History of Mechanism and Machine Science

Volume 40

This bookseries establishes a well-defined forum for Monographs and Proceedings on the History of Mechanism and Machine Science (MMS). The series publishes works that give an overview of the historical developments, from the earliest times up to and including the recent past, of MMS in all its technical aspects.

This technical approach is an essential characteristic of the series. By discussing technical details and formulations and even reformulating those in terms of modern formalisms the possibility is created not only to track the historical technical developments but also to use past experiences in technical teaching and research today. In order to do so, the emphasis must be on technical aspects rather than a purely historical focus, although the latter has its place too.

Furthermore, the series will consider the republication of out-of-print older works with English translation and comments.

The book series is intended to collect technical views on historical developments of the broad field of MMS in a unique frame that can be seen in its totality as an Encyclopaedia of the History of MMS but with the additional purpose of archiving and teaching the History of MMS. Therefore. the book series is intended not only for researchers of the History of Engineering but also for professionals and students who are interested in obtaining a clear perspective of the past for their future technical works. The books will be written in general by engineers but not only for engineers. The series is promoted under the auspices of International Federation for the Promotion of Mechanism and Machine Science (IFToMM).

Prospective authors and editors can contact Mr. Pierpaolo Riva (publishing editor, Springer) at: pierpaolo.riva@springer.com

Indexed by SCOPUS and Google Scholar.

More information about this series at https://link.springer.com/bookseries/7481

Marco Ceccarelli · Rafael López-García
Editors

Explorations in the History and Heritage of Machines and Mechanisms

7th International Symposium on History of Machines and Mechanisms (HMM)

Springer

Editors
Marco Ceccarelli
Department of Industrial Engineering
University of Rome Tor Vergata
Rome, Italy

Rafael López-García
Department of Mechanic
and Mining Engineering
University of Jaén
Jaén, Spain

ISSN 1875-3442 ISSN 1875-3426 (electronic)
History of Mechanism and Machine Science
ISBN 978-3-030-98498-4 ISBN 978-3-030-98499-1 (eBook)
https://doi.org/10.1007/978-3-030-98499-1

This Springer imprint is published by the registered company Springer Nature Switzerland AG
The registered company address is: Gewerbestrasse 11, 6330 Cham, Switzerland

Preface

In general, history is considered not only as a background but also the tracing and recognition of the evolution of a community or a discipline. In the specific case of the Mechanism and Machine Science (MMS) domain, the history of MMS addresses attention to the past personalities, achievements, and developments in machine technology in all its aspects and as such it is not only a means for tracking the MMS evolution but also the recognition of an identity of a science domain and/or its corresponding community with the aim also of being a source of ideas and inspiration for future developments.

The organization of an international symposium on the History of Machines and Mechanisms (HMM) is the main activity of the Permanent Commission (PC) for the History of Mechanism and Machine Science of IFToMM, the International Federation for the Promotion of Mechanism and Machine Science.

The first two symposia, HMM2000 and HMM2004, were held at the University of Cassino in Cassino, Italy, in the years 2000 and 2004. The third symposium, HMM2008, was held at the National Cheng Kung University in Tainan in 2008. The fourth symposium was held at Vrije University in Amsterdam, The Netherlands in 2012. The fifth symposium was held at Queretaro University in Queretaro, Mexico, in 2016. The sixth symposium was held at the Chinese Academy of Sciences' Institute for the History of Natural sciences (HINS) in Beijing, P.R. China, in 2018.

The present volume contains the proceedings of HMM2022, the 7th International Symposium on the History of Machines and Mechanisms that was held at University of Jaen and Granada University, from April 28 to 30, 2022, after a postposition of the original plan in 2021 due to the COVID-19 pandemic.

The mission of IFToMM is to promote research and development in the field of machines and mechanisms by theoretical and experimental methods, along with their practical applications. The aim of the international symposia on HMM is to create an international forum for the exploration of the history of machines and mechanisms through study and presentation of historical achievements in MMS. The scope of the symposia is wide because they emphasize the history of technical systems and their applications with historical backgrounds but peculiar attention to technical contents for understanding re-evaluates the past MMS in modern means. Relevant topics deal with the history of theories and design methods, biographies, the history of the institutions involved, the relations with other disciplines, the history of engineering education, and the social and cultural aspects of machines. Special attention in this event has been paid to past PC distinguished members and achievements in Iberoamerican contexts.

After the review process, 38 papers by authors representing 14 different countries from all around the world were accepted for publication in these proceedings of HMM2022. A selection of them was evaluated for the HMM Best Paper Awards and Springer Award according to IFToMM rules for regular and student papers. One glance at the table of contents is enough to see that we succeeded in bringing together an interesting group of scholars from a variety of disciplines with a stimulating variety of subjects. We are very satisfied with this result, and we thank the authors for their valuable contributions and for the efforts in submitting in time the final versions of the papers. Moreover, we would also like to thank the colleagues who helped us in the review process with timely feedback and careful evaluations.

This book shows the long-term development of machines and mechanisms in various contexts and reflects the uniformity and diversity of knowledge evolution in MMS domains. We believe that it will be interesting and useful for researchers, graduate students, engineers, and all others with an interest in the History of Machines and Mechanisms. We also believe that it can inspire and motivate them, not only to an attention of the history of MMS but also to conceive new results in MMS.

We would like to express our sincere gratitude to the members of the scientific committee:

M. Ceccarelli, University of Rome Tor Vergata (Italy), Chair,
I. Aslan Seyhan, Bartin University (Turkey)
T. Chondros, University of Patras (Greece)
O. Egorova, Bauman Moscow State Tech. University (Russia)
A. Gasparetto, University of Udine (Italy)
T. Koetsier, Vrije University (Netherlands)
J. C. Jáuregui, Querétaro University (Mexico)
R. López-García, University of Jaén (Spain)
R. Pisano, Lille University (France)
B. Zhang, Chinese Academy of Science (China-Beijing)
H. S. Yan, Cheng Kung University (Taiwan, China-Taipei)

We also thank the sponsors of the symposium: Spanish Association of Mechanical Engineering, University of Jaén, University of Granada, and IFToMM through the IFToMM Permanent Commission for History of MMS. Moreover, we are very grateful for the support we received from many friends and colleagues.

Without their support, we would not have been able to organize HMM2022 with a reorganization for and during the COVID-19 pandemic.

February 2022

Marco Ceccarelli
Rafael López-García

Organization

Organizing Committee

Rafael López-García (Chair)	University of Jaén, Spain
Rubén Dorado-Vicente	University of Jaén, Spain
Marco Ceccarelli	University of Rome Tor Vergata, Italy
Eloisa Torres-Jiménez	University of Jaén, Spain
Raquel Nieto Álvarez	University of Granada, Spain
Daniel López-Cano	University of Jaén, Spain
Agustín Castillo-Martínez	University of Jaén, Spain
Miguel Martinez Monedero	University of Granada, Spain

Scientific Committee

Marco Ceccarelli (Chair)	University of Rome Tor Vergata, Italy
Irem Aslan Seyhan	Bartin University, Turkey
Thomas Chondros	University of Patras, Greece
Olga Egorova	Bauman Moscow State Tech. University, Russia
Alessandro Gasparetto	University of Udine, Italy
Teun Koetsier	Vrije University, Netherlands
Juan Carlos Jáuregui	Queretaro University, México
Rafael López-García	University of Jaén, Spain
Raffaele Pisano	Lille University, France
Baichun Zhang	Chinese Academy of Science, China-Beijing
Hong-Sen Yan	Cheng Kung University, Taiwan, China-Taipei

Contents

Past PC Distinguished Figures and People

In Memory of Past PC Members 3
Marco Ceccarelli

Recent Advances and Challenges in the IFToMM PC for History of MMS 10
Alessandro Gasparetto and Marco Ceccarelli

Contribution of Professor Veniamin Goldfarb to Promotion of Gearing Science and International Cooperation 24
Evgenii S. Trubachev, Natalya A. Barmina, and Marco Ceccarelli

An Historical Review and Upload of Juanelo Turriano's Mechanism in Toledo 41
José Antonio Hernández-Torres, Angel Mariano Rodríguez-Pérez, and Julio Jose Caparrós-Mancera

The Contribution of Huseyin Vasfi Efendi from Plovdiv to Ottoman Technological Literature 49
Melek Dosay Gokdogan

The Robert H Thurston Materials Testing Laboratory Collection at Cornell University 57
Francis C. Moon

Modern Reviews of Past Works

Additive Technologies for Restoration of Kinematic Mechanism Models 71
Grigoriy Tikhomirov, Nikolay N. Barbashov, and Olga V. Egorova

3D Reconstruction of Two Medieval Artifacts by the Engineer Al-Murādī 84
Gustavo Medina-Sánchez, Jorge Moreno-Buesa, Rubén Dorado-Vicente, and Rafael López-García

Reconstruction Designs of the Automata in the Ancient Korean Water Clock "Heumgyeonggaknu" 97
Yu-Chieh Chiu, Yu-Hsun Chen, and Hong-Sen Yan

Analysis of Mechanisms 2 and 21, Included in Ibn Khalaf Al-Muradi's "The Book of Secrets" 106
Rafael López-García, Israel Fernández-Liétor, Rubén Dorado-Vicente, and Gustavo Medina-Sánchez

Analysis of the Codex Madrid I as a Compendium of Mechanisms 120
Higinio Rubio, Alejandro Bustos, Cristina Castejon, and Jesús Meneses

Ctesibius, Vitruvius and Leonardo: A Digital Reconstruction of the Water Clockwork Timeline 134
Massimo Callegari, Paolo Clini, Luca Carbonari, Renato Angeloni, Mirco D'Alessio, and Cecilia Scoccia

History of the Design of Machines and Mechanisms

The Hoist of the Aerial Rigging of the Mystery of Elche 149
Emilio Velasco-Sánchez, Sixto Manuel Marco-Lozano, and Abel Navarro-Arcas

"El Araceli" Non-destructive Test from the "Misterio de Elche" Play 160
Abel Navarro-Arcas, Sixto Manuel Marco-Lozano, and Emilio Velasco-Sánchez

Ports Design and Construction Machinery Through 19th Century Atlases 173
Agustín Castillo-Martínez

Power and Performance the Water Lifting Machines Used in Ancient Mining in the Southwest of the Iberian Peninsula 184
Angel Mariano Rodríguez-Pérez, Julio Jose Caparrós-Mancera, José Antonio Hernández-Torres, Juan Carlos Fortes-Garrido, Jose M. Dávila-Martín, and Javier Castilla-Gutiérrez

Mine Drainage and Steam Engine 197
David Calderón, Francisco Mata, Luis Mansilla, Demetrio Fuentes, Issam Hanafi, and A. Carrasco

History of Belt Conveyors Until the End of the 19th Century 210
Nenad Zrnić, Miloš Đorđević, and Vlada Gašić

Historic Development of Theories and Teaching

A Short Comparative Historiography of Science & Technology of Mechanics into Engineering—and—Architecture Literatures During the Renaissance . . . 227
Raffaele Pisano and Julie Robarts

Historical Details of Coulomb's Torsional Forces in Metal Wires . . . 241
Agamenon R. E. Oliveira and Raffaele Pisano

Revisiting Inflection Circle via Level Luffing Crane for Academic Revival . . . 255
Prashant B. Shiwalkar and Pramod M. Padole

Developing a Virtual Museum of the Escuela Técnica Superior de Ingenieros Industriales of Madrid . . . 266
Oscar Monjas Hebrero, Luisa Martínez Muneta, Antonio Carretero Díaz, and Juan de Dios Sanz Bobi

Kinematic Modelling in Greek Astronomy: The Design of Heavenly Mechanisms . . . 279
Teun Koetsier

Representation, Encoding and Geometry of Machines and Mechanisms

The Contribution of Jesuit Johann Schreck to Development of Chinese Scientific Culture in 17th Century . . . 295
Michela Cigola

Drawing the Vibrations of Sound . . . 305
Arturo Gallozzi and Rodolfo Maria Strollo

Nomograms: An Old Tool with New Applications . . . 314
Giovanni Mottola and M. Cocconcelli

The Armillary Sphere: A Representation of the Celestial Sphere and Knowledge Symbol . . . 330
Barbara Aterini

Developments of Mechanical Design and Automation

"A Great Body of Machinery": From Robots to Mecha and Back . . . 345
Nadia Ambrosetti

A Historical Development of LARM Finger Design Shape . . . 360
Emanuela Chiavoni, Michela Cigola, Alekos Diacodimitri, and Marco Ceccarelli

Evolution History of Non-circular Gears 372
Olga V. Egorova, Nikolay N. Barbashov, and Roman M. Kiselev

Dynamic Analysis of a Semi-automatic Telegraph Key 382
Cosimo Fonte and Marco Cocconcelli

Geometric Optimization of a Product Using the Design for Disassembly (DFD) Criteria 395
Pedro A. Ojeda-Escoto

Brief History of the Pulleys and Explanation of Chief Instructor Ishaq Efendi's Work on Pulleys and Pulley Systems 406
Irem Aslan Seyhan

Past Iberoamerican Distinguished Figures, Experiences and Designs in MMS

Mechanical Engineering Devices in the America Colonization 421
Rubén Dorado-Vicente, Rafael López-García, Juan Manuel Quero-Nieves, and Gustavo Medina-Sánchez

The Stamp Mill in Mexico's Antique Mines 433
Juan Carlos Jauregui-Correa

Industrial Heritage of the City of Granada: The Water Well "Santa Clara" in "La Vega De Granada" 444
A. Dávila Ponce de León, M. J. Gámez, C. González-Sevilla, Rafael García de la Mata, and Lourdes Rubio

Bartolomé Sureda: Mechanic, Inventor and Artist. A Character of Enlightened Knowledge 456
Patricia Zulueta Pérez and A. Sánchez Lite

Inheritance Value of the Machinery of the Factory "Azucarera del Pilar" in Motril and the Sugar Industry in Eastern Andalusia 470
José Antonio Saez-Calvo

Industrial Heritage in Granada and Almería - The Recovery of the Zurgena Railway Station as an Example of the Reinterpretation of the Lorca-Baza Railway Line 487
Rafael García de la Mata, C. González-Sevilla, A. Dávila Ponce de León, M. J. Gámez, and Lourdes Rubio

Correction to: Evolution History of Non-circular Gears C1
Olga V. Egorova, Nikolay N. Barbashov, and Roman M. Kiselev

Author Index 499

Past PC Distinguished Figures and People

In Memory of Past PC Members

Marco Ceccarelli(✉)

University of Rome Tor Vergata, Rome, Italy
marco.ceccarelli@uniroma2.it

Considering the generational change in progress and desirable to confirm the success of the PC for History of MMS commission, the HMM2022 symposium is also planned with particular attention to the memory and merit of the commission members who recently passed away. In particular, the HMM2022 symposium was also organized in memory of the emblematic personality of the following past PC members: Carlos Lopez-Cajun, Mexico; Stefano Paipetis, Greece; J.S. Rao, India; Alberto Rovetta, Italy; who have made notable contributions not only to the commission and IFToMM community but also in various technical-scientific fields of the MMS disciplines. Below are short biographical notes in memory and indication of their personalities as an example of inspiration for the activities within the commission and the entire IFToMM community.

1 Prof Carlos López-Cajún (1948–2020)

Prof Carlos López-Cajún has been a great IFToMMist contributing and supporting significantly activities of the Permanent Commission of History of MMS in organization of events and exploration of the history of IFToMM and MMS with very valuable publications.

He was born on January 18, 1948 in Campeche, Mexico and passed away on Sunday 20 December 2020 in hospital in Queretaro, Mexico.

Prof Carlos López-Cajún has been a great IFToMMist as a distinguished MMS scientist figure with a gentleman dynamic attitude in his ideas and proposals with a remarkable ability to connect people and to make things happen. He has been a very prolific MMS scientist with a great reputation worldwide, not only within the IFTOMM

M. Ceccarelli and R. López-García (Eds.): HMM 2022, HMMS 40, pp. 3–9, 2022.
https://doi.org/10.1007/978-3-030-98499-1_1

community, in domains of design and kinematics of mechanisms, with a great interest on the history of MMS and related people.

Prof. López-Cajún got his Bachelor and Master degrees on Mechanical Engineering from Universidad Nacional Autónoma de México (UNAM) in Mexico City and his PhD from Case Western Reserve University, USA in 1982. He worked for some years in Industry and then he moved to Universidad Autónoma de Querétaro, campus San Juan del Río, where he got the position of Full Professor in Mechanical Engineering in 1996. He has been Visiting Researcher at McGill University, Quebec, CA, and at University of Notre Dame, USA, while being Professor at the Faculty of Engineering of UNAM. He was a member of the Mexican Academy of Engineering, Member of ASME, strong promoter of IFToMM Mexico, and a Founder Member of the Mexican Society of Mechanical Engineering (SOMIM). He has been awarded with several honors by local, national and international institutions, among which it may be highlighted the Mexican National Researcher award. He served IFToMM as Member of the PC on History of MMS, Member of the Executive Council and also as Secretary General in 2008–2011. Moreover, he proposed and contributed to the organization of the 13th IFToMM World Congress held at Guanajuato, México in 2011 and he also was Chair of the HMM2016 Symposium on History of Mechanism and Machines held at Querétaro, México in 2016.

Since the beginning of his long academic career he gave courses on Mechanics of Transmissions and Mechanism Design, and many other courses always with special attention to Mechanism Science, even within invited activities all around the world so that he is remembered also as a great teacher.

He has been member of several scientific committees for international conferences and journals as result of his reputation of an intense scientific activity that is documents by papers, books (for teaching and with research results), and patents. Significant are the book on Cam Transmissions co-authored with prof Jorge Angeles and the book on Mechanism Design (in Spanish) coauthored with prof Marco Ceccarelli. He has contributed to many projects in different topics, beside Mechanism Design, like in Biomedical devices, Space systems, and History of Engineering and Machines with achievements in highly disseminated publications.

Prof Carlos López-Cajún has been a prestigious figure, who has been admired worldwide also for his unique attitude to combine friendships and scientific activity in working the true spirit of IFToMM for collaboration, sharing, and improving the technology for the benefit of the society in the welfare of human beings.

Mario Acevedo, Chair of IFToMM Mexico.

Marco Ceccarelli, past IFToMM President.

2 Stefanos A. Paipetis (1938–2020)

On May 8, 2020, Professor Stefanos A. Paipetis passed away at the age of 82 after a short fight with heart illness.

He served the University of Patras from many positions, as a member of the Rectorate Board, the Chairman of the Mechanical Engineering and Aeronautics Department Faculty Board and many other Committees. Prof Paipetis broad research reached across disciplines and captivated both students and faculty with his insight and enthusiasm. His immense and rich work has influenced the research of many of us and has shaped several generations of researchers worldwide, not only in Greece.

He was a distinguished member of IFToMM Greece and his contribution in Mechanics, mechanics of materials, fracture mechanics, Vibration, and the History of Technology is highly esteemed. He is the author of numerous papers and books on a vast area of research issues in engineering, along with a wealth of works on history, poetry, and art. A keen interest in the origins of technology in the ancient world, and the contribution of ancient inventions in the progress of human society can be traced in the books and Conference Proceedings that Prof Paipetis has edited during his lifelong career of service and leadership at the University of Patras. We are all immeasurably grateful for all he has given to us through his lifetime in the IFToMM federation since its very beginning, both institutionally and scientifically, as well as personally.

Thomas Chondros, Chair of IFToMM Greece.

Marco Ceccarelli, past IFToMM President.

3 J. S. Rao (1939–2020)

Prof. J. S. Rao made phenomenal contributions to the fields of engineering and education, including history of mechanical engineering. He was an inspiring teacher much loved by his students, and a dedicated professional immensely admired by colleagues and associates.

Jammi Srinivasa Rao was born on 27th December 1939 in Madugula village in Visakhapatnam district in the state of Andhra Pradesh in India. His father was an official at the Madugula Fort estate. He lost his mother at a very early age and was brought up by his sisters. The most significant moment of his early life was to have received the blessings of the Father of the Nation Mahatma Gandhi, at Samalkot railway station.

He received his Bachelor of Engineering degree in Mechanical Engineering with Honours in 1960 from Andhra University. Thereafter he completed his M.Tech. from the Indian Institute of Technology (IIT) Kharagpur in 1963 and obtained his Ph.D. (1965) and D.Sc. (1971 while being on the faculty from the same institute. Professor J.S. Rao became a full professor at IIT Kharagpur in 1970. He then moved to IIT Delhi in 1975, to establish the Center for Industrial Tribology, Machine Dynamics and Maintenance Engineering. He also chaired the Department of Mechanical Engineering at IIT Delhi and subsequently occupied the Bharat Heavy Electricals Ltd. (BHEL) endowed Chair. Prof. Rao served the Government of India for five years during 1981–86 as Science Counsellor at the Indian Embassy in Washington DC. He played an important role in setting up linkages in science and technology between the US and India. In the year 2000 Professor Rao moved from academics to industry on a permanent basis. He became the Chief Technology Officer at QUEST, Bangalore and played a major role in its rapid, nearly ten-fold, growth in four years. He later assumed the role of CSO at ALTAIR, Bangalore. He also provided academic services to Kumaraguru College of Technology, Coimbatore, KL Deemed University, Vijayawada and several other Educational Institutions.

Prof. J.S. Rao's research interests covered Vibrations, Blade and Rotor Dynamics, Design, Theory of Machines and Mechanisms and Thermo Fluid Mechanics, with interest on historical backgrounds. He was very active in servicing in the community at several positions and very prolific in publications and teaching.

Professor J.S. Rao was a visiting Professor at many universities around the world.

Notable accomplishments amongst several consultancy jobs were, investigation of the catastrophic failure of an Atomic Power plant in Narora, India, where the root cause was identified through successful analytical simulations, and investigations into the brake-squeal problem of Washington Metro trains. He was a full-time consultant to Stress Technology Inc., Rochester.

Prof. J.S. Rao played a significant role in advancing design culture in private industry in India, particularly in the fields of rotor and blade dynamic technologies for several multinational companies like, General Electric, Nuovo Pignone, Pratt and Whitney, Rolls Royce and GE Hydro. He contributed significantly to the design of LP Compressor for the Kaveri engine of the Light Combat Aircraft and to the rotor dynamic design of High-Speed Cryogenic Pumps for Last Stage of Geostationary Launch Vehicle for ISRO. Professor Rao was also a Director on the Board of GMR Energy Group.

Prof. J.S. Rao played a significant role in the establishment of IFToMM, right from its inception – in preparing the constitution and signing the same at the inaugural ceremony in 1969 in Poland. He was Chairman for the Sixth Congress held in New Delhi in 1983. He was a member of its executive council. He also played a major role in establishing the Rotor Dynamics Technical Committee of IFToMM and chaired it for two terms, steering it into a leading and very reputed international body; he was elected lifetime Emeritus Chairman of this committee. He was also chairman of Permanent Commission on Conferences and member of several committees over the years for IFToMM, including the Permanent Commission for History of MMS.

Prof. J.S. Rao was honoured by several bodies in India and worldwide for his outstanding scientific achievements and developing international understanding in science and technology. He was President of the Indian Society of Theoretical and Applied Mechanics (ISTAM) and founder President of the Vibration Institute of India (TVII). Prof Rao was the Chief Editor and also the Founding Editor of the journal "Advances in Vibration Engineering" (AVE) which was established in 2002. The journal was subsequently renamed as "Journal of Vibration Engineering & Technologies" (JVET) in 2014. He was honoured with the IFToMM Distinguished Service Award 2004 and in 2011 with IFToMM honorary membership. The J.S. Rao Medal in Vibration Engineering was established in 2017 in honour of Prof. J.S. Rao for his enormous contributions in the relevant field.

Prof. J.S. Rao passed away in Bangalore, in the early hours of 4th July 2020 after a brief illness.

Chitta Amarnath, Chair of IFToMM India.

Marco Ceccarelli, past IFToMM President.

4 Prof Alberto Rovetta (1940–2020)

Prof Alberto Rovetta has been a great IFToMMist of a polyhedric activity and figure with a dynamic (volcanic) attitude in his ideas and proposals. He has been a very prolific MMS scientist with a great reputation worldwide not only within the IFToMM community, with interest in the history of IFToMM and MMS documented with very valuable publications.

He was born in Brescia on 19 June 1940 and he worked out a brilliant academic career at Politecnico di Milano. He started his research activity under the guidance of Prof Giovanni Bianchi, co-founder of IFToMM and former IFToMM president, and got the full professor position for Mechanics of Machinery in 1980. Since then he gave courses on Mechanics of Robots, and many other courses in the domain of Mechanics of Systems with special attention to robots, also abroad in several periods along his career all around the world.

Prof Alberto Rovetta worked positions in several international bodies (IEEE, IFAC, ASME, NASA, NATO, UNESCO and Academies) and mainly in IFToMM for which he was member of IFToMM Executive Council and chair of the TC Micromachines, beside being member in several TCs and PCs also with an initiator activity and continuous participation. He has been member of scientific committees for international conferences and journals as result of his reputation of an intense scientific activity that is documents by more than 400 papers, over 20 books (for teaching and with research results), and several patents. He has contributed to many international projects in different topics, beside Robotics, like in Biomedical devices, Space systems, tele-surgery, Cultural Heritage, and history of engineering and technology. He has been promoter and organizer of events for scientific dissemination and promotion, among which it is to note the 1995 IFToMM world Congress that was held very successfully in Milan. His achievements range from pioneering robot designs like a very first wrapping tendon-driven hand through medical devices like innovative limb prothesis to discovering the ancient machine technologies like in the bearing design in the Tutankhamen chariot. He has been awarded with several honors by local, national and international institution, among which it may heighted the IFToMM Award of Merit in 2010.

Prof Alberto Rovetta has been a prestigious figure, who has been admired and to be taken as inspiration, also for his unique attitude to combine friendships and rigorous scientific activity in working out the true spirit of IFToMM for collaboration, sharing, and improving the technology for the benefit of the society in the welfare of human beings.

He passed on Wednesday 25 November 2020 at his home in Milan.

Alessandro Gasparetto, Chair of IFToMM Italy.

Marco Ceccarelli, past IFToMM President.

Recent Advances and Challenges in the IFToMM PC for History of MMS

Alessandro Gasparetto[1] and Marco Ceccarelli[2](✉)

[1] University of Udine, Udine, Italy
gasparetto@uniud.it
[2] University of Rome Tor Vergata, Rome, Italy
marco.ceccarelli@uniroma2.it

Abstract. The paper presents a brief account of the evolution of the IFToMM Permanent Commission (PC) for History of MMS in terms of activities and people involved in order to illustrate the evolution of the commission with its peculiarities, problems, results, and the possibility of satisfying the constitution purposes of the commission in terms of keeping memory of the history of the IFToMM and of the thematic areas of the MMS (Mechanism and Machine Science) with evaluations and interpretations of historical-technical nature. The examination of these activities is centred on the developments of the last twenty years referring to an activity of the commission with an increasing impact not only within the IFToMM community. The conference, editorial, and social activities are reported with chronicle outlines, including the interaction among the members of the commission, with particular reference to the HMM Symposium on history of mechanisms and machines started in 2000 and the book series of History of MMS started in 2007.

Keywords: History of IFToMM · History of MMS · History of PC history of MMS · History of HMM symposium · History of book series on history of MMS

1 Introduction

The history of a community, even specific to a commission, can be considered significant to outline not only the development of issues of interest to that community but also to trace the involvement of people with their specific activities at the service of the community, as stressed in (Koetsier 2000).

In general, historical studies of communities refer to much broader areas and only recently, also thanks to the IFToMM PC (Permanent Commission) on the history of MMS (Mechanism and Machine Science), attention has been directed to tracing the historical evolution of even specific communities of a more modest associative dimension.

In particular, the history of the commission was analyzed in a first work (Ceccarelli and Koetsier 2004) which outlined its characteristics and evolutionary ideas in the first years of activity up to a maturation period from which this historiographical work centers its focus. The history of the PC for history of MMS also hinges on the activities of the

M. Ceccarelli and R. López-García (Eds.): HMM 2022, HMMS 40, pp. 10–23, 2022.
https://doi.org/10.1007/978-3-030-98499-1_2

entire IFToMM world community, contributing to its identity and historical development as reported in wide-ranging papers like for example (Artobolevski 1976; Morecki 1995; Angeles et al. 2004; Ceccarelli 2015) and also in papers on the experience of individuals in various roles at the service of the technical-scientific community, such as for example (Ceccarelli 2014b and 2022).

This paper is intended to be a report of the successful activities carried out in the last twenty years by the PC for History of MMS with commitment and success by all its members, but it has also the aim of indicating the need for attention to specific aggregation for the commission future with its purposes of monitoring, research, and interpretation of the historical developments of the whole IFToMM community and of its technical-scientific issues relating to MMS. Particular attention is paid to the commission's own identity and historical development through this survey.

2 A Short Outline of the History of the PC

Figure 1 summarizes the historical development of the commission with an indication of the main facts that characterized its evolution. In particular, the historical development can be outlined in three phases, namely a first pioneering phase of initiation and consolidation of the commission within the IFToMM structure in the years 1972–1985, a second phase of rooting of the commission with a first activity in the history of engineering at the level of interest for the IFToMM community in the years 1986–2000, and finally a third phase of reached maturity of the commission with substantial activity in research and publication not only within the IFToMM community in the years 2000 onwards. The historical development of these three phases was also characterized by the leadership of the commission chairs who, with their vision and activities, allowed and guided the development in the commission as it is now recognized both at the level of a substantial IFToMM community and at the level of specific interests in the technical-scientific history of mechanical engineering with particular interest in the MMS.

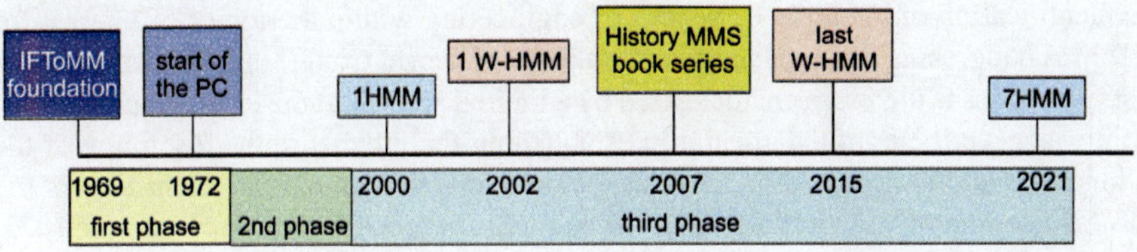

Fig. 1. A timeline of the historical evolution of the PC for History of MMS

In particular, the milestones in Fig. 1 refer to the most significant decisive events for the development of the commission. In 1972 the commission was created following the proposal of Professor Jack Phillips who, with a pioneering vision, recognized the importance of the history of the community and of the discipline that can be worked out from a technical-historical point of view, also and above all. To track the developments of the IFToMM Federation and the MMS.

After a long incubation period in which the commission chairs have rooted their interest in the history of the disciplines and of the Federation, in the year 2000 the HMM symposium series began following a documented maturity of the commission with an ever-increasing number of publications especially in the IFToMM world congresses. The HMM 2000 symposium, (Ceccarelli 2000), established a specific congress area for the commission on a regular basis and with an international organization that achieved regular success with events on all continents and with the seventh symposium the frequency of the event was reduced from four to two years. Furthermore, the success of the HMM symposium has led to the possibility of proposing and therefore launching the HMMS book series at the international publisher Springer in 2007, (Ceccarelli 2007a, b). To date, the series counts the publication of 39 volumes including conference proceedings and monographs referring to studies and historical evaluations of various disciplines of the MMS. No other milestones are highlighted in the timeline in Fig. 1, especially in order not to diminish the continuous development and obtaining concrete results that have determined a consistent community of the commission with an ever more significant commitment and impact both in the disciplinary areas of the history of science and technology as within the international community of the IFToMM Federation. In the following paragraph some of the main activities will be indicated with their most significant results.

The indications of the chair of the commission with their temporal mandate is useful to indicate the plurality and internationalization that the commission has always had since the first days: Jack Phillips (1972–1981) from Australia, Elisabeth Filemon (1982–1990) from Hungary, Teun Koetsier (1991–1997) from Netherlands, Marco Ceccarelli (1998–2002) from Italy, Hong-Sen Yan (2003–2006) from China-Taipei, Hanfried Kerle (2007–2010) from Germany, Thomas Chondros (2010–2013) from Greece, Olga Erogova (2014–2017) from Russia, and Alessandro Gasparetto (2018-present) from Italy.

In the first phase, the research activities in the history of mechanical engineering were very limited and sporadic, with one or two papers in the IFToMM world conferences, as well as in the second phase but with a greater impact also documented by the identification of the topic of history of engineering within the topics of interest for IFToMM congresses and especially in the IFToMM World Congress. The history of the first two phases is therefore characterized by a limited activity at the level of publications but it was nevertheless fundamental in establishing the interest in the discipline of the technical-scientific history of mechanical engineering and not only in the history of the IFToMM community. As initially established for the purpose of the commission in 1972, as reported in (Ceccarelli and Koetsier 2004). The third phase that can be identified with the start of the HMM symposium, IFToMM International Symposium on History of Machines and Mechanisms, in the year 2000, is characterized not only by a commission with a large number (more than 45) of active members but above all by a considerable activity both at the congress and editorial levels also with the start of the publication of a specific book series on the history of mechanical engineering.

3 Recent Advances

In particular, the aspect of growth of the consistency of the commission in terms of number of members and its international representation can be summarized considering that at the start of the third phase the PC had just 4 members (prof T. Koetsier from Netherlands, prof. A. Dimarogonas from USA, prof.H. S.Yan from China-Taipei, and prof. M. Ceccarelli from Italy) and already in 2005 it had 45 until reaching 60 members in 2015 representing almost all 44 countries that are represented in the IFToMM community (see Appendix for current members). The aggregation of people to the PC was possible due to the interest in the subjects of the activities of the PC and thanks to the regular organization of committee meetings not only as prescribed by the IFToMM constitution but also as a meeting forum for mutual knowledge and discussion of programming for the PC. Figure 2 shows photos as examples of the PC meetings to which generally not all the members have participated but to which all have referred. However, the photos shown also represent an alternation of participation by members in these meetings. In addition to the institutional meetings of the PC, since 2002 the Workshop HMMS was also instrumental with its two-year schedule and location that facilitate the participation of PC members:

- ADMONT 2002 by prof M. Husty and M. Ceccarelli
- DRESDEN 2004 by prof Corves and dr Kerle
- MOSCOW 2005 by prof Golovin
- QUERETARO 2005 by prof C. Lopez-Cajùn
- ITHACA 2006 by prof F. Moon
- BANGALORE 2007 by prof A. Ghosal and prof J.S. Rao
- PORTO ALEGRE 2009 by prof A. Oliveira
- BEIJING 2010 by prof B. Zhang
- ATHENS 2013 by prof T. Chondros
- PALERMO 2013 by prof. F. Sorge and M. Ceccarelli
- TIANJIN 2014 by prof J. Dai and T. Huang
- St PETERSBURG 2015 by prof. A. Evgrafov and O. Egorova

The HMM symposium also had a similar function of meeting, since it was celebrated regularly going around the world in different cities and in the last period, also thanks to the success of participation with authors form more than 15 countries at each event, has seen a reduction in the frequency of programming with the events listed below:

- HMM 2000 in Cassino by M. Ceccarelli, (Ceccarelli 2000)
- HMM 2004 in Cassino by M. Ceccarelli, (Ceccarelli 2004)
- HMM 2008 in Tainan by H.S. Yan, (Yan and Ceccarelli 2008)
- HMM 2012 in Amsterdam by T. Koetsier, (Koetsier and Ceccarelli 2012)
- HMM 2016 in Queretaro by C. Lopez-Cajùn, (Lopez-Cajùn and Ceccarelli 2016)
- HMM 2018 in Beijing by B. Zhang and Y. Cheng, (Zhang and Ceccarelli 2019)
- HMM 2021 (postponed to 2022) in Jaén by R. Lopez-Garcia

a)

b)

c)

Fig. 2. Examples of PC meetings: a) in 2004 at WC in Tianjin (China); b) in 2011 at WC in Guanajuato (Mexico); c) in 2018 at HMM in Beijing (China)

a)

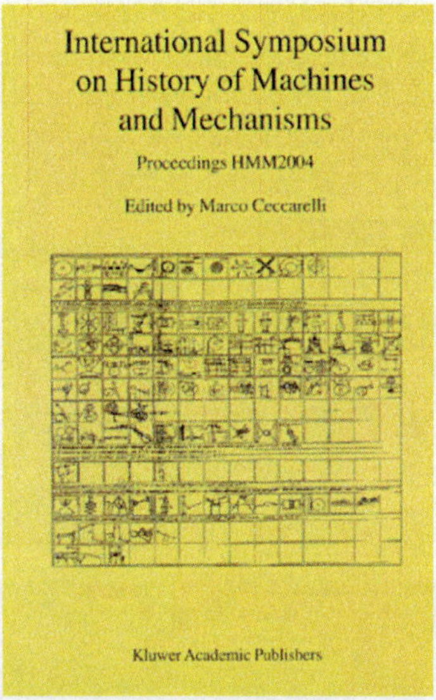

b)

b)

c)

Fig. 3. Examples of Proceedings of HMM symposium: a) in 2000 by Kluwer; b) in 2004 by Kluwer; c) in 2016 by Springer; d) last one in 2018 by Springer

Figure 3 shows the cover pages of the Proceedings of HMM symposium over the years with the last title characterization through the name 'Explorations'.

Figure 4 shows examples of participants though the group photos that are regularly taken at the HMM symposia both for historic records and personal memory of the

Fig. 4. Participants at HMM symposium over the years: a) in 2008 in Tainan; b) in 2012 in Amsterdam; c) in 2016 in Queretaro; d) in 2018 in Beijing

participants. It is to note that at the HMM symposia participants come also from other communities permitting a very multidisciplinary program dealing with historical aspects with viewpoints from mechanical engineers (as most of the PC members), engineers from other fields, historians, architects, archaeologists, philosophers and so on. Figure 5 shows a group photo of the members of the International Scientific Committee (ISC) for HMM who are PC members with the will to support and organize the event not only in their countries but also looking at new communities for sharing results for the significance of the history of MMS and mechanical engineering with technical understanding for a modern revalorization. With such a goal, Best Paper Awards were planned since the second event and even with an interest from the published Springer, as shown in Fig. 6.

Fig. 5. The meeting of ISC for HMM in 2012 (from left: H.S. Yan, M. Ceccarelli, T. Koetsier, C. Lopez-Cajun, A. Golovin, and H. Kerle)

The most significant activities in terms of impact and productivity of the PC can be summarized in the HMM symposia and HMMS book series which are the two reference initiatives, respectively in terms of conference event and editorial context. The outcome of the symposium with the first two events in 2000 and 2004 made it possible to convince the publisher Kluwer, later merged into Springer, to start in 2007 a specific book series for the subjects of history with also technical aspects that are the characterization both of the symposium and of the volumes published in the series. This characteristic differentiates the PC community from the traditional fields of history of science and technology as the interests and results of the investigations refer to complementary historical analyses when they are better focused on technical contents for a modern interpretation and re-evaluation of past achievements and figures in mechanical engineering. The HMHS series on the history of MMS was launched in 2007, (Ceccarelli 2007) with a wide-ranging vision of all issues referring to historical investigations such as theory, design, practice, and use of machines, including attention on how machines have evolved over time in the history of social and technological evolution.

The particularity of the approach for the book series is also summarized by the logo chosen for the series shown in Fig. 7 as a synthesis of the images used in the Call for Papers of the HMM symposium of 2000 and 2004. In particular, this logo represents, at this right, the image of a model for analysis and synthesis of a mechanism by Lorenzo

Fig. 6. Best Paper Awards at HMM symposium: a) the first one to prof. K. Kerle in 2004; b) the first Springer Award at prof B. Zhang in 2012; c) first student and regular Awards in 2018

Allievi in his book Cinematica della Biella Piana of 1895th is used as an idealization of the mechanisms that move progress being the evocative image of a human figure in advance and combining it with a portion of the table catalog of mechanisms proposed in Monge's analysis in Lanz and Betancourt's book of 1808.

A particular aspect has been dedicated from the beginning, as the first volume of the series demonstrates in Fig. 8a), to identifying the personalities who have contributed most to technical development also with reference to the specific IFToMM community. The aim is not only to recognize the significant figures in the fields of MMS over time, but to create a sort of specific Dictionary of Who-is-Who with special attention to the technical-scientific contributions in the fields of MMS that could be of reference and inspiration for the future generation of MMS scientist.

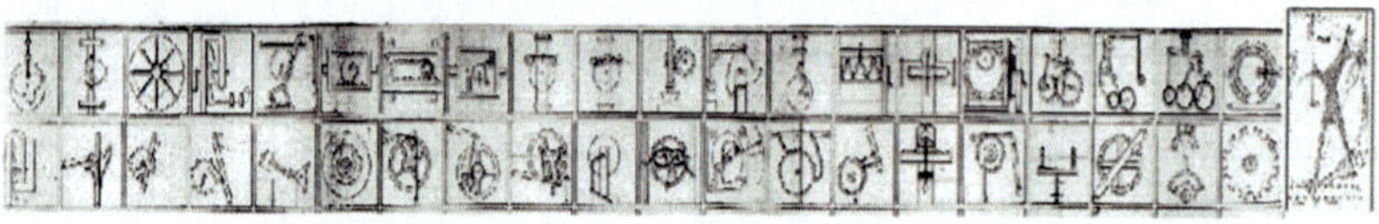

Fig. 7. Logo of the book series on History of MMS

In this context, a series of volumes has also been launched as dedicated specifically to the study and memory of those personalities (Ceccarelli 2007, 2010, and 2014a, b; Ceccarelli and Fang 2020), presenting not only biographical information that can justify the interest and passion of these figures but also addressing attention to technical investigations on the value of the contributions that they made as worthy of being a historical and technical reference for the community. An example of those volumes is shown in Fig. 8.

a) b)

Fig. 8. The series in Distinguished Figure in MMS: a) first volume in 2007; c) last volume in 2020

The book series HMMS (http://www.springer.com/series/7481) is nowadays recognized as an important frame also by other communities as proven by several volumes that are published by authors not belonging to the PC nor to the IFToMM community. Up today the book series contains 39 volumes that all are still of great interest to a wide community as demonstrated by the several citations and downloads they receive. Examples of such a variety and success can be outlined by indicating the 2010 conference proceedings celebrating 23 centuries of Archimedes legacy (Paipetis and Ceccarelli 2010), book on Homeric machines (Paipetis 2008), book on Greek and Roman machine technology (Rossi and Russo 2017), book on Chinese ancient mechanism (Hsiao and Yan 2014), and more ahead on time history the book on Russian mechanism collection (Golovin and Tarabarin 2008) up to modern time with the book on rotating machines (Rao 2011) and the book on intelligent machines (Koetsier 2019).

Summarizing, the PC has been advanced in the third phase of his history from a small group of interested MMS scientists to a community addressing and attracting interest on the history of MMS and mechanical engineering at large in all its aspects with a variety of initiatives among which the most significant ones are the HMM symposium and the Springer HMMS book series.

4 The Future of the PC

The IFToMM PC for History of MMS, as it has been highlighted in the previous sections, has an established and successful history; however, it faces major challenges in the near future. In this section we will highlight the strengths, weaknesses, opportunities and threats of the PC for HMMS. A strong point of the PC is that of being a community with a consolidated organization and a strong core, made up of researchers who are very interested in historical issues and who have been carrying out research on the history of mechanics for decades.

It should also be noted the existence of consolidated events organized by the PC (such as the HMM Symposium), as well as the contribution to renowned publication sites (such as the Springer series, Advances in Historical Studies journal).

Coming to the critical aspects, the main one is that the average age of the PC for History is high: many of its members are over 60 years old, and therefore it is necessary to foresee the entry of new forces into the PC, to replace the members who will retire in the next coming years. A campaign to raise awareness and interest for the PC among young researchers in the mechanical sector has already been launched: however, it is not always easy to find availability from young people, because the consolidated academic paths are based upon a frenetic activity of the young researchers on issues strictly inherent to their technical-scientific discipline, while the historical aspects are considered less impactful in relation to the career of the researchers themselves. To remedy this, one could consider the involvement in the PC of young researchers from other disciplines (thus not necessarily of Mechanics), who are interested in carrying out research relating to the history of machines. This interdisciplinary synergy between different disciplines would certainly enrich the PC and will allow it to recruit new members. One can think, for example, of the research sector known as "Digital humanities", where technological devices are employed to serve goals related to humanities: within this context, it would be possible to find applications (just to mention an example: robotics at the service of cultural heritage) that may be of interest to the PC.

The SWOT Table 1 summarizes strength, weaknesses, opportunities and threats of the PC for History of Mechanisms and Machine Science.

Table 1. A SWOT analysis for the PC HMMS

STRENGTH Strong core of researchers, consolidated organization, consolidated conferences and publication sites	WEAKNESSES High average age of the PC members, difficulty in recruiting young people
OPPORTUNITIES Possibility of intercepting synergies with humanistic sectors and more	THREATS Tendency of the "scientific-technical" sectors to neglect historical research in assessing the researchers' careers

5 Conclusions

The History of the IFToMM PC for History of MMS is outlined in this paper referring to the second period of complete maturity of the activities with the HMM symposium and the Springer book series on History of MMS, also considering a considerable participation of members, who have gradually increased in representation of the entire world-wide IFToMM community. In the last twenty years it can be noted that the commission has found its organizational efficiency not only in institutional roles according to the IFToMM constitution but also as a cultural reference in the history of science and technology with particular interests in mechanical engineering aspects. Despite the success that has emerged with growing results in the last twenty years, problems are emerging for the future, especially linked to the fact that the currently active community needs a generational change that hopefully will take place with consolidation and even innovation of the commission's activities.

Appendix

Members of the IFToMM OC for History of MMS as in January 2022

ACEVEDO Mario (Mexico)
ANDRIENKO Pavel A. (Russia)
ASLAN SEYHAN, Irem (Turkey)
CARVALHO M. João Carlos (Brazil)
CECCARELLI Marco (Italy)
CHEN Yu-Hsun (China-Taipei)
CHONDROS Thomas G. (Greece)
CORVES Burkhard (Germany)
CUADRADO Juan I. (Spain)
DAI Jian (United Kingdom)
EGOROVA Olga Vladimirovna (Russia)
EVGRAFOV Alexander (Russia)
FANG Yibing (China-Beijing)
FERRANDO PIERA Francesc (Spain)
FORTES GARRIDO Juan Carlos (Spain)
FRANCO Walter (Italy)
LÓPEZ-GARCÍA Rafael (Spain)
GASPARETTO Alessandro (Italy)
Gökdoğan, Melek Dosay (Turkey)
HUSTY Manfred L. (Austria)
KOETSIER Teun (Netherlands)
KOZLIKIN Denis (Russia)
KUROKI Hidetaka (Japan)
LIN Tsung-Yi (China-Taipei)
LU Zhen (China-Beijing)
MIMMI Giovanni (Italy)
MOON Francis (USA)
OLIVEIRA Agamenon (Brazil)
PAPADOPOULOS Evangelos (Greece)
PENNOCK Gordon (USA)
PISANO Raffaele (France)
RAVANI Bahram (USA)
SEO Moon Hwo (Korea)
SHIM Jae Kyung (Korea)
TARABARIN Valentin (Russia)
TOLOCKA Rymantas Tadas (Lithuania)
WAWRZECKI Janusz (Poland)
WOJNAROWSKI Józef (Poland)
YAN Hong-Sen (China-Taipei)
ZHANG Baichun (China-Beijing)
ZHANG Nong (Australia)
ZIELINSKA Teresa (Poland)
ZRNIC Nenad (Serbia)

References

Angeles, J., Bianchi, G., Bessonov, A.P., Maunder, L., Morecki, A., Roth, B.: A history of IFToMM, Chapter 2. In: Proceedings of HMM 2004 - the Second IFToMM International Symposium on History of Machines and Mechanisms. Kluwer, Dordrecht, pp. 25–125 (2004)

Artobolevski, I.I.: Past, present and future of the theory of machines and mechanisms. Mech. Mach. Theory **11**, 353–361 (1976)

Ceccarelli, M. (ed.): International Symposium on History of Machines and Mechanisms - Proceedings of HMM 2000. Kluwer, Dordrecht (2000)

Ceccarelli, M. (ed.): International Symposium on History of Machines and Mechanisms - Proceedings of HMM 2004. Kluwer, Dordrecht (2004)

Ceccarelli, M. (ed.): History of Machines and Mechanisms. Springer, Dordrecht (2007a)

Ceccarelli, M. (ed.): Distinguished Figures in Mechanism and Machine Science: Their Contributions and Legacies – Part 1. History of Machines and Machine Science, vol. 1. Springer, Dordrecht (2007b)

Ceccarelli, M. (ed.): Distinguished Figures in Mechanism and Machine Science: Their Contributions and Legacies – Part 2. History of Machines and Machine Science, vol. 1. Springer, Dordrecht (2010)

Ceccarelli, M. (ed.): Distinguished Figures in Mechanism and Machine Science: Their Contributions and Legacies – Part 3. History of Machines and Machine Science, vol. 1. Springer, Dordrecht (2014a)

Ceccarelli, M.: Twenty-five years of activity in IFToMM. J. Theory Mech. Mach. **11**(2), 3–14 (2014b)

Ceccarelli, M.: A short account of history of IFToMM and its role in MMS. Mech. Mach. Theory **89**, 75–91 (2015)

Ceccarelli, M.: Experiences in leadership IFToMM: achievements and challenges. In: Beran, J., Bílek, M., Václavík, M., Žabka, P. (eds.) TMM 2020. MMS, vol. 85, pp. 3–16. Springer, Cham (2022). https://doi.org/10.1007/978-3-030-83594-1_1

Ceccarelli, M., Koetsier, T.: On the IFToMM permanent commission for history of MMS. In: Proceedings of International Symposium on History of Machines and Mechanisms HMM 2004, pp. 10–25. Kluwer, Dordrecht (2004)

Ceccarelli, M., Fang, Y. (eds.): Distinguished Figures in Mechanism and Machine Science: Their Contributions and Legacies – Part 4. History of Machines and Machine Science, vol. 1. Springer, Dordrecht (2020)

Golovin, A., Tarabarin, V.: History of Machines and Machine Science, vol. 5. Springer, Dordrecht (2008)

Koetsier, T.: Mechanism and machine science: its history and its identity. In: Proceedings of HMM 2000 International Symposium on History of Machines and Mechanisms, HMM 2000, pp. 5–24. Kluwer Academic Press, Dordrecht (2000)

Koetsier, T., Ceccarelli, M. (eds.): Explorations in the History of Machines and Mechanisms - Proceedings of HMM 2012. Springer, Dordrecht (2012)

Koetsier, T.: The Ascent of GIM, the Global Intelligent Machine: A History of Production and Information Machines. History of Machines and Machine Science, vol. 36. Springer, Dordrecht (2019)

Hsiao, K.-H., Yan, H.S.: Mechanisms in Ancient Chinese Books with Illustrations. History of Machines and Machine Science, vol. 23. Springer, Dordrecht (2014)

Lopez-Cajùn, C., Ceccarelli, M. (eds.): Explorations in the History of Machines and Mechanisms - Proceedings of HMM 2018. Springer, Cham (2016)

Morecki, A.: Past present and future of IFToMM. Mech. Mach. Theory **30**, 1–9 (1995)

Paipetis, S.A.: Science and Technology in Homeric Epics. History of Machines and Machine Science, vol. 6. Springer, Dordrecht (2008)

Paipetis, S.A., Ceccarelli, M.: The Genius of Archimedes – 23 Centuries of Influence on Mathematics, Science and Engineering - Proceedings of an International Conference Held at Syracuse, Italy, 8–10 June 2010. History of Machines and Machine Science, vol. 11. Springer, Dordrecht (2010)

Rao, J.S.: History of Rotating Machinery Dynamics. History of Machines and Machine Science, vol. 20. Springer, Dordrecht (2011)

Rossi, C., Russo, F.: Ancient Engineers' Inventions: Precursors of the Present. History of Machines and Machine Science, vol. 33. Springer, Dordrecht (2017)

Zhang, B., Ceccarelli, M. (eds.): Explorations in the History of Machines and Mechanisms - Proceedings of HMM 2018. Springer, Cham (2019)

Yan, H.S., Ceccarelli, M. (eds.): International Symposium on History of Machines and Mechanisms - Proceedings of HMM 2008, Springer, Dordrecht (2008)

Contribution of Professor Veniamin Goldfarb to Promotion of Gearing Science and International Cooperation

Evgenii S. Trubachev[1], Natalya A. Barmina[1](✉), and Marco Ceccarelli[2]

[1] Scientific Department "Institute of Mechanics named after Professor V. I. Goldfarb", Kalashnikov Izhevsk State Technical University, Izhevsk, Russia
barmina-nat@mail.ru

[2] Laboratory of Robot Mechatronics, Department of Industrial Engineering, University of Rome Tor Vergata, Via del Politecnico 1, 00133 Roma, Italy
marco.ceccarelli@uniroma2.it

Abstract. The MMS community knows Professor Veniamin Goldfarb mostly for his activity in 2000–2015 within IFToMM – as a Gearing TC member and Chair and then IFToMM Vice-President. He had an exciting diversified life of scientist and active leader with a wide range of interests, skills and events. This paper presents a survey of his way of becoming a scientist, organizing his own scientific school, foundation of the Institute of Mechanics as a scientific department of Kalashnikov Izhevsk State Technical University, starting a unique for Russia scientific production enterprise "Mechanik", and his multifaceted university activity with his publications and patents. In this paper we would like to present Veniamin Goldfarb as a man of the world and outline mainly his educational and international activity – editing and publishing efforts, promoting the integration of the Russian gearing science into the international community, and his contribution to all-Russian and international cooperation in the field of MMS, especially in gears and transmissions.

Keywords: Veniamin Goldfarb · Gears · Gearbox production · Institute of Mechanics of Kalashnikov ISTU · IFToMM

1 Biographical Notes

Veniamin Iosifovich Goldfarb was born on February 01, 1941 in Izhevsk (Russia) in the family of medical doctors. But he chose to become an engineer and in 1962 he graduated from the Department of Instrument Design of Izhevsk Mechanical Institute - IMI (now Kalashnikov Izhevsk State Technical University - ISTU. The issues of instruments and everything connected with them, including computers and their programming were not just popular, but super popular that time. The future professor Veniamin Goldfarb was one of the best and most notable in this community. It should be said that even then he was distinguished not only by his intellect and quick wit – his charisma and leadership qualities were evident to everybody. After graduation in 1962 he started his scientific

M. Ceccarelli and R. López-García (Eds.): HMM 2022, HMMS 40, pp. 24–40, 2022.
https://doi.org/10.1007/978-3-030-98499-1_3

and educational activity. Since 1962 till 2019 the university IMI/ISTU was his only affiliation. Veniamin Goldfarb devoted his whole life to this university where he made a respectful career from the assistant, associate professor to the full professor, head of Mechanical Engineering department, director of the scientific department "Institute of Mechanics" and the head of the scientific production enterprise "Mechanik":

1957 - 1962 – Student at Izhevsk Mechanical Institute (now Kalashnikov ISTu)
1962 - Assistant of "MMS and Theoretical Mechanics" Department
1963 - Assistant of "Gyroscopic devices" Department
1965 - Senior Lecturer of "Gyroscopic devices" Department
1968 - Postgraduate course
1969 - PhD thesis on Engineering
1970 - Associate Professor of "Instrumentation parts" Department
1986 – Doctor of Engineering Science thesis
1987 - Professor of "Devices and methods of quality control" Department
1991 - Professor of "Technology of robotized production" Department
1994…2019 - Head of the newly organized scientific department "Institute of Mechanics"
1995…2014 - Head of "Design and manufacturing preparation of mechanical engineering production" Department
2014…2016 - Professor of "Design and manufacturing preparation of mechanical engineering production" Department

These biographical notes are expanded in [1].

Professor Veniamin Goldfarb was a Doctor of Technical Sciences, Professor, Honored Scientist of the Russian Federation, Director of the Scientific department "Institute of Mechanics" of Kalashnikov Izhevsk State Technical University, member of the International Federation for the Promotion of Mechanism and Machine Science (IFToMM), Chairman of the Technical Committee for Gearing and Transmissions from 1998 to 2005, member of the Executive Council from 2007 to 2011, Vice President of the Federation from 2011 to 2015, member of the International Organization for Standardization (ISO), Honorary Member of the Slovak Union of Mechanical Engineers and the Bulgarian Union of Mechanical Engineers, New York Academy of Sciences, Academician of the International Informatization Academy and the Russian Academy of Natural Sciences. Professor Veniamin Goldfarb was the Prize winner of the State Prize in the field of science and technology, Honorary Worker of Higher Professional Education of the Russian Federation, Honorary Professor of ISTU. He was awarded the Medal of Academician V.I. Vernadsky, Medal of Academician S.P. Kapitsa "To the author of a scientific invention" by the St. George Cross of the Russian Academy of Natural Sciences and the medal "Inventor of the USSR" for his merits in economics and science.

Prof. V. I. Goldfarb was the author of more than 300 scientific publications, including 10 monographs in the field of research, development and implementation of gears. For more than a quarter of a century, he led a scientific school "Institute of Mechanics" - the creative community of talented and successful, young scientists, where 16 Dr.Sc. and 38 Ph.D. theses were completed, more than 30 patents of the USSR and Russian Federation were obtained. He was a supervisor of 16 Ph.D. students and consulted 5 Doctors of

Science. A large number of gear scientists from other scientific schools got the support of Professor Goldfarb.

While working at Kalashnikov Izhevsk State Technical University, professor Goldfarb combined several activities in teaching, research, design, and technological transfer (Fig. 1).

Fig. 1. Professor Veniamin Goldfarb (1941–2019)

First of all, he was a teacher at the Mechanical Engineering department. In spite of the fact that he graduated from the Instrumentation Engineering faculty, his focus of design and research was switched to the mechanical engineering, namely, MMS and gears, immediately after the graduation. Professor Goldfarb used to present the most difficult and serious MMS issues with the help of his deep understanding of the question united with the natural skill of public speaking and extraordinary sense of humour.

The academic courses of professor Goldfarb varied much with time and academic level. His courses for future Engineers and Bachelors involved the Theory of Machines and Mechanisms (MMS), Applied Mechanics, Mechanisms and Instrument's Parts, CAD of gears. As for Master and post-graduate programs, he lectured on the Theory of Gearing, Theory of Design and Manufacturing of Spiroid Gearing, Accuracy of Mechanisms, Theory of Helical Surfaces, Applied Differential Geometry, etc.

He was so unique in his way of lecturing, addressing students and dealing with various problems that many of his students became deeply involved in the scientific research, often making then their own careers of scientists and university teachers in different engineering areas. Professor Goldfarb supervised dozens of BS and MS graduation works, 16 Ph.D. students and 5 Dr.Sc. applicants. He was a member of numerous certification committees that awarded the degrees of engineers, BS, MS, PhD and Dr. Sc. in different engineering and scientific areas.

His main scientific interests were in the field MMS-related areas including and theory of gearing, and theory and practice of design, manufacture, testing and application of gears, gearboxes, and gearmotors [2]. Being a graduate of the Instrumentation engineering department, Professor Goldfarb introduced CAD into gear research in the early 1970s which was a new trend and unstudied area in the USSR that time.

The main type of gears to be studied and further produced under the leadership of Prof. Goldfarb was spiroid gears (Fig. 2). The first information about spiroid gears appeared in the USA. Their inventor is a talented American engineer Oliver Saari (1914–2003), who got the first patent for "Speed-Reduction Gearing" in 1954 [3, 4]. Beginning from 1955 the well-known publications by F. Bohle [5, 6] and W.D. Nelson [7, 8] appeared, where the attractive name "spiroid" was mentioned for the first time, meaning the new type of a hyperboloid gear. Later the American company Illinois Tools Works (ITW) Inc., where the mentioned authors were working and which was the pioneer of mastering the manufacture and introduction of these gears into different engineering areas, patented the names SPIROID and HELICON (the variety of a spiroid gear with cylindrical worm, also invented by O. Saari) which became the trademarks of ITW [7, 8]. That is why many other companies and authors deeply engaged into investigations and manufacture of these gears began calling them other names in order to avoid conflict with ITW: face worm gear, spiroid worm gear, spirodrive, hypoid-worm gear, spiral-type gear and others.

Prof. Goldfarb's Ph.D. thesis in 1969 was devoted to studying the varieties of special and complicated for analysis spiroid gears, and DSc. thesis in 1986 was about the fundamentals of the theory of automated geometrical analysis and synthesis of general type worm gears [9], where he stated the basic results obtained by him. The main scientific developments of Professor Goldfarb consisted in the classification and study of spatial gear schemes [10] (Fig. 3), rational structuring of the process of computer-aided design of worm gears, the development of a "non-differential" method for finding surfaces formed by enveloping, and the research of a variety of spiroid gears [11, 12].

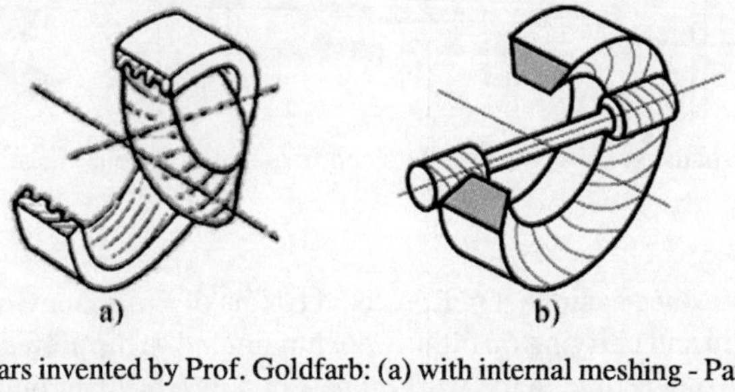

Fig. 2. Spiroid gears invented by Prof. Goldfarb: (a) with internal meshing - Patent RU № 806935 and (b) non-orthogonal gear with two meshing zones - Patent RU № 2101582.

In the last two decades of his career, Professor Goldfarb focused on the practical implementation and improvement of worm-type gears and gearboxes in pipeline valves. Under his leadership, it became possible to create a large number of competitive products and to establish their efficient production (Fig. 4, 5). The production became and in many respects remains a unique example of implementation of scientific, technical and technological developments directly by the scientific department of the University in the small innovative enterprise "Mechanik" in Izhevsk [13]. It deserves a more detailed acquaintance, which can be done according to publications made by Prof. Goldfarb and his colleagues such as [9, 13–15].

Professor Goldfarb was an organizer of the widest range of scientific events, publications and activities: USSR and All-Russian scientific conferences, seminars, meetings and symposia; international conferences on gears; editing and publishing proceedings, scientific journals and books on MMS; initiating international scientific and educational cooperation between Russian and foreign universities, their examples are: SIOMMS-2011 [16, 17], development of MMS terminology [18, 19] and the International Symposium "Theory and practice of gearing" to the 100th anniversary of Prof. Faydor L. Litvin [20, 21].

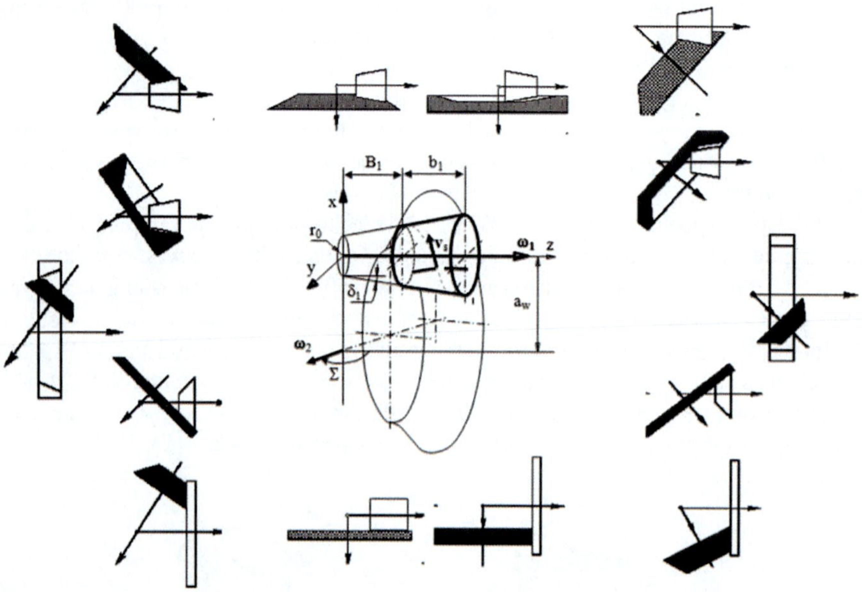

Fig. 3. Scheme of spatial gear by V. I. Goldfarb and fragment of his classification of single-stage gears [4]

We should note once again that multifaceted talents of Professor Goldfarb's personality, his leadership and unifying qualities were not limited to his professional activities. Colleagues (and many people in the world) knew him as a brilliant musician and piano player, an indispensable member of IMI/ISTU jazz ensemble, a satirical student theater of miniatures, and one of the main activists of the IMI/ISTU sports camp in on the bank of the Kama River [1]. He had an inherent depth of analysis, knowledge of a complex mathematical apparatus, the ability to single out the main scientific component, to structure new knowledge, and to formulate it clearly and beautifully, without being sprawled out on side points, which, unfortunately, are often numerous in scientific searches. He generously shared this skill both with his students and with many colleagues from other scientific schools. His unique way from a talented student to the well-known gear scientists – from the very first glance to spiroid gears to prominent scientific results, to the successfully functioning serial gearbox production, and to the position of IFToMM Vice-President – is fully described in [1].

Fig. 4. Several generations of spiroid gears, gearboxes and gearmotors produced by SIE "Mechanik" at the boundary of the 20th–21st centuries [22].

2 All-Russian and International Activities

2.1 USSR and All-Russian Conferences and Joint Projects on Gears

Along with the educational activity and scientific research, Professor Goldfarb was an active participant and organizer of different conferences and symposia on gear-related issues since the early 1960s till 2017. Soviet and then Russian gear scientists felt their unity, a number of interested young scientists increased multiply, there were many locations to meet, present their results, discuss their scientific research and find support of the

Fig. 5. Advanced version of worm-type gearboxes currently produced by SIE "Mechank" [23].

masters. Various scientific events were held in universities of the USSR with Moscow (International congress ITS-93 "Information communications, networks, systems and technologies", 1993), Saint-Petersburg (All-Russian scientific and practical conference with international participation "Gearbox engineering in Russia: state of arts, problems, prospects, 2002), Khabarovsk (Scientific conference "Problems of investigations, design and production of gears", 1974), Kurgan (The 6th international symposium "Theory of real gearing", 1997), Chelyabinsk (Scientific conference "Improvement of machine design and methods of machine parts processing", 1978), Kharkov (Scientific technical conference "Standardization and unification in the field of gearing", 1990), Odessa (All-Union scientific and technical conference "Gears: the present and progress" 1990), and Izhevsk (International conference "Theory and practice of gears", 1996) among them. Professor Goldfarb was one of the most active organizers of these events in Izhevsk. These events on the theory and practice of gears and CAD of mechanical transmissions were supported and financed by the USSR government till 1990. In spite of the fact that every scientific event was followed by the publication of a paper volume of the conference proceedings, they were all printed in Russian and with a very limited number. That is why they went practically unnoticed by the world MMS community. Unfortunately, cooperation and information exchange between Soviet and Western specialists in the mid-20th century were complicated (or too scarce), but we must say that even in the relatively young (or new) at that time Izhevsk gearing school, specialized proceedings appeared regularly starting from 1967, where works of both authoritative gear experts and young researchers were published, among whom was the future professor V.I. Goldfarb.

In 1980s the most popular topic of scientific research and discussion was the computer-aided design of mechanical transmissions. Veniamin Goldfarb was the organizer and active participant of numerous conferences on computer-aided design of mechanical transmissions, gears and gearboxes and its introduction into mechanical production in Izhevsk in 1981 (Scientific and technical conference "Development and implementation of CAD systems"), 1982 (Scientific and technical conference "Computer-aided design of mechanical gears", 1984 (Coordination meeting "Mathematical support of CAD and FMS in mechanical engineering"), 1987 (Coordination meeting "Mathematical and software support of integrated CAD/FMS systems"), 1989 (Scientific and technical seminar "Computer-aided design of gears and gearboxes") and 1990 (Scientific and technical conference "Development and implementation of CAD and CAM systems in mechanical engineering").

In difficult economic transition conditions in the 1990s, Russian scientific schools in the field of gears and gearbox engineering found themselves in an extremely difficult financial situation. Besides many difficulties, Prof. Goldfarb was struggling in many offices of Ministries of Industry, Science and Education for continuing financing scientific work. In these conditions, from 1991 to 2004, Professor V. I. Goldfarb achieved financing and initiated the joint participation of leading Russian scientific schools of 12 Russian technical universities in target programs of Russian state and interstate financing. These state target programs were a pioneer solution of supporting and financing MMS science and applying the results of scientific research in the real mechanical engineering production. The federal state programs included such programs as: "Transfer techniques", "Progressive gears", "New generation of mechanical drives", "Innovations in higher schools", "Integration" and so on.

For example, the program "Progressive gears" with the Institute of Mechanics headed by Professor Veniamin Goldfarb as the leading executor involved 17 projects and promoted mastering the production of progressive gears, development and manufacturing implementation of high-technology competitive products comprising various gears. It united the following Russian universities:

- Novouralsk Polytechnic Institute (Novouralsk) – project "Machines and mechanisms of compression and transfer of liquids and gases on the basis of progressive gears";
- Yaroslavl State Technical University (Yaroslavl) – project "Development and implementation of a series of progressive mechanical stepless regulated gears for machine drives";
- Kurgan State University (Kurgan) – project "Mastering the manufacturing support of gear production based on estimation of their load state and lifetime when operating according to of integral type strain gauges";
- Rostov State University (Rostov) – projects "Development of geometry and mastering the production of competitive gears with improved properties" and "Development of scientific fundamentals for design and production of unified series of gearboxes with increased load capacity for oil pumping units";
- Volgograd State Technical University (Volgograd) – project "Development of the theory, design methods, reserve-saving laser techniques providing the increase in the technical level and lifetime of highly stressed gears";
- Vladimir State Technical University (Vladimir) – projects "Development of actuating roller screw mechanisms of electromechanical steering devices of flying apparatus" and "Development of low-size electromechanical drive on the basis of gears with elongated roller and its production technique for nanotechnology equipment and autonomous orthopedical apparatus";
- Chelyabinsk State Technical University, Zlatoust Branch (Zlatoust) – project "Development of designs of power transmitting mechanisms on the basis of planetary gears with small difference in tooth numbers";
- Kolomna Branch of Moscow State Open University (Kolomna) – project "Development of the technique and means of production of high-accuracy spur and helical arc gears";

- N.E. Bauman Moscow State Technical University (Moscow) – project “Development and mastering of progressive designs and techniques of their production for new-generation wave gears”;
- STANKIN Moscow State Technological University (Moscow) – project “Design expert system “Expert” for computer-aided synthesis and analysis of bevel and hypoid gears”;
- Mechanical Engineering Research Institute (IMASH) of the Russian Academy of Sciences (Moscow) – projects “Development of Russian proposals on correction of ISO standards on strength analysis of gears” and “Development of heavy-loaded planetary drives for ship industry”;
- and the Institute of Mechanics of Kalashnikov Izhevsk State Technical University (Izhevsk) as the leading and coordinating organization. One of its many projects was “Development of general-purpose and special low-size competitive single- and double-stage spiroid gearboxes and gearmotors”.

2.2 International Cooperation and Membership in International Organizations and Committees

In 1990s the organizational talent of Professor Goldfarb showed itself very quickly in the diverse international cooperation between universities and scientific schools of different countries. It was gear experts from Great Britain, Czech Republic, Poland, and Slovakia who came to Izhevsk (ISTU) as the first foreign guests. Every year the geography of contacts and the number of new joint projects were expanding rapidly. The first agreements on international cooperation at ISTU with the universities of Slovakia (Bratislava, Trnava), Czech Republic (Brno), Poland (Warsaw), Bulgaria (Sofia), England (Nottingham), and USA (Los Angeles) were signed on the initiative of Professor Goldfarb.

The first International program “New Generation of Mechanical Drives” started already in 1996 and lasted till 2000; it involved scientific schools of universities from 7 countries (Slovakia, Bulgaria, Czech Republic, Poland, England, USA, and Russia).

Professor Veniamin Goldfarb made an invaluable contribution to the integration of Russian gearing science into the international scientific community and to the development of international scientific and technical cooperation in general and, in the field of gears and gearbox engineering, in particular.

The first international event on MMS was held in Russia already in 1993 exactly in Izhevsk [24] – it was the International Symposium “Development of Geometrical Theory of Gears” with foreign participants from Bulgaria, Poland, Slovakia and Czech Republic and then there were regular conferences in Izhevsk organized by Professor Goldfarb:

- International Symposium “Advanced Gears” in 1994;
- International Conferences “Theory and Practice of Gears” in 1996, 1998, 2004, 2008, 2017;
- International scientific seminar “Modern information technologies. Problems of Design, Research and Production of Gear Transmissions” in 2001 with about 70 participants including foreign ones from 8 countries already;

– International Symposium "Theory and Practice of Gears" in 2014 devoted to the 100th anniversary of Professor Faydor Litvin. About 100 scientists from 12 countries took part in this event.

During different periods of his scientific activity V. I. Goldfarb became an active participant of scientific communities. In particular, he was a member of Technical Committee TC-258 for Machine Parts and Gears of the Russian State Standardization Committee, and he was the initiator and Vice-President of Russian partnership "Association of Mechanical Transmission Engineers". Later on, Professor V. I. Goldfarb became a member of editorial boards of international journals "Mechanism and Machine Theory", "Machine Design", and a frequent chairman and member of scientific program and organizational committees of international conferences on gears and transmissions: "Power Transmissions" (Greece, 2009), "Power Gears and Transmissions" (Japan, 2009), "Fundamentals of Machine Design" (Bulgaria, 2009), "Gears" (Germany, 2010), "Terminology in Mechanisms and Machines Science" (Belarus, 2010), "Development and Study of Mechanical Components and Systems" (Serbia, 2011) and a member of program committees of IFToMM World Congresses on MMS (Finland-1999, China-2004, France-2007, Mexico-2011, Taiwan-2015).

2.3 Publication and Editor's Activity

At the end of the twentieth century it became obvious that a certain level of foreign language proficiency was necessary to initiate international scientific and educational cooperation. Still the problem of getting acquainted for Russian and foreign gear scientists remained urgent and in 1991 professor Goldfarb initiated a bilingual journal (English and Russian) "Gearing and Transmissions", which was published for fourteen years till 2004 mainly by the staff of the Scientific Department "Institute of Mechanics" of Kalashnikov Izhevsk State Technical University and was distributed for free among Russian and foreign technical universities and production enterprises (Fig. 6). It was a perfect solution for both English-speaking scientists to know about the fundamental and latest research results of Russian-speaking scientists and vice versa.

The Editor-in-Chief of this journal was Veniamin Goldfarb and the Editorial Board of the Journal covered many well-known foreign and Russian gear scientists: Prof. K. Arnaudov (Bulgaria), Prof. W. Oleksyuk (Poland), Prof. H. Winter (Germany), Prof. M. Grois (Czech Republic), Prof. A. Kubo (Japan), Prof. H. Linke (Germany), Prof. J. Mudrik (Slovakia), Prof. P. Parushev (Bulgaria), Prof. A. Siereg (USA), Prof. L. Slutsky (USA), Prof. B.-R. Hohn (Germany), Prof. D. Su and many others. Three foreign universities took an active part in editing and publishing of this journal within the international cooperation along with the ISTU: Technical University of Brno and Slovak University of Technology in Bratislava (both Slovakia) and Nottingham Trent University (UK) with the hardest contribution of Prof. Daizhong Su from the UK and Prof. J. Mudrik and Prof. S. Kral from Slovakia. After Professor Goldfarb joined IFToMM community, this journal became its official publication and was fully supported by the IFToMM Gearing and Transmission Technical Committee. It was a real breakthrough to start the international cooperation (without Internet that time!) from giving such a unique opportunity

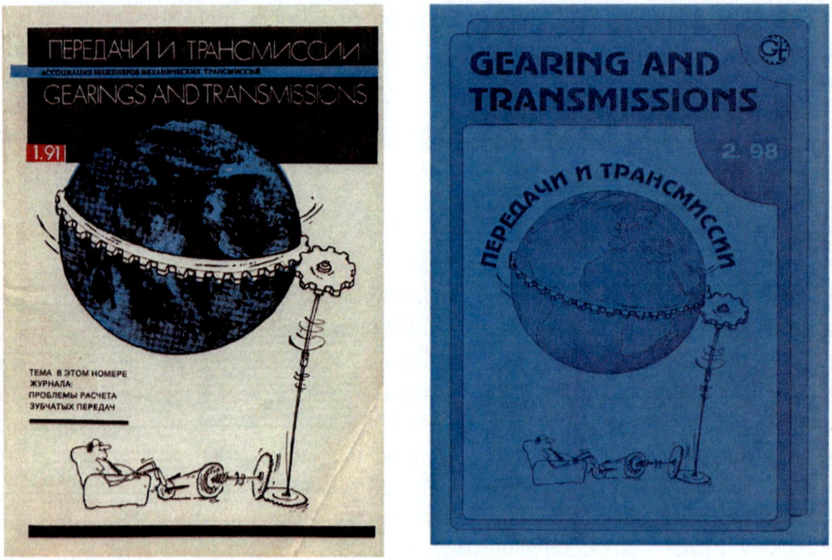

Fig. 6. Journals "Gearing and Transmissions".

for foreign and Russian scientists to know about each other and get in contact with each other.

In 2014, by the initiative of Professor M. Ceccarelli (at that time - IFToMM Past President), Professor Goldfarb started a new project - regular publication of contributed volumes on gears in the series "Mechanisms and Machine Science" in the large European publishing house "Springer". As the editor-in-chief of these books, he united fundamental and latest manuscripts of the leading gear scientists from Russia, China, the United States, Japan, Italy, Germany, Canada, Bulgaria, Belarus, and other countries. Now, with the electronic version accessible all over the world, this special series on gears is of great importance and promotes the gearing science (thousand of these books are read and sold, in accordance with the Springer statistics), encourages young researchers and skilled gear masters to go on their studies and communications and to improve their knowledge. Three contributed volumes have already been published in 2016, 2018, 2020 with Veniamin Goldfarb as the chief editor of these books and the fourth one was published in 2021 as a tribute to Professor Goldfarb after his passing away (Fig. 7) [21, 25–27].

2.4 Initiating International Student Olympiads on MMS

The first-ever international student subject Olympiad on MMS was held in Kalashnikov ISTU on April 19–21, 2011. The idea to organize the Student International Olympiad on MMS (SIOMMS) was proposed by Veniamin Goldfarb in 2009 in Mexico at the IFToMM conference. IFToMM Executive Council made a decision to organize the International Student Olympiad (SIOMMS) in 2011 in Izhevsk State Technical University, Izhevsk, Russia (Fig. 8, 9). SIOMMS-2011 was held on April 19–21, 2011 [14–16]. 17 student

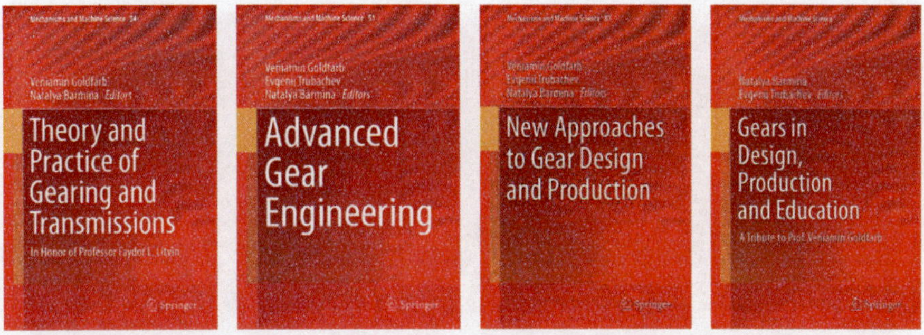

Fig. 7. Contributed volumes on gears edited by Prof. Goldfarb within MMS series in Springer Publishing House [21, 25–27].

teams from 8 countries took part in this competition. This Olympiad initiated regular Students International Olympiads on MMS which are held every two or three years in different parts of the world: 2011 (Izhevsk, Russia), 2013 (Shanghai, China), 2016 (Madrid, Spain) and 2018 (Lima, Peru).

Fig. 8. Professor Goldfarb – organizer of SIOMMS-2011 [16].

2.5 IFToMM Membership and Initiatives

The Technical Committees for Gearing and Transmissions (TC) within the International Federation for the Promotion of Mechanism and Machine Science (IFToMM) was founded in 1976 by the Initiative of the Secretary General Prof. A. Morecki (Poland).

Fig. 9. Chair persons at the opening of SIOMMS-2011 in Izhevsk, Russia.

The first Chairman of the Gearing TC was Prof. D. Dudley (USA). In 1986 he resigned and Prof. K. Stolzle (Germany) became the TC Chair till 1993. After that Prof. A. Kubo (Japan) was elected to be the new Chairman. As for Russian participation in the IFToMM, it was not very active because of the mentioned above language barriers. In 1994 Prof. Veniamin Goldfarb joined the IFToMM and was the second Russian to join the TC for Gearing and Transmissions. His main goals within the Gearing TC were as follows:

- to continue publishing the bilingual journal "Gearing and Transmissions" already with the support of IFToMM community and to distribute it among gear scientists, manufacturers and technical universities all over the world;
- to promote the knowledge and raise the authority of IFToMM among Russian scientists;
- to continue carrying out the scientific research on gears at the highest international level, presenting the results at international conferences on gears;
- to intensify communications between foreign and international gear scientists, to promote mutual design projects and participation in scientific events, etc.

In 1998, Prof. V. I. Goldfarb was elected to be the Chairman of this TC, and in 2001, after the report meeting of the Committee, his chair term was extended for 4 more years. The General Concept of the TC for Gearing and Transmissions activity within the framework of IFToMM, its main objectives, tasks, priorities were published in two languages in the international journal "Gearing and Transmissions", No.1, 1998 [28]. The Chair of the TC, Prof. Veniamin Goldfarb, outlined the urgent problems specific to this TC activities that time. This Concept was the plan which helped the Chair rationally manage the efforts to solve various problems. The Concept had been developed and first sent to all TC members after Prof. V. Goldfarb had started his activity as the Chair on Jan.1st, 1998 and approved on April 20th, 1998 in Thun, Switzerland at the Gearing TC meeting within ISO TC meeting. The Concept met the main objectives of IFToMM's activities defined in the Constitution. 7 main directions of activities were proposed: organization of international scientific technical programs in various fields of the gearing science; organization of the cooperation with TCs and PCs of IFToMM, with national and international associations of gearing specialists; publishing activity; educational activity; organization of scientific conferences; expertise; and organization of databases and informational exchange.

Professor Goldfarb regularly organized Gearing TC meetings within different international scientific events with the thorough discussion of all possible progressive international ways of cooperation between gear scientists. Numerous research and development projects were actively discussed there with initiators from Slovakia, Belarus and Russia, including the following projects:

- Development of Acoustic Emission (AE) Techniques for early diagnostics of gear failures.
- Development of transmission elements' life evaluation method with the aid of integral strain gauges.
- Development of a complex methodical and informational provision of gear and mechanical transmission reliability prediction.
- Development of simulation methods and experimental equipment for analysis of machine aggregates with new types of gears.
- Setting up Virtual Institutions.
- Development of a Multilingual Glossary of Gearing Terms.

Achievements of Gearing and Transmission TC activity stated by Professor Veniamin Goldfarb as a part of his Chairman report:

1. The committee unites the scientific potential of many countries according to one of important directions of MMS – theory and practice of gearing and transmissions;
2. Large international conference on gears are held annually with active participation of TC members as reporters and organizers;
3. National and international R&D projects are successfully implemented within the TC activity.
4. Cooperation with ISO and national associations in the field of gears is carried out, it unites scientists and practical engineers of many countries;
5. The experience was implemented of organizing scientific programs inside the committee with attraction of financial possibilities of institutions, participating in programs;
6. The precedent has been created of interaction with Reliability TC within the program "Creation of methods of diagnostics and evaluation of gearbox systems".
7. Gearing section is one of the most active in all World Congresses of IFToMM.

Prof. Goldfarb's active work in this Technical Committee was appreciated by the Executive Council of the Federation and since 2008 he became one of three Russians in the history of the Federation who was a member of IFToMM Executive Council. In 2011 Professor Goldfarb was elected as the Chairman of the Permanent Commission for Communications, Publications and Archiving (IFToMM PC CPA). The crowning achievement of Prof. V. I. Goldfarb's international activities was his election as Vice-President of this organization for the period from 2012 till 2015. Besides publication and archiving activities, during this period he was firstly dealing with promotion, administrating and updating the IFToMM webpage as the most efficient means of communication for scientists all over the world.

The main results of Prof. Goldfarb's international activity within IFToMM were also reported in many publications [16–21, 28–31], the complete list of his publications is presented in [1].

3 Conclusion

There is a famous phrase by Lion Feuchtwanger "A talented person is talented in everything". Besides being a well-known MMS scientist and researcher, a brilliant teacher and leader, a musician and sportsman, Professor Veniamin Goldfarb was a brilliant organizer who initiated a lot of interesting publications, cooperation and events. Many of these results and events seem obvious and common now (joint R&D projects, publication of scientific journals, international cooperation of Russian and foreign gear scientists, etc.), but 20–30 years ago it was professor Goldfarb who initiated this novel activity and contributed much to make it ordinary now in the Russian MMS community. In 2020 the Scientific Department "Institute of Mechanics" of Kalashnikov Izhevsk State Technical University (Udmurt Republic, Russia) was renamed into the "Institute of Mechanics named after Professor V. I. Goldfarb" to recognize his figure with his outstanding achievements.

References

1. Trubachev, E., Barmina, N., Malina, O.: Professor V. I. Goldfarb: life activity and contribution to gearing science. In: Barmina, N., Trubachev, E. (eds.) Gears in Design, Production and Education. MMS, vol. 101, pp. 1–59. Springer, Cham (2021). https://doi.org/10.1007/978-3-030-73022-2_1
2. Goldfarb, V.I., et al.: Spiroid gears for pipeline valves, Moscow, Veche (2011)
3. Saari, O.: Speed-reduction gearing. Patent USA № 2696125
4. Saari, O.: Skew-axis gearing. Patent USA № 2954704
5. Bohle, F.: Spiroid gear. Machinery **62**(2), 155–161 (1955)
6. Bohle, F.: Spiroid gear and their characteristics. Machinery **88**(6), 17–21 (1956)
7. Nelson, W.D.: Spiroid gearing. ASME Paper № 57-A-162, N.Y (1957)
8. Nelson, W.D.: Spiroid gearing. Part 1, 2, 3, Machine Design, 33(4), 136–144, (5) 93–100, (6) 165–171 (1961)
9. Goldfarb, V.I.: Fundamentals of the theory of computer-aided geometrical analysis and synthesis of general type worm gears: Thesis of Doctor of Technical Sciences (Dr. Sc. in Engineering). Ustinov (1985)
10. Goldfarb, V.I., Malina, O.V.: Skew axis gearing scheme classifier building technique. In: Proceedings of the 10th World Congress on MMS, vol. 6, Oulu, Finland, 20–24 June, pp. 2227–2232 (1999)

11. Nesmelov, I.P., Goldfarb, V.I.: Non-differential approach to the solution of the enveloping problem. In: Mechanics of Machines, Moscow, Nauka, no. 61, pp. 3–10 (1983)
12. Goldfarb, V.I., Nesmelov, I.P., Teterin, A.N.: Double-rim gear with intersecting axes. Author's certificate №1059325 dated 7.12.83, bul. N45 (1983)
13. Goldfarb, V.I., Trubachev, E.S., Kuznetsov, A.S., Makarov, V.V.: Development and manufacturing of spiroid gear drives for pipeline valves. In: Proceedings of the International Conference on Mechanical Transmissions, China, vol. 1, pp. 220–223. Science Press (2006)
14. Goldfarb, V.I., Makarov, V.V., Trubachev, E.S., Kuznetsov, A.S.: New progressive applications of spiroid gears. In: Proceedings of the 12th World Congress on TMM, Bezanson, France (2007)
15. Goldfarb, V.I., Trubachev, E.S.: Manufacturing synthesis of spiroid gearing. In: The 11th World Congress in Mechanism and Machine Science, Tianjin, China, pp. 901–905 (2004)
16. Goldfarb, V.I., Krylov, E.G., Serova, T.S: IFToMM contribution to attraction of youth to MMS development and promotion. Mechanism and Machine Theory, pp. iii–vii. Elsevier Science Publishing Company, Inc. (2016). https://doi.org/10.1016/j.mechmachtheory.2016.02.008
17. Goldfarb, V., Krylov, E., Elenski, A.: The first student international olympiad on mechanism and machine science - the challenge in MMS education. J. Mech. Eng. Autom. **3**(3), 152–158 (2013)
18. Starzhinsky, V.E., Shalobaev, E.V., Kane, M.M., Goldfarb, V.I.: Activities of Russian-speaking scientists in development of MMS terminology. In: Mechanisms and Machine Science, vol. 46, pp. 209–216. ISSN: 2211–0984, eISSN: 2211–0992. Kluwer Academic Publishers (2017). https://doi.org/10.1007/978-3-319-45450-4_21
19. Starzhinsky, V.E., Goldfarb, V.I., Algin, V.B., Shalobaev, E.V., Kane, M.M.: Participation of scientists from the USSR and CIS countries in IFToMM activities. In: Theory of Mechanisms and Machines, vol. 12, no. 2 (24), pp. 81–102 (2014)
20. Theory and practice of gearing. In: Proceedings of the International Symposium, January 21–23 2014, Izhevsk, ISTU Publ. (2014)
21. Goldfarb, V., Barmina, N. (eds.): Theory and Practice of Gearing and Transmissions: In Honor of Professor Faydor L. Litvin. vol. 34, Springer, Cham (2016). https://doi.org/10.1007/978-3-319-19740-1. ISBN 978-3-3 19-19739-5
22. Spiroid gearboxes and gearmotors. Production catalogue. Izhevsk (2001)
23. https://mipmechanic.ru/. Accessed 23 Jan 2022
24. Development of the geometric theory of gearing. In: Proceedings of the International Symposium, Izhevsk, ISTU Publ. (1993)
25. Goldfarb, V., Trubachev, E., Barmina, N. (eds.): Advanced Gear Engineering, vol. 51. Springer, Cham (2018). https://doi.org/10.1007/978-3-319-60399-5. ISBN 978-3-319-60398-8
26. Goldfarb, V., Trubachev, E., Barmina, N. (eds.): New Approaches to Gear Design and Production, vol. 81. Springer, Cham (2020). https://doi.org/10.1007/978-3-030-34945-5. ISBN 978-3-030-34944-8
27. Trubachev, E., Barmina, N., (eds.): Gears in Design, Production and Education. A Tribute to Prof. Veniamin Goldfarb, vol. 101. Springer, Cham (2021). https://doi.org/10.1007/978-3-030-73022-2. ISBN 978-3-030-73022-2
28. Goldfarb, V. I.: Concept of activity of IFToMM technical committee for gearing and transmissions. In: Gearing and Transmissions, no. 1, pp. 53–65 (1998)
29. Goldfarb, V.I.: Theory and practice of gearing in machines and mechanisms science. In: Ceccarelli, M. (ed.) Technology Developments: the Role of Mechanisms and Machine Science and IFToMM, pp. 133–139. Springer, Heidelberg (2011). https://doi.org/10.1007/978-94-007-1300-0_10

30. Goldfarb, V.I.: Directions in gear research – contribution and function of IFToMM technical committee on gearing. In: Proceedings of the International Conference on Gearing, Transmissions and Mechanical Systems, Nottingham, London, pp. 3–9 (2000)
31. Goldfarb, V.I., Mudrik, J., Kral, S., Oleksyuk, V.: Some results of joint research on the international scientific program "Mechanical Drives of New Generation". In: Proceedings of the International Seminar "Problems of the Dynamics Aggregates", Trnava, Slovakia, p.7 (1996)

An Historical Review and Upload of Juanelo Turriano's Mechanism in Toledo

José Antonio Hernández-Torres(✉), Angel Mariano Rodríguez-Pérez, and Julio Jose Caparrós-Mancera

University of Huelva, Huelva, Spain
joseantonio.hernandez@dimme.uhu.es, angel.rodriguez@dci.uhu.es, julio.caparros@diesia.uhu.es

Abstract. This work introduces the figure of Juanelo Turriano, his works and historical legacy. The main work developed by Juanelo was the hydraulic mechanism, placed in the Spanish ancient city of Toledo. The purpose of this mechanism was to provide a clean water supply to the city. Taking advantage of the study of Juanelo's main work, an analysis about the evolution of the hydraulic mechanisms, which replaced Juanelo's, and the possibilities for future developments is carried out. This study analyze the implementation of a pumping group, fueled by the energy generated by a turbine, thus it uses the power generated by the water flow through the Tajo River. This update draw from the same principles and premise, it also have a similar purpose and gets a similar result tailored to the new hydraulic urban requirements. Despite of hosting a similar population, the present city of Toledo has increased enormously its water requirements.

Keywords: Turbine · Pump · Renewable energy

1 Introduction

There are several Spanish scientists, engineers and writers, who pointed out in different fields during the XVI and XVII centuries. Nonetheless there are few literature dedicated to highlight their work and remember their figures [1].

Spain has stood out throughout history as the cradle of illustrious artists and writers who have reached high positions in all fields of knowledge on many occasions. There have also been many great inventors and scientists who had no choice but to choose the path of "exile" or to reach the end of their lives in absolute poverty [2]. The biography of Juanelo Turriano is one of the many examples that tell about people who, despite their ingenuity, had the misfortune to develop their knowledge in a place where scientific progress was viewed with envy and even suspicion [1].

What is known about Juanelo's life refers almost entirely to the stage in which he worked in the service of Carlos V, repairing and building clocks and executing automatons [3] (Fig. 1).

Juanelo Turriano was born in Cremona, Lombardy (Italy), between the years 1500 and 1511. The date of his birth has not been settle but the latest studies tend to place

M. Ceccarelli and R. López-García (Eds.): HMM 2022, HMMS 40, pp. 41–48, 2022.
https://doi.org/10.1007/978-3-030-98499-1_4

Fig. 1. "Bust of Juanelo Turriano", Pompeo y Leone Leoni, ca. 1565–1570. Museum of Santa Cruz (Toledo).

it closer to the beginning of the century [4]. Juanelo, who was born under the name of Giovanni Torresani, died in 1585. After the death of Carlos V, Juanelo continued in the service of his son, Felipe II, who asked him to build an artifact to raise the waters of the Tajo river to the Real Alcazar [3]. This mechanism is Juanelo's most known work.

From a humble family, his father Gherardo Torresani had two mills in operation on the Po River. Juanelo was not an illiterate, it could not be otherwise given his friendship with characters like Giorgio Fondulo, who, in addition to being a renowned physicist, doctor, philosopher, mathematician and astrologer, taught philosophy at the University of Pavia. The influence that Fondulo exerted on Juanelo was fundamental in his life by intervening in the early formation of the young inventor [5].

Juanelo arrived to Spain to serve the Emperor Carlos V in 1530. He worked as watchmaker for the Court [6]. His restoration of Giovanni Dondi's planetary clock and the construction of the famous "Cristalino" astronomical clock are remembered from this stage. He also stood out as engineer, mathematician and manufacturer of mechanisms, enjoying so much the favor of the monarch, that Juanelo accompanied the King in his retirement [5, 6].

After the death of the emperor, his son, Felipe II did not want to run the kingdom without his service and appointed him as Senior Mathematician, participating at this time in the reform of the current calendar, while he was "lent" to the service of Pope Gregorio XIII. During his stay in Spain. Juanelo's contributions to the Empire's heritage were numerous: construction of the Tibi reservoir dam in Alicante, the highest in the world for almost 300 years; he designs the bells, during the construction of San Lorenzo del Escorial, manufactured by his friend Juan de Herrera; he also builds mills, various automatons [7, 8].

It was in 1565 when Turriano settled permanently in Toledo. By then, the city was no longer the imperial capital, since it has been a few years since Felipe II established his Court in Madrid. In the years in which it has been, Toledo has lived through a period of splendor and demographic expansion, although it has not solved its problem of water supply that it drags [6, 9].

This task had already been tried before, the people in charge of the first attempt were two German officers whose attempt finally failed in 1526, and two Flemish engineers suffered the same fate who also tried to satisfy this demand. These previous rulings caused harsh and demanding clauses to be imposed on Juanelo. He would not receive any part of the payment until the water reached the Alcázar. The amount of water destined for the Real Alcazar was one seventh of the total to be raised. After 4 years, the work was finished, but Juanelo's success was not recognized and he did not get paid the agreed amount of money. By the side of the city of Toledo did not pay because the water reached the Alcazar in and from there it was distributed to the city and by the side of the king, because he had not been compromised to pay for it.

Frustrated and in a difficult economic situation, Turriano proposed to the city the construction of a second device. This time Juanelo would have been the one who retained the rights to the exploitation. The work was completed in 1581 and, this time, it seems, Turriano did get paid. Although his ordeal was not over. Juanelo could not cope with the subsequent costs of maintaining the mill and had to end up giving control of it to the city.

After numerous requests, only his daughter received partial compensation after Juanelo's death [3, 6].

2 Background and Its Context

2.1 The Need of Water

Advances in technology and architecture have always been essential to the formation and durability of empires throughout history, and water supply has also played an important role. During the XVI and XVII centuries, the Spanish crown became a major political and economic power in the world, so great that "the sun never set" under King Felipe II reign [9].

Water is and has always been a fundamental element in the development of any society. There is concrete news from more than 4000 years ago by which various peoples worried about finding good quality and abundant water. The existence of pipelines and deposits in Jerusalem with more than 3000 years old is known [10]. Relevants hydraulic devices were designed along the history from Antiquity (by Ctresibius, Heron, Vitruvius etc.), in the Middle Ages (Villard de Honnecourt etc.), in the Renaissance (Francesco di Giorgio, Sangallo etc.) and in the following centuries (Branca, Ramelli, Puccini, Ozanam etc.) [11, 12].

This is why the settlement of the first people of Toledo is understood to have occurred in the vicinity of the riverbed, settling on the nearby hills, having to go down to the water constantly. The clarification of the water was carried out by decantation in cisterns or reservoirs [13].

In Roman times, civilization and improvements in the life of population arrived to the peninsula. Roman Empire governors resolved the supply problem by building a dam in Alcantarilla, in the municipality of Mazarambroz [14] as shown in Fig. 2: Alcantarilla's dam engraving by Santiago Palomares. s. XVIII as well as a stone conduit with aquarium towers to bridge the unevenness. The dam was 500 m long and had a maximum height of 21 m [7].

Fig. 2. Alcantarilla's dam engraving by Santiago Palomares. s. XVIII

2.2 Juanelo's Mechanism

The Juanelo's mechanism, Fig. 3, carried 1.17 m^3/h in 1569 [8]. The population of Toledo back to this year was around 53,700 people [15]. Which means that each person consumed 0.52 L per day, this would currently be an extremely small amount of water, insufficient for current water-demanding activities. Nowadays, the average consumption per habitant entails an increase of more than the 33000% in water consumption.

2.3 Further Mechanisms After Juanelo's

The construction of the mechanism in Toledo caused a great sensation, not only within Spain, where most of the great writers of the Golden Age mention it in their works, but

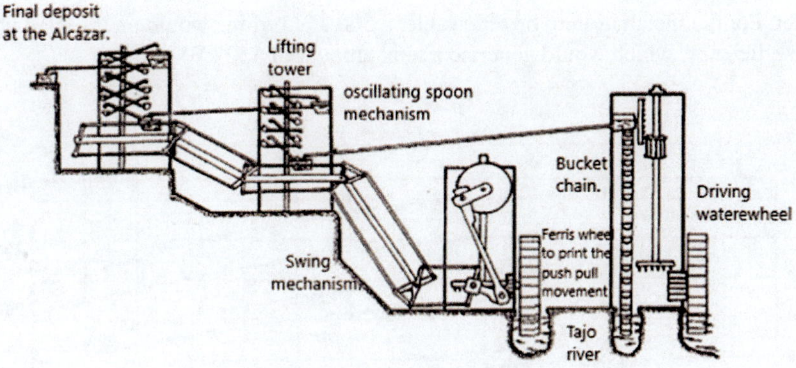

Fig. 3. Juanelo's water lifting mechanism [16].

also outside, being visited by such relevant figures as D. Juan de Austria [7, 16]. Until then nothing similar had been done in the world, since the highest elevation previously made was that of Augsburg; a tower of less than forty meters and that in view of descriptions of the time used Archimedean screws to raise the water level. Despite the success and renown he gained, Turriano would die at his home in Toledo after having been forced to hand over his artifice to the city by not being able to take charge of its maintenance [8, 16].

The machines, however, continued operating until 1639, although each time they gave less performance. By then, due to lack of maintenance and parts theft, the two machines were already in a very bad state. That year, the first was dismantled and the second was left standing as a symbol of the city. Without the artifices of Juanelo, the situation returned to normal and the water rose again to the city on the back of his donkeys. With the passage of time, little was left of the second. The looting ended up reducing it to ruins as well.

Despite the passing of the years and the physical disappearance of the machinery, the admiration for the artifice has not been lost, and the answer to the question of how the Artifice of Juanelo worked still remains a matter of controversy. Several have tried to find an explanation and several proposed models, but it is not an easy task, as no plan or drawing of the device has been preserved. The only thing they have been able to count on, those who have tried, has been the descriptions made by travelers and writers of the time.

3 Mechanism Update

The current average consumption per habitant in the city of Toledo is 132 L [17] per day, which represents a flow of 440 m^3/h. According to the capacity of the available commercial pumps, for a 50 m^3/h water flow and its characteristic curve, as shown in the Fig. 4, thus, to replace the mechanism and make it renewable today, it would be necessary to install 9 commercial pumps, model CS 50-250A from Speroni Company, with 22.5 kW power each. This group makes a total consumption of 202.5 kW of installed power. For the installation to be renewable, a 500 kW turbine would be installed with a 70% efficiency, which would generate a total amount of 350 kW [18].

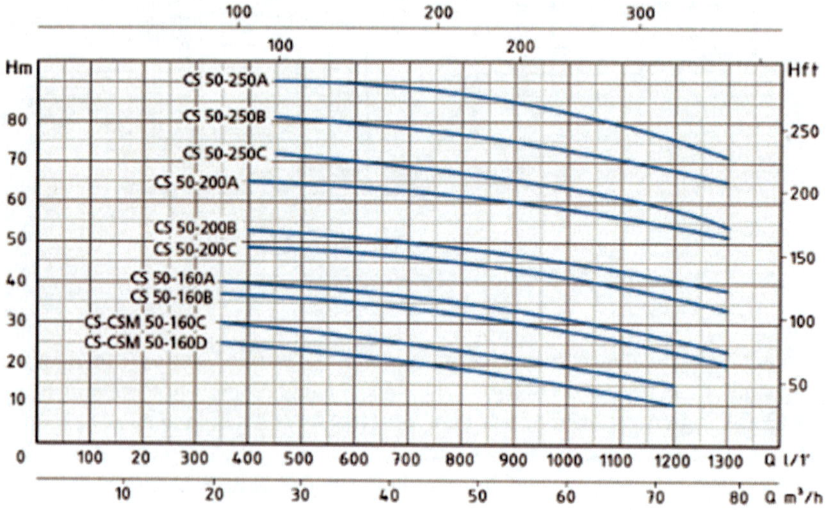

Fig. 4. H/Q Curve for the selected pump [19]

The flow that would have to pass through the turbine to generate that power would be 2 m^3/s, this corresponds to 4% of the river flow at the point where the turbine would be placed [20]. To achieve this flow, a quadrangular channel 1 m wide by 1 m high would be built. At the entrance of this channel, a gate would be placed to regulate the inflow through said channel. It is also necessary to place a grid to prevent fish and weeds from entering it. Keep in mind that maintenance would have to be carried out every so often to clean the dirt that may remain on the grill. Once it passes through the turbine, part of that water is pumped and a part of it returns to the river.

The work that would have to be carried out would be similar to the one that can be seen in the Fig. 5: Model for hydroelectric generation in a river, which is a scale model made for a civil work that is being carried out in the town of Pastores (Guatemala).

The objective of the turbine in both locations would be similar, which is to generate enough energy so that a pump can be powered. This machine would be in charge of pumping the water from the Tajo River to the city of Toledo. In the case of Guatemala, it would be the Guacalate River as it passes through Pastores. In this case, it must be taken into account that in the region where this turbine is to be installed, the majority of the population is in a situation of extreme poverty.

The estimated cost of the installation to be placed in the Tajo River would be of 2670 € per pump, 10,000 € for the turbine and 50,000 € for civil works. Which adds up to a total amount of 84,000 €.

Fig. 5. Model for hydroelectric generation in a river

4 Conclusions

The technological advances developed by Juanelo Turriano represented a great revolution in his time, despite the remaining documentation from that time, the complexity of the mechanisms developed continues to pose many unknowns. In particular, the mechanism developed for the city of Toledo is of great application today. However, social needs for water supply have changed over time. Therefore, an update of its mechanism is proposed based on available technologies and current needs.

References

1. Martínez-Pinna, J.: Juanelo Turriano, el genio renacentista español. Clío. Revista de Historia, **210**, 88–91 (2019)

2. Eamon, W.: Spanish Science in the Age of the New, de A Companion to the Spanish Renaissance, Hilaire Kallendorf, pp. 473–507 (2018)
3. Zanetti, C.: Juanelo Turriano, de Cremona a la corte. Fundación Juanelo Turriano, Madrid (2015)
4. Barisotti, R.: Janello "Torresani" Alcuni documenti cremonesi e Il Baptismum del Battistero. Bolletino Storico Cremonese, vol. VII, pp. 255–267 (2000)
5. Revuelta-Pol, B., Romero-Muñoz, D.: Juanelo Turriano. Relojero e ingeniero cremonés. de En: Realismo y espiritualidad. Campi, Anguissola, Caravaggio y otros artistas cremonenses y españoles en los siglos XVI-XVIII, Valencia, pp. 73–81 (2007)
6. Bautista, E., Muñoz, J., Echávarri, J.: Juanelo (1501–1585). In: Ceccarelli, Marco (ed.) Distinguished Figures in Mechanism and Machine Science, pp. 95–110. Springer, Dordrecht (2010). https://doi.org/10.1007/978-90-481-2346-9_5
7. de-la-Escosura-y-Morrog, L.: El artificio de Juanelo y el puente de Julio César, Madrid: MAXTOR, 1888 (Re-edited 2012)
8. Cabovolo: El prodigioso y desaparecido artificio que Juanelo construyó en Toledo. Cabovolo (2010). http://www.cabovolo.com/2010/02/el-artificio-de-juanelo-en-toledo.html. Accessed 08 Nov 2021
9. Otero, J., et al.: The twenty-one books of devices and machines: an encyclopedia of machines and mechanisms of the 16th century. In: Yan, H.-S., Ceccarelli, M. (eds.) International Symposium on History of Machines and Mechanisms, pp. 115–132. Springer, Dordrecht (2009). https://doi.org/10.1007/978-1-4020-9485-9_9
10. Paz-Maroto, J., Paz-Casañé, J.M.: Abastecimiento y depuración de aguas potables, p.14: Escuel técnica superior de ingenieros industriales (1969)
11. Ceccarelli, M., Cigola, M.: Trends in the drawing of mechanisms since the early Middle Ages. Proc. Inst. Mech. Eng. Part C: J. Mech. Eng. Sci. **215**(3), 269–289 (2001). https://doi.org/10.1243/0954406011520715
12. Cigola, M., Ceccarelli, M.: Machine designs and drawings in renaissance editions of de architectura by Marcus Vitruvius Pollio. In: Sorge, F., Genchi, G. (eds.) Essays on the History of Mechanical Engineering. HMMS, vol. 31, pp. 291–307. Springer, Cham (2016). https://doi.org/10.1007/978-3-319-22680-4_17
13. Bernabé-Merino, M.L.: El problema del agua en Toledo en la Edad Moderna. Escuela de Magisterio de la Universidad de Castilla-La Mancha, Ciudad Real (1994)
14. Porres Martín-Cleto, J.: Historia de las calles de Toledo, vol. 1, Toledo, Zocodover, pp. 47–48 (1982)
15. Kagan, R.L.: Contando vecinos: El censo toledano de 1569. de STVDIA HISTÓRICA. HISTORIA MODERNA, XII, Baltimore, Johns Hopkins University (1994)
16. Bermejo-Herrero, M., González-Conde, L., Del-Río-Cidoncha, M.G., Martínez-Palacios, J.: Reconstrucción visrtual del artificio de Juanelo Turriano para elevar agua del río Tajo a Toledo. Fundación Juanelo Turriano, Madrid (2013)
17. Instituto-Nacional-de-Estadística: Estadística sobre el suministro y saneamiento del agua. Instituto Nacional de Estadística, Madrid (2018)
18. Rodríguez-Pérez, Á., Pulido-Calvo, I.: Analysis and viability of microturbines in hydraulic networks: a case study. J. Water Supp.: Res. Technol.-Aqua **68**(6), 474–482 (2019)
19. Speroni: Speroni cat Gen 2018: Corefluid (2018). https://www.corefluid.es/product-category/bombas/bombas-centrifugas-speroni/bomba-speroni-cs/. Accessed 15 Nov 2021
20. Centro de Estudios Hidrográficos. Anuario de aforos 2017 - 2018. Estaciones de aforo (2018). https://ceh.cedex.es/anuarioaforos/afo/estaf-datos.asp?indroea=3904. Accessed 15 Nov 2021

The Contribution of Huseyin Vasfi Efendi from Plovdiv to Ottoman Technological Literature

Melek Dosay Gokdogan(✉)

Ankara University, Ankara, Turkey
mgokdogan@ankara.edu.tr, mdosay@hotmail.com

Abstract. In this paper, *History of Production, Inventions and Discoveries by Mankind* (1888), written by Huseyin Vasfi Efendi from Plovdiv where in Bulgaria today, will be presented and evaluated in terms of the history of machinery and technology of the Ottoman Empire. In this book, only inventions, discoveries and innovations are listed chronologically; their technical specifications are not disclosed. Written by the Ottomans at the end of the 19th century, this work has been studied with the idea that it can shed light on the Ottomans' understanding of industrialization and technology, despite the Europeans who lived and realized the industrial revolution earlier.

Keywords: Huseyin Vasfi · History of discoveries · Ottomans' technological understanding

1 Introduction

In this paper, information about technology, especially about machines, will be determined in Huseyin Vasfi bin Ahmed Hulkî Filibevî's work titled *Târih-i İ'mâlât ve İhtirâ'ât ve Keşfiyât-ı Beşeriye* (History of Mankind's Production, Inventions and Discoveries) written in 1888 [3; Fig. 1] and will be evaluated in terms of the history of machinery and technology of the Ottoman Empire.

Our writer Huseyin Vasfi Efendi is from Plovdiv. Plovdiv is the second most important city in Bulgaria after Sofia today (Fig. 2). It was one of the important centers in the Balkans during the Ottoman rule. This importance was due to its cultural assets and the scholars who grew up there. Governor of Rumelia, Lala Şahin conquered this place for the first time in 1361. Bulgaria was subjected to the Ottomans for nearly 500 years. In Rumelia it was a general Ottoman method to take young boys and girls from their families and make them Ottomans. These young apostates, whose background was probably influenced by the most advanced centers of Europe due to geographical and cultural proximity, played an important role in introducing modern sciences to the Ottomans. However, Huseyin Vasfi did not have any relationship with Bulgaria after that. In 1885, the newly established Bulgarian Principality captured Plovdiv. Thereupon, most of the Turkish population here emigrated [6].

M. Ceccarelli and R. López-García (Eds.): HMM 2022, HMMS 40, pp. 49–56, 2022.
https://doi.org/10.1007/978-3-030-98499-1_5

Very little is known about the life of Huseyin Vasfi Efendi from Plovdiv. He probably settled in Istanbul when Plovdiv fell into the hands of the Bulgarians or at an earlier date. He is mentioned with his two books in the references. He gave information about his profession in the preface of these works. Accordingly, he was the clerk of the indictment committee in the appellate court of the supreme Ministry of Justice [4, p. 3].

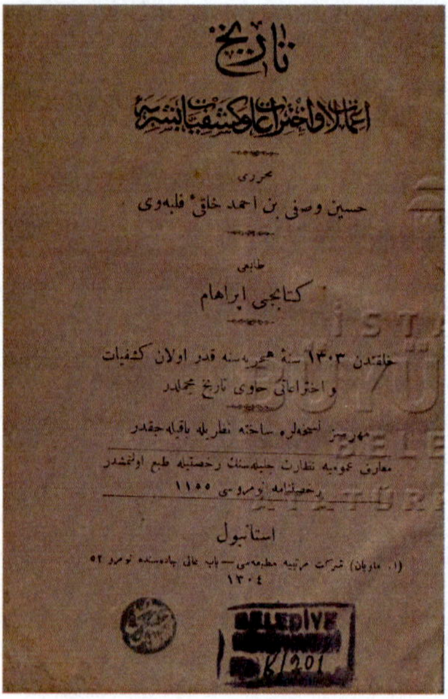
تاریخ
اعمالات واختراعات وکشفیات بشریه
محرری
حسین وصفی بن احمد خاقی فلبه‌وی
طابعی
کتابجی ابراهام
استانبول
١٣٠٤

Fig. 1. The cover page of *History of Production...*

One of his books is called *Esile ve Ecvibe-yi Hikemiye Yahut Tenvîr-i Efkar* (Questions and Practical Answers or Clarification of Ideas) and is about ethics. It consists of 67 pages. The name of this work is mentioned in the studies on ethical books in the Ottomans [For example; see 1, p.175, and 2].

Two editions of the work that we will examine here have been reached. Although their publication years and numbers of page are different, the content is the same, only the 1305 edition contains (152 pages) a list of errors. Our text on which we depend in this paper is consisting of 109 pages [3]. Existing of two editions of this book shows its popularity.

Among the simple machines and devices mentioned in this book are water and windmills, clocks (with alarm and pendulum), steamboat, knives, hammer, axes, surveying instruments. Apart from these mechanisms, all kinds of innovations related to science, technology and culture (such as the Great Wall of China, electricity and magnet, galvanometer, sewing machine, guns and their kinds, towers, some historical bridges, etymology for the word of Spain) are mentioned.

When another similar book, *Inventions and Discoveries* (Hijri 1298/AD. 1881) by M. Naim, written just before Huseyin Vasfi's book, is examined, it is seen that it is not the

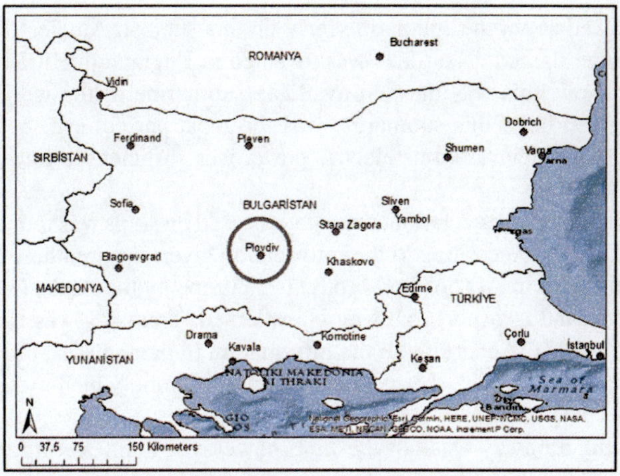

Fig. 2. Plovdiv in the map.

same in style. The subjects examined in this 76-pages of book are determined as mill, clock, steam power (steam-railway-etc.), compass, glass-mirror-binoculars, weapons, paper, measuring air, measuring temperature, balloon, lighting, electricity, telegraph, printing press, and camera [8]. Information about these is given beyond the encyclopedic nature. So it surely can be concluded that our writer (Huseyin Vasfi) did not influenced from this book.

2 History of Mankind's Production, Inventions and Discoveries

In the book, the inventions are listed chronologically, and a little more information is given about the relevant invention by opening parentheses in almost every article. The dates are grouped into three categories. The author explained this as a warning: It was divided into three classes, first creation date (actually it is a Jewish history in which the year of 2000 corresponds to 6800), second Christian era and third hijri calendar, and inventions were written in order according to these three classes [3, p. 3].

In the foreword, the author, after praising Sultan Abdulhamid Han II, the sultan of the time, and explaining that it is our duty to pray for the Sultan, who is trying to bring the Ottomans to a high position between European states and civilized nations, he says that he attempted to write this book in the hope of helping the sultan in his good works.

Here, we will give some simple tools and machines among the chronologically listed inventions in the book. The criteria in the selection of these tools and machines were determined to be comprehensible in the Ottoman text and to be compatible with the universal knowledge about these tools.

For example, the invention of the steamboat is mentioned. According to the history of creation in 2241, Prophet Noah built and invented ships. According to the detailed information given in parentheses, Denis Papin developed the steamboat by inventing a steam engine in Hijri 1114/AD 1705. The American Fulton also improved it and made his trial on the Seine River in Paris in Hijri 1218/AD 1803 [3, p. 8].

Later, he talked about the submarine ferry on this subject. An electrically powered submarine steamer named "Nautilus" was invented in England in Hijri 1303/AD 1798 (As far as we know, this was the first available submarine in the world). According to the details given here, this submarine was the most perfect ever built. It had two propellers and two rudders, and its electric power was sufficient to travel eighty miles [3, p. 107–108].

As a subject related to navigation, the compass used on ships was found in Andalusia in Hijri 683/AD 1284. According to the information given in parentheses, the compass was invented in China in AD 565 and brought to Europe by the Arabs [3, p. 78].

As related to land transport tramway is explained. Tramway was built in Istanbul in Hijri 1288/AD 1871. According to the information in parentheses, in Hijri 1290/AD 1875, the "tunnel" in Istanbul, that is, the underground tram, which moves with a belt, was put into use in Galata [3, p. 106].

Of course this tramway was not the first; however it is interesting to see that our writer is pointing to the constructing of a new vehicle in his country without mention its first appearance in England.

This tram was powered by two steam engines and by connecting the Galata Bridge to Pera Hill (Beyoglu today), it saved people from climbing uphill on steep streets where vehicles could not enter. The idea came from Eugène Henri Gavand, a French civil engineer. The "tunnel", nicknamed the "rat hole" by the city dwellers, was one of the oldest (actually second) and shortest European subways [10, p. 435].

Yet he has mentioned the invention and construction of the railway. Accordingly the railroad was invented by British engineer George Stephenson in Hijri 1230/AD 1814. According to the information in parentheses, a railway was built from Manchester to Liverpool for the first time in Hijri 1240/AD 1830 [3, p.104; 7, p. 330].

According to the date of creation, in 4070, the car was first invented as a two-wheeler in the city of Rome. According to the information in parentheses, this car was an ox cart. Four-wheeled ox carts were introduced by the famous French ruler Thierry (III) in AD 680. A type of car was invented, which later took the name Carl, and some masters called "sîrak" practiced their art with them, even in Istanbul. Some horse-drawn carriages with the names of phaetons and cups were also revealed in Hijra 1205/AD 1790 [3, p. 19].

One of his most concerned about devices is clocks. According to the date of creation in 5432, a water-powered alarm clock was invented in Rome. According to the information given in parentheses, this clock was used in the city of Rome; the current alarm clocks, on the other hand, were invented in Baghdad in Hijri 175/AD 792, used in the palace of the famous Abbasid Caliph Harun Rashid, and one of them was sent to the King of France Charlemagne by Harun Rashid with the key of the Church of the Blessed Grave in Jerusalem. The pocket watch was found in Hijri 922/AD 1516 [3, p. 44; it is said that this was a clepsydra, see; 5, p. 475].

On the other hand, a type of pendulum clock was invented in Italy in Hijri 304/AD 917. According to the information given in parentheses, Huygens, one of the famous scholars, applied the pendulum to the hanging clocks and found the use of the spring in the clocks in general [3, p.68].

As to the scientific instruments, in Hijri 1049/AD 1639, the "barometer", that is, the air measuring instrument, was invented in Florence, Italy, by docente universitario

Torricelli. According to the information in parentheses, the famous Cavendish later improved this instrument and brought it into its current form. Saussure, one of the scientist, found the device that measures the amount of expansion and moisture present in the air, and Hauksbee invented the air discharge device for evacuating the air in Hijri 1116/AD 1704 [3, p. 88].

As to the telescope, it was invented by the Dutchman Zacharias in Hijri 1175/AD 1761. According to the information in parentheses, the spectroscope instrument was also invented by applying it. Zacharias discovered that there are more than 1500 comets in the sky. Mentioning that these comets move in all directions in the planetary region and that the comet "peyl" passes so close that it leaves part of the Earth's orbit in a cloud, he said, "There is no doubt that the Earth will collide with one of them one day." Indeed, on the 29th day of October in the year Hijri 1278/AD 1832, a comet named "Gâmbâr" circulated 10 million billion meters away from the orbit of the Earth, but this star completed its cycle once every seven years. Wilson, one of England's master astronomers, discovered that each of the spots seen with a telescope on the surface of the Sun consists of a dark pit several thousand miles deep in Hijri 1185/AD 1771. Sometimes fragments of fire emanate from the surface of the Sun, reaching an altitude of 40 to 45 thousand miles. Although there were some volcanoes on the surface of the moon globe and Mercury in the past, they have completely extinguished and there are no volcanoes other than these extinct mountains in these spheres. Although the volcanic mouths on Earth only disappear and flatten over time, the fire-spreading mouths in them always maintain the state they opened. Some scholars have also discovered that there is some steam and fragments on the surface of Mars, and even that the white spots on it are snow. Newton, one of the famous English scholars, discovered that all inanimate objects in the air are suspended in the air and are in balance without ever falling by the mutual forces attracting each other. British Chamberlaine, who recently thought that all diseases are caused by some small animals called microbes, mostly occurring in water consisting of two types of gas, invented a tubular microscope instrument for this purpose [3, p. 90–92].

The invention of the printing press was a major technological advance and as a forerunner of the Renaissance, our writer naturally mentions it. In Hijri 847/AD 1443, a man named Gutenberg in the city of Mainz in Germany manufactured letters from small pieces of wood with a person named Fust and invented the printing press in this way. According to the information given in parentheses, the art of printing was first discovered in the city of Beijing in China in Hijri 300/AD 913. The wooden letters invented by Gutenberg were developed by turning them into lead and then copper in Hijri 867/AD 1463. The famous lithograph was invented in Egypt in 1215/AD 1800. Printing was heard in Istanbul in 1130/1718, and the printing house was established in Hijri 1138/AD 1726 and in 1259/1843 two types of ta'lik (a style of Arabic script) and three types of naskh (a cursive style of Arabic calligraphy which most popular for and characteristic of the Arabic language) and two types of movable letters were invented in Istanbul by masters called engineers [3, p. 80].

Another matter related to the rise of Renaissance, gunpowder was invented in Andalusia in Hijri 41/AD 662. According to the information in parentheses, gunpowder was initially used by the Arabs who besieged the city of "Nepila". In Hijri 741/AD 1341, Moroccan soldiers fired cannon with gunpowder during the siege of the castle called

"Tarife" and they invented this cannon in Hijri 762/AD 1361. Some European historians have said that the Arabs took gunpowder from the Chinese and spread it to Europe [3, p. 65].

As to the interesting invention, according to the date of creation in 5049, a scholar named Honkuçu (?) in the province of Chantuni (?) in China manufactured and invented a kind of shoe made of elephant skin. According to the information in parentheses, this person also invented a kind of sword by sharpening each side of the long bones very finely. On the other hand, shoes were found in the 17th century and later developed, and a very neat and soft kind of shoe was produced from snake skin in America. A shoe-making machine called "Neptune" was also invented in Europe [3, p. 33].

In Hijri 1062/AD 1652, the umbrella was invented by the famous British Robinson on an island where he was forced to live for twenty-eight years. According to the information in parentheses, the umbrella was subsequently improved in London. It was also spread in Istanbul in Hijri 1227/AD 1812 and a red one was first presented to Sultan Mahmud [3, p. 89].

Our writer also has been disregarded for the invention of telephone; in Hijri 1286/AD 1876, the famous American Graham Bell invented the telephone device. According to the information in parentheses, Bell carried out wireless communication via electric current between two ships placed two kilometers away and even established a telephone company in London [3, p. 106].

In Hijri 1180/AD 1766, photography, that is, the device of taking pictures with the help of light, was invented by the French people named Niepce and Daguerre. According to the information in parentheses, this instrument was later improved by the Swedish Schenil and this fine art was spread in Istanbul in 1277/AD 1860 [3, p. 93–94].

This book has a wide range of interest and arms of war were among them. According to the date of creation, in 2925, the iron mace (a mace used in battle) and sword and knife were invented by the famous Persian ruler Tahmuras. According to the information in parentheses, this ruler also found ways of riding on animals and loading things, and teaching his hawk and falcon birds to hunt. The hammer, ax and saw were invented by the ancient Mexicans [3, p. 12].

There are interesting issues within the book which are not machines, however these related to scientific approach and attitude. For example, according to the date of creation, in 5236, it was invented to use as a mirror by polishing silver and other metals in Egypt. According to the information in parentheses, the current mirror was invented in Venice in Hijri 923/AD. 1517. Glass, on the other hand, was first invented by the famous Phoenicians, who later found the crystal, from saltpeter. In Hijri 1118/AD 1706, the French King Louis (XIV) had a large and very clear mirror made in Paris with 114 fingers long and 67 fingers wide. This mirror is preserved in the Louvre palace in Paris today [3, p. 39–40].

As another example related to his own country, in AD 538, the astonishing wall surrounding the great city of Istanbul was built by Emperor Justinian to be safe from teasing by Bulgarian bandits by land and by other enemies than the sea. According to the information in brackets, unusual pictures were engraved and carved (decorated with paint) in some parts of this wall [3, p. 58].

In Hijri 905/AD 1500, King Ferdinand of Spain had a 17-m-high statue built with a cross decorated with diamonds, emeralds and rubies. According to the information in parentheses, this Ferdinand expelled the Jews living in Spain in Hijri 901 with the encouragement of his wife, and some of them preferred to settle in Thessaloniki. In 1047/AD 1637, they abolished their religion by being attached to Sultan Murat in Edirne [3, pp. 85–86].

As a last example, in Hijri 923/AD 1517, the shipyard in Istanbul was first built by the famous sultan, hardworking Sultan Selim. According to the information in parentheses, he also had the Anatolia Fortress built, and the admiralty, which means naval commander, comes from the word "emir-ul-ma" [3, p. 86].

3 Conclusion

It seems like an exaggeration to consider this book as a history of technology or machine book. It can be considered as an encyclopedic history of inventions written to inform Ottoman intellectuals at the end of the 19th century. As a matter of fact, in the last word of his book, the author states that he wrote this work in order to both serve his nation by giving a little bit of information about the progress of humanity, and to advise the young people of the country to work hard in order to complete these deficiencies in the country despite of having no his places at the education institutions such as universities, technical or professional schools [3, p. 109]. The author did not specify the reference or references. The information given in parentheses is completed with the verb "it is said". As it can be understood from this, the author has arranged the information he gathered from some foreign books in chronological order. Although his education is unknown, it is reasonable to admit that he did not have a technical education, given that he was the author of another book on ethics and held the position of court clerk. In this case, it is natural that the author does not give technical information about the subjects he mentions.

Considering the political and military downsides of the Ottomans at the end of the 19th century, it is admirable that our writer, Huseyin Vasfi Efendi wanted to inform the Ottoman intellectuals and society of the history of technical developments in the world and in his own country, by writing such a book in the face of industrialized and world power Europeans. So he exposed his desire of progression in both scientifically and technically his country.

On the other hand, it is seen that some papers on philosophy of technology and machine had been published in Ottoman literature by the time in first quarter of twentieth century. One of the most remarkable among them is the translation of *Apologie der Technik* by Richard von Coudenhove-Kalergi who was an Austrian- Japan politician and philosopher, (Leipzig 1922), translated to Turkish by engineer Bekir Sitki [9]. This book is about the philosophy of technology. Apart from that with some papers the discussions on technology and machine had been continued: "Machine and Life of Thinking" by Mustafa Şekip Tunç (1927), "La Rançon du Machinisme" by Jean Gallotti (1931), and "Let us Perform the Machine, It will Save the Humankind" by Roger Francq (1932).

Consequently *the History of Mankind's Production, Inventions and Discoveries* which I have studied here, may have contributed to the setting in which this technology and machine discussion takes place.

Acknowledgement. I would like to thank for her help for English translation of this paper to Professor Mine Dosay Akbulut. Other thanks to Associate Professor Nazan Karakaş Özür for drawing the map in Fig. 2, and to Dr. Vural Başaran for his support the technical matters of this paper.

References

1. Tahir, B.M., Rifat: "Ahlâk Kitaplarımız", Sadeleştiren: Saadettin Özdemir, Süleyman Demirel Üniversitesi İlahiyat Fakültesi Dergisi, Number **22**, pp.161–176 (2009)
2. Erdem, H.: Osmanlıda Ahlak ve Bazı Ahlak Risaleleri. Necmettin Erbakan Üniversitesi İlahiyat Fakültesi Dergisi **10**(10), 25–64 (2000)
3. Hüseyin Vasfi bin Ahmed Hulkî Filibevî: Târih-i İ'mâlât ve İhtirâ'ât ve Keşfiyât-ı Beşeriye, Şirket-i Mürettibiye Matbaası, İstanbul (H.1304). (In Ottoman Turkish)
4. Hüseyin Vasfi bin Ahmed Hulkî Filibevî: Esile ve Ecvibe-yi Hikemiye Yahut Tenvîr-i Efkâr (in Ottoman Turkish), Matbaa-yi Amire, Istanbul (H.1306)
5. Karakuş, N.: Abbasi Halifesi ve Frenk Hükümdarı Büyük Karl İlişkisi. Hitit Üniversitesi İlahiyat Fakültesi Dergisi, **16**(32), 463–486 (2017)
6. Kiel, M.: "Filibe", İslâm Ansiklopedisi, TDV, vol. 13, pp. 79–82 (1996)
7. McClellan III, J.E., Dorn, H.: Science and Technology in World History, Translated to Turkish by Haydar Yalçın, Ankara (2006)
8. Naim, M.: İhtiraat ve Keşfiyyat (Inventions and Discoveries), Mihran Matbaası, Istanbul (H.1298/AD. 1881). (in Ottoman Turkish)
9. Richard von Coudenhove-Kalergi, Teknik Apolojisi, translated by Bekir Sıtkı Bey, prepared by Remzi Demir, Muhayyel Yayıncılık, Istanbul (2021)
10. Solnon, J.-F.: L'Empire Ottoman Et L'Europe XIVe – XXe Siecle, translated to Turkish by Ali Berktay, Istanbul (2019)

The Robert H Thurston Materials Testing Laboratory Collection at Cornell University

Francis C. Moon(✉)

Cornell University, Ithaca, New York, USA
fcm3@cornell.edu

Abstract. Cornell University has preserved 19th century artifacts of significance to the history of mechanical engineering. They were designed by Robert Henry Thurston [1839–1903] who made contributions to materials engineering related to the technology of steam power. A torsional testing machine invented by Thurston uses a unique cam mechanism to automatically write a stress strain diagram. This machine led to the discovery of work hardening in metals. Another artifact relates to the testing of tribological properties of machine lubricants. This Collection received the ASME designation as a National Heritage Collection in 2019.

Keywords: Materials · Testing · Cam · Mechanism · Tribology · Robert Thurston · ASME

1 Introduction

Robert H Thurston was one of the most famous engineers in the United States in the late 19th century. [1–3] (Fig. 1) Born in Providence Rhode Island in 1839, Thurston was educated at Brown University with a "Certificate" in civil engineering. He served as a naval engineer during the Civil War at the beginning of the steam power era of the Navy. He worked in his father's steam engine manufacturing company for a short time before becoming one of the first faculty at Stevens Institute of Technology in Hoboken NJ [1871–1885]. At Stevens he began implementing his ideas for a more scientific based mechanical engineering curriculum and for developing an engineering testing laboratory. He brought these ideas to Cornell University in 1885 until his untimely death in 1903 [8].

The course of Robert Henry Thurston's career was inevitably linked to the development of steam power in the US. In 1841 his father Robert Lawton Thurston became a partner with Babcock and stared building steam engines, especially suited for marine applications. The company built a version of the Noble T Greene steam engine which was a rival to the famous Corliss engine. The flywheel and piston are a classic slider crank mechanism. In 1855 while enrolled at Brown university, RH Thurston built an operating model of a Greene engine using the facilities of his father's shops. That model is in the Cornell Collection. The Thurston company had Navy contracts at the time. Thus it was no accident that at the outbreak of the Civil War in 1861, young Robert Henry

M. Ceccarelli and R. López-García (Eds.): HMM 2022, HMMS 40, pp. 57–67, 2022.
https://doi.org/10.1007/978-3-030-98499-1_6

Fig. 1. Photograph of Robert Henry Thurston, Director of the Sibley College of Mechanical Engineering at Cornell University, 1885–1903; First president of American Society of Mechanical Engineers, 1880.

Thurston would become an engineer for the Navy servicing the new steam powered vessels. After the War he taught at Annapolis before teaching at Stevens Institute.

When Thurston was on the faculty of Stevens Tech, he was appointed to a US Board to study steam boiler accidents that were killing more than several hundred people a year. This led him to invent his autographic strength testing machine and to the discovery of work hardening. Thus the technology of steam power brought his genius into both thermodynamics and mechanics of materials. This work of Thurston on boiler safety was a forerunner of the ASME Piping and Pressure Vessel mission.

Thurston began his career in the age of steam power and steam transportation and finished his career in the age of electric power and the gasoline engines. He died a few weeks before his friend Samuel Langley and Cornell graduate Charles Manly failed in their attempt to fly a heavier than air machine and shortly before the Wright Brothers announced they had succeeded in 1903. Yet a decade earlier he had encouraged another friend Octave Chanute to organize an international meeting on flying machines at the Worlds Fair in Chicago in 1893.

One of the areas that Thurston helped change was the creation of the field of mechanical engineering both at the professional and educational levels. In 1880, John Sweet, formerly a professor at Cornell, called for a meeting of engineers in New York to create a new profession related to the technologies of steam power and transportation as well as manufacturing. Forty one year old Thurston of Steven's Tech was elected to head the new American Society of Mechanical Engineers. Beginning at Steven's and continuing at the Sibley College at Cornell University in 1885, he outlined a curriculum that would balance shop knowledge with principles of science. One of the linchpins of this curriculum was the *engineering testing laboratory* where both materials and machines were tested, based on the sciences of mechanics and thermodynamics.

Thurston's work on the thermodynamics of steam is perhaps more widely known because of his popular books on steam power [16]. He had also translated Carnot's famous treatise on entropy from French to English. But Thurston also made important contributions to mechanics of materials. In 1884 he published a three-volume book on engineering materials [14]. In this book he not only summarized the properties of materials but included chapters on how to establish those properties using scientific laboratory measurements. His 1894 book on *The Animal as a Machine,* is one of the early books in bioengineering.

2 Work Hardening of Metals

One of Thurston's major scientific discoveries that impacted mechanical and civil engineering at the time, was the discovery of work hardening of metals in 1873; that is the elastic limit was increased when the metal was initially stressed beyond its elastic limit.

James F Bell of John's Hopkins University in his 1973 scholarly work "*The Experimental Foundations of Solid Mechanics*" wrote [Thurston] "designed and developed an autographic torsion testing machine, measurements from which deeply influenced the thinking of the next forty years." "This apparatus provided the prototype for the present day standardized automated materials testing devices" [4].

This large testing machine is in effect a complaint mechanism in that the test specimen deforms elastically and plastically where the operator is able to apply a torque through a worm gear and a pendulum. [Figs. 2, 3 and 4] The unique aspect of this machine is the automatic measurement of both the torsional deformation of the test specimen as well as the torque applied. [Fig. 5] The torque is proportional to the sine of the angle of the pendulum which is translated into an angle through a cam element in the follower scribe mechanism [14].

In an obituary in the AAAS Science of November 1903, after citing Thurston's work on the US Board to Test Metals in 1972, and devising a machine for torsional tests, they wrote; "the most important [result] being that which established the fact that the elastic strength of wrought iron and steel is increased by stressing the material beyond that limit." [1] Thurston published his early work on this phenomenon in the transactions of the American Society of Civil Engineers in 1874, (pp. 239–240) after having presented his work at a meeting of ASCE in 1873.

3 Invention of a Lubrication Testing Machine

In addition to testing materials strength properties, which helped establish rational structural design, he also believed in testing the efficiency of machines. One of the important but often overlooked components of machine design is the loss of energy due to friction and lubrication. In connection with consulting work for railroad companies, Thurston invented a machine that would measure the viscous properties of lubricants. In Volume III of his tome on engineering materials, Thurston described the construction of his new machine and the mathematical and physical principles on which it works. His machine could measure the temperature of the lubricant, vary the relative speed between moving elements and control the applied normal load between machine elements. Robert

Fig. 2. Lithograph of Thurston's torsion testing machine in his three volume work, Materials of Engineering, [14].

Fig. 3. Photograph of autographic torsional testing machine in the Thurston Collection at Cornell University, Ithaca, NY. USA.

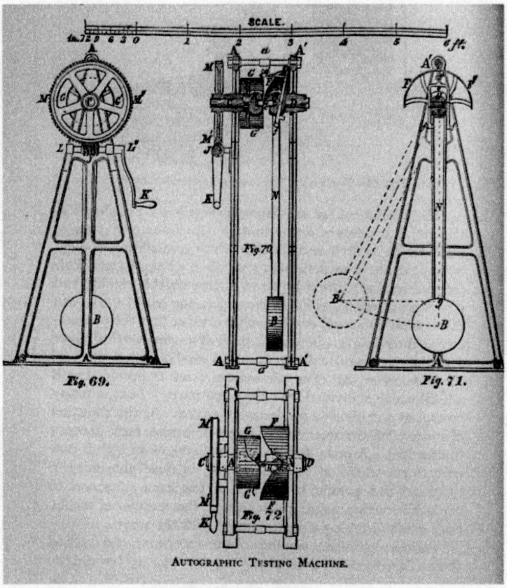

Fig. 4. Drawing of torsion testing machine showing pendulum motion and sine cam mechanism at the bottom [14].

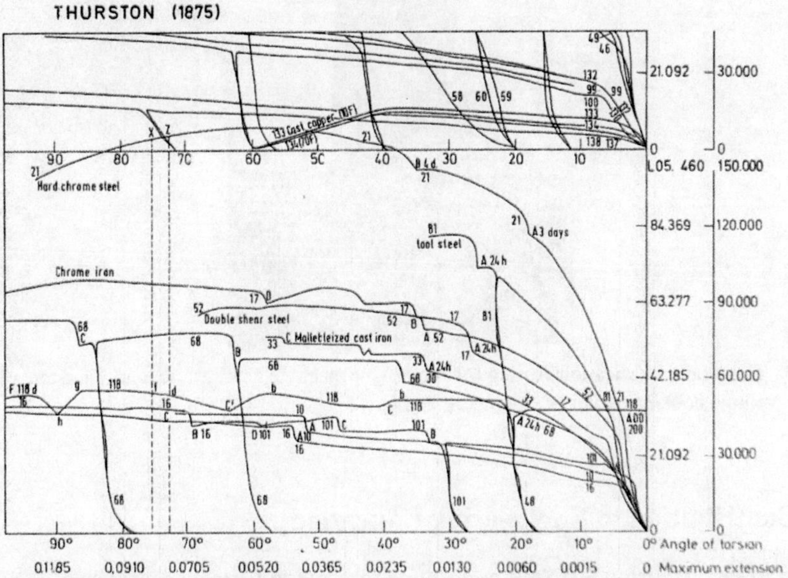

Fig. 4.24. Automatically recorded torque-angle data obtained by THURSTON from his apparatus of Fig. 4.22. Note the THURSTON effect following a pause in the loading.

Fig. 5. Torque- angle (stress vs strain) data from Thurston's Autographic Testing Machine, 1875, [14].

Thurston wrote a unique book on tribology of machines in 1885; *A treatise on Friction and Lost Work in Machinery and Millwork.* This book not only examined the physical aspects of dry and fluid friction, but also discussed the economic impact of friction losses in machinery [15].

Quoting D. Dowson of the University of Leeds; "His major contributions were three-fold; he drew attention to the importance of studying tribological concepts; he presented coefficients of friction for a wide range of machine elements and he firmly established empirical testing of lubricants." Dowson also reported;

"He appears to have been the first person to report that the coefficient of friction passed through a minimum as the load was increased." [7].

This machine is in the Thurston Collection at Cornell [Fig. 6].

Fig. 6. Thurston's Lubrication Testing Machine. This machine was described in Thurston's 1883 Three Volume book Materials of Engineering [14].

4 Contribution to Engineering Education

When Robert Thurston left Stevens Institute to come to Cornell, each school had less than ten faculty teaching mechanical engineering. When Thurston died at Ithaca in 1903, Cornell's Sibley College of Mechanical Engineering had over forty faculty in seven departments teaching approximately 1000 students. It was perhaps the largest institution of mechanical engineering in the US. AAAS *Science* magazine in their November

1903 obituary for Thurston wrote, "There can be no doubt—, that [the scientific and engineering work of Thurston] was of great benefit to mankind, for he made engineers better scientists, promoted engineering education, and helped to put engineering upon a higher professional plane." [5, 6, 9, 10].

Perhaps because of his working experience in his father's factory Thurston is quoted as writing "Manual training is done not to make [the engineering student] a good mechanic or to teach a trade, but to enable [the engineer] to design intelligently, to criticize good and bad work,"

Thurston wrote numerous articles on the engineering science nature of engineering education, one of which was obtained by the great German engineer Franz Reuleaux of the Berlin Polytechnic Institute who had Thurston's article translated into German. Thurston had visited Berlin during his appointment as US Commissioner to the Vienna World Exhibition in 1873.

5 National and International Statesman for Mechanical Engineering

Thurston was not only president of ASME but was a vice president of the American Association for the Advancement of Science (AAAS) as well as the vice president of the American Institute of Mining Engineers. In addition to his national service on the US Board to Test Metals, in 1873 he was the US Commissioner to the Vienna Exhibition or World's Fair. He sent seventeen letters for publication in Scientific American on the technical exhibits at the Exhibition. He also visited several European countries and reported on the progress of engineering education. He made two other trips to Europe in 1889, 1894. During one of his trips to Europe he visited the laboratory of the famous kinematics engineer, Franz Reuleaux, in Berlin. Cornell University has one of the largest collections of Reuleaux Kinematic Models.

As Director of the Sibley College at Cornell he invited many famous engineers to lecture to the students. The Cornell Collection of Thurston's personal papers in the Rare Manuscripts Division includes correspondence from famous engineers and capitalists at the time, such as, Thomas Edison, Andrew Carnegie, Samuel Langley, Cyrus McCormick, Alexander G Bell and Nikola Tesla, to mention the more familiar names [10].

Thurston also edited a technical publication at Cornell called the *Sibley Journal of Engineering* [1885–1919]. Up until his death in 1903, this Journal contained many articles by Thurston on the nature of engineering education and technology. In addition to all of his books, the Cornell Library has the complete collection of the *Sibley Journal of Engineering*. This Journal has articles by famous engineers at the time such as Steinmetz on advances in electrical power and the lectures of aviation engineer Octave Chanute who in 1891 lectured on the progress in flying machines a decade before the Wright Brothers flights.

6 The Uniqueness of the Collection

In 1885, the year that Robert Thurston came to Cornell to assume the directorship of the Sibley College of Mechanical Engineering and Mechanic Arts, *Scientific American* published a cover story on the Sibley College at Cornell. The article describes;

"A 'mechanical laboratory' — in which are placed a variety of testing machines for determining the strength, elasticity, ductility and 'resilience' or shock resisting power of iron and steel or other materials of construction. **Thurston's 'autographic' and lubricant testing machines,** meters, indicators, scales, dynamometers and all forms of apparatus for determining the quality of materials used by engineers and the power given by machines of all kinds and their efficiency. This department forms a very prominent part of the establishment and the course of instruction includes a considerable amount of work of this kind. This laboratory is one of the most interesting—in this great college."

Fig. 7. Bar-relief bronze sculpture of Robert H Thurston by the artist Herman Atkins McNeil in 1910. This commemorative work of art sits in the entry way to Thurston Hall, home to the teaching of mechanics and materials at Cornell University

Thurston came to Ithaca in 1885, and this article suggests that he brought with him from Stevens Tech the autographic torsion testing machine built by Pratt & Whitney around 1880. However in a query to the Archivist at Stevens' Tech they have one of his autographic testing machines built around 1871–2 at Stevens by students. This machine is on exhibit in their library. According to the historian James Bell, Thurston published his initial work on work-hardening at a ASCE meeting in 1873. So it is likely that his earliest results were performed on the Stevens built machine. However in his three-volume tome on the *Materials of Engineering* 1883–4, he presents strength data on many examples of metals and it is likely that the later machine built by Pratt & Whitney and now at Cornell was used in these experiments.

As mentioned above the Greene steam engine model was made by Thurston himself and was modeled after a full sized machine manufactured by his father's company. It is not likely that another one exactly like the Cornell model exists. Other artifacts include kinematic planimeters and devices to measure strain in metals during load tests. A list of major items in the Collection is given in the Table in this paper.

In addition to the mechanical artifacts, the Thurston Collection also contains his original annotated books and papers, [11–16].

Dedication plaque accompanying the Cornell Thurston Collection.

Mechanical Engineering Heritage Collection
Robert H. Thurston Collection – Cornell University
Early Engineering Laboratory Devices and Testing Machines

The devices in this collection, used at Cornell between 1885 and 1903, exemplify Thurston's vision of the central role of the engineering laboratory in training mechanical engineers.

Building on his work at Stevens Institute of Technology, Thurston fully implemented his pedagogical vision at Cornell's Sibley College of Mechanical Engineering and Mechanic Arts, which under his leadership became the largest and most influential mechanical engineering school in the U.S. Thurston's program achieved national and international recognition as the key to moving engineering training from shop to university.

American Society of Mechanical Engineers, 2019

Table
List of major Artifacts in the Robert H Thurston Collection

1. Autographic Testing Machine, US Patents 1874, 1880 of Robert H Thurston, Manufactured by Pratt and Whitney, Hartford Conn

2. Lubrication Testing Machine, designed by Robert H Thurston circa 1875-1880

3. Working Model of Greene Steam Engine built by R H Thurston while at Brown University c. 1855,

4. Working Model of Steam Engine used in teaching in the Sibley College, Cornell c. 1890

5. Swiss and English kinematic planimeters used to measure efficiency of steam engines, Sibley College 1890 - 1900.

6. Pressure - Displacement indicators with straight-line kinematic mechanism used to measure efficiency of steam engines in Laboratory of Sibley College c. 1890-1900.

7. Mechanical Strain measuring instruments used in metals testing laboratory, Sibley College c. 1900.

8. Bronze topographic plaque showing the effect of alloys of Copper, Lead and Zinc based on torsional strength tests with Thurston's autographic testing machine [14]

9. Personal books of Robert H Thurston [Thurston Bookplates]
a) Report on Machinery and Manufactures at the Vienna International Exhibition 1873 with an account of European Manufacturing Districts by Robert H Thurston, Member of the Scientific Commission of the United States. Washington Printing Office 1875 [includes original photographs]
b) Materials of Engineering in three Parts: Part I Non-Metalic Materials, Part II Iron and Steel 1883, J Wiley & Sons.
c) Memoirs and Professional Reports, 38 Pamphlets 1861-1903.

Acknowledgement. The author would like to recognize Professor Elizabeth M Fisher, former Associate Director of the Sibley School for organizing both material and student resources in creating the Cornell exhibit for the ASME dedication in 1919. The Cornell student section of ASME also contributed design and fabrication help.

References

1. AAAS Obituary Robert Henry Thurston, Science, p. 610, 13 November 1903
2. American Machinist: Bronze Memorial to Robert Henry Thurston, pp. 325–327, 17 February 1910

3. ASME Obituary 1049, Robert Henry Thurston, Transactions of ASME, pp. 1113–1120 (1903)
4. Bell, J.F.: The experimental foundations of solid mechanics. In: Truesdell, C. (ed.) Mechanics of Solids, vol. 1. Springer, Heidelberg (1973)
5. Calvert, M.A.: The Mechanical Engineer in America, 1930- Professional Cultures in Conflict, Part II, Chap. 5, pp. 87–106. School versus Shop: Cornell's Sibley College (1967)
6. Diederichs, H.: The curriculum in mechanical engineering. The Sibley Eng. Cornell Univ. **XLIX**(4), 71–73, 77 (1935)
7. Dowson, D.: A History of Tribology. Longman, London (1979)
8. Durand, W.F.: Robert Henry Thurston, A Biography. The American Society of Mechanical Engineers, New York (1929)
9. Hewett, W.T.: Cornell University: A History, vol. II, chap. IV, pp 303–325. Sibley College of Mechanical Engineering (1905)
10. Kimball, D.S.: The story of Sibley college of mechanical engineering and the mechanic Arts. J. Eng. Cornell. Univ. **XLIX**(4), 65–67, 80–82 (1935)
11. Thurston, R.H.: "Work of the founders; the early history of Sibley college. Sibley J. Eng. **2**(6), 255–260 (1898)
12. Thurston, R.H.: Vienna International Exhibition 1873, Report on Machinery and Manufactures with an Account of European Manufacturing Districts, Government Printing Office, Washington DC (1875)
13. Thurston, R.H.: Torsional Resistance of Materials Determined by a New Apparatus, with Automatic Registry, J. of The Franklin Institute, Reprinted in Thurston's bound Memoirs and Professional Reports (1873)
14. Thurston, R.H.: The Materials of Engineering, in Three Parts. Wiley, Hoboken (1883)
15. Thurston, R.H.: A Treatise on Friction and Lost Work in Machinery and Millwork, 7th edn. Wiley, Hoboken (1885)
16. Thurston, R.H.: Heat as a Form of Energy. Houghton Mifflin, Boston (1890)

Modern Reviews of Past Works

Additive Technologies for Restoration of Kinematic Mechanism Models

Grigoriy Tikhomirov[1(✉)], Nikolay N. Barbashov[1], and Olga V. Egorova[2]

[1] Bauman Moscow State Technical University, Moscow, Russia
tgv-22@yandex.ru
[2] V.I. Vernadsky Crimean Federal University, Simferopol, Russia

Abstract. In recent years, there has been a significant interest increase in collections of kinematic mechanism models as important objects for the study and preservation of memory of the technology history and its evolution over time. The models could be used in the educational process in the form of demonstrations of rare technical devices, including those, that posted on websites. To restore broken or damaged kinematic mechanism models the additive technologies are used, which have become widespread over the past 10 years and are among the most advanced, convenient and rapidly developing technologies in the field of machine parts production. The paper presents the main restoration norms and ethical rules that should be followed during any restoration work. The authors have split their experience of the RP-technology practice applied to the renovation of the planetary mechanism with roller wave generator and flexspline wheel. Also, modern role of the Russian collection of kinematic mechanism models is noted.

Keywords: History of mechanism and machine science · Restoration · Renovation · Reconstruction · Kinematic mechanism model · RP-technology · Additive technology

1 Introduction

The Bauman Moscow State Technical University has one of the world's best collections of kinematic mechanism models, usually called as the Russian collection, numbering more than 500 units [1]. In recent years, there has been a significant interest increase in the Russian collection. Initially, this interest arose from a historical point of view to the most ancient mechanism models (museum exhibits) as important objects of study and preservation of memory about the history of technology and its evolution over time.

The development of the Internet sparked a new interest in using the collection of mechanisms in the educational process in the form of demonstration of rare technical devices placed on websites. Given the difficulties in creating a digital space, the Bauman Moscow State Technical University (BMSTU) has created a historical database (HDB) of the collection of kinematic mechanism models with solved issues of adaptation to various modern platforms, including tablets and smartphones.

The modern multimedia space has made it possible to use the museum exhibits and the HDB during lectures, seminars, laboratory works and workshops in the study of such

M. Ceccarelli and R. López-García (Eds.): HMM 2022, HMMS 40, pp. 71–83, 2022.
https://doi.org/10.1007/978-3-030-98499-1_7

subjects as «Theoretical Mechanics», «Theory of Mechanisms and Machines» (TMM) and «Machine Parts» (MP), which form in conjunction with other subjects the technical part of education, called the «Russian method of engineers' training» [2].

To restore the broken and damaged kinematic mechanism models from the Russian collection [3], at BMSTU the additive technologies are used, which have become widespread over the past 10 years and are among the most advanced, convenient and rapidly developing technologies in the machine parts production.

1.1 A Brief History of BMSTU Mechanism Collection

Due to the rapid growth of industry in the middle of the XIX century there was a shortage of technically educated specialists, capable of designing, manufacturing and operating complex mechanisms and machines. Because of that, the higher engineering and technical education, which provided theoretical and practical training of engineers, necessarily included in the curriculum two disciplines: «Theory of Machines» and «Theory of Mechanisms». These disciplines became soon basic at all Polytechnic schools and Universities. But their quality teaching was impossible without visual aids as kinematic mechanism models. As a consequence, scientific communities began to create collections and cabinets of mechanism and machine models.

In 1850–1860s at the Imperial Moscow Technical College (IMTU), as the Bauman Moscow State Technical University was called from 1868 to 1918, the first collection of kinematic mechanism models was created. It was called the "Cabinet of Applied Mechanics" (Fig. 1).

Currently, the role of mechanism models has changed, as well as the traditional course of mechanism designing in modern higher education. With the development of new technologies the kinematic mechanism models began to lose their didactic meaning, while remaining an example of visualization in teaching of the technical disciplines.

Today, kinematic mechanism models are used not only to demonstrate the motion transformation at TMM and MP lectures or to conduct experiments in the teaching laboratory, but in fact, these models themselves have become a subject of scientific research as valuable monuments of the history of technology [4].

All historical mechanism models (museum exhibits) at the Bauman Moscow State Technical University, which were made in the 19th - early 20th centuries, we can divide into three groups:

a) Kinematic mechanism models of the second half of XIX century. The formation of higher technical and engineering education in Russia started a little later than in Europe, so its development was strongly influenced by the European experience [5, 6]. It is known that under the guidance of a professor of «Descriptive Geometry and Mechanics» department Alexander Stepanovich Ershov (1818–1867) the first kinematic mechanism models were produced at the workshops of the Moscow Higher Technical School (previous name of BMSTU from 1830 to 1868). Today these models form the basis of a unique collection, which presently containing more than 50 museum exhibits [3]. It should be noted that A.S. Ershov was one of the founders of Russian higher engineering education and played an important role as a Director of the University.

Fig. 1. Cabinet of applied mechanics [1]

b) Alexandre Claire's Collection of Kinematic Mechanism Models. At the Bauman Moscow State Technical University museum there are eight kinematic mechanism models produced in Claire's workshop. Recently, with the help of the catalog of the Crozatier Museum (France) [7], we were able to attribute another four models made in the workshop of Alexandre Claire. Figures 2, 3 and 4 show three of them. The attribution was carried out by correlating BMSTU museum's exhibits with the images from Claire's catalog. It should be noted that the models absolutely exactly matched in appearance, size and materials. Figure 2 shows the kinematic mechanism model of a universal hook. The cross-piece is installed in links in cylindrical pairs. When the input shaft rotates, its cross-piece is moving on an axis with a double frequency ratio. There are no redundant constraints in this mechanism but it is impossible to use it in practice [1]. Hook's joints are used in pairs, because only then the uniform rotation of the output shaft could be provided. To provide the uniformity, it is necessary to install forks of intermediate links in the same plane. If this condition isn't executed, the uniformity of the rotation of the output shaft is increased. It could be demonstrated by the model of a Double universal hinge (Fig. 3) [1]. Figure 4 shows the model designed to transmit and convert rotational motion between coaxial shafts.

c) Franz Reuleaux Collection of Kinematic Mechanism Models. Franz Reuleaux, a student of Ferdinand Redtenbaher, created the largest collection of kinematic mechanism models for teaching students. There are 18 models at the Bauman Moscow State Technical University museum, which are probably the only Reuleaux models left in the world [8, 9] (Figs. 5 and 6). The mechanism of an overtaking clutch (Fig. 5) is designed to connect automatically co-axial rotating shafts. It is used in various machines and devices as a clutch. Figure 6 shows a machine tool designed for the use in lathes for processing workpieces with an elliptical cross-section. The inventor of this mechanism is Leonardo da Vinci.

1.2 Application Prospects of Mechanism Models Collection

Taking into account many years of experience of the Bauman Moscow State Technical University museum, as well as global trends, we can highlight the following prospects

Fig. 2. Universal hook with cylindrical pairs (author's photo).

Fig. 3. Double universal hinge (author's photo).

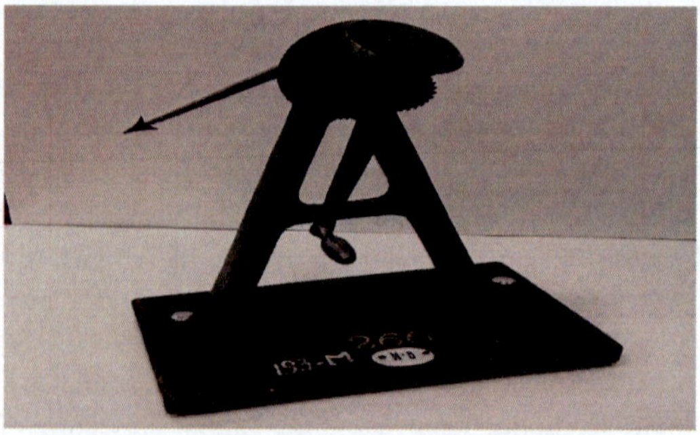

Fig. 4. Coaxial non-orthogonal bevel planetary gear (author's photo)

Fig. 5. The overtaking clutch with autoswitching or the clutch of Pouyer (author's photo)

Fig. 6. The double slider ellipsograph or the machine tool of Leonardo Da Vinci (author's photo)

and further opportunities for using of kinematic mechanism models in the educational process, in the museum space, and in the field of popularization of scientific knowledge:

a) Creation of virtual three-dimensional models of mechanisms using digital CAD/CAM/CAE packages, development of specifications and routing technologies, development of 3D-visualization (animation, rendering, etc.) for visual representation of natural laws of science. This direction has been successfully implemented since the end of the 20th century and continues to improve today [10];
b) Creating and growing three-dimensional samples of mechanism models by prototyping (or other additive technologies). This area has a great future and a wide range of application. This approach follows from the previous point in terms of preparation of technical documentation and uses the same soft environment;
c) Designing new teaching methods (distance learning, gaming, problem-based [11]) on the basis of modern communication platforms as webinars, joint projects with remote access, interactive lectures, and etc.;
d) Creation of virtual laboratories for the study of technical disciplines, when theoretical and practical training is carried out, taking into account the possibility of flexible change of the environment itself, which is impossible with the classical approach [12];
e) Adaptation of complex educational and scientific material for different age and educational groups of society as schoolchildren, students, teachers, and pensioners.

2 Aim and Research Objectives

2.1 Aim

The aim of the work is to apply in practice the acquired knowledge of kinematics and dynamics of solids and computer modeling skills, such as 3d-modeling, when restoring kinematic mechanism models including those from the Russian collection [1].

2.2 Research Objectives

a) To unite students who are interested in studying and restoring kinematic models and to create a club of like-minded people;

b) Publish the results of the work done on a specially created website to share experiences with other users of the Internet;
c) Determine the main directions of application of RP-technologies for restoration of kinematic models of mechanisms;
d) To create new methods and use new technologies to restore models.

3 Restoration and Technical Renovation of Kinematic Mechanism Models

Restoration of already existing mechanisms at the Bauman Moscow State Technical University museum, but which have lost their original appearance, requires manufacturing and installation/assembly of the lost or broken parts. Due to the fact that the museum exhibits themselves have a high historical value, the Bauman Moscow State Technical University has formed the basic principles and rules for restoration (in brief form), which are binding:

- the new part must be made as close as possible to the old one (material, accuracy, quality of the surface or coating, authenticity of the equipment and tooling used, etc.);
- the consequences of restoration and renovation must be fully reversible and must be imperceptible;
- after a new part has been installed, there must be further tests, measurements and examinations, which will make sure that the new part corresponds to the old one and is interchangeable with each other, etc.
- conditions for tests, measurements and test runs must be formed separately, for each model of mechanism, taking into account the most careful attitude to the existing exhibit.

These rules should be taken as an axiom and should be followed scrupulously when restoring or/and renovating, otherwise there is a high probability of spoiling an existing model.

The Bauman Moscow State Technical University observes the aforementioned principles and rules, which makes it possible to preserve the historical identity of the mechanism models represented in the museum.

Figures 7 and 8 show the restored models illustrating cycloids as trajectories of fixed points of a producing circle rolling without sliding along a straight line. The models are designed to demonstrate the process of forming cycloidal planar curves: cycloids and trochoids. In Fig. 7, the forming circle moves on a smooth flat surface, in Fig. 8, a rack and pinion gear are engaged. The models had broken glasses, which needed to be repaired. By measuring the geometric parameters of the circles, prototypes of cycloids [13] were accurately built by 3D modeling, and transferred to the new glasses.

4 What is RP-Technology?

RP-technologies are technologies of layer-by-layer build-up of layered materials, allowing to obtain any forms and surfaces. These technologies are most widespread for the production of single items, which have a great uniqueness. Below is a list of technologies with their brief description [14–16]:

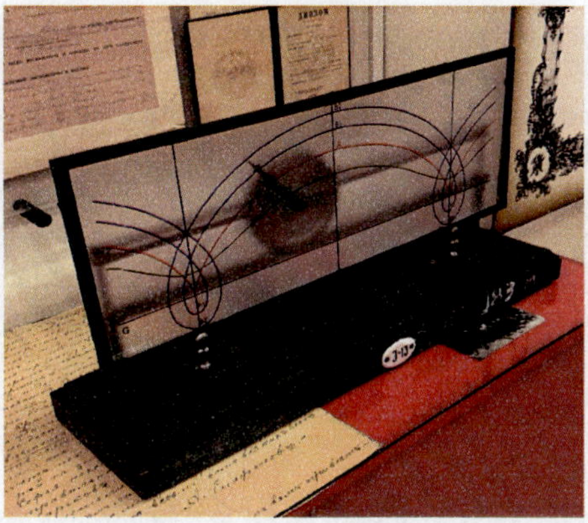

Fig. 7. Model of cycloid formation (author's photo)

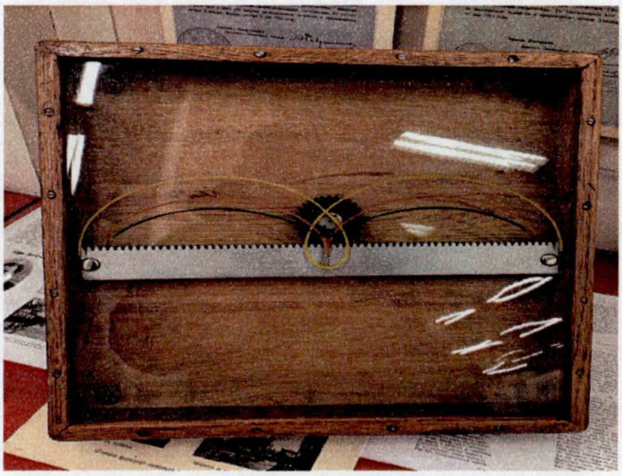

Fig. 8. Model of cycloid formation (author's photo)

- SL is a method of stereolithography. In this process, based on photo-initiated laser radiation, the desired model is grown from photopolymers in a liquid state.
- SLS is a selective laser sintering, one of the most widely used additive technologies. The principle of SLS is the point sintering of plastic powders with different components by a laser beam.
- FDM, a layer-by-layer overlay method, is an additive manufacturing process that is realized through the extrusion of materials. In FDM, an object is constructed by applying molten material according to a predetermined algorithm, layer by layer. The materials used are thermoplastic polymers and are in the form of filaments.
- DOD is a technology that is used for high-resolution printing based on pushing the drop from the chamber to the package surface by means of a piezo element, to which a voltage is applied during printing. Applying voltage to the piezo element causes a linear expansion, which in turn pushes the ink droplet out of the chamber.

In the case of using mentioned additive technologies to create new mechanism models or to recreate lost parts of old models, Francis Moon's article "3D printing of mechanism history» [16] is interesting. Francis writes: "There are several technologies for rapid prototyping, including laminated object manufacturing (LOM), selective laser sintering (SLS), photo polymerization (stereolithography, SLA) and fused deposition modeling (FDM). The FDM process was used in this study to reproduce several kinematic Reuleaux-Voigt kinematic models. This process creates a sequence of thermoplastic layers from a filament wound coil that is heated and extruded through a nozzle. The trajectory of the nozzle is derived from the triangle mesh, so as to raster scan and fill solid volumes. In order to create functioning mechanisms, a second, soluble release material is placed in the gaps between the movable parts. The basic material used was acrylonitrile-butadiene-styrene (ABS)".

One cannot ignore the facts of the financial and restorative-ethical problems faced by Moon. Thus, he wrote in his article: "These qualities of the models in this collection guarantee that their reproduction in today's market would be very expensive (the Cornell set of approximately 250 models cost $8000 in 1882 American dollars). Providing a less expensive reproduction method in cheaper materials, such as thermoplastics, is a fine goal. However basic limitations of the rapid prototyping process leads to certain necessary compromises. Specifically, we strived to make every model printable as a single, preassembled unit, so that it is fully functional right out of the printer. This necessitates careful consideration and, in some cases, redesign of some features of the model. We thus sacrifice some dimensional authenticity for functional authenticity".

From financial point of view, Moon's current estimate of $8,000 in 1882 would be approximately equal to $300,000, or about 21.5 million rubles.

Another problem is the use of plastic in restoration, because plastic has very specific mechanical and functional properties. The main disadvantages of using plastic are such properties as: its insufficient strength and elasticity, the appearance of gaps between the links and the emergence of roughness, as well as the shrinkage effect on the surfaces of plastic parts after their manufacture. Aesthetically, all-plastic kinematic mechanism models do not have a very attractive appearance compared to metal and wood, and they also break very easily. It can be concluded that the use of plastic does not have the best effect on the appearance and performance of a kinematic mechanism model, but in cases where no other alternatives are available, RP technology provides the best result.

5 Additive Technologies While Preserving Mechanism Models

Based on the analysis of modern experience of RP-production of visual and illustrative models, art objects, copies, duplicates and others [17–19], we propose to identify 3 (three) main promising areas (levels) of application of additive technologies in restoration, renovation, and transformation of kinematic mechanism models:

LEVEL 1. «Basic» application: replacement of polymeric parts with the same polymeric parts;

LEVEL 2 - Reproduction of exact copies and duplicates of real mechanism models. The method is used when restoration by traditional methods is not possible;

LEVEL 3 - Use of RP-technology as supporting and auxiliary in the processes of restoration and renovation of lost parts of mechanism models made of metal, glass, wood and other non-polymeric materials.

6 Example of a Restored Model

6.1 Restoration of a Harmonic Mechanism Model

A model of a harmonic (wave) mechanism with a roller wave generator and a flexspline wheel was made for "Theory Mechanisms and Machines" (TMM) Department at the Bauman Moscow State Technical University (Fig. 9) because in 1970, the TMM department decided to include a topic «harmonic drive» in its course of lectures.

The model (Fig. 9) is a flat one of a harmonic drive with a roller wave generator. The model was designed by the engineer T. Komarova in 1970. It consists of a rigid gear fixed to a frame, a double-roller wave generator and a flexible gear. The drive has no output shaft and clutch. Input movement is set to the shaft of the wave generator and the output movement is the rotation of the flexible gear [1]. The previous wheel was made of steel, but due to careless handling, it fell down to the floor and split.

ABS plastic was chosen as the material for manufacturing the prototype of the flexible wheel. The parameters of the flexible wheel were calculated, then using the CAD-system "Autodesk Inventor" a number of models of flexible wheel of different diameters were created, which finally resulted in a flexible wheel, which most clearly and correctly demonstrates the principle of operation of the harmonic mechanism [20].

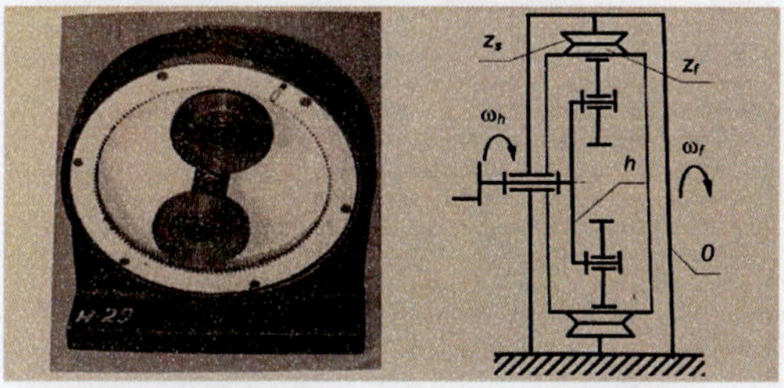

Fig. 9. Harmonic drive and its structure scheme [3].

It should be noted that manufacturing of a metal flexible wheel of this model is much more difficult process, and in practice it wears out faster than the same but a plastic one. We chose the FDM method because we had access to this equipment and the method itself allowed us to get the exact mechanical properties of this part (a flexible wheel) [14] that we needed. Figure 10 shows an image of a flexible wheel when it was printed on a 3D printer Ultimaker 2.

Figure 11 shows a photo of the restored kinematic mechanism model of a harmonic mechanism.

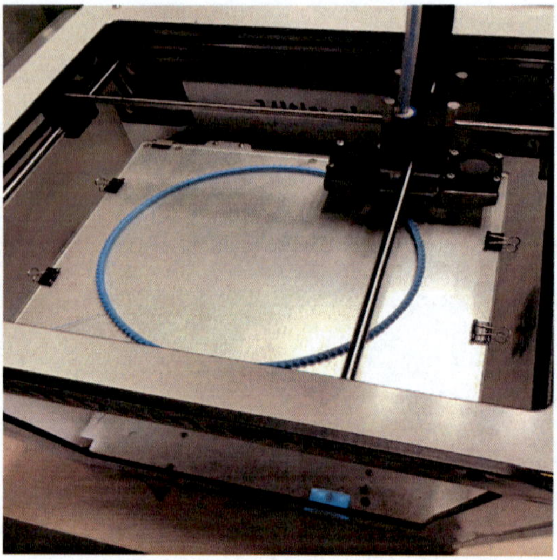

Fig. 10. 3D-printing of flexible ring (author's photo)

Fig. 11. Restored harmonic mechanism (author's photo)

6.2 Restoration Methodology for a Broken Part

When restoring a lost or a broken part, it is extremely important to determine its dimensions as precisely as possible. The use of high-precision measuring instruments made it possible to perform a 3D model with maximum accuracy and precision [21–23]. In our example, a digital caliper was used to measure the diameter of the rigid wheel, and a caliper was used to determine the dimensions of its tooth. Despite sufficiently accurate measurements, it was not possible to create a suitable size flexible wheel from the first time.

It is also necessary to choose the material, color and printing parameters on which its properties and appearance depend. Special attention was paid to this issue. For prototyping we chose ABS plastic, the most affordable and cheapest on the market. The elastic deformation of the product became the main criterion for our final choice.

6.3 *Installation and Test Run Methodology*

After manufacturing a model series of wheels with different diameters and sizes of teeth, it was necessary to establish and test the geometric and mechanical properties of the printer-printed flexible wheels [24–26]. For this purpose, we mounted the printer-printed flexible wheels, one at a time, and rotated the mechanism, checking the behavior of the wheels in operation. This technique was a forced measure, since due to the difference in mechanical properties of the previous metal wheel and the current plastic wheel, the flexible wheels took different geometric shapes. The test run after manufacturing of a broken part is the most important step, in which the main objective is to install a new printer-printed wheel as gently as possible so as not to damage the mechanism model.

6.4 Restored Harmonic Mechanism Model

The most suitable printer-printed flexible wheel in terms of geometric and mechanical parameters fully returned the model to work. Now it can again perform its main task and demonstrate the properties of a harmonic mechanism.

7 Conclusions

1. Additive RP-technologies have a high potential of application for restoration, renovation, reconstruction, and modification of kinematic mechanism models, which are in various collections all over the world, including those at "Theory of Mechanisms and Machines" Department and BMSTU museum. Despite sufficient complexity of application, the results of RP-technology fully satisfy the requirements for quality, dimensional accuracy, surface roughness, aesthetics, etc.
2. Application of additive RP-technologies for restoration, renovation, reconstruction, and etc. of museum exhibits, including kinematic mechanism models, has strict limitations, but their potential is very high and in demand today.
3. When restoring museum exhibits, the basic principles and ethical rules, which are described in this paper should be taken as an axiom and should be followed scrupulously, otherwise there is a high probability of spoiling an existing model.
4. The use of RP-technology for auxiliary purposes in no way diminishes the contribution of these processes to the overall restoration.
5. Historical kinematic models and museum exhibits from the Russian collection have themselves become the subject of scientific research as valuable monuments of the history of technology.
6. Creation of the «Student Center», which is engaged in restoration, allows students not only to apply their skills in practice, but also to preserve the historical and scientific heritage of BMSTU.

Acknowledgements. The authors would like to express their gratitude to Galina Alekseevna Bazanchuk, BMSTU Museum Director, and Sergey Vitalievich Kurakov, BMSTU Museum Engineer, for invaluable help and support.

References

1. Golovin, A., Tarabarin, V.: Russian Models from the Mechanisms Collection of Bauman University. Springer, Heidelberg (2008). https://doi.org/10.1007/978-1-4020-8776-9
2. Shaknov, V., Zinchenko, L., Rezchikova, E., Glushko, A.: An opportunity in engineering education: Russian BYOD tendencies : BMSTU case study. In: 2015 International Conference on Interactive Collaborative Learning (ICL), pp. 299–304. IEEE, Florence (2015)
3. Golovin A., Tarabarin V.: Modeli mekhanizmov russkikh uchenykh i inzhenerov v kollektsii MGTU imeni N. E. Baumana. Izdatel'stvo, vol. 300. Pervyi Tom, Moscow (2019)
4. Moon, F.: Robert Willis and Franc Reuleuax: pioneers in theory of machines. Notes Rec. R. Soc. Lond. **57**, 19–21 (2003)
5. Antsupova G.: MGTU glazami istorika. Izdatel'stvo MGTU im. Baumana, Moscow (2005)
6. Volchkevich, L., Volchkevich, I.: Moskovskii gosudarstvennyi tekhnicheskii universitet imeni N. E. Baumana. Publ. 'Istoriya i sovremennost' Moscow (2005)
7. Catalogue des Collections de Mecanique. Collection Alexandre Claire, Musee Crozatier, Le Puy-en-Velay., pp. 82–102 (1992)
8. Voigt G.: Kinematic Models After Reuleaux. Catalog, Berlin (1907)
9. LNCS Homepage. https://engineering.library.cornell.edu/kmoddl. Accessed 20 Nov 2021
10. Tarabarin, V., Tarabarina, Z.: The electronic a learning-methodical complex at the course the theory of machines and mechanisms. In: China-Russian Symposium on Problems of Development and Use of the Information Technology in Engineering Education, HPU, China, Harbin, p. 139 (2007)
11. Ceccarelli, M.: Introduction to automation and robotics. In: Fundamentals of Mechanics of Robotic Manipulation. International Series on Microprocessor-Based and Intelligent Systems Engineering, vol. 27, pp. 1–28. Springer, Dordrecht (2004). https://doi.org/10.1007/978-1-4020-2110-7_1
12. Carbone, G., Ceccarelli, M.: Experimental tests on feasible operation of a finger mechanism in the LARM hand. Int. J. Mech. Based Des. Struct. Mach. **36**, 1–13 (2008)
13. Varfel, T.: Prototyping: A Practitioner's Guide. Rosenfeld Media, Brooklyn (2009)
14. Redwood, B., Schöffer, I., Garret, B.: The 3D Printing Handbook. 3D Hubs B.V., Amsterdam (2017)
15. Grigor'yants A., Shiganov I., Misyurov A., Tret'yakov R.: Lazernye additivnye tekhnologii v mashinostroenii, uchebnoe posobie ed. by Grigor'yants A. Izdatel'stvo MGTU im., vol. 278. N. E. Baumana, Moscow (2018)
16. Lipson, H., Moon, F., Hai, J., Paventi, C.: 3-D printing the history of mechanisms. J. Mech. Des. **127**, 1029–1033 (2005)
17. Egorova, O., Evgrafov, A.: Modern view on «History of Science» teaching at technical universities. In: García-Prada, J., Castejón, C. (eds.) New Trends in Educational Activity in the Field of Mechanism and Machine Theory, vol. 19, pp. 137–144. Springer, Cham (2014). https://doi.org/10.1007/978-3-319-01836-2_15
18. Egorova, O.: Special issue on the role of history of MMS in evolution of mechanics. In: International Review of Mechanical Engineering (IREME), p. 302 (2016)
19. Egorova, O., Shcherbinin, D.: Creating technical heritage object replicas in a virtual environment. Front. Mech. Eng. **11**(1), 108–115 (2016). https://doi.org/10.1007/s11465-016-0363-4

20. Egorova, O., Timofeev, G., Grigoriev, I.: Applying modern CAD systems to reconstruction of old design. In: Wenger, P., Flores, P. (eds.) New Trends in Mechanism and Machine Science, vol. 43, pp. 323–331. Springer, Cham (2016). https://doi.org/10.1007/978-3-319-44156-6_33
21. Yanushkin, N., Melekaev, M.: Reconstruction of measuring linear and angular discplacements on the basis of the extreme precision of the measuring system. Metrology 110 (2017)
22. Yanushkin, N., Melekaev, M.: Modernization of instruments for measurement of linear and angular displacements on the foundation of a precision extremal measurement system. Meas. Tech. **60**, 1243–1246 (2018). https://doi.org/10.1007/s11018-018-1347-z
23. Kolenkin, A., Obuhov, I.: On the experimental evaluation of the components of inaccuracy measuring instruments and sample. In: 11th Russia Conference on Young Scientists and Specialists Bauman Moscow State Technical University, pp. 69–72. BMSTU, Moscow (2018)
24. Gusev, A., Buda-Krasnovskiy, S., Starodubtseva, S.: Statisctical determination of fatigue strength. Russ. Eng. Res. **39**, 303–306 (2019). https://doi.org/10.3103/S1068798X19040087
25. Timofeev, G., Kostikov, Y., Yaminski, A., Podchasov, E.: Prominent Russian constructions of harmonic drives. Int. J. Mech. Eng. Robot. Res. **9**, 775–778 (2020)
26. Timofeev, G., Kostikov, Y.: Torsional rigidity of harmonic gear drives. Russ. Eng. Res. **36**, 995–998 (2016). https://doi.org/10.3103/S1068798X16120169

3D Reconstruction of Two Medieval Artifacts by the Engineer Al-Murādī

Gustavo Medina-Sánchez(✉), Jorge Moreno-Buesa, Rubén Dorado-Vicente, and Rafael López-García

University of Jaen, Jaen, Spain
{gmedina,rdorado,rlgarcia}@ujaen.es, jmb00015@red.ujaen.es

Abstract. "The Book of Secrets in the Results of Ideas", usually called "Book of Secrets" is a codex containing drawings and descriptions of thirty-one artifacts attributed to the engineer Alī Ibn Khalaf al-Murādī, who lived in Andalusia in southern Spain at the beginning of the 11th century. This manuscript is one of the first written testimonies that describe medieval mechanisms with complex precision. The aim of this work is to reconstruct and study from a historical and technological point of view two of the ancient artifacts contained in the "Book of Secrets", the "Fortress Demolisher" and the "Magic Well". The "Fortress Demolisher" is a war machine designed to demolish the upper part of the walls or towers and consists of a battering ram mounted on a platform that can be raised several meters above the ground using a scissor mechanism. The "Magic Well" is a mechanism with several pulleys whose purpose is to raise water several meters from the bottom of a well by collecting a small portion of rope. The physical reproduction of the two artifacts was made by 3D printing, using the Fused Deposition Modeling (FDM) technology. Previously, a virtual model and a detailed study of the geometry and operation of the mechanisms have been carried out.

Keywords: History of machines and mechanisms · Digital modelling · 3D printing · Scale reconstruction

1 Introduction

Since its origin, human beings have always wanted to record their passage through history and transmit to next generations how they lived. For this reason, since the invention of writing, all cultures have given great importance to the transmission of their knowledge through the written word.

The evolution of knowledge has led to the evolution of the transmission medium, evolved from oral transmission or basic writing to the variety of multimedia and digital options of today.

In this work, an emerging technology such as 3D printing will be used for the transmission of knowledge, with the aim that the real scale reproduction of ancient mechanisms helps to better understand their operation, and thus understand the most complex modern mechanisms.

M. Ceccarelli and R. López-García (Eds.): HMM 2022, HMMS 40, pp. 84–96, 2022.
https://doi.org/10.1007/978-3-030-98499-1_8

1.1 Written Transmission of Knowledge in Ancient Times

Throughout history, the transmission of the knowledge has always been one of the priority objectives for the development of societies. This knowledge was stored since ancient times in libraries. Perhaps the earliest known library as an organized collection of documents is that of the Assyrian king Ashurbanipal, discovered during excavations in the city of Nineveh. In it, some thirty thousand fragments of clay tablets appeared buried among the remains of the Royal Palace. Later, the libraries of Alexandria, Pergamum, and different Roman libraries were famous. From those times until the advent of the printing press (Johannes Gutenberg, 1440), the fundamental supports for this transmission were tablets, scrolls and codices [1].

The tablet was an element used since ancient times in Rome, which consisted of a rectangular wooden table filled with wax. It was written on it using a wooden or metal punch, called a stylus.

The manuscript scrolls appeared in Egypt and were obtained from the Cyperus papyrus plant, very abundant in Egypt, which was cut into small strips that were later macerated and glued, overlapping the different cut layers to form the final tissue. They were stored in different temples and libraries, in clay pots, wooden boxes or leather cases [2].

The codices are one of the previous formats of the book as we know it today. They are made up of a set of folded and bound papyri or parchments [3]. They were the most used element for manuscript conservation and historical dissemination since the Middle Ages.

1.2 Science and Technology in the Medieval Arab World

The first sciences that attracted the curiosity of the learned Muslims were astronomy and mathematics, since their practical nature led them to dedicate themselves mainly to the exact sciences. Observatories were built in the most important centers of the Islamic empire, highlighting those of Baghdad, Cairo, Córdoba, Toledo and Samarkand. In mathematical sciences, many basic principles of arithmetic, geometry and algebra were discovered by Muslim scholars. In arithmetic, we still use numbers and the counting method invented by the Arabs. The invention of algebra is attributed to the Arabs. Among the most famous authors Al-Khwarizimi stands out, one of the best scientific minds of Islam, who exerted a great influence on mathematical thought throughout the Middle Ages.

The Arabs also made important contributions to physics and chemistry sciences, such as those made in the field of optics by Hassan-Ali-Aitan. In chemistry sciences, they discovered distillation and were the first to use the methods of sublimation, crystallization, coagulation, and cupellation to extract and mix substances.

After mathematics and chemistry, medicine was the science that most attracted Muslims scholars. During the first centuries of the Hegira, it was an integral part of a complete education, so the number of famous doctors and treatises on medicine is considerable. Muslim doctors played a decisive role in Western medical science. For several centuries, the works of Rhases, Ibn-Sina, Abul-Cassis and Ibn-Zohar were the basis of medical studies in all European universities. They also became interested in agronomy because of their interest in traditional herbal medicine.

Possibly the first engineering treatise written by Al-Murādī is from the Taifa era, when great interest in applied science was aroused. This manuscript is the first sample of the western Islamic mechanical tradition, which will be continued in North Africa and Europe, characterized by the use of rough mechanisms and powerful machines compared to the eastern mechanical tradition, which stands out for its delicate mechanisms and controls [4].

1.3 The Book of Secrets

The Book of Secrets (ةجتانلا راكفالا رارسأ باتك, kitab al-asrar al'afkar alnnatija), is a codex with a collection of thirty-one artifacts attributed to the engineer Alī Ibn Khalaf al-Murādī, who lived in Andalusia, probably in Córdoba or Granada, at the beginning of the 11th century AD.

In this manuscript, the first written testimonies of medieval mechanisms with complex precision are drawn. The codex has been preserved for centuries in the Biblioteca Medicea Laurenziana in Florence. However, his drawings were not interpreted until the 21st century.

The existence of this manuscript was brought to the fore by the engineer and historian Donald Routledge Hill around 1970 and was not given much attention until after about thirty years. It was then when, thanks to the economic collaboration of the Emir of Qatar Sheikh Hamad bin Khalifa Al Thani, who financially endowed the project. The facsimile reproduction of the manuscript was carried out [5], also its study both from the technological point of view and historiographic, and the reproduction of two of the inventions collected to be exhibited at the Museum of Islamic Art in Doha, in Qatar (Fig. 1).

Fig. 1. Pages from the facsimile edition of the Book of Secrets [5]

Among the thirty-one artifacts described in the codex, there are five automatons in the form of a "little theater", two calendars (one zodiacal and another astronomical), four war machines, two mechanisms to draw water from a well and, above all, there are clocks. The tools most used to create the mechanisms of the artifacts are ropes, pulleys, water tanks and counterweights. The present work has focused on recreating the 23th and 25th artifacts, a war machine called "Fortress Demolisher" and a pulley mechanism of a well called "Magic Well".

2 Description and Digital Modelling

As a previous step to the physical reconstruction of the two aforementioned artifacts it was necessary to carry out a digital modelling of them. CAD software CATIA has been used for digital modelling, which allows the design of the parts, their assembly and the conversion of each part to the STL file format needed for the 3D printing.

The original manuscript of the book of secrets contains numerous drawings and schematics of the artifacts, but they are deteriorated by the passage of time and present quite a few inaccuracies. For this reason, to design the geometry of the different pieces of each artifact, it has been necessary to take as reference the reproductions of the original work made by the engineers of the Leonardo 3 studio [5]. In these reproductions, an interpretation of the original schematics of the codex is made with 3D simulations (Figs. 2 and 3). The simulations show the operation of the artifacts with some clarity, but do not contain detailed drawings or dimensions. For this reason, in order to make a digital model that could later be printed, the design had to be reinterpreted and resized based on these simulations and the little information in the original codex.

Fig. 2. Description, drawings and interpretation of Fortress Demolisher [5].

The reproduction scale has been decided based on the limitation imposed by the size of the 3D printer's printing plate, which does not allow a printed part length greater than 200 mm. Therefore, the largest piece has been designed based on this limitation, and the rest of the pieces have been scaled to maintain the geometric proportion.

Fig. 3. Description, drawings and interpretation of Magical Well [5].

2.1 Fortress Demolisher

The Fortress Demolisher is an assault war machine similar to the typical battering ram that was used to destroy the lower parts of the walls or the gates, and to open breaches in the defenses of the enemy. The study device differs from a classic battering ram in that it is mounted on a platform that can be raised several meters high by means of a scissor mechanism in order to be able to knock down the upper part of the enemy fortifications. The original codex defines the use of the mechanism as follows: "*We want to install on top of this tower and to its left a weapon which can demolish the top of a wall or tower*" [5].

Initially, walls hide the platform with the ram. The platform is raised above the walls by the movement of four winches operated by a pair of men each. When picking up the rope with the winches, wheels on rails move and close the scissor mechanism rising the platform. (Fig. 4 (a)). Once the battering ram is positioned at the desired height, it could be moved backward with a system of ropes and pulleys, to then release it and hit the enemy wall with a strong impact (Fig. 4 (b)).

(a) Scissors mechanism (b) Elevated battering ram

Fig. 4. Details of the reproduction of the Fortress Demolisher by Leonardo3 study [5].

As there is not a clear reference to the geometry or elevation height of the Fortress Demolisher scissor mechanism in the Book of Secrets, the height has been deduced from the information in another mechanism in the book with a similar lifting system, the 22th called Mechanical Tower. After interpreting the data from this mechanism, it was deduced that the platform rise from a height of three meters to a height of eleven meters. A design of the scissor mechanism has been made based on these measurements, adapted to the reproduction scale. The rest of the elements have been designed with dimensions proportional to the scissor mechanism.

An attempt has been made to make the reproduction as faithful as possible to the original model (Fig. 5). However, because the design is oriented towards its subsequent 3D printing and manual operating, small modifications have had to be made in order to print and later assemble the parts. For example, to reproduce the movement of the scissor mechanism and that it could be carried out with only two hands, the shafts that move the pulleys have been lengthened to be able to be driven from the outside (Fig. 5 (c)).

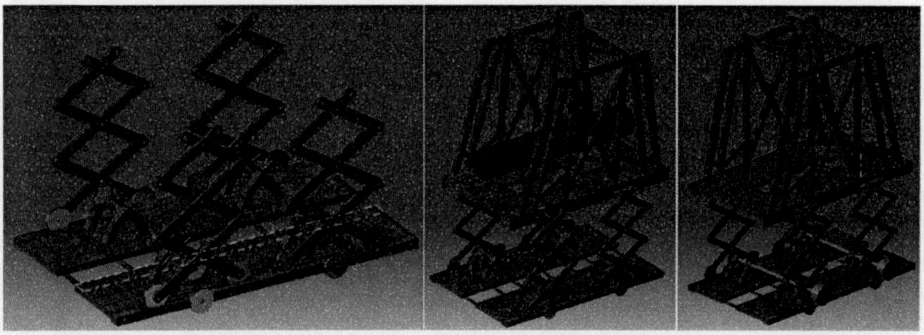

(a) Scissors mechanism design (b) Battering ram design (c) Modified shaft

Fig. 5. Digital model of the Fortress Demolisher.

To give greater realism to the designed Fortress Demolisher, its design has been rendered resulting in the image shown in Fig. 6.

2.2 Magic Well

This device is a simple multi-pulley mechanism whose objective is to be able to lift a load several meters collecting a small portion of rope. In this case, the load is a bucket of water into a well. The codex expressly indicates the purpose of the device: "*we want to take water from a well, whose depth is eight meters, collecting only one meter of rope*" [5].

This mechanism is based on the lock-and-board principle. A rope is passed through a pulley and tied to a fixed location. The other end of the rope is attached to the element to be lifted. The pulley is pulled up and a double-acting mechanism is obtained that makes the objects go up by picking up half the rope.

The original manuscript does not mention the anchoring position of the ropes, nor their length, possibly because the exact dimensions did not want to be revealed so that

Fig. 6. Digital reproduction of Fortress Demolisher.

the well did not lose its "magical" character. However, the interpretation of the drawings indicates that the length of the strings has a sequence of 1-2-4 to be able to bring the bucket from the bottom of the well to its curb. The general law that governs the operation of this mechanism is based on the distance between the supports of the pulleys and the distance to the bottom. For the device to be effective, the distances between pulleys must be in a geometric ratio of 2 (Fig. 7). With this design, the length of rope to be picked up is reduced, but the lifting force of the water is increased since the force necessary to lift each pulley is twice that of the lower pulley.

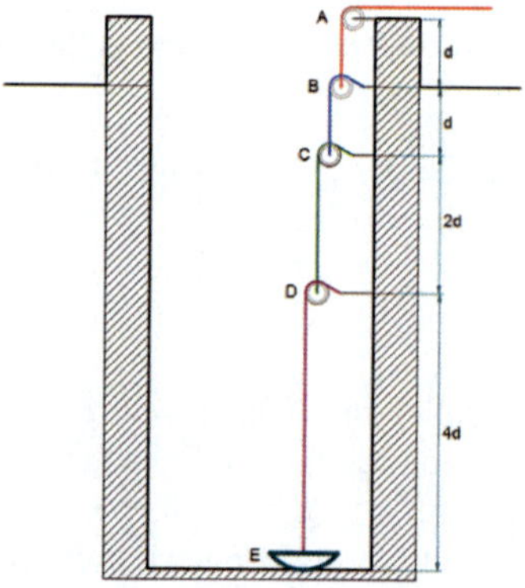

Fig. 7. Geometric relation of the Magic Well.

The dimensions of the pieces of the digital model have been designed by scaling the depth of the well with the maximum dimension available in the printer. All pieces have been scaled according to this ratio. Figure 8 shows the final digital model.

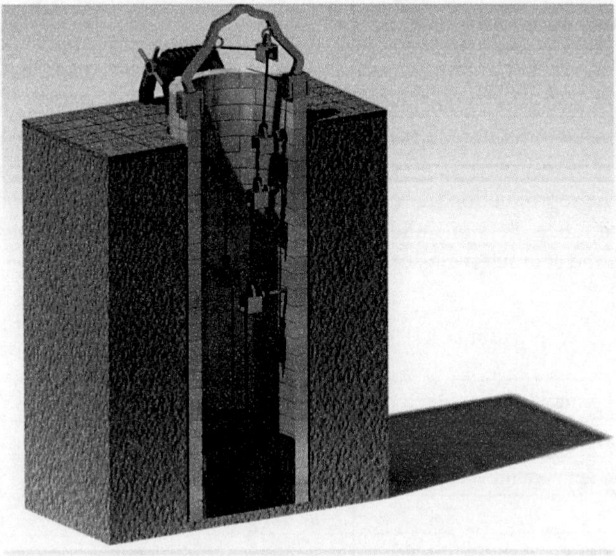

Fig. 8. Digital reproduction of Magic Well.

3 Physical Reconstruction

3.1 3D Printing

Once the digital modelling was done, the two artifacts have been physically reproduced. All the parts of the mechanisms were manufactured with a 3D printer using Fused Deposition Modeling (FDM) technology. FDM technology is based on depositing layers of a molten polymer filament according to a horizontal path followed by an extruder. The polymer is extruded at a temperature close to melting temperature, so it quickly solidifies after being deposited. The overlapping of solid polymer layers allows the generation of three-dimensional structures. The movement of the extruder is what defines the geometry of the printed part. This movement is governed by an alphanumeric code similar to the ISO code used by Computer Numerical Control (CNC) machines [6].

Starting from the digital model made with CATIA software, a discretized STL model can be obtained. This model approximates the surface of the three-dimensional object with a closed triangle mesh. The discretized file is processed with a slicer software that performs horizontal slices of the discretized model. The intersections of the triangular planes with the horizontal planes generate joined straight lines that mark the contour of the part in each plane. The coordinates of these lines are those that the extruder will follow to deposit the molten polymer.

The CURA 3.6 slicer software was used to generate the GCODE text file with the code of the paths and the conditions of the movements. This file is reproducible by the 3D printer to generate the part. This software not only allows the processing of the digital model but also manages its printing by choosing the critical printing parameters such as temperature, printing speed, layer height, wall thickness, filling type and density, etc.

3.2 Materials and Printing Set up

The material used for the printing of the mechanisms parts has been PLA (polylactic acid). The PLA is a biodegradable thermoplastic polymer obtained from cornstarch, and is widely used in molten deposition printing. PLA does not have great mechanical resistance but is widely used for prototype reproduction due to its ease of printing and low cost. To give the reproduction more realism, gray and brown filament spools have been used, which simulate the stone and wood that would be the originally used materials. Table 1 shows the characteristics of the PLA used.

Table 1. Characteristics of PLA filament

Recommended printing temperature	200/220 °C
Melting temperature	145/160 °C
Glass transition temperature	56/64 °C
Density	1,24 g/cm^3
Filament diameter	1,75 mm

An ANET A8 printer has been used for printing, which is a low cost printer with quite good print quality for its price. The printer has a square print surface of 200 mm per side, and a maximum print height of 215 mm. These printing dimensions conditioned the scaling of the printed parts. To reduce vibrations and improve the quality of the finished parts, the original design of the printer has been modified, lightening the weight of the extruder by changing the position of the extrusion motor (Fig. 9).

Fig. 9. 3D printing process.

Table 2 shows the value of the main printing parameters.

Table 2. Main printing parameters

Layer height	0,1 mm
Wall thickness	1 mm
Top/bottom layer thickness	1 mm
Infill density	20%
Infill pattern	Triangular mesh
Printing temperature	210 ºC
Print speed	60 mm/s
Infill speed	80 mm/s

3.3 Assembly and Real Scale Reproduction

After printing all the parts, they were assembled to obtain a real scale reproduction of each artifacts. The union between parts were made using a cyanoacrylate adhesive, since its result in the gluing of PLA is very satisfactory. To join the larger fixed parts, a joint system with holes and shafts has been designed, which facilitates assembly and gives greater stability to the reproduction. For the union of the mobile elements, such as those of the scissor mechanism, small shafts were manufactured and inserted between the moving parts. The end of the shafts was flattened by applying heat, acting as a rivet (Fig. 10).

Fig. 10. Detail of the assembly of the joints

To ensure the correct operation of the fortress demolition lifting mechanism, the eight trestles that house the shafts for manual lifting had to be aligned and adjusted (Fig. 11).

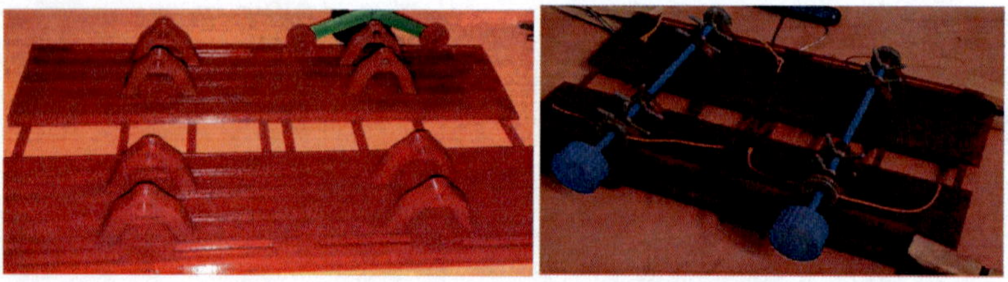

Fig. 11. Lifting mechanism shaft alignment

The only difficulty in assembling the Magic Well is determining the length of the strings. Before the printing of his pieces, a physical simulation was made with a wooden slat and four eyebolts to which the rope was attached. The length of the rope was varied until reaching the solution that was closest to reality (Fig. 12). Due to their small size, the pulleys could not be printed to the same scale as the well. Pulleys with a larger scale have been used to be able to give it a real movement.

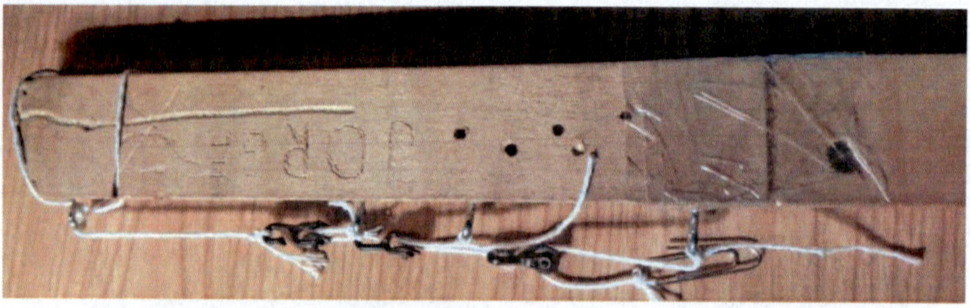

Fig. 12. Rope length test

Figure 13 shows the results of the two reproductions.

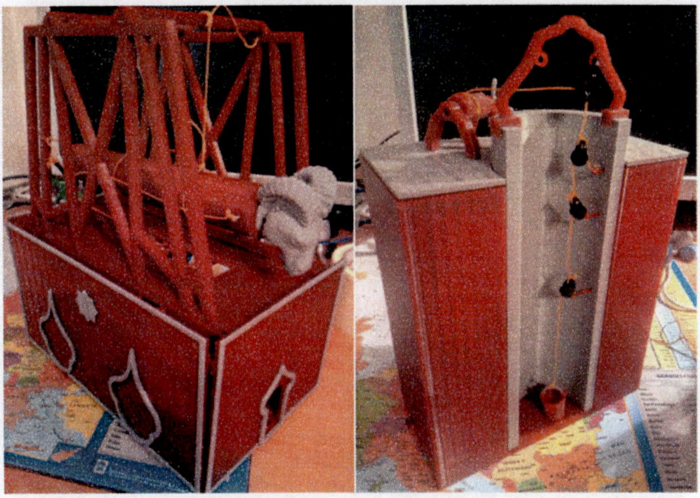

Fig. 13. Real scale reproduction of the two medieval artifacts

4 Conclusions

As a result of this work, the digital modelling and physical reconstruction of two medieval artifacts, the "Fortress Demolisher" and the "Magic Well" have been obtained. Digital modelling allows to study in detail the geometry of artifacts and to subject them to mechanical analysis using engineering software. The physical reconstruction allows to study its movement and operation in a more visual and didactic way.

The use of new technologies allows us to take another step in the transmission of knowledge. Nowadays knowledge is not only transmitted through printed writing, but it is mostly done digitally in written, graphic or audiovisual form.

The use of 3D printing in the reproduction of machinery or mechanisms, manages to go from digital information to a real three-dimensional scale model, quickly and economically. This physical reproduction allows a more direct interaction with the reproduced artifact. This can help to better understand the functioning of the mechanisms that compose it, which is a great help for students, especially in the field of mechanical engineering.

The study of historical machines allows us to deepen our knowledge of the most elementary mechanisms and better understand the evolution that has been followed up to the most complex mechanism designs.

References

1. Escolar, H.: Historia de las bibliotecas. Fundación Germán Sánchez Ruipérez, Madrid (1990)

2. De Sousa, J.M.: Pequeña historia del libro, 4th edn. Ediciones Trea, Gijón (2010)
3. Dahl, S.: Historia del libro, 2nd edn. Alianza Universidad, México (1998)
4. Viguera, J.M., et al.: Los reinos de Taifas. Al-Andalus en el siglo XI. Historia de España Menéndez Pidal, vol. VIII-1. Espasa Calpe, Madrid (1994)
5. Al-Murādī, I.K.: The Book of Secrets in the Results of Ideas: Incredible Machines from 1000 Years Ago. Leonardo 3, Milán (2008)
6. Gibson, I., Rosen, D., Stucker, B.: Additive Manufacturing Technologies: 3D Printing, Rapid Prototyping, and Direct Digital Manufacturing. Springer, New York (2010). https://doi.org/10.1007/978-1-4939-2113-3

Reconstruction Designs of the Automata in the Ancient Korean Water Clock "Heumgyeonggaknu"

Yu-Chieh Chiu[1], Yu-Hsun Chen[1](✉), and Hong-Sen Yan[2]

[1] National Taiwan University of Science and Technology, Taipei, Taiwan
yhchen@mail.ntust.edu.tw
[2] Da-Yeh University, Changhua, Taiwan
hsyan@mail.ncku.edu.tw

Abstract. *Heumgyeonggaknu* (欽敬閣漏, Clock of Respectable Veneration Pavilion) is a Korean water clock made by Jang Yeong-Sil (蔣英實) in 1438. Based on astronomical observation, the functions of this water clock include time-keeping through a water wheel and telling the hours through many automata puppets set on the five floors of an artificial mountain. However, this device was not preserved. Although the external appearance and the functions are described in ancient literature, the internal structure is unknown. Based on the development of science and technology in ancient Korea, the historical records and the modern restoration designs of *Heumgyeonggaknu* are surveyed, and the works of timekeeping devices in Korea during the subject's time are discussed. In this study, a systematic design procedure is proposed for reconstructing the ancient mechanisms with unknown parts, and the automata set on the first floor of the *Heumgyeonggaknu* are taken as a design example. At first, the design specifications of the automata in the device are concluded from the review of historical documents. Then, based on the atlas of generalized chains, specialization is implemented according to design requirements and constraints. All feasible designs that meet the scientific theories and techniques of the subject's time period are synthesized. As a result, two feasible mechanisms are generated for the automata on the first floor. Then, dimensional parameters are designed to meet the required motion, and a computer-aid virtual models is built for simulation. Finally, a prototype of the reconstructed automata is made to verify the movement characteristics and the feasibility of the presented reconstructed designs.

Keywords: Ancient machines · Reconstruction design · History of machines and mechanisms

1 Introduction

The clock is a scientific instrument for measuring time, and the automaton is a mechanical device that is able to convert the input motion to a specific output action by means of a mechanism transmission for time-telling or performance purposes. Both of them are

M. Ceccarelli and R. López-García (Eds.): HMM 2022, HMMS 40, pp. 97–105, 2022.
https://doi.org/10.1007/978-3-030-98499-1_9

important manifestations of science and technology. The scientific principles applied such as the flow rate control of water, the application of buoyancy, and the design of mechanical components are indicators of the development level in the history of machinery. Moreover, they are also the proof of the exchange between the cultures.

For thousands of years, the ancients have used long-term astronomical observations and records to develop methods for predicting the position of astronomical stars and calculating calendars, which serve as an important reference for the activities of the farming society in various seasons. In ancient Korea, King Sejong (世宗, 1397–1450) of the Joseon Dynasty (1418–1450) assigned the engineer Jang Yeong-Sil (蔣英實, 1383–1450) to make the self-striking water clock. After traveling to China and studying the designs of these kinds of devices [1, 2], Jang set up two mechanical clocks in Korea, namely the *Borugaknu* (報漏閣漏, clock of annunciating clepsydra pavilion) in 1433 [3] and the *Heumgyeonggaknu* (欽敬閣漏, clock of respectable veneration pavilion) in 1438 [4], as Fig. 1 shows. Both of the two clocks are powered by constant-speed water flow for timekeeping, and there are mechanical puppets equipped to the two clocks to tell the time. However, the records of these two devices in the literature are very different. In the former device, a rolling ball system to drive all automata is described in detail. For the latter device, the appearance, functions, and motions of automata puppets are illustrated, but just a few words related to the internal mechanical structure are presented.

(a) (b)

Fig. 1. Jang Yeong-Sil's two water clocks: (a) a modern reconstruction of the time-telling automata of *Borugaknu* [5], (b) the appearance of *Heumgyeonggaknu* [2].

According to the descriptions of *Heumgyeonggaknu* [2], the mechanical structure can be arranged as Fig. 2. Three subsystems are included, namely power source, transmission mechanisms, and automata puppets. The power source is a wheel driven by water flow. The automata puppets subsystem can be separated into five layers to compose a mountain on the pavilion, as Fig. 1 shows. From bottom to top, on the first floor (W1), twelve Gods of the Zodiac (十二神) show up at the corresponding time (Rat God for 23 o'clock, Ox God for 1 o'clock, Tiger God for 3 o'clock, Rabbit God for 5 o'clock, Dragon God for 7 o'clock, Snake God for 9 o'clock, Horse God for 11 o'clock, Goat God for 13 o'clock,

Monkey God for 15 o'clock, Rooster God for 17 o'clock, Dog God for 19 o'clock, and Pig God for 21 o'clock). At the same time, a cave behind the acting Zodiac God opens, and a fairy with an hour-displaying card steps out of the cave. On the second floor (W2), an official and three warriors standing on a stage. One warrior strikes the bell every double-hour, another warrior beats the drum at the five double-hours in the night (at 19, 21, 23, 1, and 3 o'clock), the other warrior sounds a gong every 24 min during the five hours in the night. The official turns to the warriors before they sound their instruments. On the third floor (W3), the Four Guardian Gods (四神, including Azure Dragon of the East, the Vermilion Bird of the South, the White Tiger of the West, and the Black Tortoise of the North) stand at the four corners and turn to a specific position according to the corresponding time. On the fourth floor (W4), four fairies rattle the bell at the relevant double-hour respectively. On the top floor (W5), an armillary sphere rotates along a circular ring to simulate the sunrise and sunset once a day passing over the mountain.

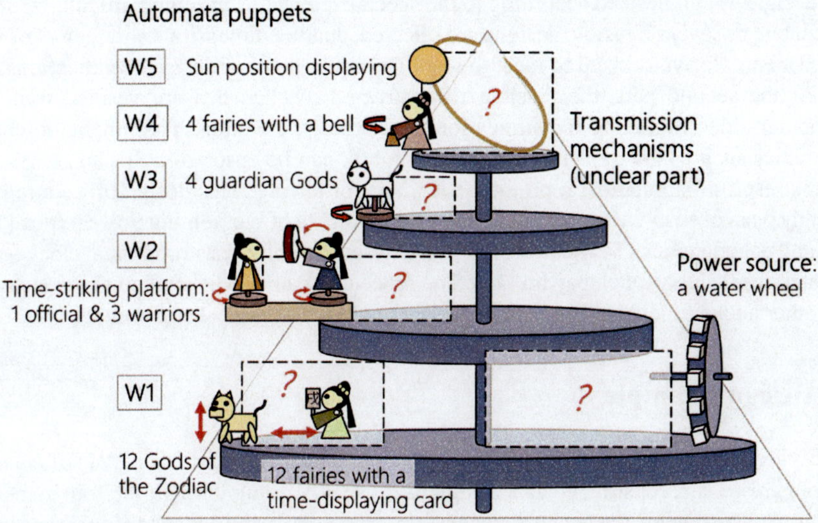

Fig. 2. Automata puppets and the unknown parts of *Heumgyeonggaknu.*

In this study, since the transmission mechanisms that drive the automata puppets are not clear in the description of *Heumgyeonggaknu* [4], a systematic design procedure is proposed to generate the feasible mechanism structures. Based on the ancient literature, the relevant design requirements and constraints are listed. By the concepts of specialization and particularization, the designs that meet the technical level of the project's time are synthesized. Finally, the feasible mechanism structures are generated, and the performance of the reconstructed mechanisms is verified through 3D modeling and simulation.

2 Design Procedure

A design procedure for reconstructing ancient mechanisms with unknown structures can be mainly divided into two parts, namely reconstructed mechanism design and reconstructed manufacturing. At the first part, the feasible structure of the lost/unknown mechanisms is synthesized through a systematic procedure [6, 7], and all feasible configurations of mechanisms that meet their design requirements and constraints with four steps. Firstly, the design specifications of the machine can be determined based on historical documents [3, 4] and modern research [1, 2, 5, 8–11], including the feasible type of links and joints, the possible number of links and joints, and the degrees of freedom. Secondly, according to the design specifications concluded in the previous step, the atlas of generalized chain is obtained. Thirdly, all feasible specialized chains are generated by assigning the types of links and joints to all generalized chains from the third step subject to the concluded design requirements and constraints. Finally, the atlas of feasible designs is synthesized according to the specialized chains through particularization. Then, one or several feasible designs are selected, and the dimensions parameters of the mechanism are synthesized to meet the motion requirement of design specifications.

At the second part, the feasible reconstructed mechanisms are verified through computer-aided modeling for simulation, and prototype testing. The kinematic characteristics such as the trajectory of the output link can be ensured in this stage. Based on the result of simulation, a prototype is made for testing. This design procedure has already been used in the reconstructions of the automata of blossoming flower clock [12, 13] and pagoda clock [14] in the Forbidden City, and a Chinese astronomical clock with a waterwheel steelyard clepsydra [15]. The procedure can also be applied to reconstruct the other ancient mechanisms with unknown/unclear mechanical structures.

3 Design Example

In this section, the mechanism of the automata puppets on the first floor (W1) of *Heumgyeonggaknu* is reconstructed as a design example. By going through the four steps of reconstruction design, the feasible mechanism structures can be generated in this section.

I. Design specifications

In the ancient literature [4], the descriptions about the first floor (W1 in Fig. 2) are as follows:

> "*At the flat ground on the first floor, the twelve gods are hided in their respective positions. ... When the ox-hour (1 o'ckock in the midnight) comes, the hole behind the ox opens, a fairy also walks outside of the hole, and the ox rises.*"

「又其下平地之上, 十二神各伏其位。...丑時至則牛後之穴自開, 玉女亦出, 牛亦起。」

According to the structural analysis in the first section and the descriptions above, the design specifications can be arranged:

1. There must be one input link (connected to the power source), two output links (one for the ox and the other one for the fairy), one fixed link (the frame), and at least one connecting link included. Therefore, this is a mechanism with at least five links.
2. Since the power source of the whole device is a water wheel, the only input of the mechanism is a rotating motion. Therefore, this is a mechanism with one degree of freedom.
3. The two output links make the ox rise/return and the fairy walk forward/backward. Both of the output motions can be analyzed as reciprocated motion.
4. With the consideration of the technical level and referring to the previous work *Borugaknu* of the inventor Jian Yeong-Sil, the type of the links can be considered as a set of linkage.

II. Atlas of generalized chain

As the design specification listed, the number of links is assumed as five, and the degree of freedom is one. For a mechanism with five links, the number of joints obtained is 6, including one joint with 2 DoF and five joints with 1 DoF. The corresponding generalized chains are listed in Fig. 3.

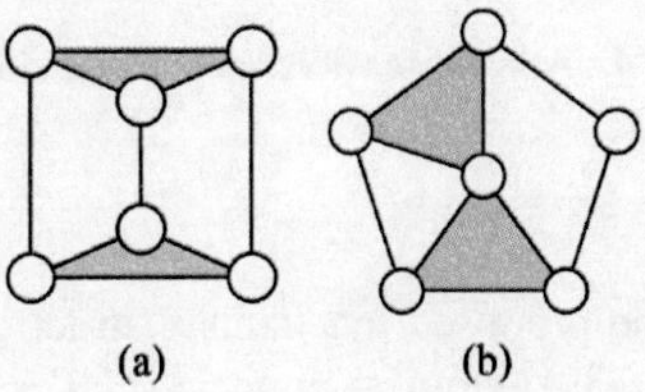

Fig. 3. Atlas of generalized chain with five links and six joints.

III. Atlas of specialized chain

According to design specifications, the obtained generalized chains are subjected to the specialized process, and the required links and joints are assigned to each generalized chain in order to generate all feasible specialization chains. To perform this step, the following design requirements and constraints are complied with as follows.

Fixed Link (Link 1, K_F)

There must be a multiple link serving as a fixed link (K_F).

Input link (Link 2, K_I)

There must be a binary link serving as the input link (K_I), and it is adjacent to the fixed link (K_F) with a revolute joint (J_R).

Transmission link (Link 3, K_T)

There must be a binary link serving as the transmission link (K_T), and it is adjacent to the input link (K_I) with a revolute joint (J_R).

Output link 1 (Link 4, K_{O1})
There must be a multiple link serving as the output link 1 (K_{O1}), and it is adjacent to the fixed link (K_F) with a revolute joint (J_R) or a sliding joint (J_P).
Output Link 2 (Link 5, K_{O2})
There must be a binary link serving as the output link 2 (K_{O2}). It is adjacent to the fixed link (K_F) with a revolute joint (J_R). Moreover, it is adjacent to the output link 1 (K_{O1}) with a cam joint (J_A).

By assigning all of the links and joints, for the generalized chain shown Fig. 3(a), there is not a loop that includes one multiple link and two connected binary links, so the first three links (K_F, K_I, and K_T) cannot be assigned in this case. Therefore, for the generalized chain shown in Fig. 3(b), two feasible specialized chains are generated as Fig. 4 shows.

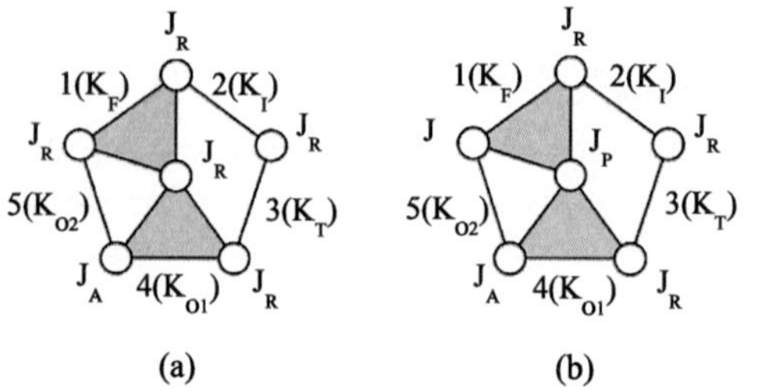

Fig. 4. Atlas of feasible specialized chain.

IV. Atlas of feasible design

The feasible designs with the proposed mechanism structure can be obtained through the step of particularization. The automata puppets such as the rooster god (one of the twelve gods) and one of the fairies with a time-displaying card are attached to the output links respectively, as the two mechanisms shown in Fig. 5.

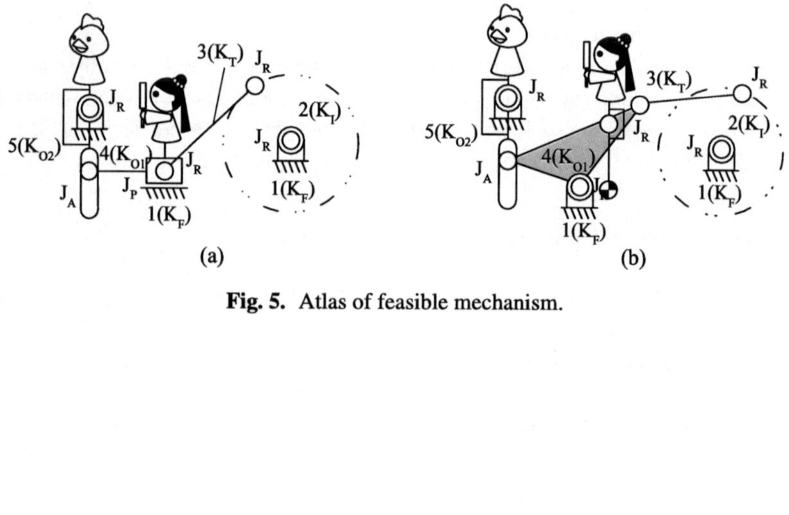

Fig. 5. Atlas of feasible mechanism.

4 Computer-Aided Modeling and Prototype Testing

Based on the feasible mechanism structures synthesized in Sect. 3, the reconstruction manufacturing is made through computer-aided modeling and prototype testing. A 3D model is built by the software SolidWorks, as Fig. 6 shows, in accordance with the mechanism shown in Fig. 6(b) and the results of dimensional parameters design to meet the motion requirements. Besides, a corresponding wooden prototype is made to verify the motion. Figure 7(a) shows the initial position of this automaton, the puppet of the twelve gods is hidden below the stage, and the fairy stays inside the cave. With the input link rotates in a half-circle, the output links arrive at their performance positions. As Fig. 7(b) shows, the puppet of the twelve gods rises to show up, and the fairy walks out of the cave to tell the time with the displaying card. Then, with the input link completing a full-circle rotation, the two puppets return to their initial position. As the result, the motion of the automata is able to meet the descriptions recorded in the ancient literature [4], therefore the feasibility of the proposed design procedure and reconstructed designs are evaluated.

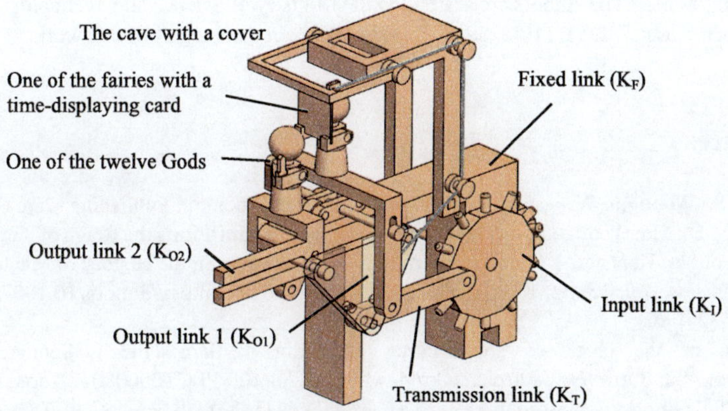

Fig. 6. A 3D model of the automata

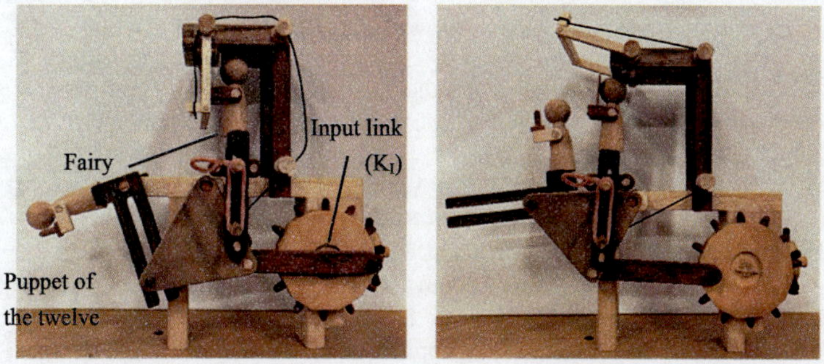

(a) (b)

Fig. 7. Prototype testing: (a) initial position, (b) performance position.

5 Conclusions

Heumgyeonggaknu is a water clock made in Korea during the 15th century. With five sets of automata puppets equipped on the five floors of the water clock for time telling, the appearance and motions for performance are clearly recorded in the ancient literature. However, the internal mechanical structure was neither documented nor preserved. In this study, a systematic design procedure to reconstruct the ancient automata with unknown structures is presented. The automaton on the first floor of the *Heumgyeonggaknu* is taken as a design example. By applying the four steps for synthesizing the mechanism structure, the design specifications are concluded, and two feasible mechanisms with five links are synthesized. Moreover, the results are evaluated through computer-aid modeling and prototype testing. A 3D model has been elaborated in SolidWorks environment for motion simulation for performance characteristics and design developments. Finally, a prototype is shown with preliminary results within a project for future developments. The design procedure is verified and it can be applied to complete the reconstruction design for the whole device of *Heumgyeonggaknu.*

Acknowledgments. The authors are grateful to the Ministry of Science and Technology (Taipei, Taiwan) under Grant MOST 110-2222-E-011-015-MY2 for the support of this work.

References

1. Baek, S.: Woongjin Wi-in Jeon-gi #11 Jang Yeong-sil. Woongjin Publishing, Paju (1987)
2. Nam, M.H.: Jang Yeong-sil: inventor of the striking clepsydra during the Reign of King Sejong in Joseon. In: Koetsier, T., Ceccarelli, M. (eds.) Explorations in the History of Machines and Mechanisms, vol. 15, pp. 83–105. Springer, Cham (2012). https://doi.org/10.1007/978-94-007-4132-4_6
3. Record of the clock of annunciating clepsydra pavilion (1433). Seongjong Sillok, vol. 82. https://www.krpia.co.kr/viewer/open?plctId=PLCT00008012&nodeId=NODE07379576#none. Accessed 8 Dec 2020. 報漏閣記 (1433). 世宗實錄, 東文選卷之八十二
4. Record of the clock of respectable veneration pavilion (1438). Seongjong Sillok, vol. 82. https://www.krpia.co.kr/viewer/open?plctId=PLCT00008012&nodeId=NODE07379576#none. Accessed 8 Dec 2020. 欽敬閣記 (1438). 世宗實錄, 東文選卷之八十二
5. Nam, M.H.: Time in Korea, Encyclopaedia of the History of Science, Technology, and Medicine in Non-Western Cultures, Springer, Cham (2016). https://doi.org/10.1007/978-94-007-3934-5_9337-2
6. Yan, H.S.: Reconstruction Designs of Lost Ancient Chinese Machinery. Springer, Dordrecht (2007). https://doi.org/10.1007/978-1-4020-6460-9
7. Yan, H.S.: Creative Design of Mechanical Devices. Springer, Singapore (1998)
8. Kim, S.H., Lee, Y.S., Lee, M.S.: A study on the operation mechanism of Ongnu, the astronomical clock in Sejong era. J. Astron. Space Sci. **28**(1), 79–91 (2013)
9. Lee, Y.S., Kim, S.H.: Structure and conceptual design of a water-hammering-type Honsang for restoration. J. Astron. Space Sci. **29**(2), 221–232 (2012)
10. Kim, S.H., Lee, Y.S., Lee, M.S., Ham, S.Y.: A study on the internal structure of Heumgyeonggaknu. J. Astron. Space Sci. **30**(2), 113–121 (2013)

11. Kim, S.H., Yun, Y.H., Ham, S.Y., Mihn, B.H., Ki, H.C., Yoon, M.K.: A study on an analysis and design of the internal structure of Heumgyeonggak-nu. J. Astron. Space Sci. **34**(2), 171–1822 (2017)
12. Chen, Y.H., Yan, H.S.: Reconstruction of unknown automata designs of blossoming flower clock in forbidden city. Proc. Inst. Mech. Eng. C J. Mech. Eng. Sci. **231**(7), 1354–1368 (2017)
13. Chen, Y.-H., Ceccarelli, M., Yan, H.-S.: Designing and prototyping reconstruction of musician automata. In: Ceccarelli, M., Cigola, M., Recinto, G. (eds.) New Activities for Cultural Heritage, pp. 22–32. Springer, Cham (2017). https://doi.org/10.1007/978-3-319-67026-3_3
14. Li, Y.Z., Chen, Y.H.: Reconstruction of unknown mechanisms of a Pagoda clock in the forbidden city. Mech. Mach. Theor. **167**, 104469 (2022)
15. Hwang, Z.H., Lin, T.Y., Yan, H.S.: Reconstruction designs of an early Chinese astronomical clock with a waterwheel steelyard clepsydra. Mech. Sci. **12**, 891–911 (2021)

Analysis of Mechanisms 2 and 21, Included in Ibn Khalaf Al-Muradi's "The Book of Secrets"

Rafael López-García(✉), Israel Fernández-Liétor, Rubén Dorado-Vicente, and Gustavo Medina-Sánchez

University of Jaen, Jaen, Spain
{rlgarcia,rdorado,gmedina}@ujaen.es, ifl00006@red.ujaen.es

Abstract. In the 11th century AD the Andalusian engineer *Ibn Khalaf al-Muradi* wrote a mechanical manuscript with the original title *Kitāb al-asrār fī natā'iy al-afkār* and called *'The Book of Secrets'*, which is preserved because a copy of it was made in Toledo in 1266 AD and is kept in the Biblioteca Medicea Laurenziana in Florence. This manuscript is of great importance in the field of science and engineering because it represents one of the earliest accounts of complex mechanisms in antiquity. The manuscript consists of 31 chapters written in Arabic, on different topics such as time-measuring machines with automata, war machines, water pumps and sundials. The present work is a historical-technological study of 2nd and 21th mechanisms. These are a theatre which works by means of a gear system and a war machine which works by means of a Nuremberg scissors mechanism. Some of these mechanisms are used in modern machines such as elevators and transmission systems. First of all, a description of the components of the mechanisms and how they work will be given. Next, a virtual reconstruction of the mechanisms will be made, using geometric modelling software such as SolidWorks, which also allows us to carry out a mechanical analysis. From this three-dimensional model, a virtual animation will be generated where the movement and operation of the mechanism can be appreciated. Finally, a scale reconstruction can be carried out by means of additive manufacturing.

Keywords: History of machines and mechanisms · Mechanical analysis · Geometric modelling · 3D printing

1 Introduction

It can be argued that the birth of the advanced technology dates back to ancient history, and we can found testimonials about inventions in ancient Asia, Middle East and ancient Greece and Rome. But Alexandria, between the 4th century B.C. and 4th century A.C., was the center of most of the ancient innovations of its time with resident likes Ctesilius, Hero and Philo of Byzantium with inventions and mechanical treatises.

At the subsequent Arabic culture, earlier scientific and technological production had been absorbed and developed. In fact, this progress was evidenced in Eastern Islam in

M. Ceccarelli and R. López-García (Eds.): HMM 2022, HMMS 40, pp. 106–119, 2022.
https://doi.org/10.1007/978-3-030-98499-1_10

treatises such as "The book of ingenious devices" written by the brothers Banü Müsa in 850 in the House of Wisdom in Bagdad which circulated widely in the Muslim world, or the posterior "The book of knowledge of ingenious mechanical devices" written by Al-Jazari in 1206.

The flowering of science and technology in Western Islam came through the influence of Eastern Islam and the support of its rulers, especially in the 9th and 10th centuries from Abd al-Radman II to al-Hakam II, which then spread to Europe, North Africa and America.

In the 11th century the engineer *Ibn Khalaf al-Muradi* wrote the only known manuscript on mechanics originally called *Kitāb al-asrār fī natā'iy al-afkār* and called *"The Book of Secrets"*, which is preserved thanks to a copy of it made in Toledo in 1266 by a translator at the court of Alfonso X the Wise. This copy is preserved in the Biblioteca Medicea Laurenziana in Florence.

The recovery of the manuscript was made possible through the patronage of His Highness the Emir Sheikh Hamad bin Khalifa Al Thani, who funded a research project that has led to its publication in Arabic and English. This manuscript is of great importance in the field of science and engineering because it represents one of the earliest testimonies of complex mechanisms in antiquity.

The manuscript consists of 31 chapters full of secrets written in Arabic, each of which describes different mechanisms; although this manuscript is not completely preserved as many of its pages have been damaged. A description of each of the mechanisms is given in Arabic and is accompanied by sketches and drawings of the mechanisms (Fig. 1).

Fig. 1. Example of one of the pages of the manuscript

In this chapter, figures 2nd and 21th of the manuscript will be analyzed and studied. Then, these descriptions of the mechanisms will be interpreted in order to understand how they work. The next step, using the software used, will be to carry out the geometric modelling, virtual reconstruction, mechanical analysis of the mechanism and possible real scale reconstruction. Finally, some current applications of these mechanisms will be discussed.

2 The 2nd Figure

Figure 2nd is a theatre in which two knights, two foot soldiers and two maidens perform a scene in which they all move around the theatre, driven by a system of gears that are moved by a rotating paddle mill driven by a stream of water.

Transcription of the description:

"We want to make a shape, which has two knights, two foot soldiers and two maidens. So let be a rectangle ABCD, in which line AD is four times longer than the line AC. On the line AC, there will be a knight, a foot soldier and a maiden. So will be the same on line BD.

In the middle of the shape and between lines AC and BD, install a six-finger-wide and three-(handbreadth)-long board. The knights run from one side to the other on this board.

The horses (of the knights) should each be installed on a circle. The two circles are separated by one handbreadth. In the center of each circle, a bar is installed. The bar extends to the head of the knight and penetrates the board below the circles. The two bars end below the board. Install a gear at the end of [...]. In the middle of the gear, install a column so that the gear turns by the turn of the column. The column ends in a board, whose width is equal to the diameter of one circle and the space between the two circles. This board lies on four orbits like wheels, which run on another board extending from the beginning of the figure to its end. The board, (on which the wheels run), has two grooves separated by a distance equal to that separating the wheels. The board, which has the wheels, also has two trays, on which the two bars turn. [...].

The columns end at the upper half of the knights if God so desires. [...] As for the [mechanism] that moves all that we have mentioned, it will be explainer as follows if God so desires.

Make an orbit IJ. If the water is scarce, make it of the kind that moves querns. Install the [orbit] on an axis KL. The axis KL is installed between the lines AD and MN. Point L is on line CB, while the axis does not touch line AD. The point K is on the line [MN] outside the figure to the side of AD. Install another orbit, BD, on axis KL. Orbit BD contains 64 teeth on one half of its circumference. Install two axes, EF and OH, parallel to axis KL. Install their ends on the lines CB and AD. Each axis holds an orbit IJ, which has teeth along its circumference and is equal in their dimensions to those of orbit BD. The diameter of orbits IJ is one fourth that of orbit BD.

Install a cross on each of the axes EF and GH. At the end of one of the crosses and at the point G, extend a rope from the cross to a pulley, R, at line AC. The rope extends from this pulley to a pulley P and ends in a cross installed on the column holding the knight. The same is to be done with the other knight. Extend another rope from cross E, in the same manner, and tie it to the front of the other side [of the cross] of each knight. Extend another rope along the cross [G] to pulley Q, which is at line AC. The rope extends from this pulley to a pulley, H, and then to the column on which the foot soldier is installed. The rope should be protected from the breeze while passing between the two pulleys H and Q by passing in a tube. Do the same to the other foot soldier. Extend a rope from cross F to a pulley, Q, at line BD. The rope then passes to a pulley [...] and ends at the bottom of the column on which the foot soldier is installed. The rope should pass in a tube between cross F, pulley Q and pulley [...] to be protected from being moved by the breeze.

Install an axis away from the orbit [BD] and install a bar on it. The bar has teeth at, one end and a weight at the other. The teeth are equal in dimensions to those of j the orbit BD. The end that is attached to a weight is also attached to a rope. The rope extends to the body of the maiden and splits into two halves, each of which passes to one of her elbows. The other maiden is moved in the same way. And that is what we wanted to do if God so desires. And this is its image."

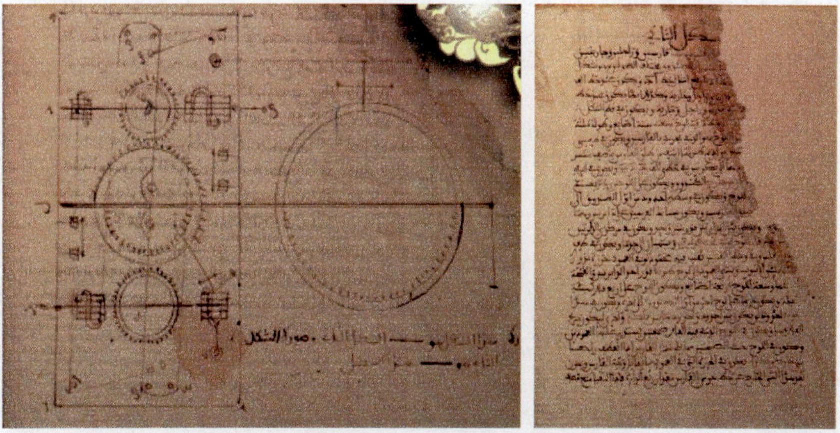

Fig. 2. Illustration and description of the 2nd figure

The figure composing a scene with two knights, two soldiers and two princesses is characterized by complicated movements, each playing a role: the knights gather for battle in the centre of the scene, while two hidden foot soldiers come out to join the battle. When they see the fight, the maidens raise their arms in the air. The mechanism used to animate the characters in this scene is similar to other mechanisms used in other figures in the manuscript.

In this figure 2nd, the main mechanism is the gear system, which is made up of three transmission shafts, on one of which is the windmill that moves the whole mechanism thanks to a current of water. On this same shaft there is a toothed wheel with straight teeth but only with teeth on half of the wheel and on the other two shafts there are two toothed wheels with teeth of the same size as the previous toothed wheel and with the same number of teeth and they have a cross for tying ropes. These ropes are then passed through rollers and tied to the figures in the theatre so that they can move.

3 The 21th Figure

Figure 21th is a war machine that is placed on top of a tower and consists of a battering ram mounted on a scissor mechanism, which is intended to bring down enemy towers and destroy fortresses.

Transcription of the description:

"We want to make a military figure 1 which is installed on top of a tower or in any other place. (The figure] can be extended twenty yards or whatever distance needed outside the tower and can throw fire which destroys whatever it meets. The figure then returns quickly to its place. So let us make a rectangle, ACBD, where line AC is twice as long as line AD. Draw lines IJ and KL between lines AC and BD. Draw two lines, GH and QT, between lines IK and JL. Draw also lines MN and MN from line IJ to line KL. Make sure that all that I describe is done with precision and accuracy. The lines intersect together with no increase or decrease.

Install on the previous figure a fixed cover which does not touch the figure and 'is similar to [...]. [The cover and the figure] are separated by nine fingers.

As for the figure mentioned before, ð is explained as I describe if God so desires. Make two (rectangles) QTTQ. The length between T and T is twelve [...] fingers and between Q and Q is nine fingers. [...] so they are four figures. Make a hole at point F of the figures [...] at line IK and install a bar (...). This bar connects the first figure and [...] the fourth figure. The distance between F and T is equal to that between the figure and the cover.Attach two ropes to the points [T]. [The first] rope extends to a pulley, M, and then is attached to a pulley [...]. The second rope extends from point T to a pulley, F, and then is attached to a cylinder O. We do the same with the ropes extending from the other point T.

Make four figures RVVR. The length of each figure between points R is six yards. The length between points V is six fingers. The thickness of the figures is six fingers each. The figures intersect with each other. Make a hole at point T and [install] an iron bar [in the hole] to conned the four figures.

Take four other figures, V, V, X and X made of wood. The length of each figure [V] is four yards, the width at points V is six fingers, and the thickness of each one is four fingers. The width at points X is four fingers and the thickness is three fingers. The figures intersect with each other and are connected with bars so that they move around these bars easily (Fig. 3).

Make two bars at point H. [Install] the two pulleys, D and D, on top of these bars. Then we make a bird's head, which is hollow and is three-yards long. The head is installed on an axis at point H and its movement is smooth. The extension of the head should not exceed the figure. Extend two ropes from points X and X to pulleys [...] then pass them to points C and C.[...] And so if we want to move it, we should turn the two cylinders together and the figure would go up. And if we want to bring it back, we turn the cylinders in the opposite direction and the figure will go back. And that is what we wanted to do if God so desires and this is its image.

The basic mechanism shown in Figure 21th is now known as the "Nuremberg scissors". It is a mechanism made up of a set of bars joined to each other at their ends and in the intermediate part of these, which allows to move from a lateral rhombus to

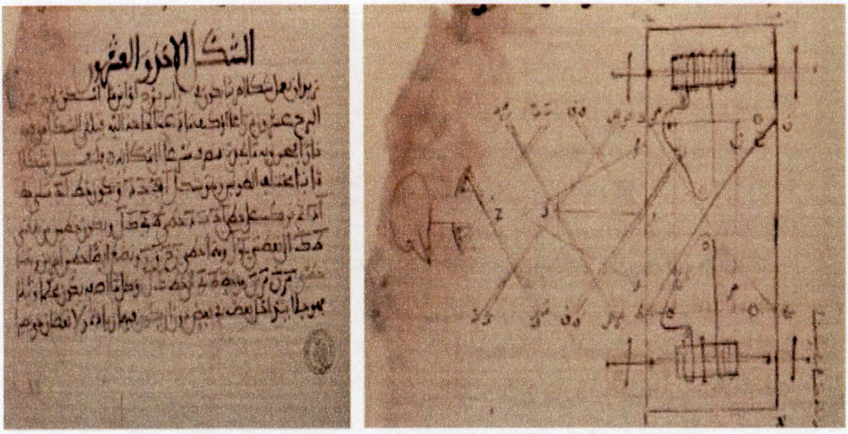

Fig. 3. Illustration and description of the 21th figure

a longitudinal rhombus, which provides a large displacement from a small space when the mechanism is retracted.

4 Geometric Modeling

SolidWorks, a CAD (computer-aided design) software for mechanical modelling in 2D and 3D, has been used for the geometrical modelling of the mechanism. The software allows parts and assemblies to be modelled and to extract from them both technical drawings and other information necessary for manufacturing. The process consists of transferring the designer's mental idea to the CAD system, "virtually constructing" the part or assembly. Subsequently, all the extractions (drawings and exchange files) are carried out in a fairly automated way.

From the modelling of simpler parts, they can be assembled into more complex assemblies using positional relationships and finally these assemblies with the Finite Element Analysis simulation complements can be meshed, forces applied and analyse them numerically, from the mechanical point of view.

The dimensions of the parts have been obtained from de few indicated in the descriptions of the mechanisms. For the rest of the elements, they have been deduced proportionally.

4.1 Figure 2nd

The modelling of the figure was based on the description and interpretation of the figure, according to the previous sections.

The main upper plate is where the movements of all the characters are made and to make it a sketch has been made with the slots for the different characters and this sketch has been extruded, then there is another small plate so that the ropes that make the system move are protected. The column that serves as a support for the characters consists of an extrusion of several circles (Fig. 4).

Fig. 4. Main upper plate and column of the 2nd figure

The arms of the maidens move by means of a rack and pinion mechanism that transform the rotary movement of the cogwheel into a linear one. As the gear only has teeth in the middle, a weight attached to a rope is used to return it to its position. The cross and the ropes attached to it are used to move the knights and soldiers. The mechanism has seven shafts, some of which are keyed, to transmit the movement (Fig. 5).

Fig. 5. Rack and cross of the 2nd figure

The main gear wheel has straight teeth, but with teeth only on half of its circumference, which engages with the two secondary gear wheels, causing half of the theatre to move and then the other half. So that while one wheel is moving, the other can return to its original position by means of a weight, so that the theatre operates continuously with the water.

To make the whole thing move, it has a wheel similar to a waterwheel with paddles that would be moved by a current of water. If there is a shortage of water, a crank can be made to move the mechanism. This piece is similar to that of a water mill. To make the piece, the two rings have been made and extruded and then one of the blades has been made and the rest has been made with a circular die.

The mechanism has several drum rollers for winding and redirecting the strings used to move the figures. In this way, the geometric modelling of the different pieces that make up the mechanism has been carried out (Fig. 6).

Next, the assembly of all the elements that make up the mechanism is carried out, by means of the position relationships between them, such as: concentric for the axles, gears, rollers; groove to place soldiers and knights; coincident to make the faces stick to each other; pinion-rack to make the gear; gears to make a position relationship between them; belt/pulley so that two separate wheels have movement (Fig. 7).

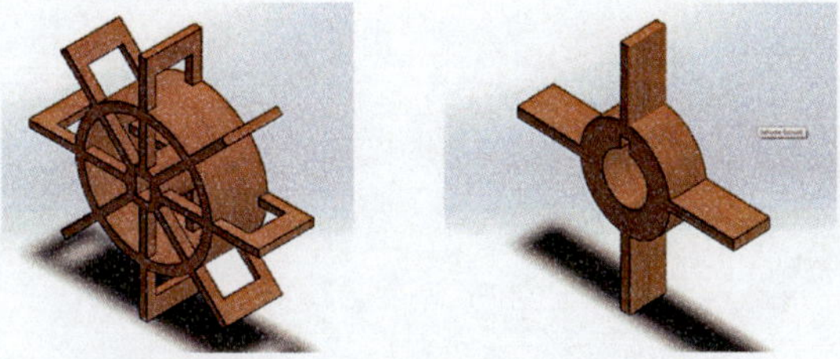

Fig. 6. Waterwheel and cross of the 2nd figure

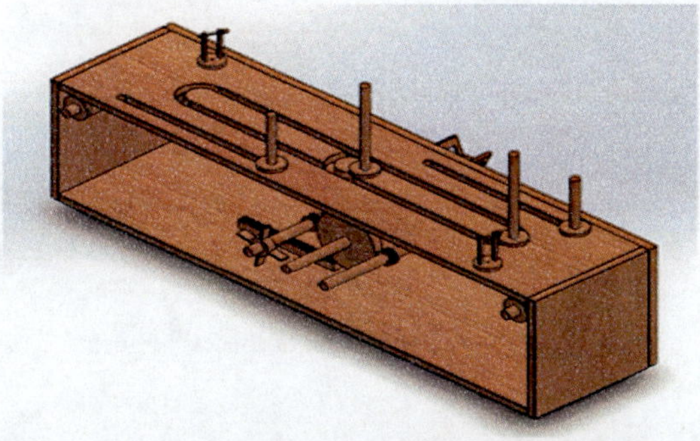

Fig. 7. Complete assembly of the mechanism of the 2nd figure

Once all the parts that make up the assembly are related, the simulation of the movement that the mechanism would have can be carried out, placing rotating, linear and trajectory motors, as well as the effect of gravity, forces on the parts, springs and contacts between parts.

4.2 Figure 21th

Following the criteria set out for Fig. 2, the geometric modelling of this figure 21 has been carried out on the basis of its description.

The dimensions of the base depend on the tower or the location where the mechanism is to be placed. This part must be very robust to be able to withstand the forces that the mechanism makes on impact (Fig. 8).

There are a total of eight bars that make up the mechanism of the scissors, of different sizes and shapes that are made by extrusion and drilling. In order for the whole assembly to move, a kind of rudder is made which is operated by human strength. It also has rollers for winding and redirecting the ropes that serve to transmit the forces that make the mechanism work (Fig. 9).

In this way, the geometric modelling of the different pieces that make up the mechanism has been carried out. The assembly of all the elements that make up the mechanism is then carried out, by means of the position relations between them, such as: concentric for the axles, gears, rollers; coincident to make the faces glued to each other; parallel

Fig. 8. Base and bar of the 21th figure

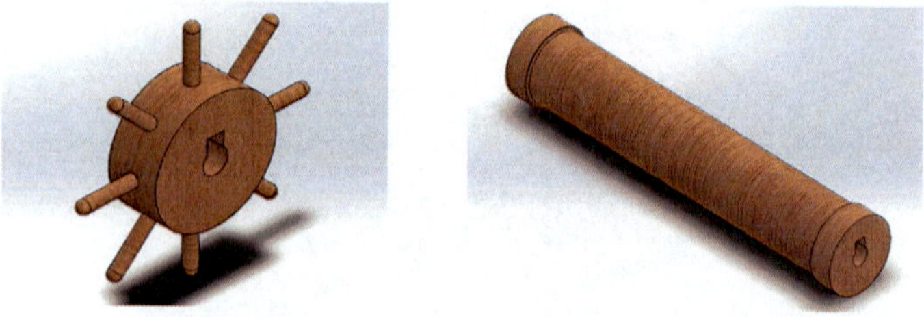

Fig. 9. Rudder and roller of the 21th figure

or distance to make the faces not glued but parallel; belt/pulley to make two separate wheels have movement.

Once all the parts that make up the assembly have been connected, it is possible to simulate the movement that the mechanism would have by placing a rotating motor on the rudder, which would be the actuator, allowing an analysis of the movement to be carried out (Fig. 10).

Fig. 10. Complete assembly of the mechanism of the 21th figure

5 Technological Analysis

The SolidWorks software has modules that allow an analysis of the movement as well as evaluating the mechanical performance, and using the position relationships of the assembly, contacts in the mechanism, forces, springs, action of gravity and motors, you can determine the movement of the assembly. Because of this, the mechanism can be analysed both from the kinematic point of view and from the dynamic point of view, and videos and animations can also be made of what the movement of the mechanism would be like.

To carry out the simulations, the geometric modelling of the 3D parts, the assembly of the set, the application of the materials, the boundary conditions, the meshing of the parts and finally the simulation were carried out.

The simplification that there is no friction between the parts and that the materials are isotropic has been made. In addition, the boundary conditions have been applied, for pins so that they remain straight, for faces to allow the displacement of the elements as a rigid solid, for contacts between parts to allow compatible deformation between the contact surfaces, and finally the support constraints.

5.1 Static *Analysis*

To carry out the simulation of the static loads, a mesh with different levels has been used, finer mesh in the areas with applied forces or discontinuities in the material and progressively coarser mesh as we move away from these points.

The FFEPlus solver has been chosen with large deformations as the dimensions of the parts are in the order of metres. The stopping setpoint is 1×10^{-8} in order to achieve a good solution and simulation result using the finite element method (Fig. 11).

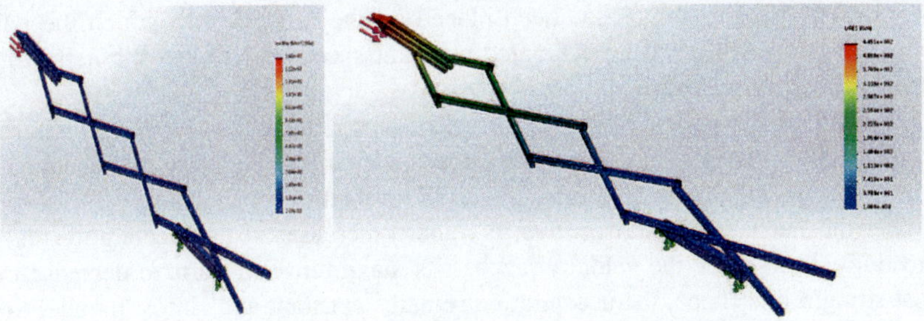

Fig. 11. Result of stresses and displacements

As the elements are joined by pins, there is no resistance to rotation at the joints of the structure and therefore no moment is produced in the sections, only axial, so it is a bar structure. It has a very slender structure, so when it is subjected to axial compressive forces, buckling occurs. The displacements that occur are large.

5.2 Fatigue Analysis

Starting from the same meshing conditions and boundary conditions of the previous section, the model is subjected to cyclic forces. Materials are assigned to obtain the S-N stress-life cycle curve, which determines their behaviour. In this case, steel and mahogany wood have been used (Fig. 12).

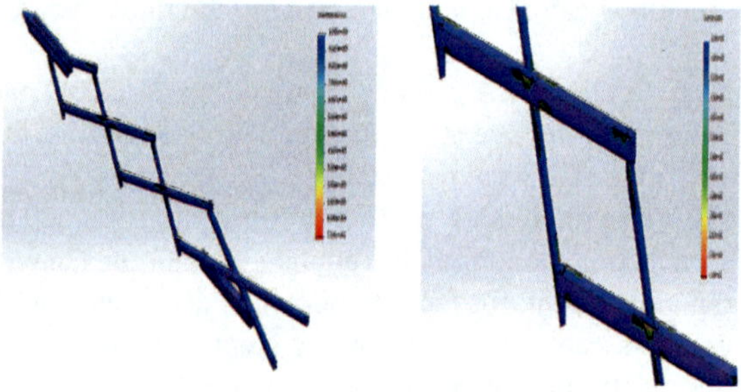

Fig. 12. Result of stresses-live

It can be seen how, as the stresses increase, the life of the materials is reduced. In the areas close to the stress concentrators, the stresses produced are multiplied, which reduces the life of the structure.

5.3 Kinematic Analysis

To carry out the kinematic analysis, another of the modules of the SolidWorks software used is used, in which the rotary and linear motors are placed to give movement to the elements. A rotary motor has been placed on the rudders with which the whole mechanism is controlled, at an estimated constant speed of 10 rpm, although it is of course modifiable, even variable.

The head of the phoenix has been tracked along the x-axis, which is where its movement takes place, in order to obtain the instantaneous velocity and acceleration curves. Subsequently, these results will be used for the impact analysis.

When the mechanism starts to move, its velocity increases, so the acceleration reaches high values. Then when the velocity reaches its maximum and starts to decrease in an almost straight line shape, the acceleration remains constant and almost parallel to the time axis (Fig. 13).

5.4 Impact Analysis

Starting from the same meshing and boundary conditions as above, it is now subjected to impact loading, which can be obtained from the velocity and acceleration, which can be obtained from the motion analysis in the previous section (Fig. 14).

It can be seen that as the acceleration increases the stress results increase, since the impact force depends on the mass and acceleration.

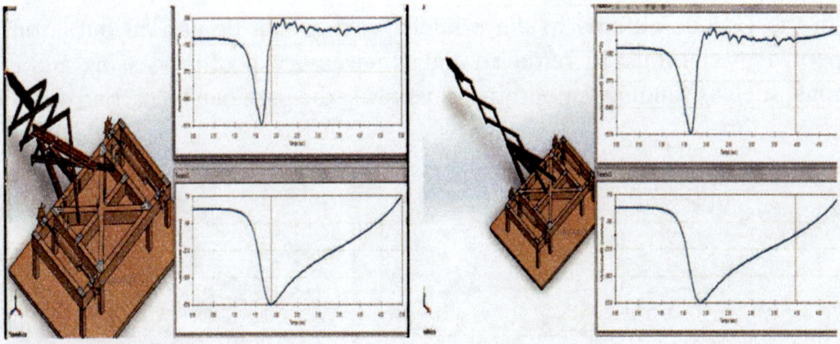

Fig. 13. Relationships between position, velocities and accelerations

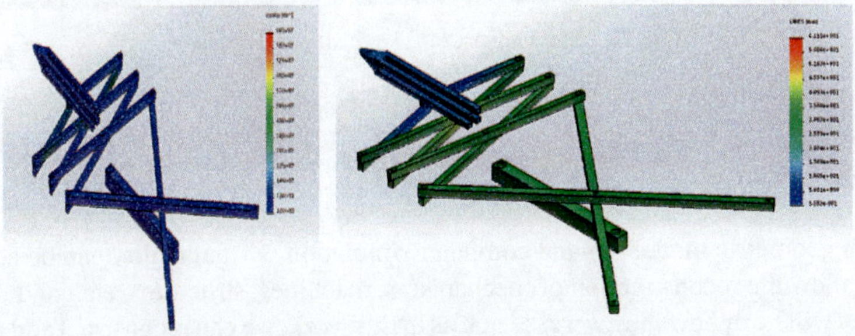

Fig. 14. Stresses and displacements on impact

6 Real Scale Reconstruction

From the geometrical modelling carried out, the real reconstruction of these mechanisms can be made. In this case, it has been done from figure 21.

An FDM (Fused Deposited Modeling) printer has been used, specifically the BQ Witbox. In this printer, a spool of thermoplastic filament (ABS or PLA, as in our case), which is supplied continuously to a nozzle, is heated and extruded in the form of hot, viscous filament. A computer controls the movement of the nozzle along the x and y axes, creating successive layers on the z axis by melting the plastic filament, which solidifies as it cools. In the latest models, a second nozzle carries the support material which can be easily removed later, allowing the construction of more complex parts.

The software used was Cura, which analyses the mesh of the geometric model made, corrects gaps or orientations, to optimise the use of material and printing time, and allows us to obtain the layers of the model and its G code.

For this purpose, the following steps have been carried out: geometric modelling; discretisation by saving the parts in STL stereophotography format; obtaining layers with the help of the Cura program which opens the STL file and obtains the ISO code or G code. Finally, the parameters of the extrusion speed, temperature, type of material, number of layers, path to be used and support material are defined.

With the G-code entered in the printer, printing can be carried out, from which the support material must be removed and, if necessary, post-processing and finishing operations, such as sanding, smoothing or painting the part, can be carried out (Fig. 15).

Fig. 15. Mechanism 21th reconstructed to scale

7 Conclusions

Through geometric modelling and computer simulation, virtual reality can be achieved as they allow the reconstruction of mechanisms, machines, structures, etc., and thus see how they will be or how they were, so that as in this work, we can reconstruct and analyse mechanisms designed by the knowledge of "engineers" ahead of their time.

The study of a theatre mechanism as in Fig. 2 and of a scissors mechanism as in figure 21, from the manuscript "The Book of Secrets", written in the 11th century by Ibn Khalaf Al-Muradi, has been carried out, which has been facilitated by the use of mechanical design and analysis software. The work required geometric modelling and analysis of the different elements of the ensemble. For the subsequent animation, it was necessary to study the movement and transmission relationships of and between each of the elements that make up the mechanism.

From the mechanical analysis it can be deduced how the sizing was carried out fundamentally by means of the trial and error process, which could allow us to redesign and optimise the mechanism.

In this type of study, knowledge of different engineering subjects must be applied, such as design, mechanics and manufacturing, and the use of CAD-CAE-CAM software, as it is not only of historical interest but also of technological interest, as we have been able to see.

This methodology gives rise to its use and application to other mechanisms that allow us to learn about and delve deeper into the History of Mechanical Engineering and Machines and Mechanisms.

References

1. Bautista Paz, E., Ceccarelli, M., Echavarri Otero, J., Muñoz Sanz, J.L.: A Brief Illustrated History of Machines and Mechanisms. Springer, Madrid (2010). https://doi.org/10.1007/978-90-481-2512-8

2. Beakley, G.C., Leach, H.W.: Engineering: An Introduction to a Creative Profession, 4th edn. Macmillan Publishing Company, New York (1970)
3. De Camp, L.S.: The Ancient Engineers. Doubleday and Company Inc., Garden City (1963)
4. Finch, J.K.: The Story of Engineering. Anchor Books, Doubleday and Company Inc., Garden City (2017)
5. Gregory, M.S.: History and Development of Engineering. Longman Group Limited, London (1971)
6. Hathaway, G.A.: Dams-their effect on some ancient civilizations, in Civil Engineering. American Society of Civil Engineers, New York (1958)
7. Al-Murādī, I.K.: The Book of Secrets. Leonardo 3, Milan (2008)
8. Montes, J.C., López-García, R., Dorado-Vicente, R., Trujillo, F.J.: Mock-up of an eighteenth-century oil mill via rapid-prototyping. In: López-Cajún, C., Ceccarelli, M. (eds.) Explorations in the History of Machines and Mechanisms. HMMS, vol. 32, pp. 65–75. Springer, Cham (2016). https://doi.org/10.1007/978-3-319-31184-5_7
9. Upton, N.: An Illustrated History of Civil Engineering. William Heinemann Ltd., London (1975)
10. Sierra, C.E.: Mecánica andalusí: otra faz de nuestra historia. Rev. Univ. de Antioq. 306 (2011)
11. Petrova, V.R.: Introduction to Static Analysis Using SolidWorks Simulation. CRC Press, Boca Raton (2014)

Analysis of the Codex Madrid I as a Compendium of Mechanisms

Higinio Rubio(✉), Alejandro Bustos, Cristina Castejon, and Jesús Meneses

University of Carlos III of Madrid, Leganés, Madrid, Spain
{hrubio,albustos,castejon,meneses}@ing.uc3m.es

Abstract. In this paper, the manuscript of Codex Madrid I, Leonardo da Vinci's workshop drawings collection, are reviewed and the main mechanisms that appear in the aforesaid codex are analysed. The works of Leonardo and the Codex Madrid I, in particular, are placed in their historical context. A compilation of the 100 main drawings of the manuscript is made. This compilation illustrates the variety of mechanical elements and simple mechanisms of Codex Madrid I, forming, as a whole, a complete treatise on mechanisms, understanding mechanisms as basic elements of machines.

Keywords: Leonardo Da Vinci · Codex Madrid I · Mechanisms · Machines · History

1 Introduction

On February 14th 1697, The New York Times published astonishing news: the American researcher Jules Piccus had found, by chance, two manuscripts with drawings, sketches and annotations by Leonardo da Vinci in the Spanish National Library in Madrid: the now named Codex Madrid I and Codex Madrid II [10]. This discovery contributed to enlarging the legend of Leonardo.

The current view of Leonardo da Vinci is that of a wise man, the prototype of a humanist man, characteristic of the Renaissance period. Leonardo is considered a great artist, scientist and inventor, though this last statement is not entirely true [7, 9]. Although Leonardo proposes the design of some original machines, their functionality is questionable and he didn't manufacture and test the aforementioned machines [4, 6, 12].

However, it should be noted all the written and, above all, pictorial material that he bequeathed as a collection of the knowledge of his time, the Renaissance [20].

In this paper, we show the main mechanisms and basic elements of Renaissance machines compiled in the Codex Madrid I, which is Leonardo da Vinci's main contribution to the history of machines and mechanisms [1, 4].

This paper aims to offer an analysis of Codex Madrid I, to relate the technology compiled in this Leonardo da Vinci's work with other authors of his historical environment, and to give a current view of the aforementioned manuscript by selecting 100 drawings

M. Ceccarelli and R. López-García (Eds.): HMM 2022, HMMS 40, pp. 120–133, 2022.
https://doi.org/10.1007/978-3-030-98499-1_11

from the main basic machine elements or mechanisms, composing a comprehensive treatise about the mechanisms of Renaissance.

The written output about the life and works of Leonardo is immense, but this paper will only cite those documents that are direct references for making this work.

2 Codex Madrid I

In 1964, two manuscripts by Leonardo da Vinci of an incalculable historical-scientific-cultural value were found in the Spanish National Library in Madrid. Both manuscripts (Codex Madrid I and Codex Madrid II) embrace almost seven hundred pages of new Leonardo writings about topics architecture, geometry, mechanics or navigation [15, 22].

Codex Madrid I can be defined as a comprehensive treatise of Renaissance mechanics, with a great graphical, descriptive and engineering ability [5]. The pagination care of the sheets and the technical-mechanical drawings presupposes that Leonardo bound this work thinking of a possible treatise.

In Codex Madrid I, Leonardo drew and studied several mechanisms to get different kinds of movement: alternating, swinging and, even, predefined movements. This paper will focus on the drawings of Codex Madrid I and we will only refer to the rest of Leonardo's work occasionally.

Although Leonardo da Vince wrote other treatises that included machine studies (Codex Atlanticus, Paris Manuscripts, ...) Codex Madrid I is unique as it establishes the first case to deconstruct machines into basic machine elements or mechanisms (Leonardo da Vinci called them "elementi macchinali") [18]. Leonardo presupposes the existence of a reader for his work and writes the scientific passage thinking about them. Had Leonardo actually published this work, it might have accelerated the development of machine design.

The Codex Madrid I contains close to 1000 drawings of machines and machine elements. Drawings are of great quality, as well as writing, the typical mannerist handwritten of Leonardo. Drawings are of high quality, clear and complete, and they contain studies about the funding concepts of modern Mechanics, it is to say, mechanisms or mechanical elements. These mechanisms are not Leonardo's machines, which are more complex and have particular usages. However, Leonardo tried to prove that by using and combining these simple elements it can be obtained machines that perform a specific function. Writing is of great quality too, the typical mannerist handwritten of Leonardo [22].

The work can be dated between 1492 and 1497, in the first period of Leonardo in Milan (1482–1499). This is when Leonardo was in the prime of his life, he was living his fourth decade [15].

Codex Madrid I is a one-volume manuscript that consisted of 192 sheets originally (only 184 sheets are preserved today), double side and dimensions 215 × 145 mm. The main body is made up of 12 notebooks, all containing 8 double sheets. The third and fourth notebooks have lost some double sheets [15].

From the point of view of its content, it is a compilation of texts mainly from other Leonardo's treatises. It also offers earlier texts about static and geometry [18].

3 Historical Scope

This paper aims to analyse the Codex Madrid I [11], placing the mechanical knowledge that contains in the historical context (the Renaissance) [14] and justifying the claim that it is the first known treatise where the main mechanisms and simple machines are compiled and studied.

The culture of the Renaissance is marked by the exaltation of the dignity of man, thought inherited from classical thinking: in Sophocles's Antigone [21], it was said: 'Wonders are many, and none is more wonderful than man'. The culture of the Renaissance took into account other topics: the exaltation of nature, the critical spirit, using reason as the fundamental instrument of any mental process, adding mathematics as a complement to literary thinking, leaving aside the principle of authority, appreciating the importance of experimental practice, etc.

Renaissance ideas were shared by the Italian intellectual elite of the 15th and 16th centuries. The Byzantine intellectuals that emigrated to Italy after the fall of Constantinople (1453) contributed to this school of thought [15].

The life and work of Leonardo da Vinci allow placing him within his historical dimension: the Renaissance spirit, both from a theoretical and practical point of view [4, 5].

Leonardo lived in one of the most cultured environments in the European Renaissance. He didn't develop original comprehensive theories in the scientific field, but he was a persistent observer and a singular exponent of a time that changed the world. His contribution to the effective progress of the science of his time was not great, perhaps due to the scant diffusion of his writings. But he left testimony of his defence of experimentation as a method of intellectual work, his claim to technique and of machines as instruments of progress and, above all, their general conception of reality governed by the laws of necessity, pragmatism, reason and proportion. Leonardo tried to demonstrate the superiority of visual creation over literary text in his work [13, 15].

One of the most important innovations of the Renaissance was to resort to drawing as an instrument of demonstration and communication. Following this idea, the graphic conventions necessary to visualize machines and suggest their features were defined. This art of describing machines became an instrument of technical-scientific study. Leonardo da Vinci was not very original but maybe he is the greatest exponent of this art.

Leonardo intended to know nature. He tried to penetrate its secrets and formulate its laws, in an attempt to solve the infinite problems posed by nature.

As we have commented, the 'Renaissance' is characterized by the emergence of the so-called "engineers" that surpasses the category of "master of war machines", as the craftsmen-builders of the Middle Ages used to be called [5, 18].

Leonardo can be classified as an 'engineer' and he knew the activities of the Engineers and Architects of ancient Rome as well as that of his predecessors in the Italian Renaissance. Analysing Leonardo's manuscripts, it can be stated that Leonardo knew the work of Vitruvius [8], Filippo Brunelleschi, Taccola [2], Leon Battista Alberti or Francesco di Giorgio Martini [3, 5, 16].

If we talk about the first works of the 'engineers', we can start from the military treatise by Guido di Vigevano as the predecessor of the notebooks of the Italian engineers of the Renaissance, as well as the military-technical writings of the German school [14, 18].

Among the Italian engineers of the Renaissance, we can speak of up to two generations of Italian engineers to reach the time personified by da Vinci. Among the authors of the first generation, we find Filippo Brunelleschi, Taccola, Leon Battista Alberti or Giacomo Fontana [5, 18]. Preceding the second generation, we can cite Roberto Valturio or Francesco di Giorgio Martini, marking the definitive transition to the second generation, of which Da Vinci was a notable representative. These two authors preceded and coincided in time with Leonardo, even Leonardo collaborated in some of his Works [14, 18].

The case of Francesco di Giorgio Martini is significant because the importance of his career is dwarfed by that of Leonardo today, who was inspired by his work and copied his drawings [3, 16, 18, 19]. However, Francesco di Giorgio was a well-known engineer in Italy in the late 15th century, whose works were known and highly regarded. Leonardo did not have as much recognition as Francesco di Giorgio in his lifetime, moreover, let us remember that Leonardo's manuscripts were not published until a long time later. In Leonardo's defence, it should be noted that Francesco di Giorgio had been inspired by Toccola's work [2, 18].

The next tables (Table 1 and Table 2) show the main authors and works about machines that came before and after Leonardo and his work. Table 1 shows the indications of F.C. Moon in [18] and Table 2 evolves the summary proposed by W. Lefèvre in [14].

Table 1. A summary of 'Theatre of Machines' books up to the 17th century.

Time	Author	Short Title
c. 27 BC	Vitruvius Pollio	*De Architectura*
1204–1206	Ibn al-Razzaz al-Jazari	*The Book of Knowledge of Ingenious Mechanical Devices*
c. 1225–1250	Villard de Honnecourt	*The Sketchbook of Villard de Honnecourt*
1335	Guido di Vigevano	*Texaurus*
1366–c. 1405	Konrad Kyeser	*Bellifortis*
1393–1455	Giacomo Fontana	*Bellicorum instrumentorum liber*
c. 1445–1516	Giuliano da Sangallo	*Opusculum*
c. 1450	Marianus Jacobus/Taccola	*De Ingeneis*
1451–1516	Bonaccorso Ghiberti	*Zibaldone*
1470–1480	Francesco di Giorgio	*Trattato di Architettura*
1472	Roberto Valturio	*De re militari*

(continued)

Table 1. (*continued*)

Time	Author	Short Title
1480–1515	Leonardo da Vinci	*Notebooks [unpublished]*
1540	Vannuccio Biringuccio	*De la Pirotechnia*
1556	Georgius Agricola	*De Re Metallica*
1578	Jacques Besson	*Theatre des instruments mathematiques et mechaniques*
1584	Jean Errard	*Le Premier Livre des Instruments Mathematiques Mechaniques*
1588	Agostino Ramelli	*Livre des diverses et artificieuses machines*
1607	Vittorio Zonca	*Novo teatro di machine et edifici*
1607–1614	Heinrich Zeising	*Theatri Machinarium*
1615	Salomon de Caus	*Les Raisons des Forces Mouvantes avec diverses Machines*
1617–1618	Jacob de Strada	*Künstlicher Abriss allerand Wasser, Wind, Ross und Handt Mühlen*
1627	Johann Schreck (J. Terentius) and Wang Cheng	*Qi qi tu shou (Diagrams and Explanations of Strange Machines)*
1629	Giovanni Branca	*Le machine*
1661	Georg Böckler	*Theatrum machinarum novum*
1672	Otto von Guericke	*Experimenta nova, ut vocantur Magdeburgica*
1695	Phillipe De Le Hire	*Traite de mecanique*

Table 2. Prominent sources of early machine drawings (1250–1650).

Time	Author	Document type
c. 1225–1250	Villard de Honnecourt	A, B
1335	Guido di Vigevano	C2
1366–c. 1405	Konrad Kyeser	C2
c. 1410–1430	German Master Gun-makers	C1
c. 1410–1450	Marianus Jacobus/Taccola	B, C2
1472	Roberto Valturio	D1
c. 1475–1490	Anonymous of the Hussite Wars	C1 (Hussite Wars 1420–1436)
1470–1480	Francesco di Giorgio Martini	D1

(*continued*)

Table 2. (*continued*)

Time	Author	Document type
1480–1515	Leonardo da Vinci	B
1486	Vitruvius	D3 (Original work c. 27 BCE)
c. 1510–1546	Antonio da Sangallo	B
1537	Niccolo Fontana/Tartaglia	D2
1556	Georgius Agricola	D1
1558	Berthold Holzschuher	A, B
1567	Giuseppe Ceredi	D1
1569	Jacques Besson	C3
1577	Guidobaldo del Monte	D2
1584	Jean Errard	C3
1588	Agostino Ramelli	C3
1588	Pappus of Alexandria	D3 (Original work c. 340)
1589	Heron of Alexandria	D3 (Original work c. 35-c. 75)
c. 1590–1635	Heinrich Schickhardt	A, B
1597	Buonaiuto Lorini	D1
1607	Vittorio Zonca	C3
1607–1614	Heinrich Zeising	C3
1615	Salomon de Caus	C3
1617	Jacob de Strada	C3
1629	Giovanni Branca	C3

A: Workshop drawings of design and construction drawings
B: Workshop drawings of sketch-books and notebooks
C: Presentational manuscripts or books of collections of drawings with or without explanatory text
C1: Practitioner booklets
C2: Representational manuscripts
C3: Theatres of machines
D: Presentational manuscripts or books of drawings in treatises and editions
D1: In technological treatises
D2: In treatises on mechanics
D3: In editions of classical sources

If we talk about the importance of Leonardo da Vinci as an engineer, it can be said that he was endowed with immense curiosity but was scattered: he started many things and later abandoned them. It should be noted his 'method' and his detailed analysis of the machines, studying the simplest elements, of which the Codex Madrid I was his main work. Leonardo bequeathed an immense number of designs and studies of machines and mechanisms to History, but his contribution of technical innovations was minimal [12]. Furthermore, he did not leave pupils and his work did not begin to be published until the 19th century.

4 The Mechanisms of Codex Madrid I

Shortly time after the Codex Madrid was rediscovered in 1965, Ladislao Reti translated the text into English and showed that da Vinci had attempted to compile a basic compendium of machine elements [11].

Reti [20] compared the drawings of machine elements and mechanisms in Leonardo's Codex Madrid I with the basic list of 22 constructive elements proposed by Franz Reuleaux in his 19th-century book on machine design "The Constructor". More recently, Moon confirmed this list [18] and made an advanced study of the machine elements and mechanisms that Leonardo described in the Codex Madrid I, including a comparison with the works of Timoshenko [17].

On the following pages, a sequence of images with Leonardo's simple machines is included, which serve as the basis for correctly interpreting machines as a set of mechanisms and machine elements. In these images, we can appreciate the great contribution of Leonardo's studies to machines engineering as a science.

In these 100 drawings (Fig. 1, 2, 3, 4, 5, 6, 7, 8 and 9), we can see 100 mechanisms and machine elements that Leonardo illustrated in the Codex Madrid I, several of which were aimed to be integrated into Leonardo's watch design Project (Fig. 10). In the Tables (Tables 3a, 3b, 3c, 3d and 3e) the drawings are very briefly described.

5 Conclusions

Leonardo was a master of painting, delved into physics and anatomy, designed machines and buildings and planned cities, and left a huge written work. However, he protected or concealed his findings led to the opinion that much of Leonardo's scientific-technical work was speculative or not original. In this paper, some of the main contributions of Leonardo da Vinci in the field of mechanism and mechanical elements have been shown and analysed.

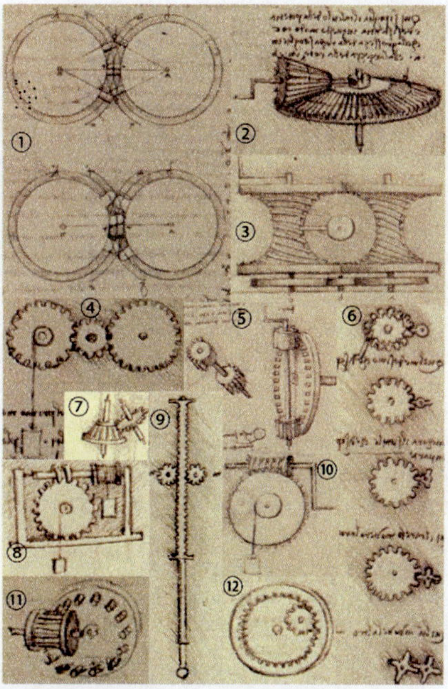

Fig. 1. Mechanisms Drawings Box 1 (MDB-1).

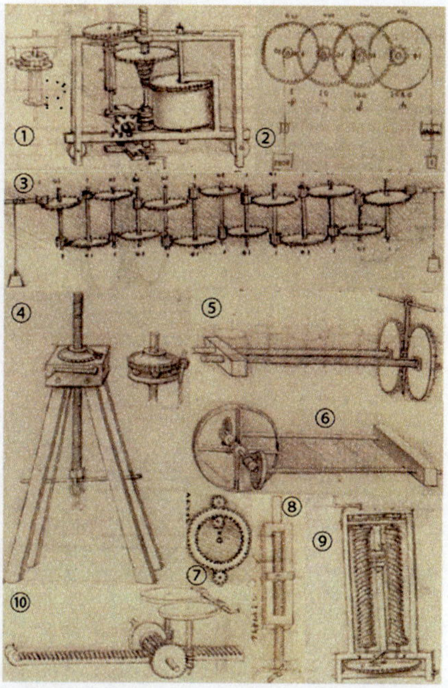

Fig. 2. Mechanisms Drawings Box 2 (MDB-2).

Table 3a. Compilation of mechanisms and machine elements from Leonardo's Codex Madrid I.

Drawing	Mechanism description	Drawing	Mechanism description
MDB-1.1	*Pins and teeth gears*	MDB-2.1	*Gear regulator device*
MDB-1.2	*Bevel gear*	MDB-2.2	*Gear reduction system*
MDB-1.3	*Double globoid worm system*	MDB-2.3	*Lantern gear reduction system*
MDB-1.4	*Spur gears*	MDB-2.4	*Screw jack with roller bearing*
MDB-1.5	*Pin-teeth rotary reversing mechanism*	MDB-2.5	*Double screw spindle*
MDB-1.6		MDB-2.6	*Double screw spindle drive*
MDB-1.7	*Crown bevel gear and lantern pinion*	MDB-2.7	*Planetary-epicyclic gear train*
MDB-1.8	*Endless screw or worm drive*	MDB-2.8	*Screw spindle*
MDB-1.9	*Rack and Pinion*	MDB-2.9	*Double helix reversing mechanism*
MDB-1.10	*Globoid worm drive*	MDB-2.10	*Rack and pinion drive*
MDB-1.11	*Crown wheel gear and lantern pinion*		
MDB-1.12	*Internal gear*		

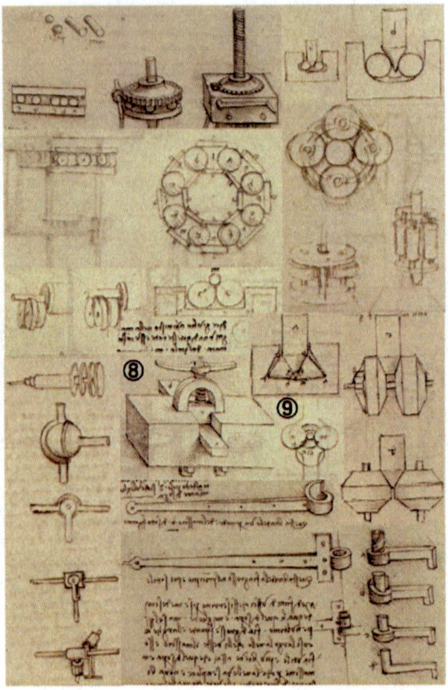

Fig. 3. Mechanisms Drawings Box 3 (MDB-3).

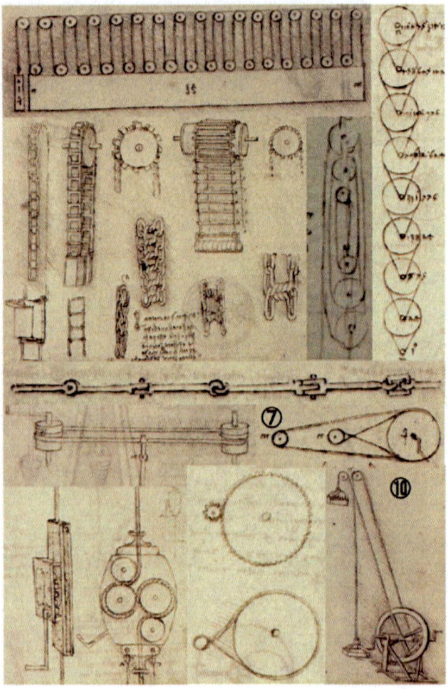

Fig. 4. Mechanisms Drawings Box 4 (MDB-4).

Table 3b. Compilation of mechanisms and machine elements from Leonardo's Codex Madrid I.

Drawing	Mechanism description	Drawing	Mechanism description
MDB-3.1	*Axial anti-friction ball bearing*	MDB-4.1	*Multiple pulley mechanism*
MDB-3.2	*Ball bearings for a conical pivot*	MDB-4.2	*Belts and pulleys reductor system*
MDB-3.3	*Radial anti-friction ball bearing*	MDB-4.3	*Models of chains*
MDB-3.4	*Rollers bearings for shaft*	MDB-4.4	*Block and Tackle system*
MDB-3.5	*Disks bearings*	MDB-4.5	*Models of chains*
MDB-3.6	*Roller bearings for a conical pivot*	MDB-4.6	*Belt and pulley mechanism*
MDB-3.7	*Universal joints*	MDB-4.7	*Open and cross belt drives*
MDB-3.8	*Spring-making machine*	MDB-4.8	*Manual lift mechanism*
MDB-3.9	*Rollers bearing for shaft*	MDB-4.9	*Analogy between gear and belt transmission*
MDB-3.10	*Standard and self-closing hinges*	MDB-4.10	*Crane with drum hoist winch*

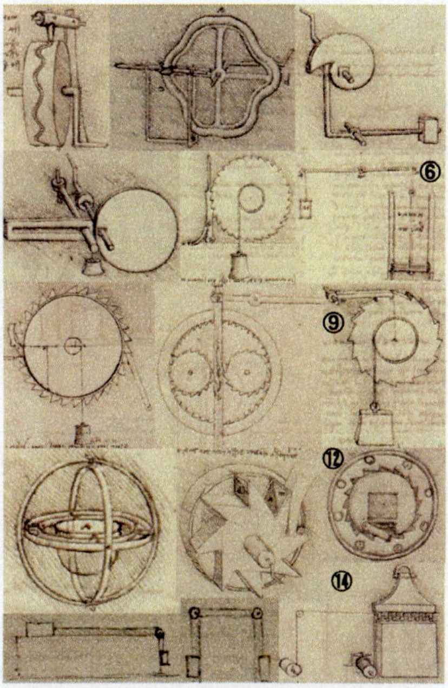

Fig. 5. Mechanisms Drawings Box 5 (MDB-5).

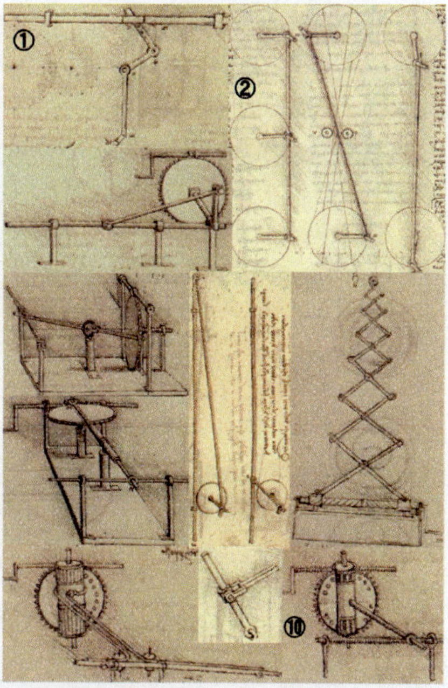

Fig. 6. Mechanisms Drawings Box 6 (MDB-6).

Table 3c. Compilation of mechanisms and machine elements from Leonardo's Codex Madrid I.

Drawing	Mechanism description	Drawing	Mechanism description
MDB-5.1	*Cylindrical cam mechanism*	MDB-6.1	*Linear motion generation mechanism*
MDB-5.2	*Cam mechanism*	MDB-6.2	*Four-bar linkage*
MDB-5.3	*Cam actuated lever*	MDB-6.3	*Linear motion generation mechanism*
MDB-5.4	*Eccentrical cam mechanism*	MDB-6.4	*Linear motion generation mechanism*
MDB-5.5	*Ratchet and pawl mechanism*	MDB-6.5	*Linear motion generation mechanism*
MDB-5.6	*Piston & Cylinder*	MDB-6.6	*Slider crank mechanism*
MDB-5.7	*Ratchet and pawl mechanism*	*MDB-6.7*	*Lazy tongs or Nürnberg shears*
MDB-5.8	*Ratchet and pawl mechanism*	*MDB-6.8*	*Linear motion generation mechanism*
MDB-5.9	*Ratchet and pawl mechanism*	*MDB-6.9*	*Three degrees of freedom joint*
MDB-5.10	*Gimbals mechanism*	*MDB-6.10*	*Linear motion generation mechanism*
MDB-5.11	*Water wheel*		
MDB-5.12	*Ratchet and pawl mechanism*		
MDB-5.13	*friction measurement system weight-driven*		
MDB-5.14	*Pulley system for weight-driven clock gear*		

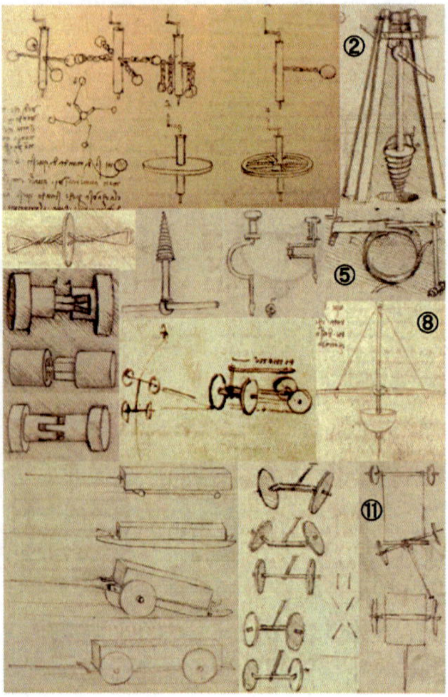

Fig. 7. Mechanisms Drawings Box 7 (MDB-7).

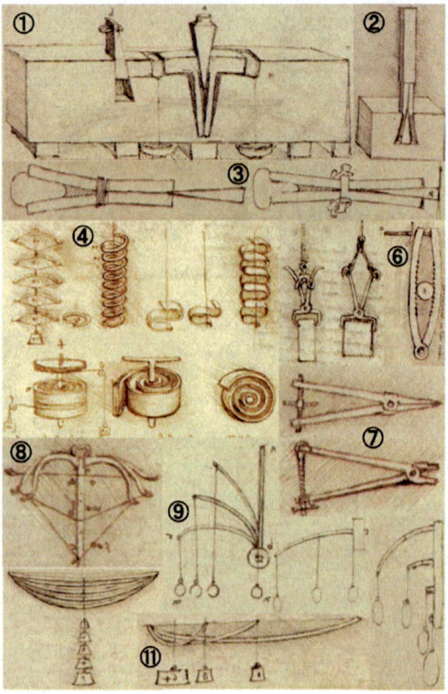

Fig. 8. Mechanisms Drawings Box 8 (MDB-8).

Table 3d. Compilation of mechanisms and machine elements from Leonardo's Codex Madrid I.

Drawing	Mechanism description	Drawing	Mechanism description
MDB-7.1	*Flywheel designs*	MDB-8.1	*Wedge mechanism*
MDB-7.2	*Drilling tool*	MDB-8.2	*Wedge mechanism*
MDB-7.3	*Rope flywheel*	MDB-8.3	*Wedge mechanism*
MDB-7.4	*Bit brace tools*	MDB-8.4	*Springs*
MDB-7.5	*Brake mechanism*	MDB-8.5	*Grippers*
MDB-7.6	*Couplings*	MDB-8.6	*Clamping system*
MDB-7.7	*Study of the curve tracing of a cart*	MDB-8.7	*Pliers*
MDB-7.8	*Manual drilling tool with rope*	MDB-8.8	*Crossbow deformation study*
MDB-7.9	*Study of weight drag*	MDB-8.9	*Deformation studies of a cantilever beam*
MDB-7.10	*Study of wheels tilt*	MDB-8.10	*Deformation study of a bi-supported beam*
MDB-7.11	*One and two-axle carriage design*	MDB-8.11	*Deformation study of bi-supported beams of different lengths*

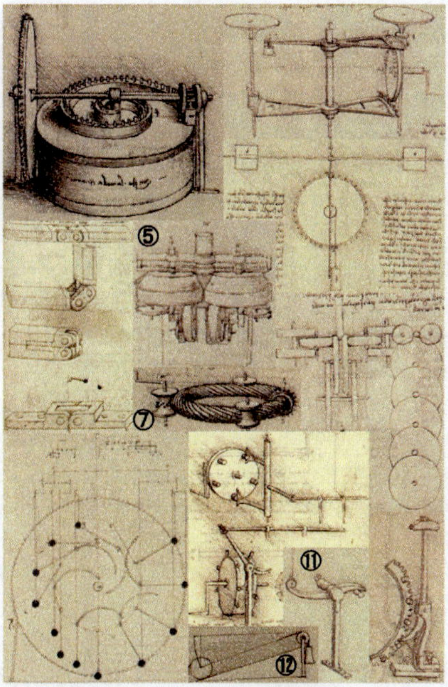

Fig. 9. Mechanisms Drawings Box 9 (MDB-9).

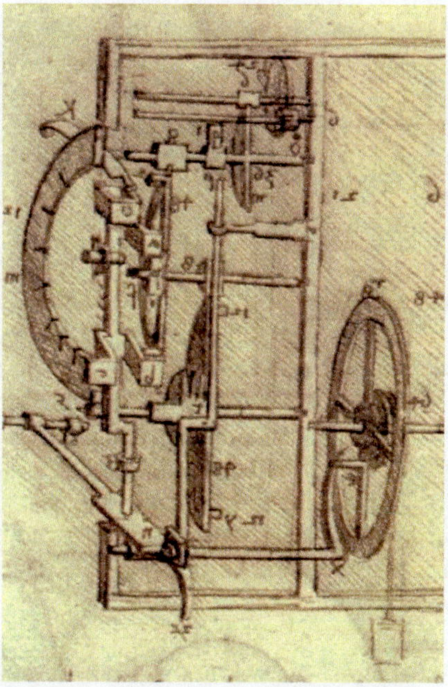

Fig. 10. Leonardo mechanical clock project.

Table 3e. Compilation of mechanisms and machine elements from Leonardo's Codex Madrid I.

Drawing	Mechanism description
MDB-9.1	*Central cylindrical drum for a clock*
MDB-9.2	*Reciprocating transformation mechanism*
MDB-9.3	*Verge clock escapement*
MDB-9.4	*positional study for a joint*
MDB-9.5	*Friction wheels*
MDB-9.6	*Friction wheels*
MDB-9.7	*Circular endless screw*
MDB-9.8	*Study of a mechanism for perpetual motion*
MDB-9.9	*Two intermittent mechanisms*
MDB-9.10	*Alternative mechanism actuated lever for a clock*
MDB-9.11	*Mechanism drive for a clock*
MDB-9.12	*Inclined plane mechanism*

After reviewing Leonardo's Codex Madrid I, this paper focuses on studying the principles of generating solutions most used by Leonardo have been studied: inspiration by nature, the organic vision of mechanisms, analysis to decompose machines and synthesis for the integration of mechanical elements and mechanisms; the analogies between different machine elements, and the economy in the design by suppressing the superfluous. All these advances allowed his mechanical designs to be better than those of his contemporaries.

The works of Leonardo and the Codex Madrid I, in particular, are placed in their historical scope and we have noted that representative genial personalities like Leonardo da Vinci and several others, could reach the heights in the Theory of Machines and Mechanisms because of contributions by many others.

In these pages, reviewing the manuscript of Codex Madrid I, the most important drawings of machine elements and mechanisms are shown and described: A compilation of the 100 main mechanism drawings of the manuscript is made.

This images compilation illustrates the variety of mechanical elements and simple mechanisms that exist in Codex Madrid I, forming, as a whole, a complete treatise on mechanisms (under-standing mechanisms as basic elements of machines), representation of the knowledge of this matter in the late fifteenth century.

Acknowledgements. Authors would like to thank Professors Emilio Bautista, Rafael López-García and Marco Ceccarelli for awakening the authors' interest in the History of Machines and Mechanisms.

References

1. Bautista, E., Ceccarelli, M., Echavarri, J., Muñoz, J.L. (ed.): A Brief Illustrated History of Machines and Mechanisms. History of Machines and Machine Science, vol. 10. Springer, Dordrecht (2010). https://doi.org/10.1007/978-90-481-2512-8
2. Ceccarelli M.: Contributions of Mariano di Jacopo (il Taccola) in mechanism design. In: XXIII Congreso Nacional de Ingeniería Mecánica. Proceedings of XXIII CNIM, Jaén, Spain (2021)
3. Ceccarelli, M.: Contributions of Francesco di Giorgio in mechanism design. In: XXII Congreso Nacional de Ingeniería Mecánica. Proceedings of XXII CNIM, Madrid, Spain (2018)
4. Ceccarelli M.: Contributions of Leonardo da Vinci in mechanisms design. In: XXI Congreso Nacional de Ingeniería Mecánica. Proceedings of XXI CNIM, Elche, Spain (2016)
5. Ceccarelli, M.: Renaissance of machines in Italy: from Brunelleschi to Galilei through Francesco di Giorgio and Leonardo. Mech. Mach. Theory **43**, 1530–1542 (2008)
6. Cerveró-Meliá, E., Ferrer-Gisbert, P.E., Capuz-Rizo, S.F.: Functional feasibility review of the technical systems designed by Leonardo da Vinci. In: Proceedings of 22nd International Congress on Project Management and Engineering, Madrid (2018)
7. Cerveró-Meliá, E., Capuz-Rizo, S.F., Ferrer-Gisbert, P.: Leonardo da Vinci's contributions from a design perspective. Designs **4**, 38 (2020)
8. Ceccarelli, M. (ed.): Distinguished Figures in Mechanism and Machine Science. Their Contributions and Legacies, Part 3. HMMS, vol. 26. Springer, Dordrecht (2014). https://doi.org/10.1007/978-94-017-8947-9

9. Contreras M.A.: Leonardo da Vinci: Ingeniero. Ph.D. thesis. University of Málaga, Spain (2015, in Spanish)
10. Da Vinci, L.: (c. 1504). Códice de Madrid. Madrid. Biblioteca Nacional de España. http://leonardo.bne.es/index.html
11. Da Vinci, L.: (c. 1500) Codex Madrid I or Tratado de Estatica y Mechanica en Italiano, 1493, Madrid. Facsimile Edition: The Madrid Codices, National Library Madrid, No. 8937. Translated by L. Reti. McGraw-Hill Book Co., 5 vols. (1974)
12. Gille, B.: Les ingénieurs de la Renaissance. Edited by Seuil (1978, in French)
13. Innocenzi, P.: The Innovators Behind Leonardo. Springer, Cham (2019). https://doi.org/10.1007/978-3-319-90449-8
14. Lefèvre, W. (ed.): Picturing machines 1400–1700. Massachusetts Institute of Technology (2004)
15. Leonardo's imaginary. Codices Madrid of the BNE. Edited by the Ministry of Culture, National Library (2012, in Spanish)
16. Merrill, E.M.: Francesco di Giorgio and the Formation of the Renaissance Architect. Ph.D. Thesis. University of Virginia (2015)
17. Moon, F.C.: History of dynamics of machines and mechanisms from Leonardo to Timoshenko. In: Yan, H.S., Ceccarelli, M. (eds.) International Symposium on History of Machines and Mechanisms, pp. 1–20. Springer, Dordrecht (2009). https://doi.org/10.1007/978-1-4020-9485-9_1
18. Moon, F.C.: The Machines of Leonardo da Vinci and Franz Reuleaux. History of Mechanism and Machine Science, vol. 2. Springer, Dordrecht (2007). https://doi.org/10.1007/978-1-4020-5599-7
19. Moon, F.C.: The kinematics of Francesco di Giorgio Martini and Franz Reuleaux. In: Proceedings of the Third International Workshop on History of Machines and Mechanisms, Moscow (2005)
20. Reti, L.: The engineer. In: Reti, L. (ed.): The Unknown Leonardo. McGraw Hill, New York (1974)
21. Sophocles (495-405 a.C.), Antigone. Ed. y trad. Hugh Lloyd-Jones, pp. 334–335. Harvard University Press, Cambridge, London (1994)
22. Two rediscovered manuscripts of Leonardo da Vinci. The Unesco Courier. Published by UNESCO, October 1974

Ctesibius, Vitruvius and Leonardo: A Digital Reconstruction of the Water Clockwork Timeline

Massimo Callegari[1], Paolo Clini[2], Luca Carbonari[1], Renato Angeloni[2(✉)], Mirco D'Alessio[2], and Cecilia Scoccia[1]

[1] Department of Industrial Engineering and Mathematical Sciences (DIISM), Polytechnic University of Marche, Ancona, Italy
m.callegari@univpm.it
[2] Department of Construction, Civil Engineering and Architecture (DICEA), Polytechnic University of Marche, Ancona, Italy
p.clini@univpm.it, r.angeloni@staff.univpm.it

Abstract. Over the centuries several physical models of Leonardo's machines have been reconstructed to solve a fascinating question: visionary and prophetic intuitions or real projects corroborated by analytical studies on their actual possibilities of use? Taking advantage of an interdisciplinary approach and exploiting the potential of digital design, this paper presents the reconstruction of one of the most complex machines represented by Leonardo and first by Ctesibius and Vitruvius: the water clockwork. Starting from the analytical study of Leonardo's drawings, the workflow proposal leads to a digital reconstruction of this machine, made available through an Augmented Reality mobile app developed for the exhibition *Leonardo e Vitruvio. Oltre il Cerchio e il Quadrato.* A digital solution to share this remarkable Cultural Heritage, making the complex language of Leonardo's drawings accessible to a wider audience.

Keywords: Leonardo · Water clock · Digital reconstruction · Augmented Reality

1 Introduction

Over the centuries several physical models of Leonardo's machines have been reconstructed starting from his drawings. The limits of these reproductions can be exceeded by new digital versions able to simulate their functioning and to easily compare different possible settings, helping understand Leonardo's intention: designing visionary and prophetic intuitions or real projects corroborated by analytical studies on their actual possibilities of use?

Transforming this extraordinary heritage into a 3D digital database also turns "hard to read" images into something that can be enjoyed by a wider audience. Virtual Reality (VR) can generate thus new forms of communication, promoting Cultural Heritage (CH) fruition in spectacular ways, both evocative and didactic. Therefore, it means more than simply elaborate scientifically validated digital twins, but generating a new educational

M. Ceccarelli and R. López-García (Eds.): HMM 2022, HMMS 40, pp. 134–145, 2022.
https://doi.org/10.1007/978-3-030-98499-1_12

repertory, bringing Leonardo's machines from physical to digital reality. A solution that opens up new opportunities in designing museum and knowledge paths, making Leonardo's mechanics easily accessible to a wider audience thanks to digital solutions.

From this perspective, the 500th anniversary of Leonardo's death represented a starting point, with great exhibitions and scientific projects based on digital technologies. This paper fits into this scenario presenting part of the scientific research carried out for *Leonardo e Vitruvio. Oltre il cerchio e il quadrato* [1], an exhibition that took place in Fano (Italy) aiming to explore the connection between Leonardo and Vitruvius, the closest ancient author to the Tuscan genius as expression of a multiform and encyclopaedic wisdom. A bond that goes beyond the well-known subject of the *Vitruvian man* and that was explored focusing on the field of mechanics and measurement. Disciplines to which Vitruvius dedicates the *Book IX* and the *Book X* of the *De Architectura*, two essential books to the history of Machines and Mechanics. The *Book IX* describes the water clock of Ctesibius whereas the *Book X* the construction of a ballista and an odometer. Three machines to which Leonardo devote the greatest attention and which mention a deep connection with Vitruvius, highlighted in three sheets of the *Codex Atlanticus* exhibited in Fano [2, 3]. This paper is specifically focused on the water clock, Fig. 1, describing its reconstruction hypothesis and AR fruition, an effective replicable solution also to disseminate other machines history and functioning.

2 Water Clocks in Antiquity

The hourglass was called by the ancient Greeks κλεψύδρα (klepsýdra) which means "water thief" since the passage of time was measured by the constant flow of water in or out of a vessel.

The oldest hourglass of which there is physical evidence dates back to c. 1417–1379 B.C., during the reign of Amenhotep III where it was used in the Temple of Amen-Re at Karnak. The oldest documentation of the hourglass is the tomb inscription of the 16th century B.C. Egyptian court official Amenemhet, which identifies him as its inventor.

In Greece the use of the hourglass was introduced in 325 B.C.: a vessel had a hole in its side near the base through which the water was drained slowly and evenly out of the container; the level of the water gave an idea of the time elapsed. However, this "clock" was not accurate because the water pressure decreased as the liquid level dropped, making the flow progressively slower.

In the first half of the third century B.C. Ctesibius of Alexandria transformed this apparatus into a water clock, that is into a real measuring instrument, thanks to various devices that were described many centuries later by Vitruvius.

The main novel idea was to keep constant the pressure at the exit hole, allowing the water to flow from a vase in which the water level was kept constant as well. Thus, the flow of water, which depends on the pressure, became constant too. Ctesibius was also the first to describe the use of a float to indicate the level of water in the basin and to operate the ring: both these elements, as well as the constant level of the head of water, are present in Leonardo's machine.

Ctesibius also turned the water clock into an automaton, able to take into account the variable duration of the hours at the different periods of the year; in fact the ancient

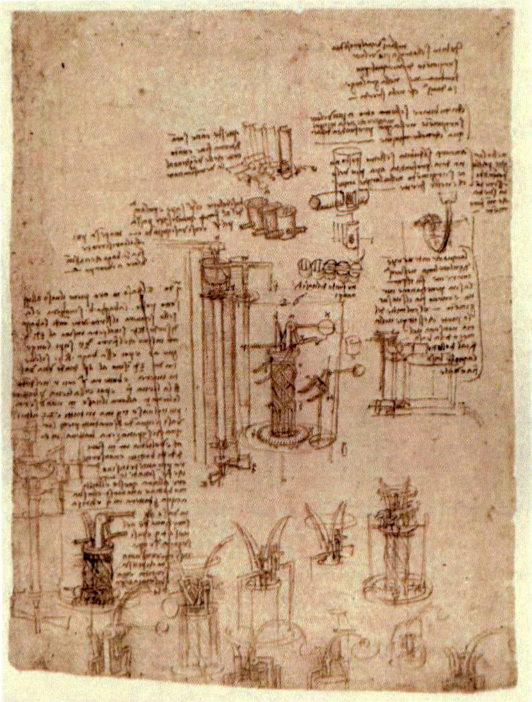

Fig. 1. Codex Atlanticus, f. 943r, around 1508–1510, pen-and-ink, 259 × 196 mm, Milan, Veneranda Biblioteca Ambrosiana

Greeks and Romans had twelve hours from sunrise to sunset and since summer days are longer than winter days, summer hours were longer than winter hours.

Going into the details of the clock, Fig. 2, a constant flow of water pours into a vessel, inside which there is a float that holds an indicator: as the float rises, the indicator slides along a graduated cylinder, pointing to the current hour for the day. When the vessel is filled up, a siphon flushes down the water which in turn rotates a wheel divided into six spokes: a series of gears connects such wheel to the graduated cylinder, so that it completes a full rotation in 365 days, showing day by day the correct time.

Based on Ctesibius' concept, various mechanisms have been conceived in the following centuries. A notable example is the Tower of Winds, Fig. 3, which is located in the Agora of Athens: it was constructed by Andronicus Kyrrhestes of Macedonia during the Roman period in the second quarter of the first century B.C.

Built of marble and octagonal in shape, it is 43 feet (13 m) tall. In its interior, time was determined by a water clock, driven by water coming down from the Acropolis.

In the first century B.C. Vitruvius in book IX of De Architectura describes the first precision hourglasses designed by great Greek mathematicians of the past, such as Ctesibius. In the book, Ctesibius' water clock or clepsydra, is described as the first to have a regulator that would maintain a constant head of water in the effluent part of the clock in order to improve the accuracy.

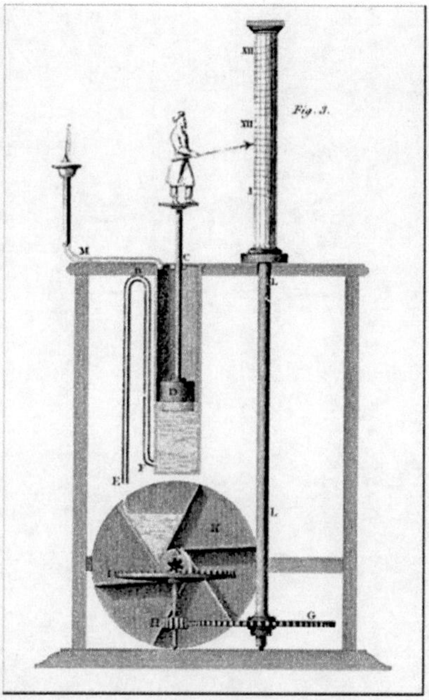

Fig. 2. An early 19th-century illustration of Ctesibius's clepsydra by John Farey, Jr. (1791–1851)

During Renaissance many artists showed a strong interest in ancient scholars and turned to the work of Vitruvius about architecture and painting techniques. Leonardo, however, was also interested in ancient technological world therefore drew on Vitruvius' work on issues related to hydraulics, machines, measurement instruments, materials, and so on [4–6].

Leonardo developed his studies on hourglasses during his service with Charles of Amboise, French governor of the state of Milan: at that time the treatise of Vitruvius was already printed but hardly accessible, moreover it was written in Latin and without illustrations.

The only attempt to represent Ctesibius's water clock in figures could be found in Roberto Valturio's De re militari. Some key features of the hourglasses described by Vitruvius have been reinterpreted by Leonardo in an original way, such as how to obtain a uniform flow of water.

After the fall of the Roman Empire, the tradition of water clocks was preserved by the Arabs with an entire book of these devices written by Al-Jazari in the thirteenth century with the "Book of Knowledge of Ingenious Mechanical Devices".

3 Leonardo's Bell Ringer

The water clock has been the subject of many studies by Leonardo, which led to different conceptual solutions, therefore it appears in various sheets of the Codex Atlanticus and in the Windsor collection.

Fig. 3. The Tower of the Winds

Folio 943 recto of CA, which was on display in the Fano Exhibition, contains many sketches used by Leonardo to assess different viable alternatives, so that it is possible to follow his entire line of thoughts during the redesign of Ctesibius' machine as described by Vitruvius.

The final design of Leonardo is sketched in the central part of the folio and is partly reproduced in Fig. 1: a description written in Leonardo's classic mirror handwriting appears in the upper left-hand side of the sheet. It is a very complex machine and the precise working flow is not yet understood in all details, also because it was probably only sketched. The scientist drew different concepts for the main components of the clock, at least as many in another part of the sheet not shown in the figure. An informed review on the studies of Leonardo upon this complex mechanism can be found in [7] while Rosheim [8] managed to build a prototype of the water clock.

Leonardo conceives an hourglass supplied by a canal which continuously pours water into a container hinged to the frame. The container can rotate to fill one after the other 24 vessels (called "bottini") representing the hours of the day: only one of them is shown in the drawing. At the end of the day all the containers are full and can be emptied.

The indexing motion of the container is actuated by the central body of each vessel, which is a complex mechanism shown with details in Fig. 4: key elements of its functioning are the helical guides that wind around the central column and two floats; one of them is spherical while the other is toroidal and is called "baga" by Leonardo.

When the vessel is filled with water to its maximum level, the spherical float rises and triggers the release mechanism that frees the horizontal bar: such bar is the upper part of a metal body passing through the central column with helical guides and is connected to a stop for the lower float. The baga can now rise up following the guides which impose a motion that is rectilinear for the first part and then helicoidal. This helical motion causes the horizontal rod to beat against two levers that rotate the container to feed the subsequent vessels; the same mechanism opens up a valve that flushes down the water of the vessel.

In addition, fork levers have been added, perhaps as counterweights, to make sure that the spherical float can lower again.

The Vitruvian mechanism, so ingeniously reinterpreted by Leonardo, would be so precise and automatic that it would never need to be tuned or reloaded. On top of the entire mechanism, Leonardo imagined installing a large bell beating the number of chimes corresponding to the hour of the day. This should be either a living being that would simply observe how many tubes were full, or some sort of automaton, almost a robot.

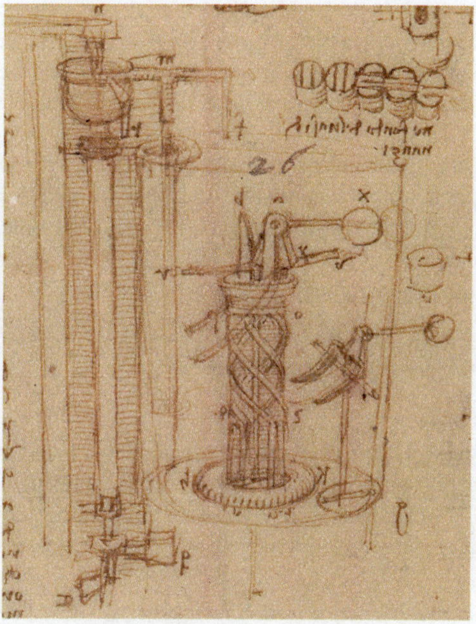

Fig. 4. Water clock sketch according to the original description by Vitruvius

4 Digital Mock-Up

The digital reconstruction of the water clock and the related animations allowed the Authors to experiment the feasibility of the concept and the Exhibition's visitors to understand its complex functioning; a sample animation can be accessed on the internet at the web address: https://www.youtube.com/watch?v=2omu1jSofUU&ab_channel=MeccanicadelleMacchine%40UnivPM.

The virtual prototype realized by the Authors allowed to assess that the mechanisms of Leonardo's clockwork and its workflow are sound despite the fact that he never realized a physical prototype. However, some issues coming from the actual realization of the clock can be pictured by the perspective of modern technology. Rosheim in [8] managed to build a small-scale clock facing many technical challenges never mentioned by Leonardo. Such issues mainly involved the materials and the surfaces finishings, which strongly affected the forces mutually exerted by the components.

Let's take the baga as an example: its material affects both the upwards and the downwards strokes. In fact, such component should be light enough to float and at the same time heavy enough to drag the whole mechanism while going back to its home position; solid wood can be an option in such sense.

Moreover, the ability of the baga to move along its helical guide depends on the quality of the surfaces in contact: the friction developed by the bodies could easily jam the whole mechanism. Such hypothesis is more than reasonable, especially if we think that the contact among two bodies (the baga and its helix) takes place under water where the use of a lubricant is virtually impossible. Moreover, the physical properties of the wood change when underwater, thus making the sliding of the components even more difficult.

This section is a brief explanation of the digital mockup that was animated during the Exhibition and was able to interact with the visitors. Figure 5 shows one of the 24 vessels, connected by pipes to transport water from one to the other. Each vessel has a complex mechanism that drives the filling and flushing of the water from the container and is composed by the following main parts.

4.1 Hours Basin and Water Supply System

The water flow fills the upper vase coming from the top left and comes out through the side duct to fill the hours vessel. When the latter has filled up, and therefore an hour has passed by, a mechanism (explained later on) rotates the water supply duct, which then feeds the next vessel. At the same time, a valve on the bottom flushes the water down. The stroke of the hour is marked by the ringing of a bell, not shown in the sheet exposed at the Exhibition.

4.2 Float Release Mechanism

When the water level in the hours vessel reaches the spherical float located in the upper part of the drawing, the float rises due to Archimedes' buoyant force and therefore it tilts the circular sector, which finally unlocks the small horizontal rod.

4.3 Central Column (Valve/Actuator)

While the water is filling the hours vessel, the toroidal float placed at the bottom receives an upward thrust but cannot rise since, being integral with the central shaft and the horizontal rod, it is blocked at the upper end by the circular sector previously described. When the water reaches the level necessary to free the central shaft, the horizontal rod, pushed by the lower float, starts rising upwards. The components of the central column have a shape such that during the rise the motion is initially straight but from a certain point onwards it becomes helicoidal, as there is also a rotation of the block around the vertical axis. In the final phase of the movement the small horizontal rod gets in contact with the adduction duct and makes it rotate of a certain angle, so that it goes to feed the aside basin of the following hour. The rotation of this duct also opens the drain valve, flushing the container (Fig. 6).

Fig. 5. Rendering of the interactive model of the water clock

5 Digital Technologies for Cultural Heritage Experience

Cultural Heritage (CH) is defined as set of elements which, for historical, cultural or aesthetic importance, are of public interest, representing a community and its cultural wealth. CH is therefore, by its very definition, of great interest to the public, who must be allowed to enjoy it in a proper way.

In order to achieve this goal, it is necessary to go beyond a static relationship between CH and its user. Overcoming the need for a rational systematization of CH expressed by the Museum Institution of late eighteenth-century, new needs are emerging today: to stimulate a dynamic and interactive perception of CH, developing a process of continuous learning is just one of the challenges to be faced shifting the attention of Museum Institution from the object to the observer.

The physical presence of CH is no longer enough to spread culture, it is necessary to build a relationship between the work of art and its observer, to know its contents,

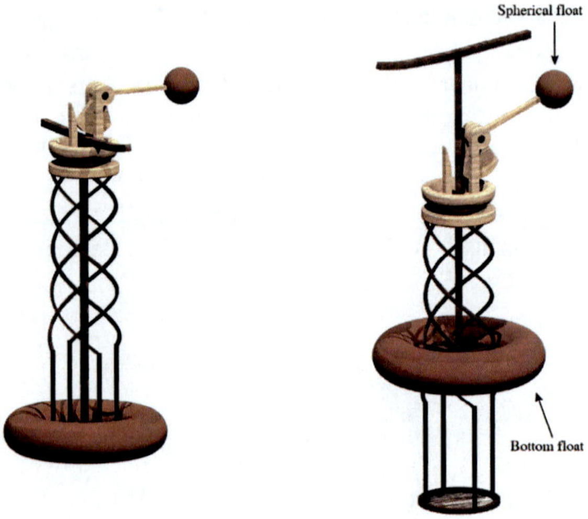

(a) (b)

Fig. 6. Details of the central column commanding the filling/flushing of water through the hours vessel; the bottom float is locked in (a) and it is released and rising in (b)

its techniques, its expressive and aesthetic meaning [9]. From an educational point of view, works of art are carried out in a complete way only when it successfully reaches its recipients, that is, when the communicative act implicit in it is completed [10].

In this scenario the use of technology as an intermediary between CH and the user, as a tool of transition from passive contemplation to a new dynamic fruition, represents an effective solution [11–14]. It must be considered that the users who enjoy CH are the same who use the screen of a PC, tablet or smartphone as privileged "window" to the world every day. Therefore, these tools have to convey cultural contents, making also CH a daily presence.

However, making fruition "technological" is not enough, technology must be used in a conscious way, avoiding, as often happens, to aim more at entertainment than education. These two components must be balanced in a learning by playing approach to involve users, designing cultural and multimedia operations that solicit connections [15, 16].

This strategy makes the technological medium and their potential benefits effective. Digital tools allow us to customize CH experience and therefore to communicate it, placing it at the user level thanks to different depths of contents and linguistic register.

The awareness of the effectiveness of smartphones and tablets for CH experience has led to a wide use of these devices. Several museums offer apps containing sections on the masterpieces preserved in their galleries or offering the possibility of virtual visits. These experiences show how the use of new media improves the fruition CH, encouraging its diffusion to a wider audience and even at a higher quality level. From new thematic paths to the enhancement of single works of art, from the 3D reconstruction of existing museums to the setting up of only virtual ones, different technological solutions can be exploited for CH fruition. In any case, a user-centered fruition that makes CH more accessible by improving its perception must be the goal, regardless of whether enhancing the vision of the real object (Augmented Reality), combining real and virtual (Mixed Reality) or creating a completely virtual environment and interaction (Virtual Reality).

6 LeonARdo: An AR App for Leonardo's Machines

The AR mobile app developed for Leonardo's water clock and the other drawings exhibited in Fano allows to integrate original drawing, one of major expressions of human inventiveness, with their digital 3D reconstruction. It has been developed for Android OS using the Unity game engine. To be used as a target, the images of the original drawings were uploaded on the Vuforia platform, where they have been analysed identifying their features to be recognized when the app is running the camera of user's device. Once the quality of the image has been validated, a dataset was automatically downloaded, providing all necessary for target recognition. This dataset and all the other elements for the

Fig. 7. User Interface. After selecting the language (1), the camera is activated. Automatically detecting the depicted machine on the framed drawing, the app generates the 3D model (3). The user is then allowed to activate the mechanisms and access in-depth folders about the representation (4), with the possibility to explore its HD digital reproduction (5). Further insights about the machine, its sub-assemblies (6) and history, are also available.

development of the App were imported into the Unity environment: the image-target and the dataset from Vuforia, the ARcamera that simulates the camera of the user's device, the 3D models of Leonardo's machines with related materials and an illumination and shadows system to make their visualization more real. Finally, the app was built after scaling and positioning the 3D models to make them appear from the original drawing as they were generated by the lines on the paper.

Concerning the user interface, it was designed to offer an immediate fruition of the contents. Once the app is open, users only have to frame one of Leonardo's drawings that was previously uploaded as target, making the corresponding 3D model appear on the screen. Once the model has been visualized, it can be investigated independently from the drawing, its components can be observed, and the functioning of the machine can be simulated. It is also possible to access a section about the original drawing, with textual information and high-resolution image, and one about the choices made for the digital reconstruction analysed from the mechanical point of view, Fig. 7.

7 Conclusions

The exhibition held in Fano has shown that is desirable to generate a digital heritage of Leonardo's inventions able exploits the possibilities granted by digital technologies for new forms of investigation, fruition, narration and understanding. A digital approach solves problems related to the complexity of the original drawings and it helps to better understand the delicate step from the concept to the representation. The physical models of Leonardo's machines are in fact not able to fully describe this passage, not allowing variations or simulations that digital reconstructions enable.

Coupling digital design and digital fruition also allowed to define methods and best practices for the transition from the cognitive model to the communicative one, generating a dynamic interaction with drawings that are by their very nature static representations. The AR application was, in fact, designed to return those drawings in the form of the artist's thought, recreating a suggestion able to penetrate the mind of the greatest genius who ever lived.

Acknowledgments. The Authors wish to acknowledge the Municipality of Fano, sponsoring institution of the Exhibition *Leonardo e Vitruvio – Oltre il Cerchio e il Quadrato.*

References

1. Borgo, F. (ed.): Leonardo e Vitruvio. Oltre il cerchio e il quadrato, Marsilio (2019)
2. Brillarelli, S., Callegari, M., Carbonari, L., Clini, P.: Digital experience of the work of Vitruvius and Leonardo. In: IOP Conference Series: Materials Science and Engineering, vol. 949, no. 1, #012041 (2020)
3. Callegari, M., Brillarelli, S., Scoccia, C.: The odometers of Marcus Vitruvius Pollio and Leonardo da Vinci. In: Niola, V., Gasparetto, A. (eds.) IFToMM ITALY 2020. MMS, vol. 91, pp. 75–82. Springer, Cham (2021). https://doi.org/10.1007/978-3-030-55807-9_9
4. Moon, F.C.: The Machines of Leonardo da Vinci and Franz Reuleaux. Kinematics of Machines from the Renaissance to the 20th Century. Springer, Dordrecht (2007). https://doi.org/10.1007/978-1-4020-5599-7

5. Cigola, M., Ceccarelli, M.: Marcus Vitruvius Pollio, history of mechanisms and machine. Science **26**, 307–344 (2014)
6. Cigola, M., Ceccarelli, M.: Machine designs and drawings in renaissance. Editions of de Architectura by Marcus Vitruvius Pollio. Hist. Mech. Mach. Sci. **31**, 291–307 (2016)
7. Zanetti, C.: Ctesibio, la clessidra ed il salvagente. In: Leonardo e Vitruvio: Oltre il cerchio e il quadrato, Marsilio, pp. 118–123 (2019)
8. Rosheim, M.E.: Leonardo's bell ringer. In: Rosheim, M.E. (ed.) Leonardo's Lost Robots, pp. 115–160. Springer, Heidelberg (2006). https://doi.org/10.1007/3-540-28497-4_4
9. Dewey, J.: L'arte come esperienza, La Nuova Italia (1951)
10. Antinucci, F.: Comunicare nel Museo, Gius. Laterza & Figli (2014)
11. Bekele, M.K., Pierdicca, R., Frontoni, E., Malinverni, E.S., Gain, J.: A survey of augmented, virtual and mixed reality for cultural heritage. J. Comput. Cult. Herit. **11**(2), 1–36 (2018)
12. Tscheu, F., Buhalis, D.: Augmented reality at cultural heritage sites. In: Inversini, A., Schegg, R. (eds.) Information and Communication Technologies in Tourism 2016, pp. 607–619. Springer, Cham (2016). https://doi.org/10.1007/978-3-319-28231-2_44
13. Ferretti, E., Turrini, D.: Visualizing Leonardo and Michelangelo through digital humanities, reconstruction and interaction design. SCIRES IT **10**(1), 5–18 (2020)
14. Gimeno, J., Portalés, C., Coma, I., Fernández, M., Martínez, B.: Combining traditional and indirect augmented reality for indoor crowded environments. A case study on the Casa Batlló museum. Comput. Graph. **69**, 94–105 (2017)
15. Infante, C.: Imparare Giocando, Edizioni Bollati Boringhieri (2000)
16. González Vargas, J.C., Fabregat, R., Carrillo-Ramos, A., Jové, T.: Survey: using augmented reality to improve learning motivation in cultural heritage studies. Appl. Sci. **10**(3), 897 (2020)

History of the Design of Machines and Mechanisms

The Hoist of the Aerial Rigging of the Mystery of Elche

Emilio Velasco-Sánchez[1](✉), Sixto Manuel Marco-Lozano[2], and Abel Navarro-Arcas[1]

[1] Miguel Hernández University of Elche, Elche, Spain
{emilio.velasco,abel.navarro}@umh.es
[2] Board of Elche Mystery, Elche, Spain

Abstract. In the town of Elche, during the 14th and 15th of August and without significant interruption since the last third of the XVth century, festivities are held in honor of its patron saint, the Virgin of the Assumption. This theatrical performance was declared Intangible World Heritage in 2001. The representation of the final days, death, assumption into heaven and coronation of the Virgin Mary is the highlight of the performance that takes place inside the temple of the Basilica of Saint Mary. This representation, unique in the world, is performed in song after converting the whole of the interior of the temple into a huge stage, with its corresponding rigging fitted with composite machinery, a double hoist, which allows lowering and raising to a height of 27 m. The aerial devices are occupied by singers, sometimes as many as five, reaching a weight of up to 600 daN, and even the simultaneous maintenance of two of the devices in the air. The lifting machines installed have undergone various modifications over time. The last of these was undertaken by the architect, Marcos Evangelio, in 1760 and, except for maintenance repairs required over the years and, despite a fire at the church in February 1936, they remain in a similar state to how the architect designed them. Studying them allows us to assess the mechanical assembly in general as very primitive technology, but which, after being used and perfected over the centuries, has allowed it to reach a high degree of precision and efficiency. This paper shows the main mechanical characteristics of the hoist, which is 300 years old.

Keywords: Historic winch · World Heritage · Mechanical lifting devices

1 Introduction

In the town of Elche, during the 14th and 15th August, and without significant interruption since the last third of the XVth century, festivities are held in honour of its patron saint, the Virgin of the Assumption. The representation of the final days, death, assumption into heaven and coronation of the Virgin Mary, is the highlight of the performance that takes place inside the temple of the Basilica of Saint Mary. This theatrical work has outlived others of its kind that were celebrated in the Christian world centuries ago, despite the clauses established by the Council of Trent that banned them inside temples. It was protected by a papal decree granted in 1632 by Pope Urban VII and the extraordinary thing about this event is that the performances have lasted until the

M. Ceccarelli and R. López-García (Eds.): HMM 2022, HMMS 40, pp. 149–159, 2022.
https://doi.org/10.1007/978-3-030-98499-1_13

present day -and during five centuries- and have only been interrupted by cases of force majeure (wars, epidemics, etc.). This representation, unique in the world, is performed in song after converting the whole of the interior of the temple into a huge stage, with its corresponding rigging fitted with composite machinery, a double hoist, which allows lowering and raising to a height of 27 m. The aerial devices are occupied by singers, sometimes as many as five, reaching a weight of up to 600 daN, and even, as we can see in Fig. 1 (a), the simultaneous maintenance of two of the devices in the air. Therefore, this extraordinary set will consist of two parts: one at ground level, known as the lower rigging, and another -the aerial one-, known as the upper rigging or sky in the dome, the two spaces being connected by means of three aerial devices—la Mangrana, el Araceli and the Santísima Trinidad. As the main component—see Fig. 2 (b)—there is a hoist responsible for moving the aerial devices between the two rigs, separated by a height of more than 24 m and a platform, by way of a balcony, which extends from the inner walls of the drum towards its centre and allows easy and convenient access to the aerial devices.

The lifting machines installed have undergone various modifications over time. The last of these was undertaken by the architect, Marcos Evangelio, in 1760 and, except for maintenance repairs required over the years and, despite a fire at the church in February 1936, they remain in a similar state to how the architect designed them. Studying them allows us to assess the mechanical assembly in general as very primitive technology, but which, after being used and perfected over the centuries, has allowed it to reach a high degree of precision and efficiency.

2 The Aerial Rigging

The different elements that make up all the necessary machinery for the representation of the Mystery have already been commented on, below its parts and mode of operation are described, for this let us observe Fig. 1(b) where a longitudinal section of the temple and the disposition of this.

As can be seen in Figs. 1 (a) and 1 (b), next to the machinery and as an essential element, is a circular canvas covering the base of the dome, the part that is visible to the audience decorated with blue-coloured paintings, clouds and musical angels; in the script of the play, it represents the sky. In the part located below the working platform, there is a square opening covered by two sliding doors, which open to allow the aerial elements to pass through. From the lower terrace to the canvas of the sky there are 27 m of height.

It works as follows: the ropes start from their respective reels and by means of pulleys known as "de remonte" (lifting pulleys)—see Fig. 2-b (n° 2)—they change their trajectory from horizontal to oblique. After reaching the existing pulleys at the vertex formed by the pillars—see Fig. 2 (a)—they turn 214° to reach a vertical position. Their ends, under the pulleys and on the platform, are attached to the apparatus, remaining suspended and ready for its occupants to get on. When the time comes for the apparatus to enter the stage, the doors open and allow it to pass through. The winch releases or recovers rope from its reel and the apparatus descends or ascends—see Fig. 1 (b).

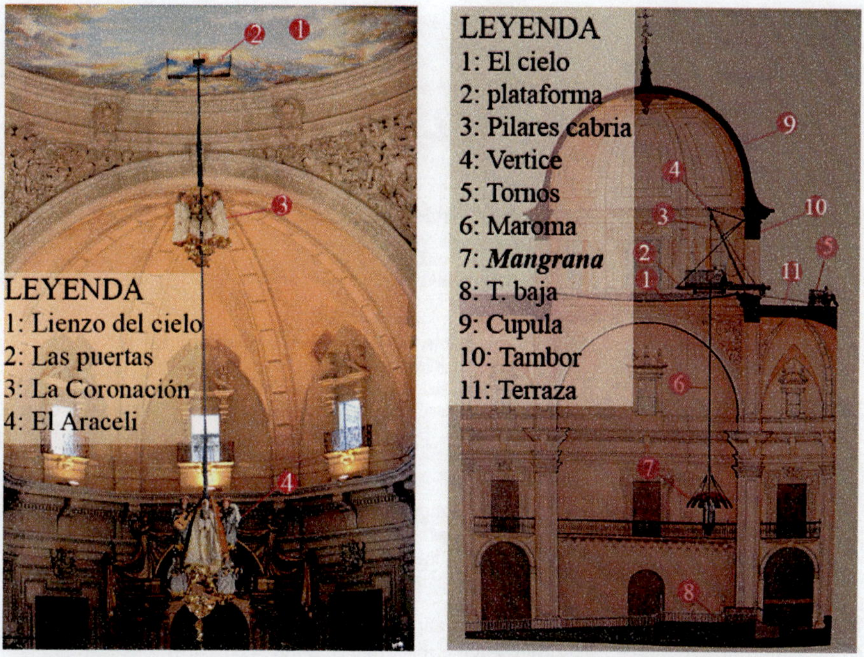

(a) LEGEND
1. Canvas of the sky
2. The doors
3. The Coronation
4. El Araceli

(b) LEGEND
1. The sky
2. Platform
3. Pillars, hoist
4. Vertex
5. Winches
6. Rope
7. ***Mangrana***
8. Lower terrace
9. Dome
10. Drum
11. Terrace

Fig. 1. A composite figure of a snapshot of the performance with two devices in the air and above, part of the canvas of the sky with their doors referenced as (a) and the longitudinal section of the temple with the arrangement of their main parts referenced as (b) [5].

3 The Hoist of the Aerial Rigging

Consisting of two pillars assembled at an acute angle, joined together at their apex and resting on the main ring, they are inclined towards the interior of the temple. A set of eight braces hold them in position, which is regulated by the length given to them. Two manually operated winches with parallel axes, each equipped with gear reduction systems, located on the terrace and separated from the pillars, are responsible for providing the force to lower and raise the devices; a force that is transmitted by means of thick hemp ropes guided by the pillars—see Figs. 1 (b) and 2 (b).

The whole assembly has, as a common support, the part of the building in which they are located: the terrace of the presbytery and the interior of the dome on the fourth floor of the basilica. These are sufficiently rigid to transmit the forces between the different members of the machine, and support the loads required, without any relative movement between their parts.

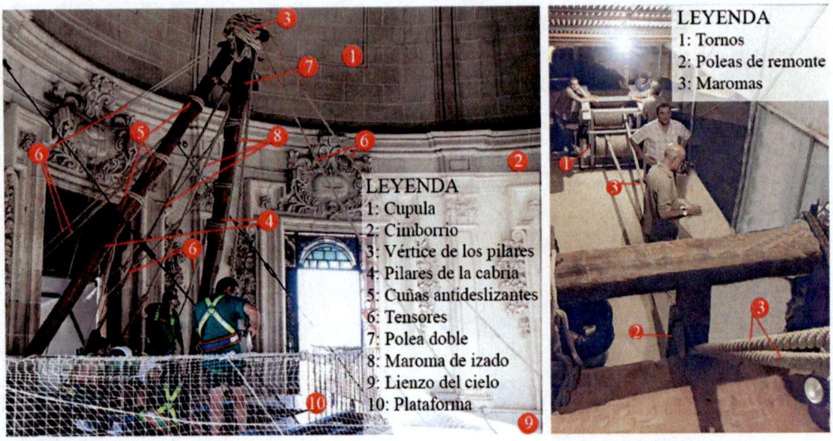

(a) LEGEND

1. Cupola
2. Dome
3. Vertex of the pillars
4. Pillars of the hoist
5. Non-slip wedges
6. Tensors
7. Double pulley
8. Lifting rope
9. Canvas of the sky
10. Platform

(b) LEGEND

1. Winches
2. Lifting pulleys
3. Ropes

Fig. 2. Showing the hoist with its components and its location within the dome referenced as (a) and the path followed by the ropes after leaving the winches referenced as (b).

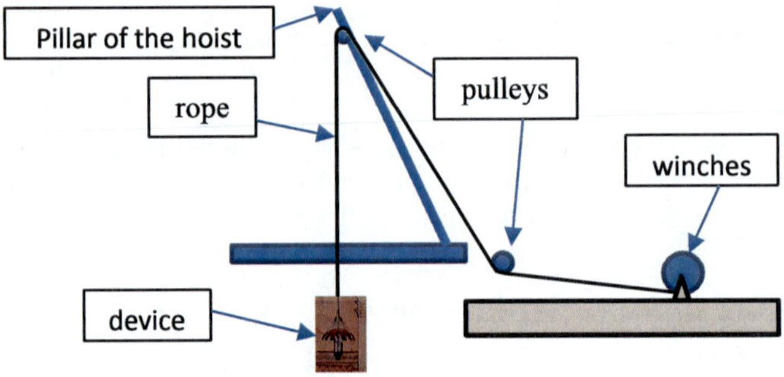

Fig. 3. Mechanical schemes of operation

3.1 The Pillars

As can be seen in Fig. 2 (a), two 6.46 m long masts form the beams of the hoist. At the tops of the two masts, there is a truncated cone-shaped cap which, as well as reinforcing the wood, holds the assembled parts together. It is customary to cover the metallic joint system of the vertex, with what would once have been its fastening system and formed by a tie with ropes, which is what can be seen in the image. Next to and below it, the two pillars are threaded by a steel bar that is responsible for holding the set of pulleys. Along the length of each mast, there are three groups of wedges, attached to it and separated from each other, which prevent the tethers from slipping downwards.

3.2 The Ropes

Twelve ropes of different diameters and lengths complement the pillars, except those used for the lifting equipment, all of which are made from industrial hemp (Cannabis sativa) and have traditionally been the only material used. Recently, those used for lifting, which are hidden from view, have been reinforced with synthetic fibres.

The ropes that function as tensors are made up of eight ropes tied at one end to different points of the beam and, at the other, to as many firm points distributed both on the inside and outside of the dome (Fig. 2a (6)). This is done symmetrically. The seventh is made from the vertex to a ring anchored to the upper part of the lintel of the window by means of a tackle that allows its inclination to be regulated. Finally, the eighth is made by means of a pulley attached to the vertex and a rope that starts and ends on the opposite side of the drum where the pillars are mounted. This counteracts the pull on the vertex caused by the lifting ropes and the opposition caused by the passive forces on the vertex pulleys.

The lifting ropes are the thickest and strongest of the whole rigging. They are corked in such a way that they have been reduced to a third of the length of the ropes, which gives them an optimum resistance value, with an apparent external diameter of 52 mm. Under these conditions, for hemp, a working load of 1912 daN and a breaking strength of 10000 daN, are estimated. With regard to the core, made of synthetic material, the data provided by the manufacturer is for the rope section given a breaking strength of 8000 daN, which gives a total breaking load for the rope of 18000 daN, a figure which, of course, is conditional on the state of the hemp.

3.3 The Pulleys

The pulleys are an indispensable complement to the ropes; fourteen of them are used in the hoist. The main ones are the two sets used to guide the lifting ropes from the winch to the apparatus: the lifting pulleys (Fig. 2b-2) and the vertex pulleys (Fig. 2a-7). The first two, which are located on the sill of the window that connects the inside of the drum to the terrace where the winch is located, consist of two bronze eye pulleys mounted in parallel with a shaft and a common steel frame (Fig. 2b-2). Those at the vertex are particularly unique, being a pair of pulleys, made of wood, mounted on top of each other, the one with the largest diameter on top intended for heavier apparatus, and both in the same frame, also made of wood and all reinforced with forged steel (Fig. 2a-7).

3.4 The Winches

The force applied to the ropes has always been with manually operated winches. The current ones entered into service in 1971 and replaced others with similar dynamic characteristics, which entered into service in 1761. After they were removed, they were moved to the Casa de la Festa, where they can be seen today. Comparing the two machines, we understand that it would be appropriate to present the hoist with the old winches given their special singularity.

The first image we have of them is the one in Fig. 3 (a) corresponding to the year 1899, Fig. 4 (b) shows its current state and corresponds to the result of a restoration that was carried out so that it could be exhibited in the aforementioned Casa de la Festa.

We can see in Fig. 4 (b) that these are two independent winches, different from each other and mounted on a common frame made from profiles of forged steel. This set was accompanied, as can be seen in Fig. 3 (a), by a wooden box that contained it and which had, among others, the following characteristics: its shafts are parallel horizontally, they are manually operated with two input cranks per winch and the multiplication of the force is provided by two chains of gears and this chain means that the number of gears is the main differentiating element between the winches, one having three pairs and the other, two.

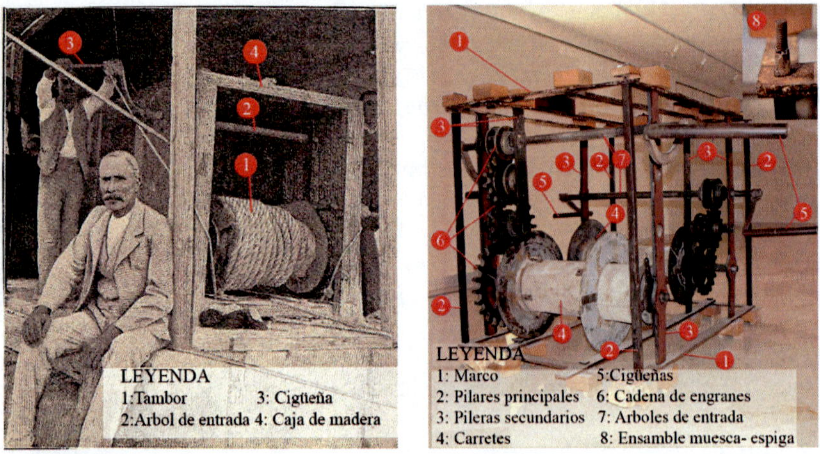

(a) LEGEND
1. Drum
2. Input shaft
3. Crank
4. Wooden box

(b) LEGEND
1. Frame
2. Main pillars
3. Secondary pillars
4. Reels
5. Cranks
6. Gear chain
7. Input shafts
8. Notch-tenon assembly

Fig. 4. Showing the hoist with its components and its location within the dome referenced as (a) and the path followed by the ropes after leaving the winches referenced as (b) [5].

They are mounted opposite each other on the frame, with the output drums at the same height, the gear trains positioned on opposite sides and the two cranks corresponding to the chain with the highest number of gears positioned higher than the others.

Until 1971, the whole assembly was anchored to the terrace of the basilica and positioned in such a way that its axes of rotation were orthogonal to the trajectory of the ropes. In this location, the winch with the chain with the highest number of gears, the most powerful one, was located at the part furthest from the dome and which we will consider as the rear part throughout this paper. As for the least powerful, the smallest and intended to lift the Santísima Trinidad, it is located, according to the reference taken, at the front.

Originally, the frame consisted of a metallic structure reinforced on the outside with a wooden box. After the box was burnt in the 1936 basilica fire, what has survived is the internal part. Due to the poor condition it was left in after being dismantled at the terrace of the basilica, prior to being exhibited, some parts had to be reconstructed, which is why, today, many of them are not the originals, which are easy to identify as they are made of laminated steel.

In order to describe it, we will consider that it has a main part, made up of two rectangular frames, one located at the base and the other at the top, their vertices joined together by four vertical columns. This main part is joined by accessory parts such as: two longitudinal crossbars in each frame, formed by rectangular plates identical to that of the frame, and also has two sets of three secondary columns formed by plates that are the support points to the shafts. To this end, at their resting point, the plates, while maintaining their thickness, produce a widening so that the hole for the bearing does not weaken the column. At the outer limits, the assembly resembles the geometric figure formed by the twelve edges of a straight parallelepiped or orthohedron with external measurements of: 1,556 mm long, 1,095 mm wide and 1,199 mm high.

The different parts that make up the structure, except for the frames, which are fastened with rivets, are joined using the system known in carpentry as notch-tenon—see detail in Fig. 4 (b)—on the upper right-hand side. Notches were made in each frame, four at the vertices and two on each side; the ones at the vertices being square and the rest, rectangular. The ends of both the main and secondary pillars, end in a tenon and, as an extension to it, has a threaded bolt to hold and support the wood of the box that encloses it. After tightening the nut to the bolt, the frame and the wood of the box would be securely fastened. The two remaining pillars are fastened using the same system as the four crossbars attached to the frames.

As we have just seen, the metallic part of the frame is entirely made up of parts arranged vertically and horizontally, without the presence of struts, cross braces, braces or crosspieces capable of triangulating the structure, a function that was assigned to the parts of the wooden box.

The groups of secondary pillars are arranged with their broad faces parallel to each other; each group having one pillar contained in the plane of each of the side faces of the orthohedron. The distance that remains from the axes of each group with respect to the plane of the rear part is 227 mm for the first and 1,177 mm for the second. The central pillar of the first group is separated from the left side faces by a distance of 426 mm, the second being separated by the same distance, but from the right side plane. Each of these pillars is attached in the same way as the side pillars, but on each of the crossbeams arranged in the upper and lower frames.

All the bearings are friction bearings and are attached to the secondary pillars. They are flanged bushings, made of bronze, on which the different shafts of the winch rest. Originally, the bushings of the hub were square on the outside and the flange was circular. After being repaired, some of them were replaced with bushings with a circular outer section.

The shafts start from a forged square section and have rods, both front and transverse, for both winches, which are roughly the same. The input shafts have a length of 1,546 mm and a square section at their ends, intended for coupling the cranks. On the output side, those supporting the drums have a maximum length of 1,122 mm and those corresponding to the intermediary gears of the kinematic chains have a maximum length of 300 mm.

As for the cranks, they consist of a main bar with a rectangular cross-section in the form of an arch. The length of the arch cord that forms the profile is 350 mm for the two corresponding to the small winch and 450 mm for the large one. All four have a square tube at the end intended to be attached to the shaft which coincides with the square shape given to the four ends of the two input shafts. At the opposite end, the one intended to be gripped by the hands of the operators, the plate was bent into a right angle. This end was forged into a cylindrical shape. These days, this cylinder has a steel tube on the outside that rotates concentrically on the forged cylinder.

The drums, both of which are the same, are where the ropes are wound. Attached to the shaft that supports them is the wooden matrix that forms a revolving cylinder on which the ropes are wound. To contain them within the cylinder, two circular crowns are fastened to it, the planes that contain them being parallel to each other and perpendicular to the axis of the cylinder. Due to the arrangement of the gear train of the small winch, only two thirds of its total length can be used on the drum of the large winch, which means it requires greater effort. The torque is provided by the force from the kinematic gear chain applied to the last gear, which is attached to the drum shaft.

The transmission of forces between each input shaft and its corresponding output shaft is effected by means of two ordinary straight gear trains located between each of the central columns and the corresponding one on the outside; to the left of the output drum in the large one and to the right in the small one.

The gears, formed from cast bronze, have turned side faces and the teeth, which are fitted by hand, are of the oblique, or so-called double point, type. The pinions are all lantern, and the head of the other gears describe the curve that results from tracing the envelope of the positions of the roller when its centre travels through the positions of its profile—see detail of Fig. 5 (a).

For the large winch, the kinematic chain is made up of three lower pairs and four members; the first is attached to the input shaft and the last to the output drum and the two intermediate ones, are attached to two shafts which rest on the central and outer column of the frame and are held in place by friction bearings housed in holes made in the columns.

The input shaft is fitted with a lantern pinion made up of eight rounds of steel attached to bronze crowns. It engages with a 14-tooth cast bronze sprocket that is closest to the outside and to this wheel, an 8-tooth cage pinion identical to that of the input shaft. The wheel and the pinion were attached to the shaft by inserting the shaft through the square holes made in them and tightening them together.

The second shaft, similar to the one described above, is located further down and its position is reversed. The wheel, which has 18 teeth, is at the same distance as the previous pinion so that the mesh between them is perfect. The pinion is in the right position to engage with the 24-tooth wheel attached to the output shaft.

The chain ratio is therefore 8/14 for the first pair, 8/18 for the second and 8/24 for the third, which means a total gear ratio of 1/0.0847; in other words, for every 11.8125 turns of the cranks, the drum does one turn. The module is slightly variable for the different gears, being roughly 18, the lanterns are all the same, with a primitive diameter of 140 mm and a pitch of 55.4 mm. Regarding the wheels, there is a primitive diameter of 447 mm with a circumferential pitch of 55 mm for the 18 teeth; a primitive diameter of 245.8 mm and a circumferential pitch of 64.4 mm for the one with 14 teeth. The thickness of the wheels is 28 mm and the geometry of its teeth is of the so-called point type, in which the point has been replaced by a cylinder [1]. They are made of cast bronze, with the teeth filed down and the faces turned.

As for the small winch, the kinematic chain comprises two lower pairs and three members with similar characteristics to the chain of the large winch. The location of the chain in this machine is to the right of the drum and the ratios for the pairs are 8/14 for the first one and 8/18 for the second, which means a total transmission ratio of 1/0.253968; in other words, for every 3.9375 turns of the cranks, the drum does one turn.

Regarding the large winch, in the case of having the drum width limited to 464 mm, a width that has always been used since the date it was installed in 1761, even without the delimiting disc as can be seen in the first photographs we have of it, the force on the handles from each of its operators varies from a minimum of 8.78 daN, corresponding to the turns of the third layer, to a maximum of 10.72 daN, corresponding to the turns that are placed on the fourth layer and the torque required in the input shaft for the first situation is 16.17 daN-m, and a maximum value of 19.74 daN-m.

If the large winch were to work with the full width of its drum, with a width of nearly 700 mm, with only two layers of rope and only two riggers, the values would be between 12.94 daN for the minimum and 14.66 daN for the maximum, and for the first situation, the input shaft required a torque of 11.90 daN-m and a maximum of 13.60 daN-m.

These days, and with the new winches, the time it takes to raise el Araceli is 7 min, 48 s, to travel a distance of 27 m at a speed of 0.05756 m/s.

With regard to the work carried out by the riggers during the lifting of el Araceli, when the whole drum is not used, a workload value of 270 kJ is obtained and the power developed by each rigger would be 144.22 W. Under these conditions, the efficiency of the winch would be 0.58.

(a) (b)

Fig. 5. Figure composed of the 1761 winch gears referenced as (a) and a gearing system, from the same era as the previous one, comprising a wooden wheel with wooden teeth and a lantern with steel cylinders attached to wooden discs reinforced with steel straps and attached to a wooden shaft and belonging to a windmill from the Cartagena countryside referenced as (b).

If the whole drum were used, with only two layers of rope, and two riggers, the workload would be 247.2 kJ and the power developed by each rigger would be 263.55 W. Under these conditions, the efficiency of the winch would be 0.59.

In order to evaluate the power supplied by a single rigger, we have consulted a specialised book on the subject [2] and it establishes average power values by age groups in different sports. It gives values for different sports and ages, and they are quite a lot higher than this figure. They did not have to be athletes, were generally young people, so we understand them to be figures, for both strength and power, so that, with only two people, the most arduous operation involved in the Mystery of Elche could have been carried out.

4 Conclusions

The most striking thing about the machine is, of course, the absence of design criteria, an aspect typical of constructions prior to the XIXth century, and the next is that the date of construction does not correspond to the high level of technology with which it was built. This leads us to understand that the machine is ahead of its time and, above all, in relation to the geographical place where it appears: the town of Elche in the year 1761 And what is so striking about it for us to say this? Quite simply, the technology of the metallic gears and their manufacturing process.

As for the construction process for the frame, there is nothing special about it, given that they are forged pieces with a better or worse finish that, in each case, depended on the skill of the blacksmith who made them, and, with regard to the shafts, the situation is similar to that of the frame, although whoever made them must have taken much more care given the straightness required and the tolerance in their measurements.

However, for the construction of the gears, the truncated cone-shaped parts that reinforce their position on the shaft and the shaft rods, not only was special care required on the part of the blacksmith, but, above all, it was essential to have special machining equipment that was certainly not available in the town of Elche or in the surrounding area. The gears were cast, the teeth were then adjusted manually and the sides of the discs along with the cone-shaped trunks that reinforce their position, were finished by means of a turning process—see Fig. 5 (a).

The technology that was being developed around the town of Elche was very different to that used in the construction of the winches and we can see this in the wind and water flour mills of the Cartagena countryside, where the designers of the winches came from, and which underwent a period of expansion and reached its maximum splendour in this same era [3].

Cartagena, after being named the headquarters of the Maritime Department of the Mediterranean in 1728, under the new Bourbon dynasty, must have been one of the places in Spain, and, above all, in these parts of its geography, with the greatest concentration of technology at that time. After its nomination, the region underwent, among other things, significant demographic growth, and with it, the need to feed its new inhabitants, as well as the occupants of the king's galleys [4]. It was traditional in those parts to grind wheat using windmills and given the growth of the population, in keeping with tradition, the number of these wind machines increased. This increased need brought about a growth of up to a total of 81 mills in the area.

A flour mill has a chain of gears that transmits the force from the sails to the grinding wheel, or via water extraction, and the mills of that era have a frame, which is the building itself, with wooden parts reinforced with steel and gears formed with wooden parts, also reinforced with steel—see Fig. 5 (b). It was not until well into the XIXth century that flour mills began to be constructed with parts of steel.

In light of all this, the question arises: where was the winch built? As a hypothesis, we can point to places where, at this time and much earlier, this technology was already in use. For example, in cathedral bell towers, where at the end of the XIVth century, as in the case of Wells cathedral in England, which are on display at the London Science Museum, similar mechanisms were already being used to operate the chimes and to support the corresponding weights.

References

1. de Martinez, A.L., de Corral Sìz, A.: Cinemática y Dinamica de Maquinas, E.T.S. Ingenieros Industriales, Servicio de Publicaciones, Madrid, p. 242 (1969)
2. Hawley, J., Burke, L.: Rendimiento deportivo máximo, Paidotribo, Barcelona, Tabla 4.7, p. 88 (2000)
3. López, F.J.M., García, M.D.A., Díaz, I.G.: El molino de viento en Cartagena durante el siglo XVIII, Universidad de Murcia, p. 148 (2009)
4. Romero, C., Galana: El molino de Zabala. III Jornadas nacionales de molinología, p. 1 (2001)
5. García, J.G.: La Festa o Misteri d'Elx. Patronato del Misteri d'Elx (2008)

"El Araceli" Non-destructive Test from the "Misterio de Elche" Play

Abel Navarro-Arcas[1](✉), Sixto Manuel Marco-Lozano[2], and Emilio Velasco-Sánchez[1]

[1] Miguel Hernández University, Elche, Spain
{abel.navarro,emilio.velasco}@umh.es
[2] Board of Elche Mystery, Elche, Spain

Abstract. In the city of Elche, every year, on August 14 and 15, it is celebrated a sacred musical drama of the death, the passage into heaven (known as the Assumption) and the crowning of the Virgin Mary. This event, known as "Misterio de Elche" play is unique in Europe. Since the mid-fifteenth century it has been performed in the Basilica of Saint Mary and in the streets of the old city of Elche, situated in the region of Valencia. The stage has two levels: the horizontal "terrestrial" stage and the vertical "celestial" stage, characteristic of the medieval mystery play. Ancient aerial machinery is used to enhance the spectacle by means of special effects. From the upper level, several aerial devices are raised and lowered in a controlled manner. These devices are called "La Mangrana", "El Araceli" and "La Santísima Trinidad". To give context to the reader, it is important to highlight that "El Araceli" is a device prepared to support five people situated in a special layout. The original supporting structure of this apparatus was made up of forged steel, wood and hemp ropes, but with the passage of time, some successive renovations were required. In 2010, wear was detected in some parts by the technical department that could cause security problems. Due to this circumstance, it was ordered a deep study. This paper analyzes a non-destructive test that was carried out on "El Araceli". A data acquisition system customized to the device was used to verify that the apparatus was suitable for the loads for which it was designed.

Keywords: Non-destructive · "El Araceli" · "Misterio de Elche" · Heritage

1 Introduction

In the town of Elche, during the 14th and 15th August, and without significant interruption since the last third of the XVth century, festivities are held in honour of its patron saint, the Virgin of the Assumption. The representation of the final days, death, assumption into heaven and coronation of the Virgin Mary, is the highlight of the performance that takes place inside the temple of the Basilica of Saint Mary [1–4].

This theatrical work has outlived others of its kind that were celebrated in the Christian world centuries ago, despite the clauses established by the Council of Trent that banned them inside temples. It was protected by a papal decree granted in 1632 by Pope Urban VII [5]. This act has remained unchanged for five centuries and has only been

M. Ceccarelli and R. López-García (Eds.): HMM 2022, HMMS 40, pp. 160–172, 2022.
https://doi.org/10.1007/978-3-030-98499-1_14

interrupted by cases of force majeure, such as wars or epidemics like the one we are currently experiencing.

In this work, unique in the world and a UNESCO World Heritage Site, the inside of the temple is transformed into a huge stage. Its aerial rigging system is equipped with a double hoist that allows the aerial artefacts to be lowered and raised to a height of 27 m.

Figure 1 (a) shows 2 of the artefacts suspended by the same rope. Traditionally, a maximum weight of 6000 N has been allowed [6].

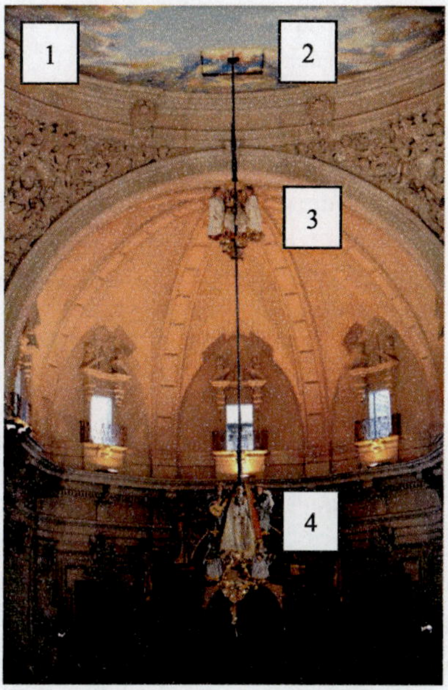

(a) Snapshot of the performance with 2 devices in the air. Canvas in sky (1), the doors (2), "The Coronation" (3) and "El Araceli" (4)

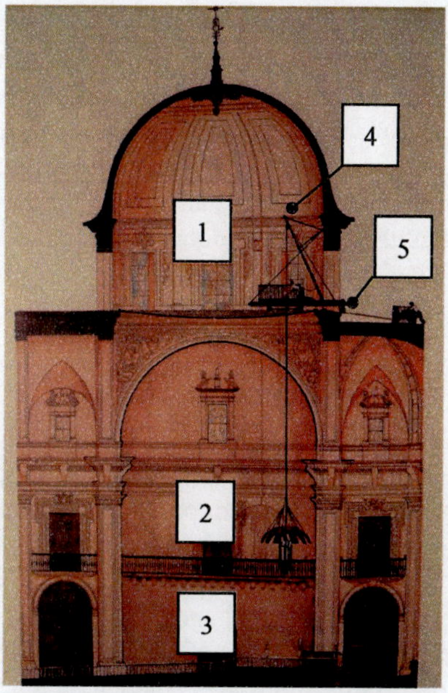

(b) Longitudinal section of the temple with the arrangement of the main parts of the aerial rigging. (1) High rigging, (2) "La Mangrana", (3) Lower rigging, (4) Hoist, (5) Terrace

Fig. 1. Site of the performance. The Basilica of Saint Mary.

The lifting machines installed have undergone modification over time. The last modification, carried out in 1760, was undertaken by the architect, Marcos Evangelio, and, except for maintenance operations, these devices remain in their original state. Despite their "primitive" technology, these machines achieve a high degree of precision and efficiency.

1.1 The Aerial Rigging

Figure 1 (b) shows the set, which is made up of two different parts: one at ground level, known as the lower rigging, and another aerial rig, known as the upper rigging or sky in the dome. The upper devices are operated with the help of a hoist that facilitates the lowering and raising of the apparatus. The 3 main aerial artefacts involved in the performance are called: “La Mangrana”, “El Araceli” and “La Santísima Trinidad” [7–12].

1.2 Manoeuvres in the Upper Rigging

The aerial stage, shown in the upper part of Fig. 1(a), is formed by a large circular canvas that covers the base of the dome. This canvas is decorated with paintings depicting the sky with musical angels. Slightly offset from the circular perimeter of the dome and aligned with the longitudinal axis of the temple, there is a square opening covered with two sliding doors, which open to allow the passage of the aerial apparatus.

Figure 1 (b) shows the general operation, where the rope starts from a reel and, by means of a series of pulleys, changes its trajectory until the rope is in a vertical position. The apparatus is attached under the upper pulley and on the platform. In this position, the actors are given access and exit to the aerial artefacts. At the moment that the apparatus enters the scene, the doors of sky open and allow it to be lowered down. The winch unwinds or collects the rope from its reel and the apparatus is lowered or raised [13].

2 “El Araceli”

The structure of “El Araceli”–shown in Fig. 3 – is the aerial apparatus where two of the most important scenes of “la Festa” take place. On the one hand, the assumption of Mary into the sky and, on the other hand, “La Coronación” (The Coronation) by “La Santísima Trinidad” (The Holy Trinity) [6, 7, 13].

Figure 2 shows the artefact in operation in 2 of its interventions during the performance. The lateral ends are occupied by angels carrying stringed instruments, who remain kneeling, throughout their intervention, on ledges arranged in each house. The two lower positions are occupied by two children who pretend to play a guitar without strings. The two upper positions are occupied by two adults. The central location may be represented by the High Angel, played by a priest, or the Virgin of the Assumption, depending on the act of the play.

2.1 Structure of “El Araceli”

The whole structure consists of a frame that is roughly symmetrical in its vertical axis, built with rectangular steel profiles, as shown in Fig. 3. These profiles are shaped in the form of a parabola and made up of sections of two superimposed profiles: two forming an arch and two more for each of the branches. The ends of the branches are joined by a crossbar (lower part).

At its upper vertex, it has an eyebolt, which can be seen at mark 1 of Fig. 3, for coupling the apparatus to the rest of the system. This ring is attached to the outer arc of the parabola by means of a single rivet. The apparatus has a second reinforcement, parallel to the crossbar that closes three quarters of the vertex, where the central platform (mark 6, Fig. 3) rests. From the central ledge and downwards, two struts (mark 8, Fig. 3) extend, and upwards, four columns which, fastened together with two parabolic arches, form the harness of the central ledge.

(a) "El Araceli" with the High Angel. Descent

(b) "El Araceli" with the High Angel. Ascent

(c) "El Araceli". The Assumption of Mary

(d) "El Araceli". "The Coronation"

Fig. 2. "El Araceli" in its two interventions during the performance.

The exterior of the frame has four side ledges: two lower ones, where the whole structure is closed, and two located at mid-height. Each exterior platform has an articulated bar which, together with a leather belt, as a preventive element to tie the occupant of the ledge to the frame. Each support is reinforced at the bottom with a strut (Fig. 3).

As for the main dimensions of "El Araceli", it is 2475 mm high, 1510 mm long and 600 mm wide. The free space of the lower side platforms limits the height of the actors to 1.170 m and is therefore intended for use by children. The central and upper side ledges situated at mid-height are intended for use by adult actors as they are not limited by their upper part.

Its unladen mass is 130 kg and its maximum theoretical load weight assigned on stage is 6000 N. It should be noted that its maximum weight in service is currently 4550 N.

2.2 Restoration of "El Araceli" in 2010

In July 2010, and during the assembly of the high rigging, certain problems were detected in "El Araceli". In the lower part of the gospel's side, signs of decomposition were observed. On the terrace of the temple itself, the action required was taken to guarantee its correct operation in safe conditions for the tests planned. After the analysis, the notch-tenon-wedge joints were perfected. Figure 5 (a) shows the previous design of the wedges that hold the tenons. Figure 5 (b) shows the redesign made, which consisted of opening the wedge at its end, which ensured that it was attached in the right position.

After the site tests were completed, the event organisers commissioned a major review. The apparatus was transferred to a suitably equipped workshop, where photographs, detailed sketches, functionalities of its parts, etc., were recorded in order to study it in detail.

From this situation, the need to perform a series of tests arose, which gave rise to the study presented here.

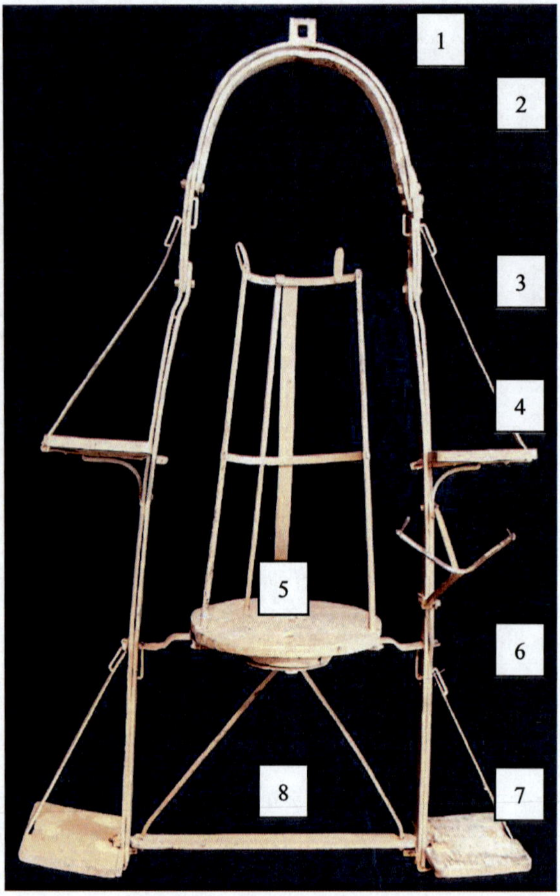

Fig. 3. Structure of the main elements. (1) Eyebolt, (2) Parabolic arches, (3) Branch, (4) Upper side ledge, (5) Central ledge, (6) Crossbar, (7) Lower side ledge, (8) Lower struts.

(a) Front view (b) Rear view

Fig. 4. Close-up of the eyebolt fastening rivet located at the upper vertex

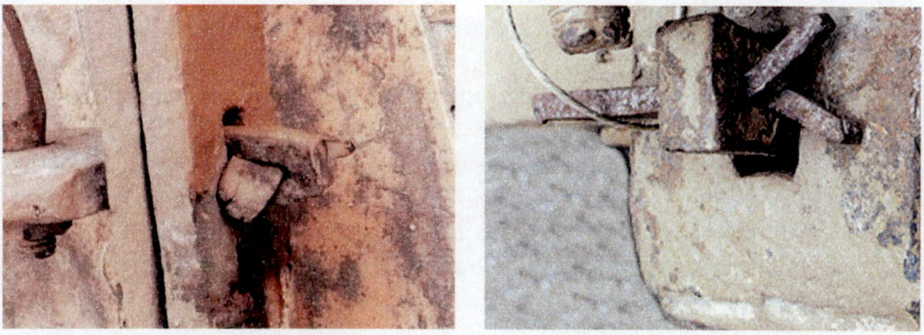

(a) Traditional wedge (b) Redesigned wedge

Fig. 5. Notch/tenon/wedge fastening system belonging to the device's locking crossbar.

3 "El Araceli" Test

These days, any machine or mechanism is subject to compliance with strict regulation in terms of safety, use and periodic maintenance. These regulations are even stricter in the case of apparatus for lifting persons, which are designed with redundant safety systems and with significantly high design safety coefficients.

To put this in context for the reader, we believe it necessary to mention that the elements that form part of the "Misterio de Elche" are exempt from the application of current regulations [14, 15], due to the fact that it is a device for transporting actors during artistic performances and because it is an artefact made up of parts that are more than 550 years old. The conservation of these elements was one of the values that led to "la Festa" being proclaimed a World Heritage Asset by UNESCO.

Historically, the inspection protocol for machines and apparatus used in the performance of this type of event was limited to a visual inspection [16]. In 2010, the possibility of performing non-destructive tests on the most critical elements that could affect the safety of people, was raised. Information is available on non-destructive or semi-destructive heritage assets [17, 18]. Most of the structures studied are made of wood although not much information exists about structures similar to those analysed in this paper.

3.1 Justification of the Test and Types of Load Applied

Although the current regulations on types of apparatus do not apply, the "Misterio de Elche" technical team studied the possibility of performing a non-destructive test. It was decided that it should be of the non-destructive type and its main objective was the integrity of the device and people. This course of action was well received by the trustee, who is responsible for its proper conservation.

It was determined that it would be advisable to subject the mechanism to a non-destructive, loading the apparatus to roughly 1.5 times the maximum load mass allowed.

The test was performed in conditions similar to real life. Figure 6 shows the apparatus attached to a frame by the upper eyebolt. This was fitted with analogue dials, on which deformations in the x and y axes, according to the load conditions, were recorded. As a safety measure, the support point was secured with a brace in the event of a hypothetical failure of the eyebolt rivet, as shown in Fig. 7.

The test performed was of the static traction type [19]. On the one hand, the apparatus was pulled by its own weight and the loads superimposed. On the other hand, it could be assumed to be static since the operating speed of the apparatus is roughly 0.35 m/min, a speed which, for the purposes of the test, was deemed not to be significant. Traditionally, this speed was used to eliminate possible adverse effects of uncontrolled inertia. It is worth mentioning that this particularity also exempts it from regulations relating to lifting apparatus [16].

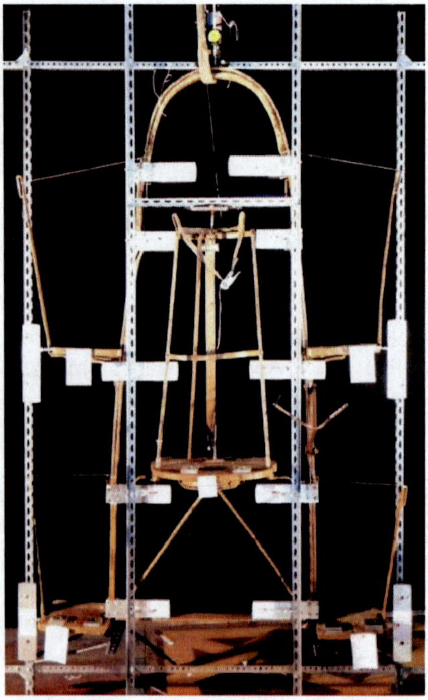

(a) Structure of "El Araceli" and reference frame

(b) Close-up of the upper eyebolt and the rivet fastening

(c) Close-up of the analogue dial

Fig. 6. Notch-tenon-wedge fastening system belonging to the device's locking crossbar.

Regarding the load for the test, it should be noted that the device was originally designed with 5 places: for two children and three adults, but no weight limit either in total or per actor had been established. It was only recommended that a certain symmetry in weight and volume be maintained to ensure the balance and aesthetics of the staging. Thanks to data from previous experiences, the maximum masses of the actors were set: 42 kg for a child and 80 kg for an adult. Taking into account the proposed safety coefficient (1.5), the total value of the load on the device should not be less than 4770 N.

To carry out the test, concrete parallelepipeds or load units (hereinafter referred to as l.u.). These parts fitted into the dimensions of the ledges and had a unit weight of 162 N. The weight of the l.u. was quite illustrative as 3 l.u. added up to 486 N (slightly more than the weight of a child) and 5 l.u. added 810 N (roughly the weight of an adult). From these data, it can be deduced that, in order to ensure the load conditions expected, the ledges reserved for the children had to be loaded to 4 l.u. and those for the adults to 8 l.u. All this is shown in Table 1.

At the top of the apparatus, the upper eyebolt was fitted with a touch probe with a scale division of 0.01 mm and a dynamometer to record the load exerted.

Table 1. Test loads and load units.

Position	Maximum nominal mass	Maximum nominal weight	Equivalent load units	Minimum test load (× 1.5)	Load units	Load on apparatus
Type of load	(kg)	(N)	(l.u.)	(N)	(l.u.)	(N)
Central figure	80	785	5	1178	8	1296
Adult left	80	785	5	1178	8	1296
Adult right	80	785	5	1178	8	1296
Child left	42	412	3	618	4	648
Child right	42	412	3	618	4	648
Total	324	3179	21	4770	32	5184

3.2 Test Procedure and Measurement on Control Points

The test was conducted rigorously in order to ensure the integrity of the apparatus. A loading system was devised so that the test could be stopped as soon as permanent deformation was detected. "El Araceli" was loaded progressively according to the following diagram:

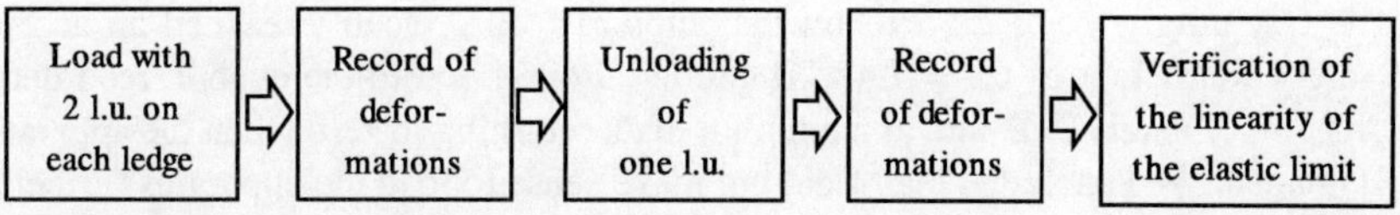

Before loading the apparatus, the dials on the control points of interest were set to zero. These dials had a scale division of 0.5 mm.

The acquisition of the data of greatest interest, due to the apparent deformation, were the ends of the upper and lower side ledges and the central seat, points 1, 2, 3, 4 and 11. Other unremarkable displacements were also recorded in the recesses of the side ledges and around the arch and branches. The deformation in the fastening rivet of the upper eyebolt was obtained by means of a touch probe with a scale division of 0.01 mm as shown in Fig. 7.

(a) Apparatus subjected to maximum test load

(b) Close-up of the eyebolt fastening, touch probe and safety system

Fig. 7. Notch/tenon/wedge fastening system belonging to the device's locking crossbar.

4 Results

Table 2 shows the displacements of the points of greatest interest as a function of the load conditions applied. Once the maximum load has been reached, according to the procedure detailed in the previous point, the apparatus was unloaded in reverse order until it was no longer under load.

The fastening rivet of the eyebolt that supports "El Araceli", reached an acceptable safety coefficient of Cs = 1.45. The touch probe fitted to the eyebolt recorded a deformation of barely 0.02 mm at maximum load, returning to zero when the apparatus was unloaded. We concluded that it did not make sense to load the apparatus further to verify a higher safety coefficient for the rivet. Loading the apparatus further could lead to the deterioration of other parts. To ensure the resistance at this point, the apparatus was reinforced with bolted braces by way of a bracket and re-secured with four 10 mm lengths of hemp rope tied to the arches of the parabola, bridging the eyebolt. All this is shown in Fig. 6 (b).

In Table 2, it is worth noting the last row, which shows the value of the residual displacements when the apparatus was unloaded. The value (0.0) in all the records indicates that there was no permanent deformation and, in no case, was the elastic limit exceeded.

Table 2. Test loads and displacements at points of interest.

	Total Weight Increase	*Point 1. C*			*Point 2. A*			*Point 3. B*		
Order	*(N)*	*q1*	*x*	*y*	*q2*	*x*	*y*	*q3*	*x*	*y*
0	0	0	0	0	0	0	0	0	0	0
1	810	16.2	0	-1	16.2	0	-1	16.2	0	0.4
2	1620	32.4	0	-2	32.4	0	-2	32.4	1	-7
3	2430	48.6	0	-4	48.6	0	-4.5	48.6	1	-10
4	2916	48.6	0	-3.5	64.8	0	-6.5	64.8	1	-12
5	3402	48.6	0	-4	81	0	-8	81	1	-15
6	3888	48.6	-1	-4	97.2	0	-10	97.2	1	-16
7	4692	64.8	-1	-5	113.4	0	-11	113.4	1.5	-18
8	5178	64.8	-1.5	-6.5	129.6	0	-12	129.6	1.5	-18
0	0	0	0	0	0	0	0	0	0	0

	Total Weight Increase	*Point 4. D*			*Point 11. E*		
Order	*(N)*	*q4*	*x*	*y*	*q5*	*x*	*y*
0	0	0	0	0	0	0	0
1	810	16.2	0	-1	16.2	0	0
2	1620	32.4	0	-2	32.4	0.5	0
3	2430	48.6	1	-3	48.6	0.5	0
4	2916	48.6	1	-3.5	64.8	1	0
5	3402	48.6	1	-4	81	1	-1
6	3888	48.6	1	-4	97.2	1	-4
7	4692	64.8	1	-4	113.4	1	-4
8	5178	64.8	2	-4	129.6	1	-4
0	0	0	0	0	0	0	0

The graphical results of the displacements as a function of the load applied, experienced by the side ends of points 1, 2, 3, 4 and 11 corresponding to the ledges A, B, C, D and E, are shown below.

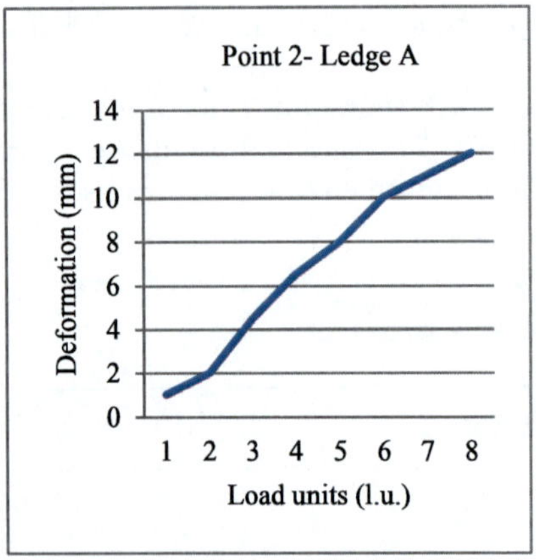

(a) Point 2 – Ledge A

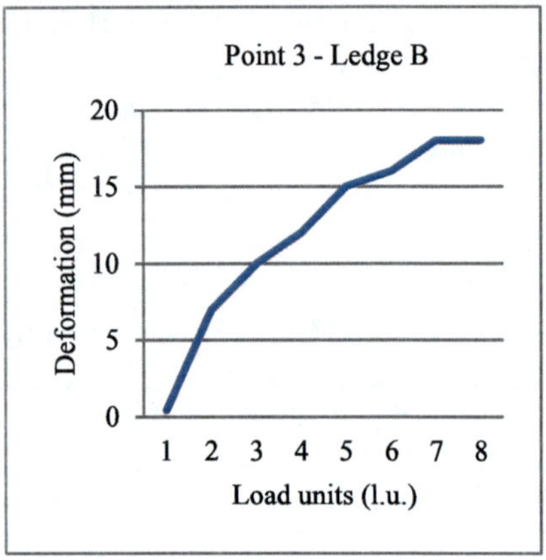

(b) Point 3 – Ledge B

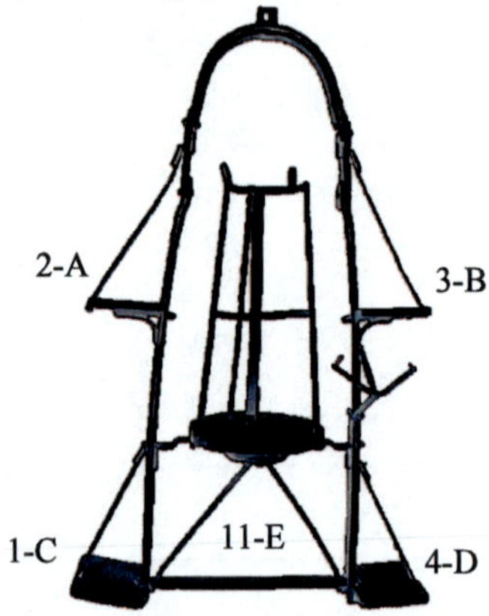

(c) Location of the points studied

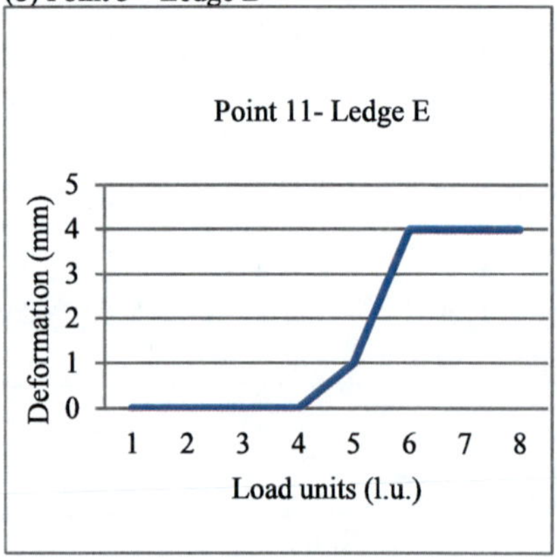

(d) Point 11 – Ledge E

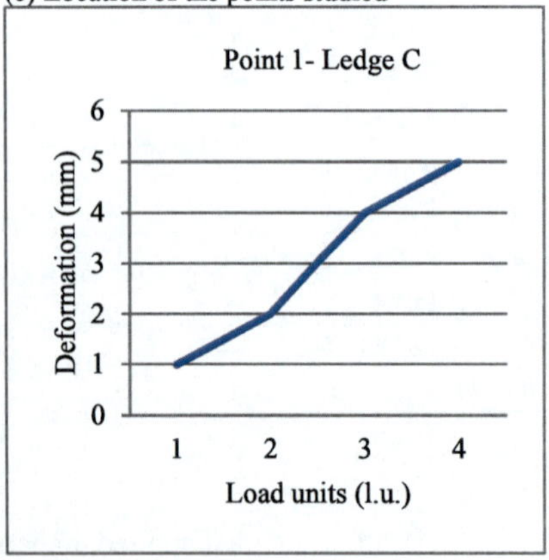

(e) Point 1 - Ledge C

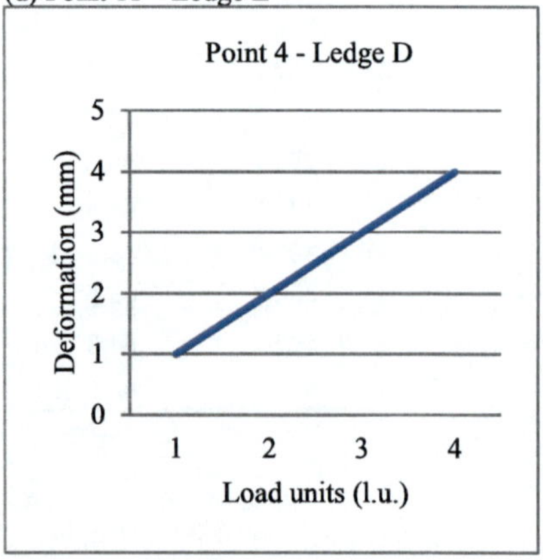

(f) Point 4 – Ledge D

Fig. 8. Deformation/Load graphs at points 1, 2, 3, 4 and 11

As can be seen in Fig. 8, points 2, 3 and 11, intended for one adult, were loaded with 8 l.u.'s or 1296 N. The minimum safety coefficient obtained was 1.62. The weight of an adult has been assessed to be 800 N according to the Occupational Risk Prevention Act.

Points 1 and 4 shows the behaviour of the structure intended for children. These platforms were loaded with 4 load units, corresponding to 648 N. Compared to the nominal weight requirement for a singing child, 421 N, a safety coefficient of 1.57 was obtained.

5 Conclusions

It is important to note that the apparatus that is the object of this study includes parts from the XVth century. The greatness of "la Festa" is recognised for the conservation and maintenance of its elements, including the music, staging, rigging system, and artefacts.

Until 2010, the inspection of the apparatus before use was based on a visual examination. During 550 years of use, no accidents worth mentioning have been recorded. Such antiquity makes the apparatus used non-destructive relics today.

From an engineering perspective, a replica of "El Araceli" could be reproduced with better mechanical characteristics. The organisers of the Mystery of Elche deem keeping the artistic legacy alive to be very worthwhile, and therefore insist on the use and maintenance of the traditional artefacts.

Once the test results were analysed, it is concluded that:

The apparatus demonstrates sufficient rigidity and elasticity to guarantee its continued use under traditional test conditions and suitable safety conditions for people have been established.

The minimum safety coefficient achieved is 1.57, which is deemed to be a sufficient guarantee.

The eyebolt from which the apparatus is suspended was tested until a safety coefficient of 1.42 was obtained. This point was reinforced with bolted braces by way of a bracket and completed with four 10 mm lengths of hemp rope tied to the arches of the parabola. All this ensures improved mechanical resistance. The detail is shown in Fig. 6 (b).

Lower value load units could have been used to obtain better graphic resolution and to enable behaviours not picked up in this study to be appreciated. This aspect will be taken into account in future analyses.

6 Future Work

The future lines of research proposed by this study group are:

To analyse, study and improve the mechanical system for the assembly of the aerial rigging stage [20]. 5 prototypes of this system have been made, and the most suitable solution is being sought.

To study the possible restoration of the metallic parts not subject to stress by applying filler for formal reintegration.

Given the antiquity of the metallic parts that form the structure of "El Araceli", it would be interesting to characterise the steel used for future renovations.

References

1. Gironés, G.: El Misterio de Elche. Patronato del Misteri d'Elx, Elche (2008)
2. Castaño, J.: Repertori bibliogràfic de la Festa d'Elx. IVEI – Ajuntament d'Elx, Valencia (1994)
3. Massip i Bonet, F.: La Festa d'Elx i els misteris medievals europeus. Institut de Cultura Juan Gil-Albert - Ajuntament d'Elx, Alicante (1991)
4. Ramos, R.: La leyenda del Misterio de Elche. Graficas Asín, Madrid (1956)
5. Pérez, J.A.: El rescripto del Papa Urbano VIII sobre la Festa o Misteri d'Elx. Tirant lo Blanch, Valencia (2008)
6. Marco, S.M., Velasco, E.: La cabria de la tramoya aérea del Misterio de Elche. In: Actas del XX Congreso Nacional de Ingeniería Mecánica, Málaga (2014)
7. Lozano, S.M.: Análisis de la tramoya aérea de la Festa o Misterio de Elche. Tesis Doctoral, Universidad Miguel Hernández, Elche, (2014)
8. Caprietti, E.: Le machine scenotecniche per le sacre rappresentazioni. Brunelleschi, Giunti Editores, Milán (2003)
9. Massip i Bonet, F.: La ilusión de Ícaro: Un desafío a los dioses. Comunidad de Madrid-Consjería de Educación y Cultura, Madrid (1997)
10. Mc Evoy Bravo, R.: El Misterio de Elche y sus antecedentes en el arte, la cultrua y la escenoténica italiana: Filippo Brunellechi y los ingenios aéreos florentinos. Festa d'Elx, pp. 159–176 (2013)
11. Pomares, J.: La "Festa" o Misterio de Elche. Patronato del Misterio de Elche, Elche (2004)
12. Massip i Bonet, F.: El Teatro Medieval. Voz de la divinidad, cuerpo de histrión. Montesinos, Madrid (1992)
13. Marco, S.M.: La restauración de "El Araceli o Rescèlica. Regidoria de cultura, Ajuntament d'Elx, Elche **59**, 217–226 (2018)
14. Directiva 2006/42/CE del parlamento europeo y del consejo de de 17 de mayo de 2006 relativa a las máquinas y por la que se modifica la Directiva 95/16/CE (refundición) (2006)
15. Real Decreto 2291/1985, de 8 de noviembre, por el que se aprueba el Reglamento de Aparatos de Elevación y Manutención de los mismos, sus ITC y actualizaciones (2013)
16. Riggio, M., D'Ayala, D., Parisi, M.A., Tardini, C.: Assessment of heritage timber structures: review of standards, guidelines and procedures. J. Cult. Herit. **31**, 220–235 (2018)
17. Kloiber, M., Drdacky, M., Machado, J.S., Piazza, M., Yamaguchi, N.: Prediction of mechanical properties by means of semi-destructive methods. A review. Constr. Build. Mater. **101**, 1215–1234 (2015)
18. Niemz, P., Mannes, D.: Non-destructive testing of wood and wood-based materials. J. Cult. Herit. **13S**, S26–S34 (2012)
19. Ramírez, F., Fernández, M.A., Alonso, A., Delojo, G., Valdecantos, C., de los Ríos, J.M.: Métodos de ENSAYOS NO DESTRUCTIVOS, Tomo I. Inta, España (1996)
20. Díez, R., Marco, S.M., Navarro-Arcas, A., Velasco, E.: Diseño de sistema mecánico para el montaje de la jácena de la tramoya aérea del Misteri d'Elx. In: Actas del XXI Congreso Nacional de Ingeniería Mecánica, Elche, pp. 452–458 (2016)

Ports Design and Construction Machinery Through 19th Century Atlases

Agustín Castillo-Martínez(✉)

Mechanical and Mining Engineering Department, University of Jaén, Edificio Departamental, Campus Científico-Tecnológico, Ronda Sur, S/N, 23700 Linares, Jaén, Spain
acmartin@ujaen.es

Abstract. During 19th century, ports and harbours design and construction techniques experienced an exponential growth in the Western countries. The new scientific paradigm brought deep knowledge about wave mechanics, and at the same time the steam technology deployed new machinery to address some of the construction issues troubling European engineers for decades. This paper shows some examples of the new machines and mechanisms developed in 19th century Europe for ports design and construction, with special interest in the *École nationale des ponts et chaussées* of Paris research. The graphic definition of such machines through engineering atlas volumes is also addressed.

Keywords: Ports design · Construction methods · History of construction · 19th century machines

1 Introduction

For many centuries, port and harbour construction techniques evolved very slowly. Some artificial harbours built during the Ancient Age were great examples of engineering quality techniques, even showing great size achievements – the civil and military ports of Cartago being extraordinary examples of both [1]. Nevertheless, during 19th century the construction techniques evolved much more rapidly than in previous times. The reasons for such a quick growth are diverse but interrelated.

First of all, the scientific corpus of mathematics and physics –and specially wave mechanics—had grown considerably. The works of Christiaan Huygens (1629–1695), Isaac Newton (1642–1727), Gottfried Wilhelm von Leibniz (1646–1716), Daniel Bernoulli (1700–1782) and Leonhard Paul Euler (1707–1783), among many others, became more accessible and familiar in many academic settings. At the middle of the 18th century, the "great schools" of engineering were founded in France, creating a model that was imitated in some other countries of Europe for many years. The *École royale des ponts et chaussées* was founded in 1747, and started researching about ports and harbour construction, as well as many other civil engineering endeavours [2]. These schools syllabuses brought to engineering the latest knowledge about mathematics and physics. In 19th century England, this trend was heavily supported by Osborne Reynolds (1842–1912) [3], who studied mathematics in Cambridge before becoming professor of engineering.

M. Ceccarelli and R. López-García (Eds.): HMM 2022, HMMS 40, pp. 173–183, 2022.
https://doi.org/10.1007/978-3-030-98499-1_15

A second reason is the birth of new engineering materials, such as modern hydraulic cements and steel. At the beginning of 19th century, Louis Vicat (1786–1861) and Egor Cheliev (1771–1839) had some success developing some modern hydraulic cement recipes in France and Russia, respectively [4, 5]. In 1822, James Frost (1780–1840) obtained a patent for his *Portland cement* or "British cement", as he would call it [6]. In 1855, sir Henry Bessemer (1813–1898) started the modern era of steelmaking with his industrial method to create steel from pig iron [7]. The new era of construction materials had started.

But the third reason is far from being the lesser of them –the steam revolution and the mechanical improvements that were brought to engineering. Traction and power became available everywhere through many mechanical advances, and railroads starting arriving to every corner of Europe, improving accessibility to these new mechanical inventions.

Some of the new machines were used for new scientific measurements related to wave mechanics and weather behaviour data collection. Others were needed for the new construction techniques, improved harbour operations or railroad integration in European ports. The objective of this paper is to acknowledge some of these machines, describing their operations and graphic representation through engineering atlases used in the *École des ponts et chaussées* of Paris. A descriptive methodology will be used, and the physic definition of the machine will be followed by a full description of uses, construction methods or other operations related to each of these mechanisms.

2 Description of the Machines and Their Graphic Representation Through Atlas Volumes

At the end of the 19th century, Médéric-Clement Lechalas (1820–1904) – general inspector of France Civil Engineering Corps- founded an editorial collection of engineering treaties, by the name of *Encyclopédie des Travaux Publics.* He was awarded the title of Officer of the Legion of Honour of France, and the gold medal of the International Exposition of Paris (1889) for exceling doing the coordination of these publications, as well as due to his presentations during the exposition and career.

As these works exemplary show the state of the art by the end of the 19th century engineering, they will be referred henceforth. Most of these books were comprised of some text volumes with descriptions and explanations for the students of engineering, as well as an atlas volume for graphic definition of machines and work procedures. The most important of the ports and harbours construction manuals were *Ports Maritimes* and *Travaux Maritimes, Phénomènes Marins, Accès des Ports,* written by professor F. Laroche, and edited from 1891 to 1893 by *Baudry et Compagnie.*

2.1 Machines to Measure Wave Physic Characteristics and Weather Extreme Conditions

Design characteristics for ports should be measured in extreme weather conditions. Stability and durability as well as finesse are required in such measure machines. In France, most of the extreme winds have a west or southwest origin (facing east) [8]. By the end of 19th century, the Robinson anemometer design was rather frequently used (Fig. 1). Wind pressure was usually calculated from average speeds registered for ten or more minutes, although some peak values could be obtained considering just extreme gusts lasting a few seconds. This machine could register pressures of 275 kp/sq mtr[1] consistently, and even 678 kp/sq mtr during an isolated event, registered in 1877, January the 30th [8]. In the atlas volume, the machine is determined by elevations, plan and perspective figures, shown below [9].

Other important values needed for ports design are related to the geometrical shape of the incoming waves during extreme weather conditions. Wave significant height and wave length are especially important to be determined, and both are related to the geometry of the incoming waves. Admiral F. E. Páris (1806–1893) invented a device to record the shape of waves going through his machine. It was based in a very long wooden pole, immersed vertically in the sea, remaining substantially immobile. A wooden plate of annular shape was then attached to the extreme of the pole, connecting the upper end with a suitably stretched elastic rubber thread, hence attached to instruments that registered the vertical movement of the pole. The atlas depicts a full elevation of the machine, as well as some details of the mechanisms making functional the whole system (Fig. 2).

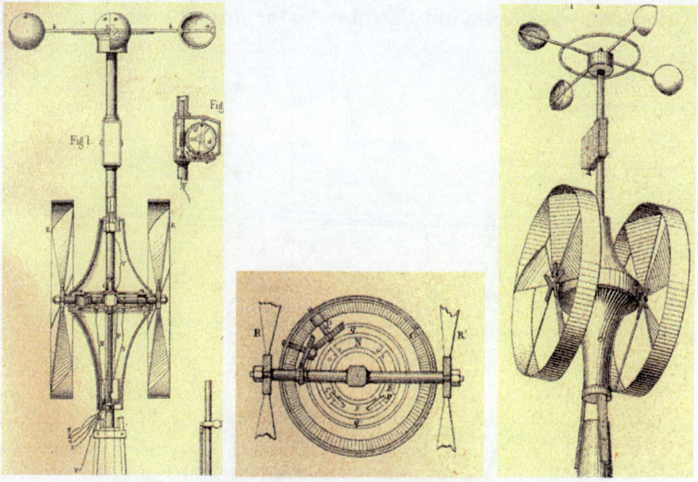

Fig. 1. Elevations, plan and perspective of the Robinson anemometer [9]

Another ingenious machine was created in 1842 by M. Aimé in Algiers harbour to recognize the depth under a vessel by the degree of agitation caused by the waves. It had to be installed in a well sized ship floating in a place where we could register water

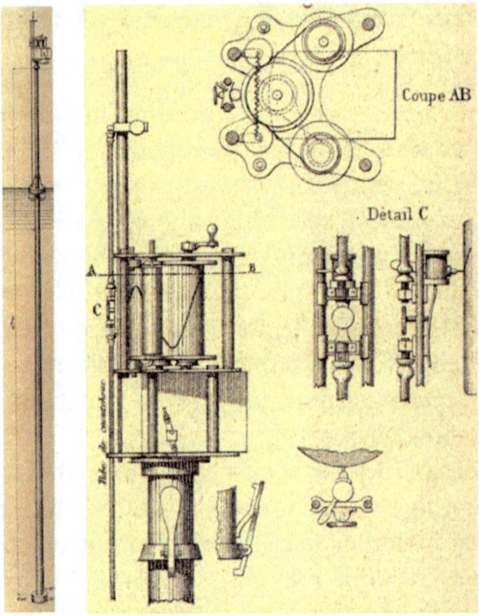

Fig. 2. Elevation and details of the admiral Páris machine [9]

agitation levels. A conic shaped piece, a little lighter than water, should be attached by its top using a short wire to a horizontal plate, usually covered in lead material. Iron weights were hanging from the corners of the plate, according to the perspective shown in Fig. 3. As the agitation registered by inclination of the conic piece decreased, the depth under the ship was known to be greater. So the machine would alert of less deep

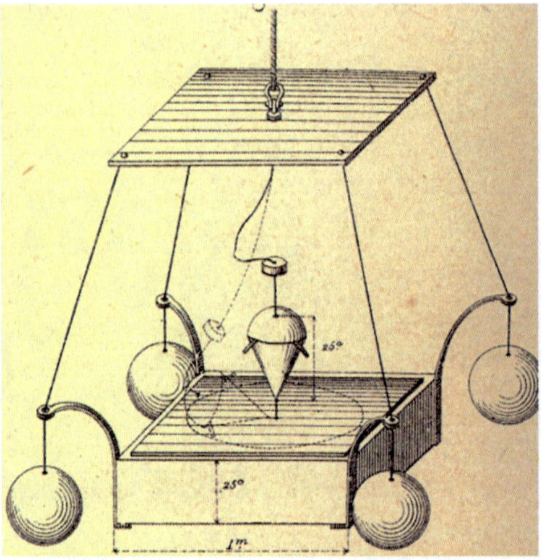

Fig. 3. Perspective of the Aimé machine [9]

areas showing greater inclinations of the conic shaped piece, according to tables created by Mr. Aimé.

2.2 Machines Used During Ports and Harbours Construction

The steam technology became extremely important to deploy traction power to the materials transportation, being probably the greatest of the 19th century innovation in pier construction engineering techniques. We will bring examples of transportation of different sizes of material used in pier building.

First, some techniques used sand or other granular materials to fill the territory where the pier was projected to be built. In these situations, little locomotives where used to transport loaded hopper buckets. The schematics of the construction methodology used for the construction of the Wicklow harbour in Ireland can be found as Fig. 4.

Fig. 4. Elevation of filling construction system using locomotives in Wicklow (Ireland) [9]

For bigger sized materials transportation, such as in the case of big stones, railroad-guided cranes were built. The design (Fig. 5) considered the bending moments balance of the whole transportation system, so the crane would not overturn by lack of equilibrium. The following machine was used during the construction of the port of Brest, in Brittany.

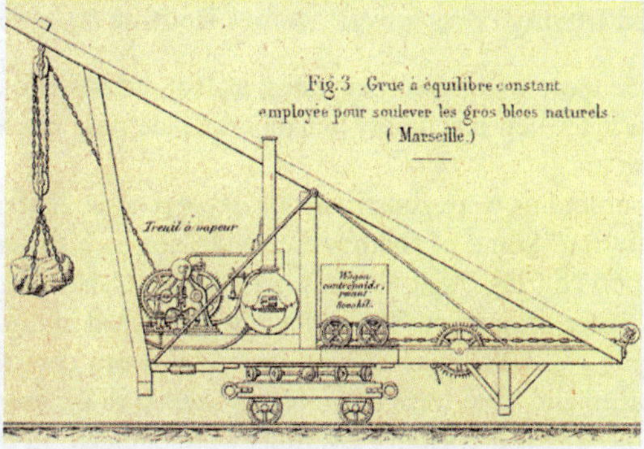

Fig. 5. Steam-powered railroad crane in Port of Brest [9]

The biggest sized blocks were transported using complete railroad rolling systems, including bigger sliding cranes (Fig. 6). The following machinery system was used to move and place into position the blocks needed for the piers of the Leixòes port, in Portugal. The graphic description in the atlas includes plan, elevations and several details of the wheeled transportation system, as well as the rotation mechanisms to allow the crane to change positions as needed. As the weight of the iron cast machinery was rather high, some of the details show the arch reinforcements needed for the whole structure to behave solidly and in equilibrium under all circumstances of load.

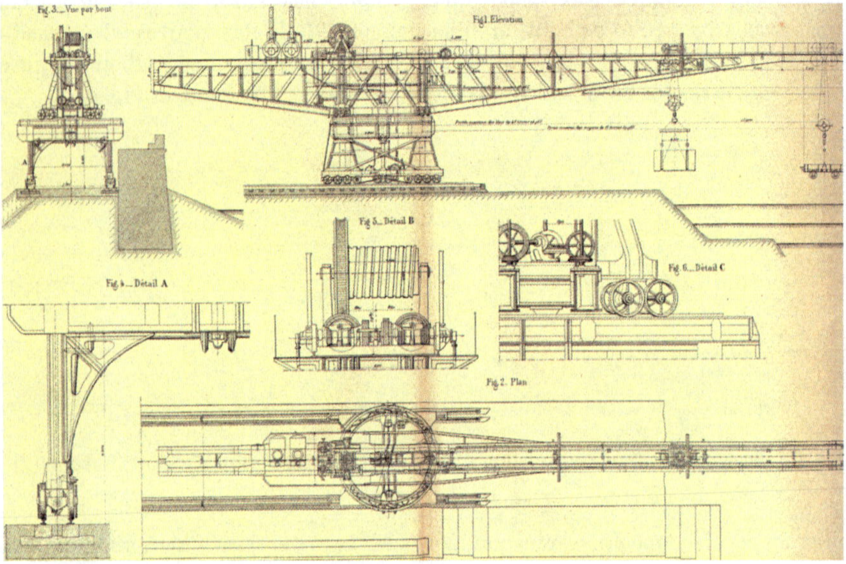

Fig. 6. Railroad big blocks transportation crane system in Leixòes (Portugal) [9]

2.3 Machines Used During Ports and Harbours Routine Operations

Although some of the more traditional machines, as load cranes, are described in these atlases, we will focus in the new machinery that modern materials and steam technologies created in 19th century.

Big breakwater structures were constructed to protect some harbours from extreme weather wave generation. Some of them were made not to last, just as an auxiliary structure to protect ports during construction works. Other breakwater structures stayed as permanent water surge protection. The following were used in the port of La Ciotat, next to Marseille (Fig. 7). Lattice-reinforced structures were connected by chains to avoid relative displacements, and were anchored to seabed in different points to assure the system as a whole was static. Both elevations were depicted in the atlases, as well as a plan distribution of the structures and anchor points (Fig. 8). Finally, a complete perspective of one of the breakwater structures was shown (Fig. 9). The complexity and weight of these structures were so high that they could only be made possible with an extensive use of steam-powered transportation systems.

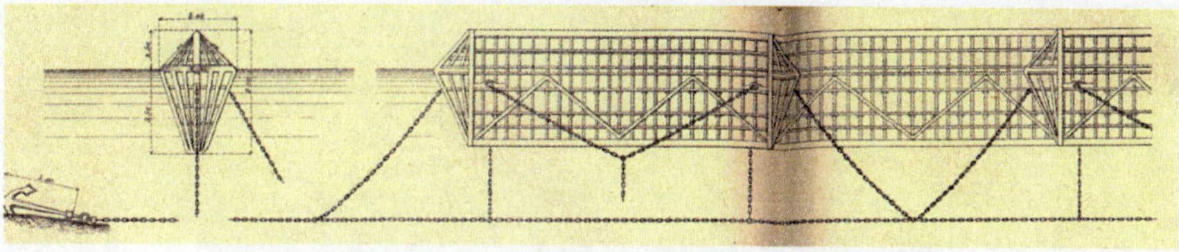

Fig. 7. Breakwater structure in La Ciotat (elevations) [9]

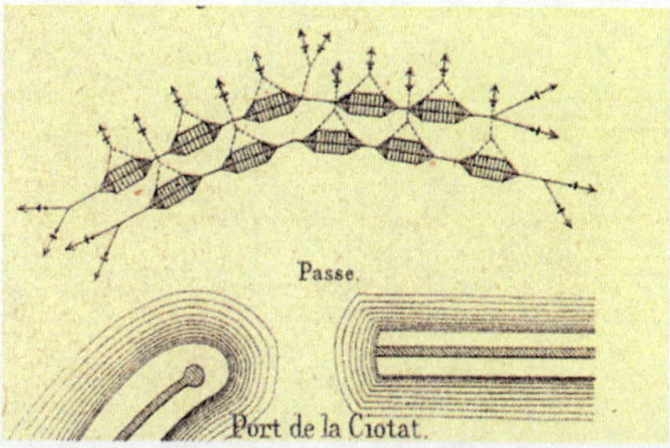

Fig. 8. Distribution of breakwater structures and anchor points in La Ciotat (plan) [9]

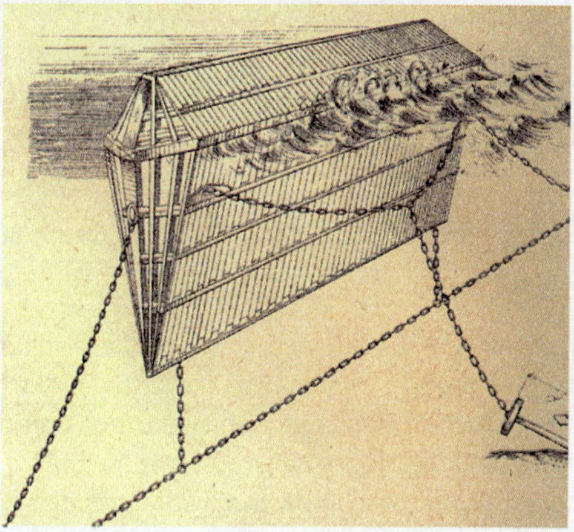

Fig. 9. Perspective view of breakwater structures in La Ciotat [9]

Probably, the floodgate systems were under their greatest changes during the 19th century. Some harbours could be improved, and some others fresh built as the floodgates measures were made much bigger in size. Bold designs were created in some English ports, Liverpool being one of the most important due to the paramount industrial activity taking place in the harbour areas. The following design was constructed in Liverpool

(Fig. 10), with a height of 11,51 m and a width of 30,48 m. Elevation, plan and three sections are shown.

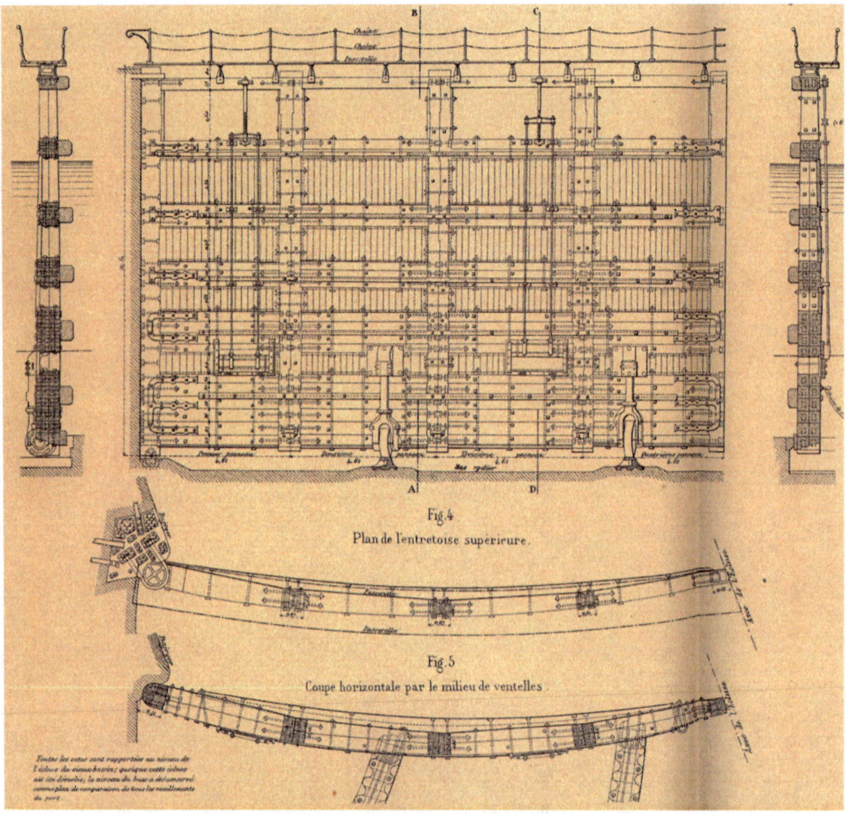

Fig. 10. Plan, main elevation and sections of Liverpool port floodgate [10]

The new steam traction systems allowed some of the bridges needed for road or railroad access to ports to be mobile. Henceforth, rotational and wheeled mobile bridge designs were built in the ports like Caen and Saint-Nazaire (France), respectively [11]. The possibility of making mobile bridges allowed for more bold designs, allowing also the displacement of access structures to allow navigation through floodgates or specially located port entrances.

The Caen rotational bridge is graphically defined by plan, main elevation and details of the rotation pivot structure, and was specially built over the Orne floodgate to allow navigation while vehicles access was not needed (Fig. 11).

The Saint-Nazaire wheeled mobile bridge could be slowly displaced using a steam boiler room installed in the vicinity. The main elevation and plan are referred (Fig. 12), as well as the steam industrial machinery attached to it (Fig. 13). Compression pumps chambers and boilers were included in the facilities.

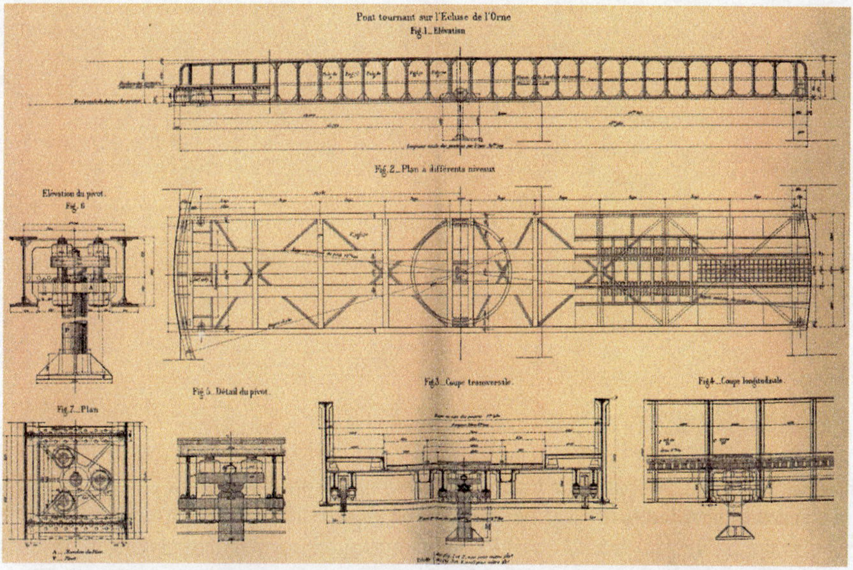

Fig. 11. Plan, main elevation and details of Caen rotational bridge [10]

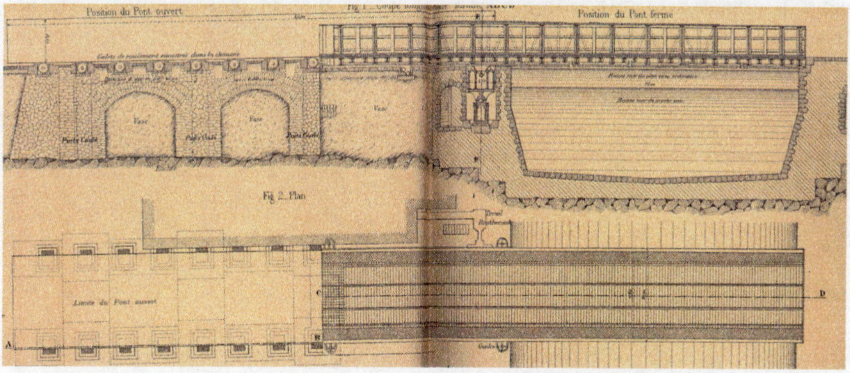

Fig. 12. Plan and main elevation of Saint-Nazaire mobile bridge [10]

As the technology became more usual, other ports installed these mobile bridge designs, as Le Havre in Normandy. Other interesting designs that can be found in the atlases are steam bells to be used as alarms in harbour installations [11], dry docks, railroad passenger and merchandise terminals, warehouses and coastal protections for beaches and military or strategic places of interest. Very advanced floodgate systems were installed in the Calais port, used by France and England as main commerce exchange international gate.

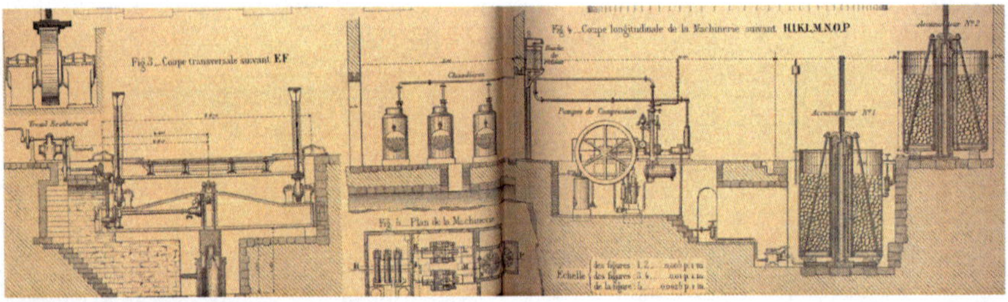

Fig. 13. Sections of the steam machinery for Saint-Nazaire mobile bridge, including boilers and compression pumps [10]

3 Conclusions

The new materials, construction techniques, steam-powered industrial designs and general scientific knowledge created an exponential growth for ports construction engineering techniques. New machines were designed to acquire environmental data, to improve construction methodologies or to allow better and easier ports routine operations. These designs were registered in engineering treaties used in the Great Schools of Engineering across the whole continent, including atlas volumes. The digitalization of these documents is essential to fully comprehend the importance of these facilities, and how to preserve the industrial remains of the steam era as part of the civil and industrial engineering history.

Acknowledgments. The author wants to acknowledge the support received from Mr. Miguel Giménez-Yanguas, retired professor of University of Granada, as the monographies quoted and digitalized in this paper belong to his own personal library.

References

1. Decret, F.: Carthage ou l'empire de la mer (Series: Points Histoire). Ed. De Seuil (1977)
2. Chatzis, K.: École des Ponts et Chaussées, in: Encyclopaedia Universalis, Ed. Encyclopaedia Britannica, Paris (2019)
3. Allen, J.: The life and work of Osborne Reynolds, in: Osborne Reynolds and Engineering Science Today. Manchester University Press, Manchester (1969)
4. Coriono, G.: 250 ans de l'École des Ponts et Chaussées en cent portraits. Presses de l'École Nationale des Ponts et Chaussées (1997)
5. Znachko-Lavorski, I.L.: Очерки истории вяжущих веществ. От древнейших времен до середины XIX века. Ed. Издательство Академии Наук СССР (1963)

6. Pasley, C.W.: Observations on lime, calcareous cements, mortars, stuccos, and concrete, and on puzzolanas, natural and artificial. Ed. John Weale (1838)
7. Boylston, H.M.: An Introduction to the Metallurgy of Iron and Steel. Wiley, Hoboken (1936)
8. Laroche, F.: Travaux Maritimes, Phéhomenes Marins, Accés des Ports, vol. 1 (Series: Encyclopédie des Travaux Publics). Baudry & Compagnie (1891)
9. Laroche, F.: Travaux Maritimes, Phéhomenes Marins, Accés des Ports, vol. Atlas (Series: Encyclopédie des Travaux Publics). Baudry & Compagnie (1891)
10. Laroche, F.: Ports Maritimes, vol. Atlas (Series: Encyclopédie des Travaux Publics). Baudry & Compagnie (1893)
11. Laroche, F.: Ports Maritimes, vol. 1 (Series: Encyclopédie des Travaux Publics). Baudry & Compagnie (1893)

Power and Performance the Water Lifting Machines Used in Ancient Mining in the Southwest of the Iberian Peninsula

Angel Mariano Rodríguez-Pérez(✉), Julio Jose Caparrós-Mancera, José Antonio Hernández-Torres, Juan Carlos Fortes-Garrido, Jose M. Dávila-Martín, and Javier Castilla-Gutiérrez

University of Huelva, Huelva, Spain
angel.rodriguez@dci.uhu.es, julio.caparros@diesia.uhu.es, {joseantonio.hernandez,jcfortes,jmdavila, javier.castilla}@dimme.uhu.es

Abstract. One of the main problems that were faced by former miners was the extraction of water in mines. When Romans arrived at the Peninsula, they introduced new technology that gave solution to the problem. However, this technology was not created by them, but this had already used in the development of other activities in the Hellenistic world in the Eastern Mediterranean. Main drainage's systems for indoor works in the mine were: Archimedes' screw, the waterwheels, bucket pulley and Ctesibius' pump. Machines are analysed from their main features and operating way. All of that without forget the most important discoveries in the southwest of the Iberian Peninsula, one of the places that biggest mining use has had throughout history. This work thus provides a qualitative and quantitative comparison of the main mechanisms for mining in ancient Huelva, supposing a basis for directing research in this area.

Keywords: Iberian pyrite belt · Roman times · Water wheel · Ctesibius pump · Archimedes' screw

1 Introduction

The mining activity in the southwest of the Iberian Peninsula, where Huelva is located nowadays, started on the third millennium B.C. with the extraction of copper and precious metals [1, 2]. First mining was approached in superficial way and using primitive technology. It was not until second century B.C., with the arrival of the Romans to the south of the Peninsula, when the mining activity begins to be noticed [3, 4], introducing new techniques to solve the problems that originate the exploitation of this sector at greater depth. One of the main problems that had to face the mining exploitation was the drainage of the water, because of the arrival of new techniques of deepening. The extractive capacity was raised, and in many cases the miners exceeded the water tables. This could even cause the inundation of underground works, with subsequent of exploitation cessation.

M. Ceccarelli and R. López-García (Eds.): HMM 2022, HMMS 40, pp. 184–196, 2022.
https://doi.org/10.1007/978-3-030-98499-1_16

At that time, the different drainage some systems were simple and elementary methods, such us manual extraction through buckets, rope baskets or cubes. Also more elaborated methods were used like the conduction of water through wooden channels and drainage galleries, or the use of machinery to lift them abroad. When the configuration of the terrain prevented the use of the tunnel's systems and galleries, it was necessary to drive the water to a specific point of the mine, or elevate it outside. Lifting mechanisms were used, less stressful and more effective than manual transportation.

The first drainage mechanisms in mining were found in Egypt during the Ptolemaic era. Engineers from Alexandria's School designed some mechanisms made of wood, like a wheel to evacuate water from the inside of the mine. Later, the Romans used the same system in mining works in Britannia (nowadays Great Britain), Dacia (nowadays Romania) and the southwest of Hispania (Riotinto, Tharsis and Sâo Domingos, Spain and Portugal).

The main relevant sources of knowledge in mining tasks are the X books written by Vitrovius, "Architecture of Vitrovius", (Roman architect, I B.C.) along with the work of J. Gonzalo and Tarin (1888) and R. Palmer (1927) in which they noted and graphically described the remains of ancient mining found in Huelva's. However, these devices were expensive and could only be used where the ore was rich [5, 6]. The main drainage machineries documented in Roman times were, Archimedes' screw, Ctesibius' pump and the water wheel [7]. All these water devices were previously used in the Hellenistic period, but the Roman technicians introduced them in mining, also improving their effectiveness. Therefore, the technical contribution of Romans was based on the usage of machines.

The main objective of this article is to review the mechanical systems of water extraction in the SW mines of the Iberian Peninsula in Roman times, an area where there was a lot of mining exploitation. Once the main methods are described, a comparison between the systems is carried out, comparing efficiency, depending on their lifting capacity, flow and energy consumed for its operation.

2 Materials and Methods

Several documentary resources were used to locate the bibliographic documents. Historical Mining Archive of the RioTinto Foundation, in Huelva, provided a massive amount of information, saved by the companies that exploited the Riotinto mines and other mines of Huelva. The Andalusian Historical Heritage Institute (IAPH) was also consulted about the mines.

A recent bibliographic research was made in various databases such us Scopus, Google scholar, Web of Science, Journal citation reports using the descriptors: Water lifting machines, Iberian pyrite belt, Water wheel, Ctesibius pump, Archimedes Screw. A search was also carried out in 'Google scholar' search engine with the same terms. The record obtained were sifted to the Roman times, and ranged between 1850 and 160 records after the combination of different keywords. Documents were selected about relevant information of the water lifting machines used in the SW mines of the Iberian Peninsula in Roman times, its design, construction, applications and lifting capacity.

2.1 Archimedes' Screw

It was invented in the third century B.C. by Archimedes, from which it receives its name, although some similar devices were found in the ancient Egypt. Archimedes based his design on theory, which allows a rational usage [8, 9]. This device is a helical gravimetric machine used to lift water, flour, cereals or excavated material and was based on a hollow cylinder and an endless screw, Fig. 1.

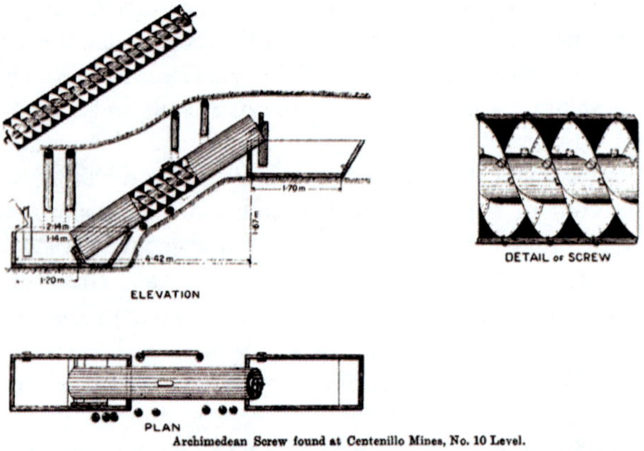

Fig. 1. Description of the composition and operation of the Archimedes' screw [10].

Generally, it was constructed of wood, although in some cases the internal helix was made in metal. Eight parallel lines, equidistant from each other, were drawn along in a trunk. These lines were cut by others, also equidistant, and separated by an eighth of circumference of the axis. At the intersection points of some lines and others were passing some slats, of flexible wood, fixed firmly and glued together with pitch. These wood's strips were those that being fixed in spiral on the trunk, formed the snail or screw. Finally, this piece was covered with wooden slats (convex tables) forming envelope cylinder of the previous piece. The whole set was tarred and solidly held with a rolled rope. It did not raise much, but it could move a large amount of water.

Vitruvius recommended eight partitions for his device, the ends of his axis had to be of iron and once mounted and his casing had to be covered in pitch to avoid leaks. The best inclination according to Vitruvius was based on the Pythagorean Theorem, being about 37°. They were placed in rows, Fig. 2, and they loaded the water from a lower tank to an upper tank, from where it was covered by another screw.

It was activated by human force. A slave worker used a bar placed on top, providing movement, Fig. 3.

From this mechanism, three devices were found in Huelva, all of them associated with a bucket pulley, near the pit of San Juan in Sotiel Coronada's mine (Calañas, Huelva) [13]. Of these three, only one in the Huelva Museum is relatively intact. From the others, pieces are preserved in the Museum of Transport in Glasgow, the British Museum, the Musée des Arts et Métiers in Paris and the Provincial Museum of Huelva. Some copies

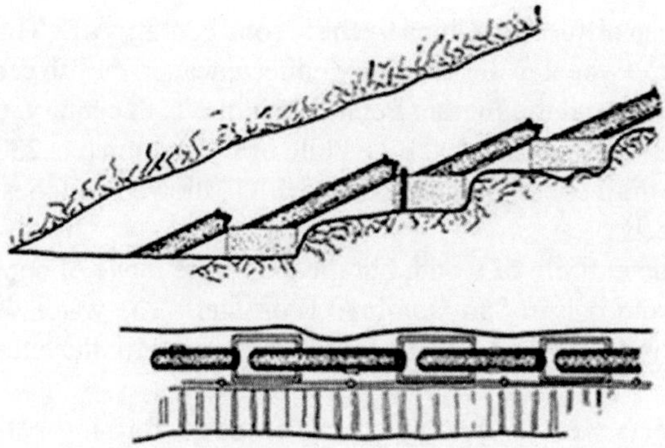

Fig. 2. Archimedes' screw battery [11]

Fig. 3. Operation of the Archimedes' screw [12].

of this mechanism were also located in the Cerro Muriano's mine, in Cordoba Copper Co. And also in Centenillo's mine (Jaen) in Centenillo and Castulo Group, where the best preserved screw can be found. The one that is in the museum has a different size than Vitruvius. This one has a ratio of 1/15, the partitions that form the spiral are three instead of eight and are formed by twenty-five layers of wood sheets in 12 m of height [12]. These plates are fixed with bronze nails greased on the cylinder, and it is covered in pitch together with a rope covered in pitch too, to secure its sealing. Its inclination was 17º.

2.2 Water Wheels (Rotate Aquae)

The water wheel is a machine used to raise water, formed by a series of buckets attached to a large wheel. It raises the water from the bottom to the top. A wheel of this type is similar to a carriage wheel, with larger dimensions and with the addition that it has buckets to catch water.

Several civilizations claim the invention of the wheel. There are Indian texts dating from 350 B.C.; Joseph Needham believed that the wheel was developed in India during the fifth or fourth century B.C. [14]. He assumed that it spread to the west by the first century B.C. and then diffused to China by the second century A.D. This was followed by widespread use of the wheel in the Eastern Mediterranean in the 5th century A.D., before reaching North Africa and the Iberian Peninsula in the 11th century. Other possibilities of its origin include the Near East 200 B.C. Philo of Byzantium (ca. 230 B.C.), a Hellenic engineer of the late third or early second century B.C., showed sketches of several distinct types of water wheels.

They were made entirely of wood, but the axis were made of copper (Fig. 4a). The wood was taken from oak in Sâo Domingo (Portugal). The water wheel preserved in Huelva, has three types of wood, walnut for the cube, pine for the buckets and fir for the radios.

The water wheels were preferably placed in pairs, forming batteries. They were placed in different steps. They lifted the water from one step to the next, communicating with each other through short galleries and channels. The couples of water wheels turned in the opposite direction (Fig. 4b), so they put the water in the cannel always with the same direction. Its assembly was made in situ due to its large diameters.

(a) (b)

Fig. 4. **(a)** Finding and reconstruction "in situ" of the Roman water wheel found in Riotinto in 1886. (Archivo Histórico Minero Fundación Río Tinto. Minas de Riotinto, Huelva). **(b):** Water wheels with compartments rim; representation based on archeological findings in Spain (adapted from [14])

Some boards were fixed on the running surface, joining the ends of spokes. These tables were used as steps receiving the impulse of the foot of the worker who operated it, sitting at a higher height on the wall in front of the wheel. This was helped by another worker, located in the part of the wheel's axis, which moved the spokes with ropes [12]. Their performance was relatively high, since they could raise large amounts of water to a considerable height (3–4 m).

The water wheels found in the Iberian Pyrite Belt, have been defined as "Hispanic type" and are characterized by being taller, slender and narrow. They have a diameter that varies between 3,60 and 4,65 m. They have a bucket between 20 and 30 cm. Their weight ranges around 200 kg and their number of buckets and spokes are between 22 and 30 [13]. Between 1850 and 1860 ten wheels were found in Sâo Domingo's mine (Mértola, Portugal) the wheel is currently kept in the Musée des Arts et Meters in Paris [15].

About 50% of water wheels discovered in the Iberian Peninsula have been discovered in Huelva. Ten water wheels were discovered in 1875 in Filón Norte's mine. Two fifths of the wheel are preserved in the Museum of Transaport in Glasgow. Later in 1886 at the Lagunazo's mine (Alosno), a pair of wheels similar to those was found in Tharsis [13].

About 50 wheels have been found in Riotinto, but only two are preserved and only one third of them. Also a bronze axis is preserved in the British Museum.

A wooden axis is preserved in the Riotinto Mining Museum too, and eight bronze axis, three wooden discs and a complete wheel found in Masa Planes are showed in the Museum of Huelva (Fig. 5) [16].

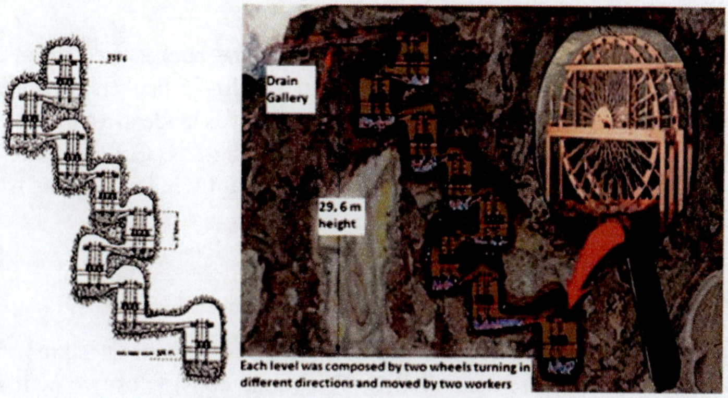

Fig. 5. Roman wheels in Riotinto [16]

Apart from these remains that have been maintained, several pairs of wheels have also been found, as the battery of eight pairs found in Filón Sur or Nerva. These 16 wheels allowed them to move the water 29,6 m from level 309 to 338,9 above sea level.

We need to know the volume of each bucket, the number of buckets, the diameter of the wheel and rotation speed for the calculation of the flow of a water wheel. It is a variant of the previous one, used when the wheel was not effective due to the depth or because the loading area of the wheel had a low level of water, which prevents the buckets to load water correctly. It consists of a hydraulic wheel with a metallic double chain on which were hanging some bronze buckets with a capacity of 3,5 L. Only a few copper buckets found in the Cabeza del Agua (Riotinto) and in Sotiel Coronada's mine (Calañas) are preserved nowadays. This device was found combined with a set of Archimedes' screws.

2.3 Bucket Pulley

The bucket pulley is a variant of the wheel for situations in which the amount of the water extracted was very small, Fig. 6.

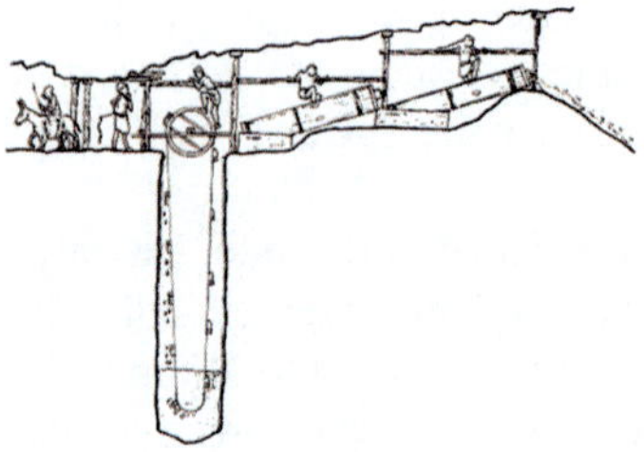

Fig. 6. Draining system combination, Bucket pulleys and Archimedes' screws (Sotiel Coronada's mine, Huelva). [12].

It must be taken into account that the volume of the bucket used to be 3,5 L. In addition, the few machines that are still preserved had 20–25 buckets. Despite the few pieces that are preserved, Sotiel Coronadas' bucket pulley is made in a single piece when it is usually made in two. You can also see two engravings: "Q.CORNELLIV (s)" and "L.VIBI. AMARANTI. P. XII S." [12]. It is believed that these inscriptions were made by the pulleys manufacturers.

2.4 Ctesibius' Pump

According to Vitruvius, his invention is attributed to the Hellenistic Ctesibius of Alexandria in the middle of the third century B.C. This pump consists of two cylinders with pistons that were moved by means of connecting rods attached to opposite ends of a single lever [17]. These pistons converge in a common chamber, in whose entrances there are non-return valves, allowing the water to take a single direction. The central chamber is attached to a nozzle through which the driven water comes out. The nozzle can be oriented in the desired direction and increase the water speed about 25 times, due to the narrowing section. These can be made of copper or lead. To manufacture the perfectly cylindrical pistons they used the procedure of lost wax.

The pump was moved by hand, operating a wooden lever with a swinging motion. Then the water ascends by the piston. The other piston impelled the water contained in the central pipe. Securing in that way a continious flow. By their operation mode, they worked only with clean water.

One of the two pumps of this type that are preserved in Spain, were discovered with a masonry work on the third floor, sleeper 25, Sotiel Coronada's mine in Calañas, Huelva, Spain, [13]. This is the most important device preserved of the Roman world (Fig. 7). It is currently located in the Archaeological National Museum of Madrid.

This pump found in Huelva was made in bronze and its size was 0,95 m high and 0,41 m wide. It had a flexible tube with the form of a T placed at the end of the main tube. Due to their characteristics, it is believed that the main use of this pump was not to

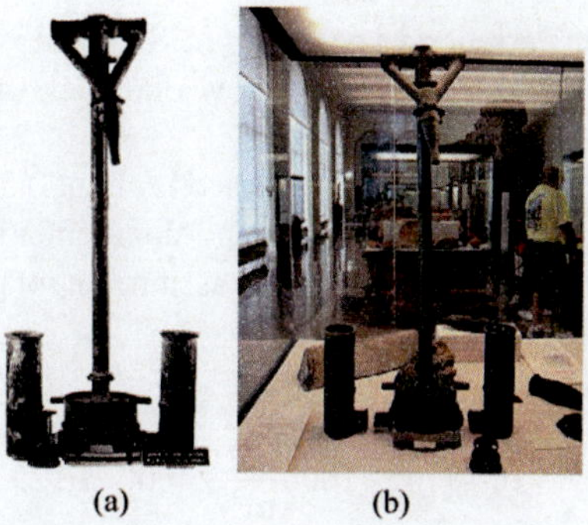

Fig. 7. Sotiel Coronada's pump, Calañas a) Virtual Archive archaeological national museum and b) Archaeological national museum, (Madrid)

drain water but to pulverize it and cool the pyrite in the demolition system of the hard rock, where the action of the fire was complemented by the water [12].

3 Results and Discussion

Once we have seen the main water extraction machines Roman times mines in the SW of the Iberian Peninsula, we are going to see their performance and capacity and we are going to compare them to each other.

Humans have a limited physical power output, which can be in the range of 0,08 to 0,10 HP. This power can be used to raise water (0,08 to 0,1 HP gives between 60 and 75 watts).

- **Archimedes' Screw**

Talking about size proposed by Landels J. [14] and in the formulation of helical screws mentioned in "Elevators: Principles and innovations" [18] the main operating parameters of Archimedes' screw can be calculated.

The following data are useful: hb = 1.16 m (Lifting height), L = 2.4 m (Screw length), $\varphi = 37^\circ$ (Inclination angle). With these values and the following expressions we can obtain the water flow that the Archimedes' screw was able to raise.

$$Q = V \cdot S \cdot \rho \cdot 3600 \cdot i \tag{1}$$

$$S = \frac{\pi \cdot D^2}{4} \tag{2}$$

$$V = \frac{P \cdot N}{60} \tag{3}$$

Where:

Q = Flow (l/h); V = Peripheral speed (m/s); S = Area (m^2); ρ = Water density (kg/m^3); i = Inclination parameter; D = Screw diameter (m); P = Step (m); N = Angular speed (rpm).

The usual diameter of similar machines studied is about 0,5 m, the distance between blades can be assumed approximately 0,25 m and the inclination parameter is 0,25 for 40°. This machine was activated by people so we assumed a rotation speed of 10–14 rpm. We are going to use the Eqs. (4), (5) and (6).

The results obtained are these.

$$Q = 0,058\frac{\mathrm{m}}{\mathrm{s}} \cdot 0,19\,\mathrm{m}^2 \cdot 1000\frac{\mathrm{kg}}{\mathrm{m}^3} \cdot 3600 \cdot 0,25 = 9918 l/\mathrm{h} \tag{4}$$

$$S = \frac{\pi \cdot 0,5^2}{4} = 0,19\,\mathrm{m}^2 \tag{5}$$

$$V = \frac{0.25\,\mathrm{m} \cdot 14\,\mathrm{rpm}}{60\,\mathrm{s}} = 0,058\,\mathrm{m/s} \tag{6}$$

These results are like Landel's 10.000 l/h. Landels used the same size so we can consider it a valid calculation method [14]. Using these values, the theoretical power can be calculated and compared later with the other devices studied using the following expression:

$$Pt = \rho * \mathrm{g} * \mathrm{Q} * \mathrm{hb} \tag{7}$$

Where:

ρ = Fluid density (kg/m^3); g = Gravity (m/s^2); Q = Flow (m^3/s); hb = Lifting Height (m).

$$Pt = \rho * \mathrm{g} * \mathrm{Q} * \mathrm{hb} = 1000 * 9,81 * 0,0027 * 1,16 = 30,69\,\mathrm{W}$$

For water density of 1000 kg/m^3 and 9,81 m/s^2 The power value is 30.69 W. Finally, we use the expression (8).

$$\mathrm{Pr} = \frac{Pt}{\eta} \tag{8}$$

Where:

Pr = Real power (W); Pt = Theoretical power (W); η = performance.

This machines were made in wood and knowing that their power came from human people the performance was too low. In this case we assume a performance of 70%, like other writers did [19]. So the real power developed in this conditions and calculated by (8) has a value of 43,84 W.

- **Water Wheels**

According to the biography "Las ruedas de achique romanas de Riotinto" (P. Manzano, et al.) the usual volume of a bucket reached a capacity of 7 l. and the number of buckets, 30 units. The total volume will never be covered so a coefficient of 0,75 is applied. [6] "Note on some Ancient Mine Equipments and Systems".

$$Volumen_{wheel} = volumen * 0,75 = 5,25l \tag{9}$$

$$Volumen_{wheel} = Volumen_{bucket} * N^{\circ}_{bucket} = 5,25 * 30 = 157,5\,l \tag{10}$$

In a complete round we would have 157,5 L of water. Knowing that the diameter of the wheel was approximately 4,5 m we can assume that the drive speed per person should not be too high. For this study one complete lap per minute has been assumed.

$$Q = \frac{157.5\,\mathrm{L}}{1turn} * \frac{1\ \mathrm{turn}}{\mathrm{min}} = 9450l/h \tag{11}$$

We obtain a value a little higher than the proposed one by Palmer but of the same order of magnitude [6]. The difference is made due to the performance used by the author or to various losses that are not discussed in this article.

Knowing that the lifting height for a wheel is 4,5 m and using the expression (7), You get a power of:

$$Pt = \rho * \mathrm{g} * \mathrm{Q} * \mathrm{hb} = 1000 * 9,81 * 0,0026 * 4,5 = 114,77\,\mathrm{W}$$

In the same way, and with the same performance in the section of Archimedes' screw (8) the real power of a water wheel is calculated. The value obtained is 163,95 W.

- **Bucket Pulley**

Using the same formulation as in the wheel (9), (10), (11) and assuming a diameter of the wheel of 1 m. driven by a person at a rate of 3–4 turns per minute and with a lifting height similar to 4,5 m and with about 25 buckets:

$$Q = 65,625l * 3\ \mathrm{turn} = 11812,5\frac{1}{\mathrm{h}}$$

Using (7) the power is:

$$Pt = \rho * g * Q * hb = 1000 * 9,81 * 0,0032 * 4,5 = 141,26\,\mathrm{W}$$

Again using (8) to calculate the power of a pumping machine, with a performance of 70% we obtain 201.8 W.

- **Ctesubius' Pump**

In this case for the calculation of the real and theorical powers, we begin with the flow data previously established by Martínez Luengo [20] These values are 3320 L/h for the flow and 2 m. of lifting hight (Boletín Arkeolan, 16, 2009–2010).

With these values we obtain the theoretical power from the Eq. (7).

$$Pt = \rho * g * Q * hb = 1000 * 9,81 * 0,00092 * 2 = 18\,\mathrm{W}$$

Applying again with a performance of 70% the real power obtained is 25.71 W.

If the rocker arm operates at 15 cycles per minute, the pump supplies 23 L. per minute and requires an input power of 54 watts. Working with a rocker, a man can probably produce between 40 and 50 watts [19], so it would be difficult for a man to handle this pump. However, two men could do it easily.

However if it is assumed, and they can control the machine, they could drive it faster (for example 20 cycles per minute). The output power will increase to 31 L per minute; and the power requirement increases to 71 watts.

With all this data we are going to try to evaluate which device was the most effective for the water extraction from the mines of the Iberian Peninsula.

Although, there is a question that admits many variants and special circumstances, the best answer we can reach is to compare the options visually.

Below we present a scoring method of flow, lifting height and real power.

A score of 10 is assigned to the best invention in each of the properties that are studied. We will use a rule of three to establish a realistic proportion between the results obtained. In this way, the highest lifting height will have a score of 10 and the rest will be weighted according to the lifting height they raise. In order to compare the three instruments studied, a weighting has been carried out base on the most effective values for each evaluated property. The highest values in lifting height, flow and lower values in power consumed will have a value of 10.

The following table shows the experimental values to be compared in a clear way, Table 1.

Table 1. Devices experimental comparison

	Water wheel	*Archimedes' screw*	*Ctesibius pump*	*Bucket pulley*
Flow (L/h)	9450	9918	3320	11812.5
Lifting Height (m)	4.5	1.16	2	4.5
Power (W)	163.95	43.84	25.71	201.8

Performing the predicted calculations one by one, the following weighted scores have been obtained, Table 2.

Table 2. Devices weighted scores

	Wheel	Screw	Ctesibius	Pulley
Flow	8	7	3	10
Height	10	3	4	10
Power	2	6	10	1
Total	20	16	17	21

4 Conclusions

The results obtained show that the best performing devices are the pulley and the wheel. This is because they move a large amount of water and they have a high lifting height. Their scores are ballasted by their power consumes. However at that time, the people who moved the device were not taken into account, so it should not be an engineering problem. In a second step are clearly the Ctesibius' pump and the Archimedes' screw. The height they are able to overcome makes them ideal for other applications in mining, such as mineral extraction in small galleries, but in terms of pure performance they are behind the wheel and the pulley.

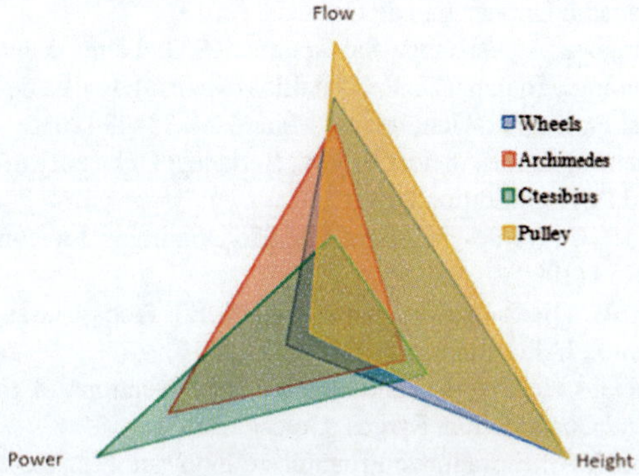

Fig. 8. Results graph

In this graph (Fig. 8) you can clearly see the result. It must be taken into account that the power is considered something negative in this study. The element that consumes the most power has obtained the least score. In this way you can see the result with the blink of an eye (the triangle with more area) and you can select the most efficient devices for water extraction in mines during the Roman era in the Iberian pyrite belt.

References

1. Nocete, F., Alex, E., Nieto, J., Sáez, R., Bayona, M.: An archaeological approach to regional environmental pollution in the south-western iberian peninsula related to third millennium BC mining and metallurgy. J. Archaeol. Sci. **32**(10), 1566–1576 (2005)
2. Nocete, F., Saez, R., Bayona, M., Peramo, A., Inacio, N., Abril, D.: Direct chronometry (14C AMS) of the earliest copper metallurgy in the guadalquivir basin (Spain) during the third millennium BC: first regional database. J. Archaeol. Sci. **38**(12), 3278–3295 (2011)
3. Mateo, M.O.: La minería antigua de Riotinto (Huelva), *De re metallica (Madrid): revista de la Sociedad Española para la Defensa del Patrimonio Geológico y Minero,* no. 3, pp. 23–32 (2004)
4. Rothenhoefer, P.B.M.a.H.N.: Old finds-new insights: remarks on two roman lead ingots from minas de riotinto (Huelva, España), Onoba. Revista de Arqueología y Antigüedad **4** (2016)
5. Gonzalo, J., Tarín: Descripción física, geológica, y minera de la provincial de Huelva., Madrid: Imp M Tell, p. 1331

6. Palmer, R.E.: Notes on some ancient mining equipmentsand system. Trans. Inst. Mining and Metallurgy, 1926–27
7. Fortes Garrido, J., Sabido Castilla, L., Dávila, F.B.J., Salmerón, I.: Máquinas de desagües utilizadas por la minería romana en la provincia de Huelva. Una Apuesta por el Desarrollo Local Sostenible., Huelva: Universidad de Huelva. (2010)
8. Ceccarelli, M.: Renaissance of machines in Italy: From Brunelleschi to Galilei through Francesco di Giorgio and Leonardo. Mechanism and Machine Theory **43**, 1530–1542 (2008)
9. Chon, T.G.: Archimedes life works and machines (2010)
10. Tamain, G.: Las minas antiguas de El Centenillo (Jaén). Oretania pp. 23–24 (1966)
11. García-Romero, J.: Tesis doctoral: el papel de la Minería y la Metalurgia en la Córdoba Romana., Córdoba: Univ. De Córdoba. (2002)
12. Arboledas-Martínez, L.: Tesis Doctoral: Minería y Metalurgia romana en el Alto Guadalquivir., Granada: Universidad de Granada. (2013)
13. Domínguez-Domínguez, A., Guisande-Santamaría, P.S.G.: La noria del museo de Huelva, un unicum en la arqueología romana. De Re Metallica (Madrid): revista de la Sociedad Española para la Defensa del Patrimonio Geológico y Minero **24**, 33–43 (2015)
14. Landels, J.: Engineering in the Ancient World., Barkeley: University of California, Barkeley, Edited in 1981 and Revised Edition 2000)
15. Domergue, C., Binet, C., Bordes, J.: La roué de Sâo Domingos. La revue., Musée des Arts et Nétiers **27**, pp. 49–59 (1999)
16. Manzano-Beltrán, R., Ojeda-Calvo, R., Rodríguez, E.: Trobajo, Las Ruedas de Achique Romanas de Riotinto., IAPH. Junta de Andalucía. (2010)
17. Kotsanas, K.: Ancient Hellenic Technology — The Inventions of the Ancient Hellenes: Research, Study and Construction, Pyrgos, Greece (2013)
18. Miravete, E., Larrodé, A.: Elevadores: Principios e Innovaciones., Reverté. (2007)
19. Stein, R.: Roman Wooden Force Pumps. Use and Performance. Brun J.P, Fiches (2007)
20. Martínez Luengo, L.: Revista Anales, 2012. https://revista-anales.icai.es/web/n_14/seccion_4.html

Mine Drainage and Steam Engine

David Calderón[1], Francisco Mata[1](✉), Luis Mansilla[1], Demetrio Fuentes[1], Issam Hanafi[2], and A. Carrasco[1]

[1] University of Castilla-La Mancha, Almadén, Spain
francisco.mcabrera@uclm.es
[2] University Abdelmalek Es-Saadi, Aloucema, Morocco

Abstract. The mine can be regarded as a testing ground for many technological advancements. Mine drainage is one of the basic tasks to improve the exploitation.

Traditional drainage systems are reviewed and then reference is made to the use of steam engines more specifically. There is no doubt that drainage is one of the earliest applications of steam engines, and mines, a favorable environment for its evolution and development.

Keywords: Mining · Drainage · Steam engine

1 The Work of Drainage in the Mines

As the mining work deepens and drops below the groundwater level, water begins to seep into the rocks, making the work difficult and even flooding the working area. To avoid having to stop the area exploitation, we need to bring the water to the surface by what is called drainage. This work should be carried out continuously, draining water as it was produced to avoid accumulation, and always keeping in mind the possibility of increased flow and bigger problems due to rainfall.

It was such the importance of drainage, that the engineers were forced to find or create new ways to carry out this work as economically as possible. The invention of the steam engine and its application represented a turning point in solving this problem. There are two types of drainage: natural drainage, for uneven land, where the water leaves the mine all by itself, taking advantage of the uneven plot or artificial drainage, which uses machines to drain the water out of the mine [1, 2].

Natural drainage can only be carried out in mines where the topography of the land allows sinkholes or drainage ditches to be formed. The shaft or passage was built with a very gentle slope for the water to leave the mine by gravity. Water was channeled through ditches or gutters carved in the ground. In cases of very high circulating water volume, a corridor was specially constructed for the draining pipe separated from the main sink.

The artificial drainage pipe uses a machine to pump water to the surface of the drainage hole or well according to the needs of the mine. The machinery used to drain water is very similar to the machinery used to extract ore and is sometimes even used to replace the container. In the case of very abundant water, the machine is dedicated to this function, so as not to damage the rest of the work. To reduce the height of pumping

M. Ceccarelli and R. López-García (Eds.): HMM 2022, HMMS 40, pp. 197–209, 2022.
https://doi.org/10.1007/978-3-030-98499-1_17

to the outside, water tanks or boilers are built on different floors inside the mine, close to the vertical line of the well, where the water from the lower floors was directed. The water was extracted by lifting water from one reservoir to another in a staggering manner until it was brought to the surface. Until the 18th century, techniques inherited from the Romans were used for drainages, such as wheels, Archimedes screws, pulleys, or hand winches. From these techniques on, it began the use of winches, pumps, and finally the steam machine or water columns [3].

When compared to the low performance attained by traditional methods involving a substantial amount of labor, higher performance was reached with the use of steam. Not only was it able to diminish it, but it was also able to drain from larger depths thanks to the steam. Although the genuine beginning point is the Newcomen machine in 1712, the earliest usage of a steam engine for lifting water in a mine is dated in 1611 by the hand of Ayanz. Steam is fast gaining popularity in England (1722), Slovakia (1726), Sweden (1726), Belgium (1726), and France (1732). The use of steam for drainage was used in all of these circumstances, despite the steam engine's significant reliance on fuel and water for the creation of steam [4, 5].

The steam engine was mostly used in coal mines, where the inexpensive cost of fuel allowed them to be used despite the large amount of energy consumed by the initial models. The advancements in the design of the steam engine resulted in a considerable reduction in fuel consumption, making this type of equipment suitable for use not just in coal mines but also in other types of mines [6]. After the Guadalcanal mines (1611), the Almadén mercury mines are the next known use in Spain (1785–1805).

Despite the significant improvements in drainage obtained with steam engines, the problem could not be said to be solved until the arrival of electricity in the twentieth century, with centrifugal pumps or submersibles.

In Figs. 1 and 2, some diagrams of the inventions used for the drainage of mines, at different times, are reproduced.

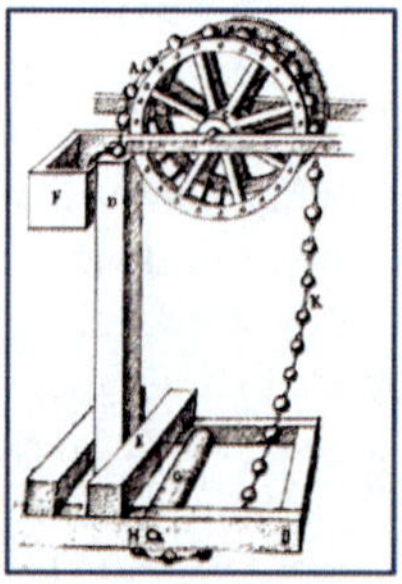

Fig. 1. Archimedes screw [7] and bucket wheel [8]

2 The Steam and the Movement

With the invention of the steam engine, human and animal powers were replaced as a source of energy for movement. Its many uses made it the most recognizable element of the Industrial Revolution.

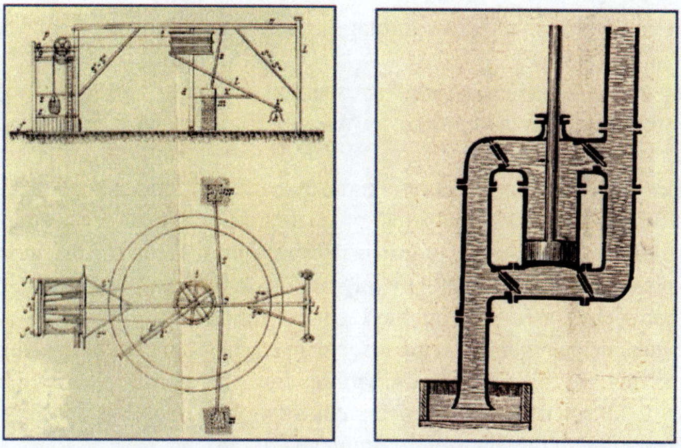

Fig. 2. Scheme of a horse "malacate" [9] and Impeller aspirant pump double effect [10]

The first evidence of the use of steam to generate movement can be found in ancient Greece. In a book from the second century B.C., Hero of Alexandria recalls the employment of a contraption capable of using steam to open and close the Temple's doors with a burning fire on its altar. Heron also referenced Heron's Eolipila, a steam turbine that was the first attempt to employ steam for motion generating [11].

What we don't know is whether these innovations were Heron's own, as he stated in his work that he aimed to collect both known and invented machinery. However, because they lack a practical application and some of them presumably did not go beyond a simple printed sketch, these devices have been seen as nothing more than simple gadgets or games [4, 12].

Figure 3 shows the Hero de Alexandria steam engine.

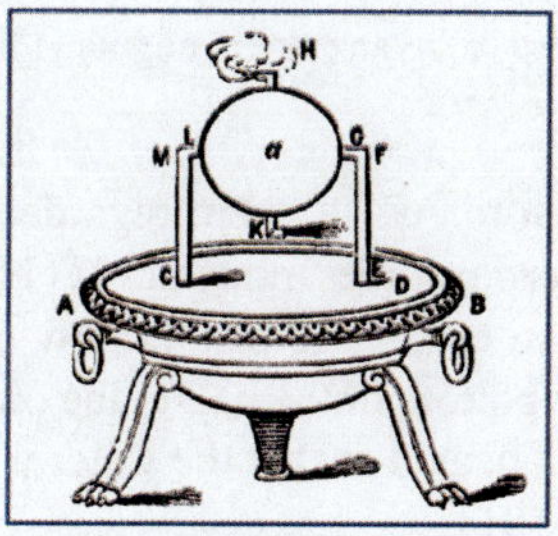

Fig. 3. Hero of Alexandria steam engine [12]

3 The First Steam Engines

There is no proof of steam use in mines until the 17th century. Thurston (1886) [12] notes that there are indicators that steam knowledge was not lost for such a long time and that the temporary gap on the beginnings of the steam force may be due to historians lack of interest at the time.

For many years, the steam engine was assumed to have originated in England, however other authors claim that Jerónimo de Ayanz, a Spaniard, was the true creator [3, 13, 15–17].

Ayanz received patent permission for more than fifty inventions by a Royal Certificate in 1606, the first of which was the patent of several types of machinery claimed to have been invented during the English Industrial Revolution. It should be mentioned that the majority of his innovations were not merely basic sketches, but he was able to put ideas into effect, being compared to Leonardo Da Vinci for this.

The one that can be regarded as the first steam engine is one of these inventions. This mechanism is thought to have been used in the drainage of Guadalcanal's silver mines since around 1611 when Ayanz founded a corporation to exploit it after being inactive for several years due to the inability of drainage [13, 14].

This machine (Fig. 4) consisted of a sphere-shaped boiler connected to two closed water tanks into which pressured steam flowed through valves and forced the water from the tanks to the surface. This functioning was similar to the Savery pump, which was invented in 1698 [18].

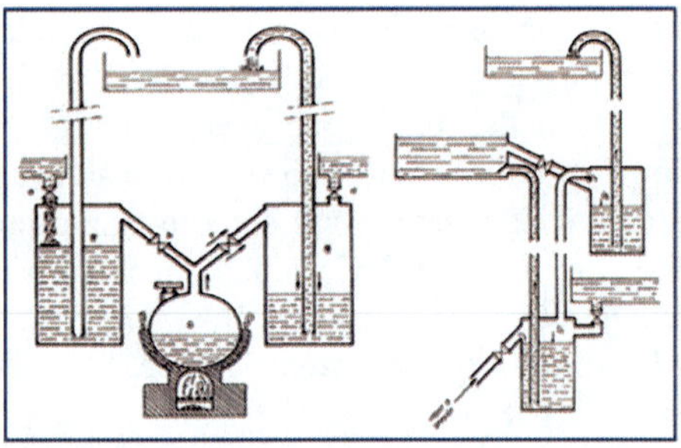

Fig. 4. Ayanz steam engine [17]

Other inventors, in addition to Ayanz, were regarded as forerunners of the steam engine. In his work "Les raisons des forces mouvantes" (1625), the Frenchman Solomon de Caus describes a mechanism consisting of a hollow copper ball with a vent and a tube that run from virtually the bottom of the ball to the outside. The ball was filled with water through the vent and then heated so that the water was raised to the outside by the tube using the force of steam [19].

Another steam engine is depicted in Giovanni Branca's work "Le Machine" from 1629. The machine (Fig. 5) was made up of a boiler in the shape of a human head that released steam through its mouth into the blades of a rotating horizontal wheel, which then moved a set of mechanisms that raised and lowered some mallets. Although this machine may be considered a precursor to the impulse turbine, it would be little more than a speculative machine due to the difficulties of making it work as intended [17].

Fig. 5. Giovanni Branca steam engine [12]

In 1630, Charles I of England gave David Ramsey a new patent for many ideas including the use of steam. An invention to raise water using fire is cited among them, but without a description, it is impossible to tell whether or not it was a real steam engine.

In 1663, Edward Somerset published a description of a way of raising water using a steam engine in his book "A Century of the Names and Scantlings of such Inventions" [20]. Figure 6 shows a illustration of Edward Somerset steam engine.

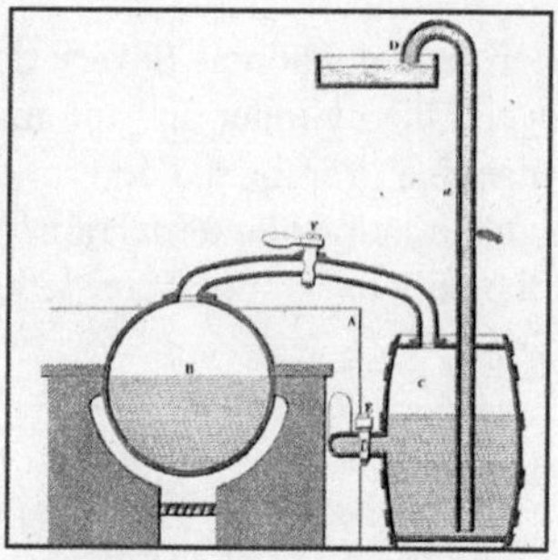

Fig. 6. Edward Somerset steam engine [12]

Another example of an invention that is not described is that of Samuel Morland, who created a system for producing water in 1682 which is quite similar to Edward Somerset's. In his work "Principles of a New Power of Fire, developed by Knight Morland in 1682, and delivered to H.M. Cristiansima in 1683," he made it obvious that he was well-versed in the "force of steam." [12].

4 Designs from the Late 17th Century

In 1690, Dionisio Papin, a French physicist, conducted experiments to prove that by introducing and condensing water vapor inside a cylinder containing a plunger with an upper face at atmospheric pressure, an alternative action of rising and falling happens when steam enters and condenses. However, Papin's suggestion did not apply to a usable machine because the cylinder itself worked as a boiler, and it was essential to have a heat source beneath the cylinder to heat the water inside it and generate the steam that elevated the piston to achieve piston movement. To cool the cylinder, the heat source had to be removed, and the cylinder had to be removed from this form (Fig. 7).

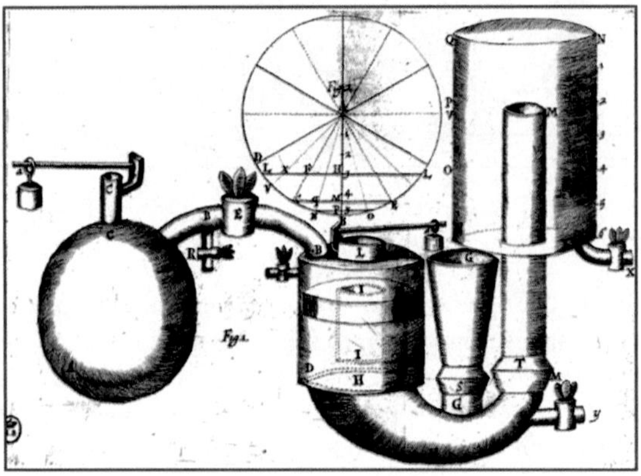

Fig. 7. Dionisio Papin steam engine [21]

In 1698, a military engineer named Thomas Savery designed a steam engine that used pressure differences between the cylinder and the mine's bottom to pump water (Fig. 8). The advances to this machine over the previous ones were the ability to achieve condensation in the glass through the action of internal rain and the application of external cold, as well as a way to feed the boiler with hot water rather than cold water.

Fig. 8. Thomas Savery steam engine [20]

This machine had some problems, such as a limit on water elevation of 12 m unless the steam pressure was dangerously increased, posing a risk of the boiler exploding, which made it unsuitable for mine drainage; or the requirement for the steam cooling and condensation process to be carried out in each operation, resulting in the loss of the steam generated [22].

In 1707, Papín developed a machine similar to this one, but on a slightly different principle by mounting it on a boat with wheels, but unfortunately when it arrived in England, it was wrecked, and Papín was never granted a patent [23].

Throughout the 18th century, experts became increasingly interested in harnessing steam energy to solve the problem of water elevation. According to, the Savery machine's performance was significantly superior to that of the Edward Somerset machine, and this machine was quite similar to the one already patented by Ayanz, despite minor variations [24].

5 The Atmospheric Steam Machine

In 1705, Thomas Newcomen gained patent rights for an atmospheric steam machine, often known as a firebomb, in England with the help of John Calley and Savery. This new machine used lower working pressures, and the boiler was a separate component of the machine.

The machine was made up of a steam boiler, a vertical cylinder with a piston, a rocker, a counterweight, and several operating components (Fig. 9).

In 1712, this machine was working at full capacity in several English mines, doing drainage operations, delivering water to some cities, and assisting with agricultural irrigation.

It was such a robust machine that, although being technically superseded by other models, some were still in operation in the nineteenth century [20].

Fig. 9. Newcomen steam engine [23, 25]

The success of the Newcomen machine drew the attention of mechanics and scientists, who studied it to see how they could improve it. This was the case with Enrique Beighton (1717), Jonathan Hulls (1736), Juan Payne (1741), Francisco Blake (1751), Polzunov (1763), and John Smeaton (1765), who studied the proportions of the elements and determined the forces of the machines, with some of them even going so far as to make changes [20, 26–28]. Polzunov machine most likely could be considered as the first known steam engine which does not require side actuation for vapor exhaust mechanism. Despite the several improvements made, Newcomen's machine still had some problems with high fuel consumption compared to the amount of labor that can be done.

6 The Simple Effect Steam Engine

James Watt will not enhance the steam engine until 1769 when he incorporates a condenser that permits the condenser outside the cylinder to retain the temperature, a greater operating rate, higher performance, and fuel economy close to three-quarters of the Newcomen machine (Fig. 10).

Watt's first experimental machine was built with the help of John Roebuck, with whom he shares the 1769 patent for a "method of lowering steam consumption, and therefore fuel expenditure, in fire machines."

The key upgrading to this process was the employment of a pump to achieve a good evacuation of the water and vapor from the condenser, as well as condensing the vapor into a separate vessel [12].

Roebuck had to sell his patent rights due to financial difficulties, and Watt tried to obtain money through his partnership with Mathew Boulton. Under the signature Boulton&Watt, they made a few more adjustments and enhancements to the initial machine.

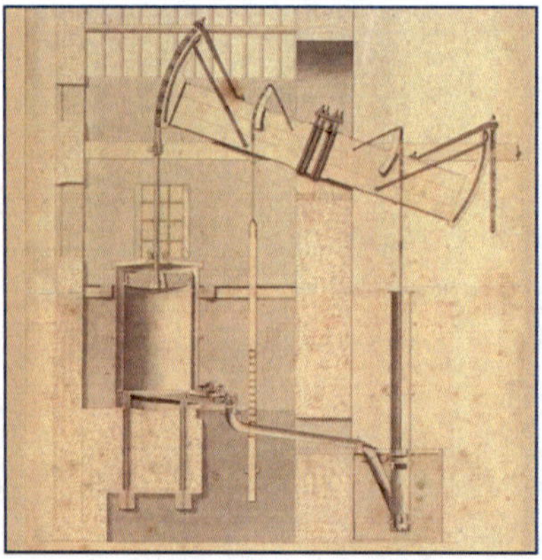

Fig. 10. Boulton & Watt steam engine. Coal mine of Hawkesbury, Bedworth. 1776. [29]

Watt also asked for John Wilkinson's help in manufacturing the cylinders for his machines. This condition would make Wilkinson a significant figure in the history of the steam engine, as he would manufacture many of the machine parts that Boulton & Watt traded in English industries and that would expand throughout the world over time [30, 31].

7 The Double Effect Steam Engine

Watt patented several steam engines types in 1782, including the double-acting machine, which could use steam on both sides of the plunger and perform work on both the rising and lowering of the piston. As a result, the force created was nearly uniform, resulting in a smaller boiler and machine.

Watt also introduced the "sun and planet" system to convert the vertical movement of the plunger into a rotation movement, avoiding the use of the already patented crank-crank system, as well as an inertia steering wheel to maintain a uniform speed; the "articulated parallelogram" to transmit the plunger movement to the rocker arm; and the "ball regulator" or centrifugal governor, which was used to regulate the steam imputation [23] (Fig. 11).

Watt's changes allowed the steam engine to be used in various industries such as mining and transportation, laying the groundwork for the Industrial Revolution. The success of Watt's machine drew the attention of new researchers, who were constrained by the Boulton & Watt firm's patent rights.

Fig. 11. Boulton & Watt double effect steam engine. New River Head, London. 1785. [32]

In 1788, Agustin de Betancourt was appointed director of Spain's Royal Cabinet of Machines, with the mission of compiling a collection of machines that were at the cutting edge of technology at the time. To accomplish this, he would travel to England in 1788. His goal was to create a model of a steam engine that incorporated all of the technological advancements of the day [33, 34].

Betancourt even built his double-acting steam machine, based on Watt's design but with a modified valve arrangement (Fig. 12). In 1789, he gave the Academy of Science in Paris his "Mémoire sur la machine à vapeur à double effect." Betancourt would also sign a huge number of innovations in addition to this contribution [35].

Fig. 12. Betancourt double effect steam engine [36].

Another of Watt's contemporaries was British engineer Richard Trevithick, who, with Edward Bull, improved the direct-action steam engine for the mine drain, reducing fuel consumption and making it a competitor to Watt's machine.

8 The High-Pressure Steam Engine

The development of high-pressure machines, which do not produce condensation and allow steam to escape directly into the atmosphere, would be the next technical triumph. Leupold was the first to think of a machine that could work at high pressure, about 1720, and Watt was the second, around 1769, but neither of them was able to put it into practice. Building a boiler that met all of the necessary safety requirements for working with steam at high pressure was out of reach for metallurgical workshops at the time.

The first high-pressure steam machine was designed by Richard Trevithick and his associate Andrew Vivian in 1802. In 1804, the machine was put into action, resulting in the world's first steam locomotive [37]. In the same year, American Oliver Evans built the first high-pressure steam engine in the United States (designed and patent in 1790) [38], which was quite similar to Trevithick's.

9 The Steam Engine Composed of Half Pressure

Around the same time, British engineer Arthur Woolf [39], badapts Watt's engine to incorporate Hornblower's idea of several cylinders, resulting in the first half-pressure composite steam engine (Fig. 13). According to Gutierrez Medina (1994) [40], this medium-pressure system had a remarkable economy in comparison to the Watt machine, as well as a more regular swing action.

The historical significance of steam is a reality that should not be overlooked. Even though steam has been largely replaced by electric motors in industry and internal combustion engines in transportation, steam turbines, which were invented almost simultaneously by Swedish engineer Gustavo de Laval in 1882 and English engineer Charles Parsons in 1884, are still used to generate electricity.

Figure 14 shows the evolution of steam engine since 1600 to 1900, with the most representative designs.

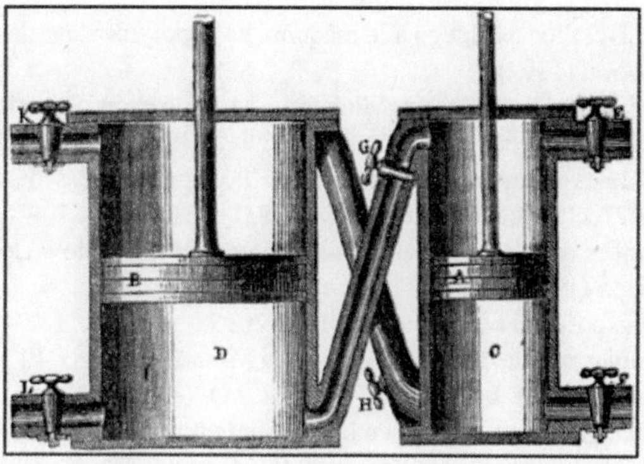

Fig. 13. Scheme of several cylinders of Woolf machine [41]

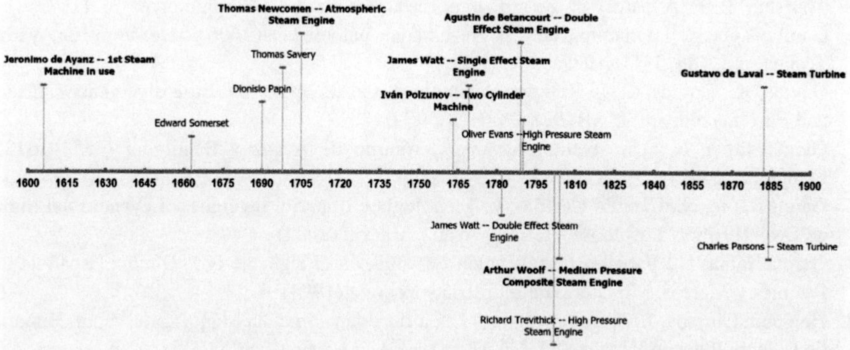

Fig. 14. Timeline with the evolution of steam engine [42]

10 Conclusions

The steam engine replaced all existing systems for mine drainage. This paper reviews the evolution of the steam engine, from the first developments to the machine that allowed the industrial revolution to take place.

The steam engine aided the development of mining by making it possible in many circumstances, and the difficulties of mining operations surely served as a constant incentive for steam engine models to evolve and improve over time.

References

1. Ezquerra del Bayo, J.: Elementos de Laboreo de Minas precedidos de algunas nociones sobre geognosia y la descripción de varios criaderos de minerales, tanto de España como de otros reinos de Europa, 2º ed. Imprenta de Don Salvador Albert, Madrid (1851)
2. Malo de Molina, M.: Laboreo de Minas, vol. 2. Establecimiento tipo-litográfico de Marcial Ventura, Cartagena (1891)

3. García Tapia, N.: Del dios del fuego a la máquina de vapor. Instituto de Ingenieros Técnicos de España, Valladolid (1992)
4. Gay, A., Dovis, S.: Máquina a vapor. Ministerio de Educación, Ciencia y Tecnologia de la Nación, Instituto Nacional de Educación Tecnológica, Buenos Aires (2006)
5. García Tapia, N.: Les premières applications de la vapeur: le cas de Jerónimo de Ayanz. Relations Science-Technique, San Francisco (1993)
6. Anónimo.: Lecciones del Laboreo de Minas. Imprenta de Cornelio y Compañia, plazuela de la Fortaleza, Oviedo (1966)
7. http://yessidp.blogspot.com.es. Accessed 1 Dec. 2015
8. Ugalde, T.: Máquinas de Elevación de Agua en la Minería Romana. El ejemplo de los pozos de la mina de Belbio (Irún). In: Públicas, F.d.l.I.T.d.O. (ed.), V Congreso de Obras Públicas Romanas. "Técnicas y construcciones en la ingeniería romana", pp. 327–345. España (2010)
9. Moncada, G.: Elementos de laboreo de minas. Imprenta José Requena, Cartagena (1992)
10. Guillman, F.: Elementos de Minería o Laboreo de Minas. Gras y Compañia, Editores, Madrid (1885)
11. Müller, E.A.: Termodinámica Básica: máquinas de vapor. Publidisa S.A., España (2004)
12. Thurston, R.H.: A history of the growth of the steam-engine. D. Appleton & Co. (1875)
13. García Tapia, N.: La máquina de vapor inventada y patentada en 1606 por Jerónimo de Ayanz. Tecnica Ind. **186**, 3–11 (1987)
14. Davids, K., Davids, C.A.: Religion, technology and the great and little divergences: China and Europe compare, ISNB: 978-90-04-23695-0 (2012)
15. García Tapia, N.: Un inventor navarro. Jerónimo de Ayanz y Beaumont (1553–1613). Pamplona (2001)
16. García Tapia, N., Carrillo Castillo, J.: Tecnología e Imperio. Ingenios y Leyendas del Siglo de Oro: Turriano, Lastanosa, Herrera, Ayanz. Madrid (2002)
17. García Tapia, N.: Patentes de invención española en el siglo de oro: Oficina Española de Patentes y Marcas. Ministerio de Industria y Energía (1994)
18. Helguera Quijada, J., Torrejón Chaves, J.: La Introducción de la Máquina de Vapor. Historia de la Tecnologia en España, vol. I. VALATENEA, S.L., España (2001)
19. V.V.A.A.: El Instructor o Repertorio de Historia, Bellas Letras y Artes, vol. 4. Londres. En casa de Ackermann y Comp. Repositorio de artes, 96.: En la imprenta de Carlos Wood, Strand (1837)
20. Tredgold, Th.: Tratado de las Máquinas de Vapor, y de su Aplicación a la Navegación, Mina, Manufacturas, etc. (G. d. l. Escosura, Trans.). Imprenta de D. León Amarita, Madrid (1831)
21. Papín, D.: Nouvelle maniere pour lever l'eau par la force du feu. A Caffell, Londres (1707)
22. Güémez, J.: Máquina Térmica de Savery. Departamento de Física Aplicada: Universidad de Cantabria (2004)
23. Figuier, L.: Los Grandes Inventos Antiguos y Modernos en Las Ciencias, La Industria y Las Artes. Imprenta y Libreria de Gaspar y Roig, Madrid (1867)
24. García Tapia, N.: Ingeniería e invención en el Siglo de Oro: El caso de Jerónimo de Ayanz. In: Paper presented at the Los origenes de la ciencia moderna. Actas años XI y XII, Canarias. Fundación Canaria Orotova de Historia de la Ciencia (2004)
25. Rolt, L.T.C., Allen, J.S.: The Steam Engine of Thomas Newcomen. Science History Publications, New York (1977)
26. Lefer, D., Buckland, G., Evans, H.: They made America: From the steam engine to the search engine: two centuries of innovators. ISBN: 9-316-07034-8 (2009)
27. Schmadel, L.: Dictionary of minor planet names. Springer, ISBN: 3-540-00238-3 (2003)
28. Hill, M.: Polzunov´s engine: an example of innovation in eighteenth century Russia. Icon **8**, 107–139 (2002)
29. http://www.search.birminghamimages.org.uk. Accessed 1 Dec. 2015

30. Poveda Ramos, G.: La tecnología mecánica y su ingreso a Colombia. Rev. Acad. Colomb. Cienc. **25**(95), 253–267 (2001)
31. Cunningham, R.E.: Historias del vacío: la ingeniería y la nada. Petrotecnia, pp. 92–105 (2009)
32. https://www.search.birminghamimages.org.uk. Accessed 1 Dec. 2015
33. González Tascón, I., Fernández Pérez, J.: Memorias de las Reales Minas del Almadén: Secretaria General del Plan Nacional de I+D. Comisión Interministerial de Ciencia y Tecnología (1990)
34. Rojas-Sola, J.I., et al.: Agustin de Betancourt´s double-acting steam engine: analysis through computer aided engineering, Appl. Sci. (2018)
35. Cioranescu, A.: Agustin de Betancourt, su obra técnica y cientifica. La Laguna de Tenerife: Consejo Superior de Investigacines Científicas. Instituto de Estudios Canarios. Universidad de la Laguna (1965)
36. http://fundacionorotava.es/betancourt. Accessed 1 Dec. 2015
37. Nieves Medina, A.: De Locomotoras de Vapor, vol. 13. Benemérita Universidad Autónoma de Puebla, Mexico (2006)
38. Ferguson, E.S.: Oliver Evans: genio inventivo de la revolución industrial estadounidense, Wilmington, DE: Fundación Eleutherian Mills-Hagley, ISBN: 0-914650-18-1 (1980)
39. Jenkins, R.: A Cornish engineer: Arthur Woolf. Transactions of the Newcomen Society, Taylor-Francis (1932)
40. Gutierrez Medina, M.L.: La España Industrial, 1847–1853. Un modelo de innovación tecnológica. Universidad de Barcelona, España (1994)
41. Canalejas, Casas, J.: Curso elemental de mecánica. Teoría y aplicada. Carlos Bailly-Bailliere, Madrid, España (1864)
42. Calderón Herrera, D.: La Tecnología del Vapor aplicada en las Minas de Almadén. Desde su origen hasta el Consejo de Administración. Universidad de Córdoba (2017)

History of Belt Conveyors Until the End of the 19th Century

Nenad Zrnić(✉), Miloš Đorđević, and Vlada Gašić

University of Belgrade, Belgrade, Serbia
{nzrnic,mddjordjevic,vgasic}@mas.bg.ac.rs

Abstract. The paper discusses the history of belt conveyors over the years until the development of the modern belt conveyor at the end of the 19th century. The roots of conveyor systems and principles of operation go back to the ancient times, with a first trace of primitive belt conveyor found in the beginning of the 17th century. The early belt conveyors can be found in the late 18th century due to the invention of Oliver Evans in USA and in the 19th century, by Lyster in England and Lopatine in Russia. The first modern solution of the heavy-duty belt conveyor was proposed by Thomas Robins in the last decade of the 19th century, with the basic principles of operation which remain unchanged up to now.

Keywords: Belt conveyor · History · Ancient conveyors · Oliver Evans · Thomas Robins

1 Introduction

Nowadays material handling equipment and systems are integral part of intralogistics (previous term commonly used was internal transport) which include organization, control, execution and optimization of in-plant material and information flows, and of goods transshipment in industry, distribution and public sector facilities [17]. Two main groups of material handling equipment classified by their design features are lifting equipment and conveying equipment. Conveying machines may have no lifting gear and move loads and people in a continuous flow [33]. Those machines with continuous action include various types of conveyors with or without pulling member, ropeways, pneumatic and hydraulic transport installations and some types of transfer equipment [13, 25]. Conveyor systems are mostly used in mining applications, depositories of bulk materials, in production lines, including their specific usage inside production machines [18]. Belt conveyors belong to the class of high-performance machines with continuous action and present the backbone of surface mining and conveying systems [3].

A belt conveyor is the most common means of transportation of bulk materials. It is capable of carrying a greater diversity of products at the rates of thousands of tons per hour in a continuous and uniform stream over long distances than any other kind of continuously operating mechanical conveyor [27]. It is used for transporting materials, generally horizontally, from one location to another. It is made up of an endless belt on top of idlers (rollers). The idlers are arranged in a trough on the carrying side and flat on

M. Ceccarelli and R. López-García (Eds.): HMM 2022, HMMS 40, pp. 210–223, 2022.
https://doi.org/10.1007/978-3-030-98499-1_18

the return side. The driving unit moves one or more pulleys which, in turn, move the belt. The whole belt conveyor is supported on a steel (rarely wood) structure, which supports and keeps the idlers in-line, as well as supports the pulleys and the drive. On the troughed side, a tripper is located wherever it is necessary to discharge material at an intermediate location. Today's conveyors consist of dozens of other components, each one specially designed for moving different materials. Belt conveyors are capable of handling a wide variety of bulk materials in open-cast/underground mines, steel plants, beneficiation plants, ports, thermal power plants, cement plants, chemical plants, fertilizer plants, etc. [11, 27]. Wherever continuous material transportation is involved, belt conveyors form the main artery of the transportation system. Belt conveyors have the adaptability to work in different types of terrain, different gradients, have flexible path of travel, and the length of the routes can be extended repeatedly as per the requirement. The lowest transportation and maintenance costs, low labor and low energy requirements are the exceptional features with belt conveyors as compared to the most other means of transportation of bulk materials [11].

2 Earliest Examples of Conveyors Since Ancient Times

Rudimentary conveyor systems, for instance screw conveyors, date back to ancient times. Ancient conveyors were much simpler, and could not perform like today's modern conveyors. However, even different types of conveyors have some basic common operating principles. Ancient machines laid the foundation for later inventions and improvements. The first machines ever built by a human were more than probably labor-saving devices for the conveyance of water [31]. Greek technology was at its zenith during and after some time of the period of Archimedes (300 BC–150 AD). The important innovations in this period include cranes, screws, gears, organs, odometer, wheelbarrows and roof tiles. Undoubtedly the first conveyor as such was designed by Archimedes (287 to 212 B.C.)—Greek mathematician, physicist and inventor—for removing water from the hold of a ship built for King Hiero of Syracuse. Later, a screw water machine of this kind was used for water supplying the city of Rome, which was regarded as the ancestor of today's helical screw conveyor, as shown in Fig. 1 [24]. The application of the principle of the Archimedes' screw is seen in many modern machines. For example, screw conveyors and rotary feeders are used to transfer both solids and liquids from one end of the conveyor to the other in industrial applications [6, 30].

Of course, there are some more analogies between modern and ancient conveying equipment. Another device which is the progenitor of the bucket elevator as a continuous machine is related to chain water pump or water elevator with buckets [31]. The mechanism shown in Fig. 2 [20, 21] (possibly dating from the first century BC), consisted of a chain of buckets that was set to motion by a horizontal shaft via pedals attached to it. It can be seen that the pedals were used by two men to rotate the shaft which is attached to a horizontal bar in the main frame of the machine. This action activates the wheel that moves the buckets which raise not only water but also sand or earth as required [21].

Additionally, from the works of Philo of Alexandria, one learns that in his era (1st century AD) in Alexandria, Egypt, there was a mechanism for lifting water from the river

Fig. 1. Archimedes screw water machine as ancestor of screw conveyors (left) [24] and a modern screw conveyor (right) [33]

Fig. 2. Chain water elevator (From the Thien Kung Khai Wu) [20]

with continuous action, i.e. conveyor, and according to the principle of operation resembles at first glance the bucket chain elevator [33]. The bucket elevator from Alexandria and today's bucket elevator for grain material are shown in Fig. 3 [33].

3 Early History of Belt Conveyors

It is obvious from the previous chapter that even before the first belt conveyor was invented people have used the principles of conveyors for thousands of years [19]. The history of primitive belt conveyors began in late 16th century [4] or beginning of the 17th century [4]. However, according to the available literature and lack of documented data, it seems that the invention of those conveyors can neither be assigned to an exact date nor to a specific inventor. In fact, there are various technical innovations that could be seen as trailblazers for the modern conveyor belt. Probably, the first representation of the primitive conveyor belt was made by Hieronymus Megiser in 1614, Fig. 4 [5].

As the Industrial Revolution took hold in Great Britain during the 18th century, and then a little bit later in America, manual labor in many production environments began to be phased out in favor of steam-powered machinery. Since the industrial revolution,

Fig. 3. Bucket elevator from Alexandria (left) and modern bucket elevator (right) [33]

Fig. 4. Conveyor belt for transporting soil from the moat [5]

people have made increasing use of mechanical methods of handling materials. The most important literature discussing the history and development of belt conveyors was written by Frederic Hetzel (1870–1946), American engineer and holder of many patents [12]. He mentioned that the earliest reference to the use of belt conveyors in American practice is in Oliver Evans' "Miller's Guide" published in Philadelphia in 1795, with several corrected later editions until the last one in 1860, Fig. 6 [7]. Oliver Evans (1755–1819) was an American inventor, engineer and businessman born in Delaware and later took root commercially in Philadelphia. He was one of the first Americans who built steam engines, and also a pioneer in the fields of automation and materials handling. Going into business with his brothers, he worked for over a decade designing, building and improving an automated flour mills without the employment of manual labor using various devices such as bucket chains (elevators) for raising vertically and conveyor belts for moving horizontally [10]. Evans was one of the most creative and influential inventors in the early years of the United States. His revolutionary book [7], written in plain language with easily understood plates, became one of the most heavily used technical handbooks of the entire 19th century, Fig. 5.

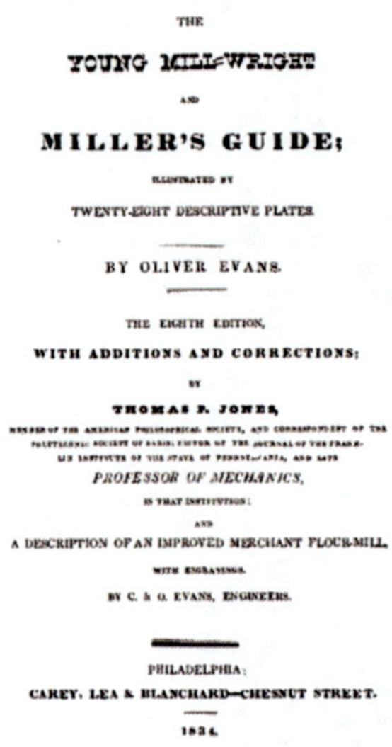
THE

YOUNG MILL-WRIGHT

AND

MILLER'S GUIDE;

ILLUSTRATED BY

TWENTY-EIGHT DESCRIPTIVE PLATES.

BY OLIVER EVANS.

THE EIGHTH EDITION,

WITH ADDITIONS AND CORRECTIONS;

BY

THOMAS P. JONES,

[illegible]

PROFESSOR OF MECHANICS,

IN THAT INSTITUTION;

AND

A DESCRIPTION OF AN IMPROVED MERCHANT FLOUR-MILL,

WITH ENGRAVINGS.

BY C. & O. EVANS, ENGINEERS.

PHILADELPHIA:

CAREY, LEA & BLANCHARD—CHESNUT STREET.

1834.

Fig. 5. Front cover of Oliver Evans' "Miller's Guide" (left) [7] and Oliver Evans portrait (right) [10]

The book [7] describes and illustrates a flat belt receiving material on its upper run and discharging over the end; Evans calls it "a broad endless strap of thin pliant leather, canvas or flannel revolving over two pulleys in a case or trough", Fig. 6 [7]. Although this machine would work by gravity, even when the descent was small, yet Evans recommended that power should be applied to it where practicable; and when driven in this way it became the prototype of the belt conveyor of the present day, which are generally used for the horizontal movement of grain in large quantities. In the flour and grist mills built by Evans and his successors some of these conveyors were probably used, but it was more common in those days to use screw conveyors to convey grain as well as the lighter mill products [12].

Before the Industrial Revolution in England and implementation of steam powered machines primitive belt conveyors in the 1700s were made of short leather belts running on wooden rollers and were powered with hand cranks and a series of pulleys, moving goods onto ships when they were in port. With the onset of industrialization in the second half of the 18th century in England many industries and companies – such as abattoirs and bakeries – were equipped with timesaving and money-saving transport system. Probably, the first steam operated conveyor belt had been set into operation in 1804 by the British Navy to produce ship's biscuits, Fig. 7 [15]. It was used to transport special long-life biscuits from a bakery to the storage and packing area.

According to [12] belts sliding in troughs, Fig. 8 (left), were used before 1840 to convey materials which were not suitable for screw conveyors, such as clay, shavings,

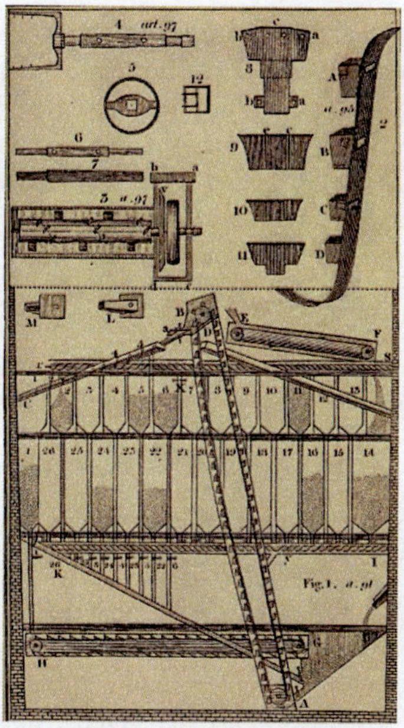

Fig. 6. Conveyor belt, Plate VI, Fig. 1, E-F [7]

saw-mill refuse and crushed stone. Sometimes the return run slid back; sometimes it was carried on rollers. In some cases, the sides of the trough were made with hinged sections which swung inward to stand diagonally across the belt and thus discharge material along the run of the conveyor [12]. When such conveyors were used for hard and gritty substances, the belt did not last long. In clay conveyors, the clay stuck to the sides and bottom of the trough, hardened there, and wore out the belt rapidly.

Fig. 7. Steam powered conveyor belt, 1804 [15]

Some of these troubles were avoided in the design shown in Fig. 8 (right) [12], which represents a form of clay conveyor in extensive use about 1870s. This construction substituted sliding friction on dirty rough surfaces for rolling friction in lubricated bearings; it saved power and helped to preserve the belt, although the material could not be carried without some leakage under the skirt-boards or trough sides, and the edges of the belt were still subject to injurious wear [12]. The further development of conveyor belts is strongly connected to the development of grain elevators. For instance, in Southwark Foundry in Philadelphia, between 1859 and 1863, wide composite belts consisting of two parallel leather belts to which bent iron bars or spreaders were riveted at intervals to support a trough of canvas which carried the grain, were used.

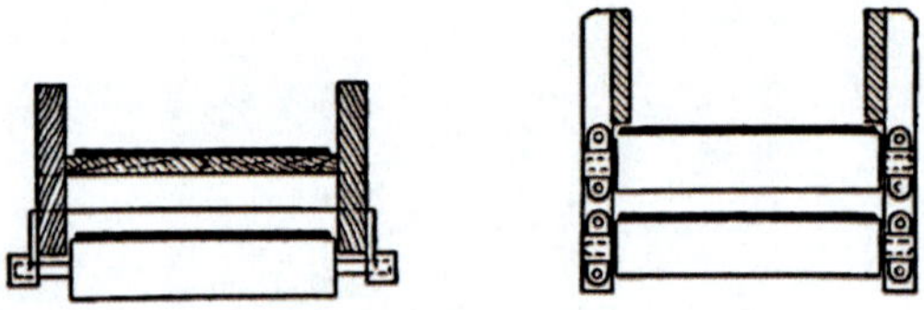

Fig. 8. Conveyor belt sliding in wood trough, 1830 (left) [12] and clay conveyor belt supported on rollers, 1870 (right) [12]

Historically, the first conveyor belts were made from layers, or plies, of canvas or woven cotton fabric, termed cotton duck, with natural rubber interlayers and natural rubber covers. In 1844, inventor Charles Goodyear patented vulcanized rubber, for which he received patent number 3633 from the United States Patent Office [14]. This innovation made rubber a much more stable substance that didn't react negatively to temperature changes. And while some sources report that rubber had been used as a conveyor belt material in earlier days, it certainly was not ideal. It became hard and rigid in cold environments and tended to melt and get sticky when warmed. About 1865, P. G. B. Westmacott and George Fosbery Lyster, engineers for the Birkenhead and Waterloo Docks at Liverpool, England, made experiments with a 12" (ca. 300 mm) belt which showed that belt conveyors carried more grain with less power than screw conveyors. In 1866 they installed a system consisting of a skip hoist for elevating the grain with belts for distributing it. The belts were 18" (ca 450 mm) 2-ply rubber, running at 450 to 500 ft per minute (up to ca. 2.5 m/s) and supported every 6 ft on straight wooden rolls [12]. It is unknown how much Westmacott and Lyster knew about American practice at that time. In 1868, Lyster described to the British Engineers Society his work on conveying bulk materials by endless belts made of two plies of canvas with a facing of rubber. However, these belts had short lives, not because of their construction but because of their use with rollers that had flared ends to form a trough. This resulted in the belt contacting the rollers at different linear speeds and rapidly wearing out [26]. Flat rollers were then used, giving greater belt life but lower conveyor capacity. In [26] it was reported that, because of the high price of rubber, it became impossible to produce rubber faced canvas belting economically and lower cost "solid woven" cotton belts, i.e., without plies or interlayers, were used for many years. These belts were said to have the disadvantages of being disposed to damage, the absorption of moisture and being less flexible than the ply belts previously used. Lyster's first idlers were spool shaped, but discarded for straight-faced wood idlers because the flared ends wore the edges of the belt, as they do today. Lyster experimented with two-roll troughing idlers and found that it was impossible to prevent the belt from centering on one roll or the other. He devised and used the three-way tripper and the revolving cleaner brush; it seems that Lyster invented many of the features of the belt conveyor of today [16, 26].

In the similar period, in Russia the first indications on the use of conveyor belts in the mining industry were published by Alexander Lopatine industrialist in Yeniseysk [8]. An article about the invention of an endless transport machine, called by its inventor "sand conveyor", was published in the government bulletin of Yeniseysk, No. 47, 1859. After several years of work, the inventor succeeded in developing, in July 1859, a machine operating as sand conveyor. The author of the invention described the machine as follows: The "sand conveyor" machine aims to substitute the means of delivery of sand with gold content and the removal of washed sand to the landfill and thus reduce the cost of horse and human labor. The main component of the machine is an endless movable belt that turns over two drums, fastened at two opposite ends, Fig. 9 [8]. The load placed at one end of the belt, by moving the belt comes to the other end and is unloaded at the designated place. When changing the direction of movement and lifting the load to a height, several spans are needed, so that the load is transferred from one to another. The belt is powered by water, steam or horses [34]. In July 1860, A. Lopatine put into service, in Pétropavlosk, three conveyors transporting gold-bearing sand. The periodical press of Eastern Siberia takes a keen interest in Lopatine's sand conveyor. Irkutsk government bulletin published four articles on this subject in 1860 [8]. Lopatine's sand conveyor exhibited all parts essentials of the modern conveyor belt. It included an endless belt, a drive head pulley and a tail pulley, a transmission, a frame and most importantly, idlers. It was the first conveyor belt in history, intended to transport rocks both in surface works and in the mining industry in general. Thus, A. Lopatine was the first to apply complex conveying in the practice of mining works [8, 34]. Lopatine's sand conveyors were employed for at least ten years in fifteen locations in the Eastern Siberia. In the early 70s of the 19th century, the sand conveyors are beginning to be replaced on the sites by the transport by wagon pulled by cable, undoubtedly due to the cessation of activity of the inventor. Nevertheless, the owners of the gold mines of this time retained interest in conveying. For example, in 1873, Michel Koouzov was granted a patent for the implementation of a system of conveyors in Siberia individually controlled by steam [8]. This invention was awarded a prize by the Russian Technical Society. Therefore, only the technical backwardness of tsarist Russia of that time did not allow to organize the production or acquisition of rubber bands, being the real reason that slowed down the use of conveyor belts in mines.

With the development of conveying machinery between 1880 and 1890 belts were used for handling coal, ore and other materials heavier than grain. Designers followed grain elevator practice for belts and idlers. One of the largest installations at that time was the ore concentrating plant of the New Jersey and Pennsylvania Concentrating Co in Sparta, New Jersey [12]. On this site (Ogden Mine) in 1891, Thomas Edison developed an enormous complex of mines, crushers, separators, and subsidiary buildings that came to be known as "Edison" or, more properly, "The Works of the New Jersey and Pennsylvania Concentrating Company" [1]. This company produced briquettes of concentrated iron ore. It had over 50 cotton belt conveyors with continuous skirt boards along the carrying run to keep the load from rolling off the belt, ranging from 20" to 30" wide and up to 500 ft long [1, 2]. As it was stated in [9], Thomas Edison assisted in the design of the

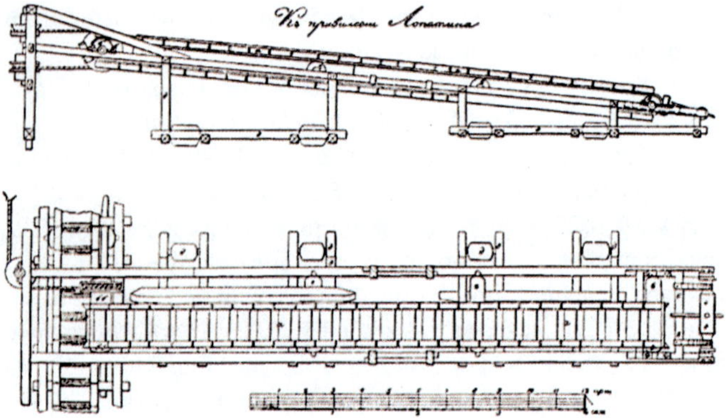

Fig. 9. Belt conveyor, by A. Lopatine [8]

troughed idler. Those idlers used there about 1891 were cast-iron spool idlers, but it was a bad solution, because in order to get the greatest capacity from the belts, the troughing was deep and the spools were much larger at the ends than at the middle. These were heavier than the spools used on grain conveyors; the work was continuous, the place dusty, and the belts wore out rapidly. When the plant was rebuilt in 1893, the idlers of the type shown in Fig. 10 (left) replaced the spool idlers [12]; the new idlers were practically the same the flat rolls with concentrators used on grain conveyors, Fig. 10 (right) [12]. The main trouble at Edison and at other places doing similar work was with the belts. The sharp pieces of heavy ore cut the fabric and the stitching of canvas belts and loosened the plies; the same thing happened with rubber belts. We know now that much of the difficulty was due to improper delivery of material to the belts and to improper construction of the belts themselves.

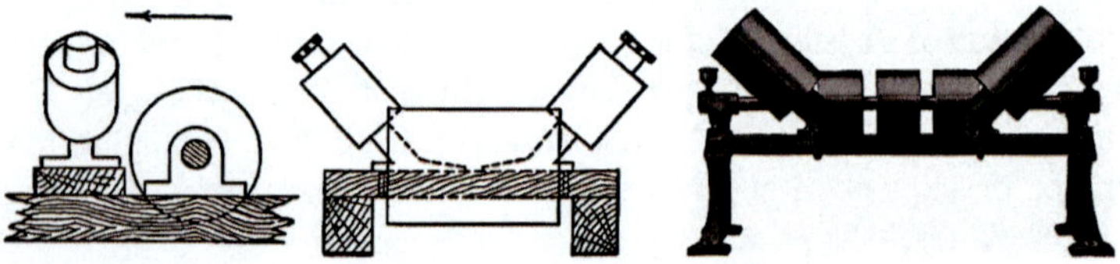

Fig. 10. Troughing idler for ore conveyor, 1893 (left) [12] and concentrators mounted on a stand with carrying pulleys, Webster Manufacturing Co. (right) [12]

However, in [26] it was stated that Robins invented the troughing idler in 1885, but without any proof or explanation about the mentioned year. Thomas Robins (1868–1957) was an American inventor and manufacturer educated at Princeton University (1887–1889), Fig. 11 [28]. In 1892 he began a series of inventions which led to the development of a belt conveyor system used for carrying coal, ores and other products [22].

According to [12] Robins, at that time in the business of making rubber belts, visited the plant about 1891 and noticed the following points: the thin layer of rubber which covered the belt resisted abrasion much longer than did the duck which formed the body of the belt; each layer of duck wore out faster than the one preceding it, showing that as the belt wore thin and the tension on the threads increased they were cut more easily; belts wore out in the middle and split longitudinally while the edges were still good [12]. From these observations Robins concluded that the sole function of the fabric in a rubber belt should be to give the belt tensile strength and that it should be protected from injury by a cover of rubber compound that would resist abrasion better than the cotton threads or the thin layer of friction rubber that covered them. After experiments with various rubber compounds, he furnished belts with a thick face of rubber that proved to be much more durable than those previously used, lasting years where others lasted months. In 1893 he patented the belt shown in Fig. 12 (left) [12], which had an extra thickness of rubber in the middle to resist the abrasion which comes from feeding material from a comparatively narrow chute; it resisted abrasion very well but the belt was so stiff that it would not conform to the shape of the troughing idlers, which, as shown in Fig. 10 (left), had the side pulleys inclined at 45° [12].

Fig. 11. Thomas Robins, developer of belt conveyors, in 1915 [28]

Additionally, Thomas Robins, seemingly independently of Lyster, followed analogous lines of investigation and eliminated the difficulty inherent in the two-pulley troughing idler by adding a horizontal pulley just back of each inclined pair [16]. He achieved this with the invention of the three-in-line-unit idler and prevented oil from the bearings from leaking onto the belt by means of hollow roll shafts with grease cups on the outer ends and radial ducts along the shaft, incidentally securing the advantage of exuded grease seals to prevent ingress of dirt into the bearing. His work in belting led him to construct belts that had two or more plies with rubber covers. Then he reduced the thickness of the underside and increased it on the carrying side to secure longer life. About 1896, he introduced the stepped-ply belts, in which the plies stepped off toward the center with corresponding increasing thickness of rubber in the central zone [2, 16], Fig. 12 (right) [12].

The space left by the omission of the plies was filled with the rubber of the top cover, making the extra thickness of protective cover there equivalent to that shown in Fig. 12 (right), but using less rubber where the abrasion was not so great. The omission of the plies at the middle made the belt more flexible in cross section, allowed it to conform more closely to the shape of the troughing idlers and at the same time left a thick stiff edge to bear against the side guide pulleys, Fig. 13 (left) [23]. These pulleys

Fig. 12. Patent of conveyor belt by Thomas Robins, in 1893 (left) [12] and Robins' stepped-ply belt, 1896 (right) [12]

were necessary to keep the belt straight on 45° troughing idlers, especially when the belts were comparatively narrow and stiff and when the inclined idlers came every 4 or 5 ft. Idlers inclined at 45° were in successful use on grain conveyors, but there the belts were relatively wide and thin and the troughing occurred only at the loading points or at intervals not closer than 8 or 10 ft. Under these conditions the grain belt had a good guiding contact with the horizontal pulleys and did not usually require side guide idlers to keep it straight; but with stiff and narrow belts on 45° idlers, the guiding contact on the horizontal pulleys was slight, the side guide pulleys were necessary and the belt had to have a thick stiff edge; otherwise, the edges were chafed, bent over or rubbed off and the belt destroyed [12, 23].

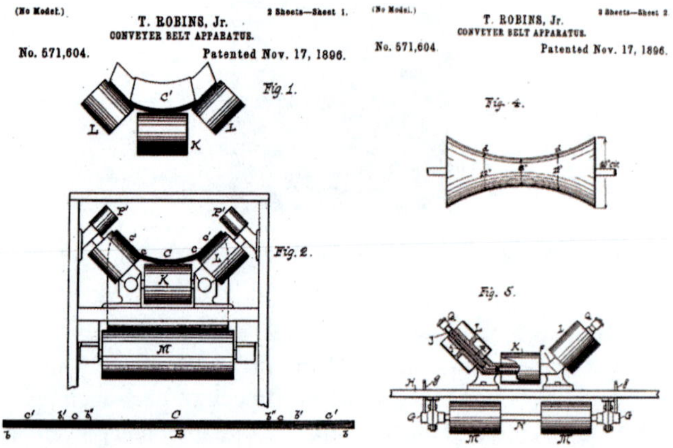

Fig. 13. Stepped-ply belt on 45° troughing idlers with side-guide pulleys, 1896 (left) [23] and Robins' three-pulley idler, 1896 (right) [23]

In the early days of the belt-conveyor business, before 1900, comparatively few belts wider than 24" were used. Troughing was considered a means to permit a narrow belt to carry what would otherwise require a wider and more costly belt, and the tendency was to use belts narrower than would now be considered good practice. The Robins patent (571604, Nov. 17, 1896, [23]) covers also the idler construction shown in Fig. 13 (right) [12]. Also, Robins Company realized the solution with six rollers, three at the top and three at the bottom belt, Fig. 14 [32].

There was a great expansion in the use of all kinds of conveying machinery between 1895 and the early 1900s and the belt conveyor had its share of the growth. Thomas

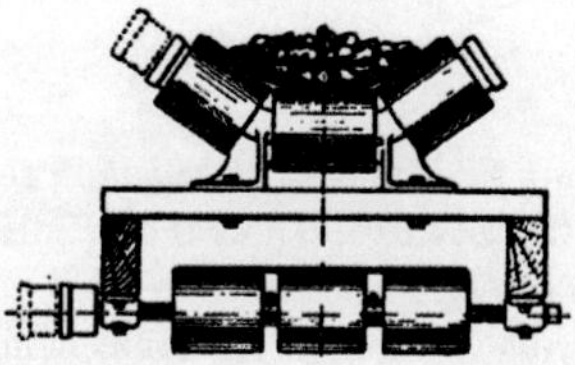

Fig. 14. Solution for the support of belt conveyor by Robins Conveying Belt Company [32]

Robins and his company (Robins Conveying Belt Company, today Hewitt-Robins Company) did a pioneer work in extending the use of belt conveyors to the handling of materials other than grain and in laying the foundations for the engineering knowledge of the business [29]. His conveyor belt received the grand prize at the Paris Exposition in 1900, and first prizes at the Pan-American Exposition and Saint Louis Exposition [22]. As usual other companies entered the market; some of them with devices intended to circumvent the Robins patents, others with improvements suggested by the growing experience of manufacturers and users of belt conveyors.

4 Conclusions

No history of the last two centuries, or of the current one, would be complete without a history of the belt conveyor. Conveyor systems have been an integral and inevitable part of material handling for around 200 years, and their origins can be traced back even further than that. Belt conveyors have their ancestors in ancient times. People have used some principles of belt conveyors operation for thousands of years. After analyzing different sources, it seems that it is very difficult to assign very strictly the belt conveyor invention to an exact date or to a specific inventor, though the most of the references named Americans Oliver Evans and Thomas Robins as the inventors of belt conveyors, while some named English engineer George Lyster and Russian entrepreneur Alexander Lopatine, as the inventor of belt conveyors in mines. Lopatine's contribution to the development of belt conveyors isn't recognized by most of the authors because his impact is highlighted merely by Russian authors. However, it is obvious and in accordance with the relevant literature that the first "real" belt conveyors showed up around the late 1700s to early 1800s. The earliest primitive belt conveyors were constructed with wood and leather and were operated by hand. The Industrial Revolution brought with it steam-powered conveyor belts at the beginning of the 19th century, taking much of the manual labor out of belt operation. The first modern belt conveyors were developed in the late 19th century. Moreover, since that time their basic principles of operation have not changed until now. However, by permanently improving features of belt conveyors over the years, their capacity rating and the length over which material can be conveyed have increased considerably, together with the power inputs, the size of components and the degree of sophistication.

Acknowledgments. This work is a contribution to the MESTD of Serbia funded project "Integrated research in the fields of macro, micro and nano mechanical engineering" (Contract number: 451–03-68/2020–14/200105).

References

1. Agreement, Edison Ore Milling Co Ltd, New Jersey and Pennsylvania Concentrating Works, December 31st, Microfilm ID, 144:614, Thomas A. Edison Papers, School of Arts and Sciences, Rutgers University (1890)
2. Bernstein, W.J.: Belt conveyors in modern mine mechanization, master thesis, Colorado School of Mines, Golden, Colorado (1950)
3. Bošnjak, S., Zrnić, N.: Dynamics, failures, redesigning and environmentally friendly technologies in surface mining systems. Archives of Civil and Mechani. Eng. **12**(3), 348–359 (2012). https://doi.org/10.1016/j.acme.2012.06.009
4. Dartnall, J.W.: A general approach for introducing materials handling topics in mechanical engineering degree course, Proc. 2005 ASME Int Mech Eng Congress Expo, November 5–11, Orlando, Florida USA, IMECE2005–82004, pp. 1–9 (2005)
5. Deutsche Fotothek, df_tg_0001790, author Hieronymus Megiser (1614)
6. Dixit, U.S., Hazarika, M., Davim, J.P.: A Brief History of Mechanical Engineering, Springer International Publishing, Switzerland (2017)
7. Evans, O.: The Young Mill-Wright and Miller's Guide, the 8th edition with additions and corrections by Thomas Jones. Carey, Lea & Blanchard-Chestnut Street, Philadelphia (1834)
8. Faddeev, B.V.: Documéntation - Historique de l'utilisation des convoyeurs à bande dans les travaux miniers. In: Revue d'histoire des sciences, tome **24**(1), 61–66 (1971)
9. Fayed, M.E., Skocir, T.S.: Mechanical Conveyors - Selection and Operation, CRC Press (1996)
10. Ferguson, E.: Oliver Evans – Inventive Genius of the American Industrial Revolution, The Hagley Museum, Greenville, Delaware, USA (1980)
11. Fruchtbaum, J.: Bulk Materials Handling Handbook. SSBM, New York (1988)
12. Hetzel, F.: Belt conveyors and belt elevators. John Wiley & Sons, New York (1922)
13. Hoffmann, K., Zrnić, N.: A Contribution on the History of Ropeways. In: Koetsier, T., Ceccarelli, M. (eds.) Explorations in the History of Machines and Mechanisms, pp. 381–394. Springer, Heidelberg (2012). https://doi.org/10.1007/978-94-007-4132-4_26
14. https://patentimages.storage.googleapis.com/1a/13/8e/18751846ded131/US3633.pdf. Accessed 17 January 2022
15. https://www.timetoast.com/timelines/significant-milestones-in-development-of-conveyor-belts. Accessed 15 November 2021
16. Hudson, W.: Conveyors and related equipment. 2nd edition, John Wiley & Sons (1950)
17. Kartnig, G., Grösel, B., Zrnić, N.: Past, state-of-the-art and future of intralogistics in relation to megatrends, FME Trans. **40**(4), 193–200 (2012)
18. Langerholc, M., Zrnić, N., Đorđević, M.D., Jerman, B.: Conveyor design optimization as the provision of sustainability. Technical Gazette **20**(5), 837–846 (2013)
19. Lovren, N., Debelić, D.: About the history of belt conveyors. Eng. Rev. **28–2**, 111–118 (2008)
20. Needham, J.: Science and Civilization in China. Cambridge Univ. Press, Cambridge (1975)
21. Paz, E.B., Ceccarelli, M., Otero, J.E., Sanz, J.L.M.: A Brief Illustrated History of Machines and Mechanisms, SSBM B.V. (2010)
22. Edwin, R.G.: Robins, Thomas, Encyclopedia Americana, ed. (1920)
23. Robins, T.: Conveyor belt apparatus, No 571,604. Patented Nov. 17, USA Patent Office, (1896)
24. Rossi, C., Russo, F., Russo, F.: Ancient Engineers' Inventions: Precursors of the Present, SSBM (2009)
25. Spivakovsky, A., Dyachkov, V.: Conveyors and related equipment. Peace Publishers, Moscow, USSR (1969)
26. Streets, H.: Sutcliffe's Manual of Belt Conveying. W. & R. Chambers Ltd (1956)

27. Subba Rao, D.V.: The Belt Conveyor, CRC Press/Balkema, Taylor & Francis Group, London, UK (2021)
28. Robins, T. Jr.: in 1915, Library of Congress, file created 17 December (2013)
29. von Hanffstengel, G.: Die Förderung von Massengütern: I Band, Bau und Berechnung der stetig arbeitenden Förderer, Julius Springer (1908)
30. Zhang, C., Yang, J.: A History of Mechanical Engineering, Springer Nature Singapore (2020)
31. Zimmer G.F.: The early history of mechanical devices, Trans. Newcomen Soc. **2**(1) (1921)
32. Zimmer, G.F.: The Mechanical Handling of Material. Crosby Lockwood & Son, London (1905)
33. Zrnić, N., Hoffmann, K.: History of moving people: Lifting and conveying equipment. Inte. J. Terraspace Sci. Eng. **5**(1), 55–71 (2012)
34. Грудачев, А.Я.: К 150-летию создания ленточного конвейера, Текст научной статьи по специальности История и археология, Горный информационно-аналитический бюллетень (2013)

Historic Development of Theories and Teaching

A Short Comparative Historiography of Science & Technology of Mechanics into Engineering—and—Architecture Literatures During the Renaissance

Raffaele Pisano[1(✉)] and Julie Robarts[2]

[1] IEMN, Lille University, Villeneuve d'Ascq, France
raffaele.pisano@univ-lille.fr

[2] Centre for Early Modern Studies, Australian National University, Canberra, Australia

Abstract. The Historiography of science deals with methods, and so interpretations of facts and events within the discipline of history. It also represents a style of method of historians in developing their specific studies. This discipline also allows historians of science to refine the borders internal or external to their historical methods. During the Renaissance, a culture of machines (including fortifications and weapons) paralleled the development of science, from the science of weights until the birth of modern science. Was it possible to build a machine without knowing Aristotle's and Euclid's sciences? Thus, *Science* on one side and (military) *Science and Technology* on the other almost independently developed themselves; in some special cases, they crossed over each other. A comparative historiography of science and technology (artisanal crafts produced by non-scientists like engineering and architecture, including applied mechanics in some cases) in the Renaissance is presented. But, how was it possible to build (in the Renaissance) a machine without knowing Aristotle's and Euclid's sciences? Do Historical (scientists, philosophers, facts and theories) or historiographical (methods, events and interpretations) elements matter? No theory—no machine? Historiography can influence the vision—and—image of a subject within history; i.e., is the case of Koyré's Newtonian studies and of Kuhn's paradigm. Thus, in this paper, we review the role played by the major contributions to the historiography of science and technology from the Middle Ages to the Renaissance as presented in the past century by Singer's authors within the first three of seven volumes of *A History of Technology* (1954–1958). This provides perspectives both on the enduring influence of this early historiographical (on the birth of Mechanism in the Renaissance) tradition and describes the ways it differently accounted for variation in unlike countries and consequently how its results remain, still nowadays, a debated question. Due to constraints preventing an extensive exploration of this amazing historiography, we focus our comparative analysis to the books cited above.

Keywords: Historiography · History of science and technology · Mechanics · Singer · Briggs · Derry · Williams

M. Ceccarelli and R. López-García (Eds.): HMM 2022, HMMS 40, pp. 227–240, 2022.
https://doi.org/10.1007/978-3-030-98499-1_19

1 Singer on Science and Technology

The historiography on Charles Joseph Singer (1876–1960) reveals how the English historian of science and medicine masterfully distinguished the history of things that men did from the history of men's ideas. In this sense, the vast history of science and technology of Singer (edited together with other scholars) is unprecedented in the history of scientific literature[1] (of the twentieth century) both for the contents and the scope of this encyclopaedic work [14]. In the third of his eight volumes, titled *From the Renaissance to the Industrial Revolution* [14, III] Singer – and his co-editors (Holmyard, Hall and Williams) – aimed to understand the role of technology and science up to that significant happening that represented the birth of modern science, while clarifying how this subsequent story was extraneous to their research. For example, at the beginning, Singer specified that the study of the birth of modern science does not fall within the limits of their historiographical project on the history of technology [14, preliminary matters: Vols. I-III]. Together with the history of technology, the geographic and cultural background of the period is also analysed. According to Singer, this is a period in which Europe is able to procure the necessary raw materials by means of new navigation techniques, thus avoiding dependence solely on pre-existing cultures: both exporting and asserting technological supremacy. In addition, at the disappearance of the city-state, economic competition begins to assume a political-economic role within the nascent nation-state. In more detail, starting from the fact that at the end of the Middle Ages the technical scientific expertise did not always progress parallel to the theoretical, the text identifies the early and persistent role of the natural philosopher, dedicated and attentive observer of nature, from the technical pragmatism of some artisans. These artisans, not being scientists, perfected their techniques because they were driven by their trades and necessity, according to Singer. The knowledge of these craftsmen was mostly transmitted verbally, and this was enough to ensure the success of their workshop. Singer does not even associate the technician-craftsman with a primitive (scientific) theory through which they could have come to be curious to find explanations for certain events – according to Singer – "Only in the seventeenth century [...] was it realized—and even then by few—that science and the crafts were alike concerned with natural phenomena and could aid each other." [14, III, pp. v-vi]. Therefore, the development of technology up until industrial organization was mainly the inventions of artisans, as well as the results of pure research. He adds, however, that "[...] the marvels of the age—the great wooden ships [...] the ponderous machines for raising water [...], the massive stone buildings [...] and the intricately woven tapestries [...]—diminish in importance as compared with the first humble offshoots of scientific research. [14, III, p. vi]. Now, if from the previous

[1] A larger—and—update of Singer's *History of Technology* is forthcoming next years by Bloomsbury (London, UK). One of us (RP) is guest editor (and author) of third volume titled *A Cultural History of Technology in the Age of Encounters (1450–1650)* [13]. This is extremely topical, because even today some *History of science* studies are separated from the *History of technology*. A combined–integrated methodology is necessary. Otherwise it is not clear, for example, why a cannon placed at 45 degrees has a maximum range (*History of science*, i.e., Tartaglia's *Quesiti* 1535–1554) regardless of its geometric shape and the material used (*History of technique and technology*); or, why a 3-phases steam engine does not work (Watt, late 17th-early-18th, *History of technology*) but rather 4-phases are required (*History of science*, i.e., Sadi Carnot, 1824).

passage we interpret Singer's term "first humble offshoots of scientific research" with *scientific theories*, then we can consider including within this discourse the first scientific theories from the early 1500s. Particular examples include those, e.g., of Tartaglia: on numbers, on the dynamics of projectiles, on science of weights [5–7, 13] on statics, and on the resolution of algebraic equations; those of Guidobaldo del Monte (1545–1607) on the laws of statics; until the theory of Galileo Galilei on the inclined plane [12, 15]. In Singer's history of technology, if the practice of building bridges [14, III, pp. 417–437] was not indebted to the principles of science, the practical technologies of navigation and industrial chemistry [*Ibidem*] were, with the subsequent *abstract* process of scientific knowledge significantly contributing to the transformation and improvement of some of their techniques [*Ibidem.*]. Even for the construction of bridges the inability of the builders to provide a static theory sufficient to the needs of the time became clear, thus the whole was strongly marked by common sense [Cfr. 14, III, p. 432]. Singer's historical work also highlights how applications of certain scientific principles supported the development of navigation technology. In particular, the concepts of pure geometry for cartography and those of pure mathematics and physical instruments (e.g. magnetic needles) for surveying techniques were of great interest. In fact, it was customary to include topics of mathematics and practical geometry in scientific treatises. This is what Tartaglia did - wrote Singer [14, III, pp. 538–539] - who in his *Quesiti et invention diverse* (1546; 1554) also dealt with instruments of magnetic needle detection aimed at measuring angles for horizontal positions [16, ff. 55r-55v; 7]. But Singer, demonstrating his need to occasionally involve even pure science in his treatment of technology, presents other surveying instruments proposed by various scientists. One is the instrument of *Triangulation* [14, III, pp. 503–504; pp. 539–541] developed by the mathematician Gemma Frisius (1508–1555), professor of mathematics at Leuven. The instrument was based on a mathematical method capable of detecting angles of position, called *orientations*, placed at both ends of the chosen baseline [14, III, pp. 503–541]. Another instrument described by Singer were *Mathematical Tables*. They are a compendium of mathematical rules that assisted surveyors in the measurement of land and landholdings. Very important in this regard is *Surveyor* (1616) by Aaron Rathborne (1572–1618; 14, III, pp. 541–542]. According to Singer, the mathematical-technical relationship within the technique of levelling developed thanks to the work of the English mathematician Edward Wright (ca. 1558–1615) who also associated this technique with *perspective glasses* to establish a greater accuracy of the position to be measured. On navigation, Singer presents other instruments and rules of measurement of the time: the *quadrant*, the *nocturnal*, the *nautical astrolabe* and the *cross staff*, etc. We conclude this excursus on Singer's thought, taking the opportunity to remind the reader not to miss the opportunity to read Singer's complete historiography in his *A short History of scientific ideas to 1900* [17].

2 Briggs, Derry and Williams on Architecture

Martin Briggs (1882–1977) also collaborated on the immense work of history directed by Singer that we analysed above [14, III, *Material Civilization, Building Construction* and *Town-Planning from the Ancient World to the Renaissance* [14, III, pp. 245–299]. As an urban planner and architectural historian, he helped to explain the mixing of art and science in architecture, interweaving the organization of the city, in every aspect and variation. In his specific essay *The architect in history* [18] Briggs gives an overview of the history of urban settlements and the ways of building houses from antiquity to the Renaissance. For the British school we will refer mainly to the work of Thomas Kingston Derry (1905–2001) and Trevor Illtyd Williams (1921–1996) on the history of technology, and in particular on the construction of buildings [14, III; 19]. More details are below, in the next section. According to Briggs' reconstruction, in the early Neolithic ages, the concept of dwelling was relegated to the idea of a human grouping in a certain area, but without the houses having a visibly recognizable geometric organization: the home was essentially based on individual needs, for example, the so-called lake villages built on stilts for protection and survival. At this time, it seems that there were already wooden and/ or stone buildings for various religious rites. Among the most famous is the work of Stonehenge for the English religious site of the same name. The builders of Stonehenge, in particular, would have used the *tenon–mortise* interlocking technique to assemble the huge blocks of stone. The monument has a circular shape, and along its circumference are placed large erect stones, better known as megaliths. In a later period, known for its role as a city of workers, is the community of Kahun (ca. 1900 BC) in Egypt, (today) just under a hundred kilometres south of Cairo. The role of this community was to house the men who were dedicated to building the pyramids. In fact, we only have conclusive information on this place for the period of 21 years, that is, the time needed to build a pyramid. In terms of urban planning, this small town was arranged with a checkerboard geometry on an area of approx. 80 ha. In spite of the huge number of people needed to work on the pyramids, the city contained only 300 dwellings, each with 4–5 rooms [14, III, pp. 269–299]. Of course, slaves were excluded from these houses. These were relegated to an area that today we would call the periphery of the city. According to Briggs, Kahun's urban structure consisted of straight streets between the square-shaped neighbourhoods. A drainage channel ran between the houses "[…] providing the earliest known example of street drainage [14, III, p. 270]. From the point of view of materials, the Middle East had little timber available, but plenty of cane and palm leaves. The dwellings were built of dry clay (in the villages to the east of the Tigris) using these same reeds and palm leaves for supports and decorations. In Jericho you can find examples of walls made of mud bricks. In particular, holes were found in the bricks to confirm that the architectural scheme also included a planned structure of supporting beams [Cfr. 14, I]. Given the type of climate, with these techniques, the dwellings lasted for a long time; in this sense, clay was the main construction material for these peoples. However, in the Nile Valley, while clay had become the material used by all, the only uses of stone seem to have been controlled by the sovereign. Another civilization, however, quite well-known today was the Babylonian along the ancient and well-known river system. The brick construction system was decidedly different from those described above. Without going into great detail, suffice to say that the upper part of their construction was slightly convex and sometimes the bricks were wedged at an angle, while other rows of brick in the shape of spines angled in the opposite direction. A religious building par excellence was the Ziggurat (ca. 2000 BC) with the emblematic narrowing towers along several terraces. The city of Babylon, crossed by the Euphrates, was much larger than Ziggurat. It is known in the literature from the description of Herodotus of Babylon (active V sec. BC) that archaeologists consider valid though sometimes confused. Herodotus presents a description of the clay material used, and the method for the fabrication of the glazed bricks [Cfr. 14, III]. The latter represent the highest form of construction of that period in Mesopotamia. Derry and Williams [Cfr. 14, I; 19] offer us an important technological insight into the Babylonian people. For example, on the Ishtar Gate [Cfr. 14, I] we can see some representations of animals on the mammoth Babylonian gate. These, according to English scholars, were first modelled on a clay plane subsequently worked onto baked bricks, and eventually glazed. Derry and Williams also emphasize the most interesting aspect of the decoration: the same representations are also found underground. In fact, underground the foundation walls are equally highly worked, both at the entrance to the city and along the perimeter walls. But the focus on the mystery and strangeness of this also concerns the fact that those abundantly decorated foundations, as they were conceived, were (to the modern eye) *among the weakest points of the buildings.* [Cfr. 14, I]. On the subject, according to Briggs, however, it was starting with Greek cities that a well thought out planning scheme began to be achieved in relation to the environment in which these communities would have arisen. With a greater consolidation of sovereignty, one could also obtain well-thought-out means of defining a city plan. Briggs, in fact, considers crucial the realization according to which art and science can intertwine in a product as long as it is visible, concrete and immediately useful to man, and able to satisfy his everyday state of well-being. *But what exactly does Briggs historiographically mean by art and science in reference to the organization of a city?* In effect two methodological approaches were possible: "As a science, it consists in preparing plans to regulate the lay-out of a town with a view to making good use of the natural advantages of a site, and to securing favourable conditions for housing, traffic, industry, and recreation. As an art, it seeks to create an effect of dignity, harmony, and beauty." [14, III, p. 269]. In this analysis of Briggs' work, considerations emerge around urban planning and the history of architecture and science, through the second of the two approaches he has described: "Here we are concerned with the former aspect, and how it was practised in the ancient empires." [14, III, p. 269].

Derry and Williams, also analyse ancient civilizations using art and science as investigative categories that seem to find greater application [*Ivi*] in the flourishing Hellenic civilizations (compared to the earlier Egyptian and Babylonian communities). Already noted by Aristotle (384–322 a. C.) as having designed the fortified structure of Piraeus, the Greek Hippodamus of Miletus (V c. BC.) was the one who introduced the principle of building straight and wide streets, and, first of all the architects, made the project of a suitable grouping of dwelling houses and also paid special attention to the coordination of the different parts of a city in a harmonious whole, centered around an *agorà* (Cfr. Aristotle's *Politics*: II, 8 1267b, 22 sgg.). The structure of the Greek temples was built with mud bricks with a timber roof covered with straw. In agreement with Briggs, it should be noted, however, that Roman architecture is better known than Greek because, except for Herodotus, Greek commentaries and writers seemed to show no great interest in the magnificence of the architecture of their time. It is known that Hippodamus – commissioned by Pericles (fl. 495–429 BC) – in addition to designing Piraeus also planned the city of Turi. In particular, Briggs stresses the original transformation of the Greek port into square blocks with roads arranged in such a way as to avoid the obligatory passage to the *agora*. The town plan (if we can already call it this) was very different from that of Athens where instead the beauty of the buildings on the Acropolis contrasted with most of the houses of the *populus*, interspersed with narrow and winding streets; all without thought-out urban organization. Briggs' reflection is reinforced by his continued reference to the relationship mentioned above between science and art. In fact, the *agora* while geometrically central and sometimes surrounded by colonnades of rare beauty, at the same time showed all its functionality when it also served as a market. A place where stalls and temporary shelters apparently clashed with the magnificence of the geometries that surrounded, it served instead the most important component of the city—the people and their daily needs –. But it was also a place which at short notice had to be cleared for public assembly, which in turn, could be a requirement that benefitted the community. In one leap we pass to the Middle Ages and then on to the Renaissance to try to answer the following question: *are art and science are still valid categories* [20] *to investigate the history of architecture between the Middle Ages and the Renaissance*? Following on from Roman artistic transformations are those in urban planning, with the first great projects towards the master plan of Pope Sixtus IV (1471–1484) and Pope Sixtus V (1585–1590). Most notable is the transformation of the main streets of Rome at the hands of the Neapolitan architect Domenico Fontana (1543–1607) up until the works of Gian Lorenzo Bernini (1598–1680) who glorified Rome (1667) with his sculptures for all posterity. But, some architects also possessed scientific-technical inclinations, such as Leon Battista Alberti (1404–1472) who in his *De re aedificatoria* described new techniques in architectural design; or Buonaiuto Lorini (1540–1611) who in *Del-Le Fortificationi* (1596–1609) emphasizes the importance of the scientific-technical relationship [4, 5]. Most architectural historians agree that the designs of Alberti and Lorini are influenced by the attempt to join science and technology, with the latter sometimes seen as art. In his idea of the central importance of urban planning, Briggs points out [14, III, pp. 282–283] that from the fall of the Roman Empire until the Renaissance there was very little room for a dedicated discourse on the urban planning of a city. Therefore, throughout the Middle Ages the relationship between science-technique and art was barely considered to be of importance to a builder. Communities of the early Middle Ages (XIII-XIV centuries) were the product of the expansion of small villages (from Roman sites) that were not accompanied, however, by an equally thought-out design nor by a choice of urban regulation. It was all guided by the concern for defence, so the urban agglomeration took refuge within the boundaries of gigantic castles. This situation, according to historians, also produced limitations on *liveability*, as the whole logically had to be contained within a place that, for military reasons, offered little space for architectural creativity. The streets were increasingly narrow, and the center passed from being the *agora*, the heart of Greek socialization, to the congested medieval market square. [14, III, pp. 283–284]. However, some examples of thought-out medieval urban organization were already beginning to appear in France and England. Edward I (1272–1307) had 50 fortified cities built, while the French organized themselves through a network of fortified cities (*bastides*), similar to the extremely regular lay-out of Monpazier [8, 12]. It is a rectangular plan (with some diagonal variations) that arrives at the church and the market-place. The perimeter walls were made up of 10 gates that skirted the *pomerium* opening onto the road that ran the entire perimeter of the site; each house had a garden. In 1281 Edward I commissioned a project for a new plan of the city. In a partial reconstruction of this plan [14, III, p. 283] Briggs also considers the geometry of the problem, that it is presented as being absolutely regular composed of 39 housing blocks of square shape (*insulæ*): a housing system that also took into account the needs of the Franciscan friars to whom was assigned a large area that directly overlooked the church, and was easily accessible as the road was devoid of urban settlements. Briggs also focuses on Hull, with a checkerboard layout, on the port of York, and some German cities such as the beautiful Rothenburg ob der Tauber. In general, public and private buildings were mostly made of wood, and brick making was a forgotten practice. Those works in stone or brick were attributed to a few expert and trusted Italian or French masons [Cfr. 14, I]. The Romanesque style crossed the Alps also because of the activity of senior prelates who during their travels (not always official) needed to stop at certain abbeys and monasteries. In particular, the churches, as well as the castles were characterized by semicircular barrel vaults of Roman origin. The thrust was supported by the introduction of a second vault or by an increase in wall thickness. Around 1400 (ca.) the walls were replaced by stained glass windows while the thick walls and support pylons were replaced by thin walls that made better sense of the geometric space available [*Ibidem*]. In this period the stone was worked and shaped directly in the quarry in order to lower the transport costs. Everything revolved around the figure of the mason with some variety of specialization: a) the *labourer* masons that transported the material; b) the *basic* masons employed only to cut the stone, c) the *specialized* masons employed in the subsequent processing of the stone according to the design requirements. The figure of the mason was flanked by that of the carpenter who was charged with building, for example, the beautiful oak roofs of English Gothic buildings. The technique of supporting rafters from outside of the stone vault, not suitable for sloping roofs, was replaced by a complex carpentry work which gave birth to trusses with beams arranged in hammer configuration. For added safety the pitched roofs of the buildings were usually covered with lead. In this period there were remarkable developments around navigation, including the construction of bridges and locks to promote trade along rivers, the return of boats to the mainland. From the point of view of urban planning, according to Briggs it was the Italian Renaissance, certainly the richest example of the science-art relationship, which produced the first examples of urban planning. Here this relationship manifested itself in line with its character according to which "On the one hand are a number of architects and other writers who developed theories of ideal town-planning in sundry treatises and projects; and on the other is the endeavour to apply such theories in practical examples." [14, III, pp. 285]. Although it is likely to appear rhetorical, among Renaissance scholars it must also be recalled that Leonardo da Vinci (1452–1519), in addition to works of mechanics, during his stay in Milan (1481–1499) was also involved in urban planning as we witness in his notes between 1483 and 1518. Briggs devotes significant space to this aspect of Leonardo da Vinci's work focused on the relationship between science (mechanics) and architecture [21, 22]. In particular, the English scholar interprets, from the point of view of aesthetics, some of Leonardo's notes on how he, as an architect, intervenes in the liveability of the city both during the day and at night. Other notes relate to techniques and theories used for defence along river courses from both enemies and floods. According to Briggs, in some passages of his notes, Leonardo anticipated both a modern town-planning as "[...] let the street be as wide as the universal height of the houses [...]" and a *modern principle of dispersing the population of overcrowded cities into satellite towns* [14, III, p. 285; 23]. In effect, Leonardo's mental representation of building an *ideal city* with a rationalised distribution of traffic on two levels was an idea possessing a certain depth. But it was also the sign of progress (today we would say) in a city's zoning plan: upper roads for pedestrians (accessible via streets at the edge of the city), and lower roads for means of transport [23]. The two levels were connected by flights of stairs. In Briggs' work there is no lack of references to Leonardo's annotations about the design of canals and inland waterways in the city. The scientist worked on irrigation (Vigevano) and it seems that he also invented lock gates set at 45-degree angles. Another passage mentioned by Briggs in which the technical (-art)-science relationship is important concerns the architect Francesco di Giorgio Martini (1439–1502), who published famous drawings on an ideal city with a polygonal and fortified shape. The city plans that appear in the works of Francesco di Giorgio Martini have geometric shapes in radial or convex forms and are mostly solutions for cities that must be built on high plateaus. This choice also derives from a certain rationalization of the terrain's orography. The city of Francesco di Giorgio Martini follows the architectural conception typical of the *ideal city* of the fifteenth century. It consists of a polygonal and regular perimeter with appropriate battlements. At the end of his work Briggs reviews some Italian cities including Turin [16, 8]. It is well known that during the Renaissance the reinterpretation of the classical style was the greatest influence on architecture of the time. According to Derry and Williams, but in accordance with most of the conclusions of architectural historians, it is possible to identify the birth of Renaissance architecture with the construction of the *brick dome that Brunelleschi erected in Florence in 1420, with two brick shells, connected with 24 stone ribs along the sides of the octagonal vault* [Cfr. 14, I]. Considered in more detail, the large dome, or cupola, of Brunelleschi in Florence is formed, in turn, by two brick domes arranged in cells, that is, separated by an intermediate space. The brick structure forms an octagonal pavilion vault. Briggs precisely summarized Brunelleschi's account (1420; 14, III, p. 248]. Regarding the cupola of Florence, Briggs cites the publication (1820) of an early drawing which showed a scaffolding for the cupola held up with supports that started from the frame at the base of the dome. According to Briggs, as this was *somewhat similar to the seventeenth-century drawing by Carlo Fontana of the centinatura used in San Pietro in Rome* [Cfr. 14, III, pp. 246–248] it can be assumed that Brunelleschi may have used the same method. In fact, it is well known that the construction of the dome of Saint Peter's of Michelangelo is rather similar to that of Brunelleschi, especially in that it does not represent significant structural advances compared to the model used by Michelangelo. Although – as Briggs rightly points out – the cupola of Florence is much steeper and therefore the overall strength-weight of the lantern is transferred to the supports, that is structurally is closer to the vertical. Briggs, also to investigate his relationship between art and science in the Italian Renaissance, identifies two lines of development that – according to him – circulate in this period among Italian architects: a) A purely architectural one, made up of projects and measures; b) One made up of building builders, bricklayers and skilled carpenters [Cfr. 14, III, pp. 245–247]. But in order to understand the structural changes, one must first understand the *sweeping revolution* that manifested itself in the front-line field, that is architecture [Cfr. 14, III, p. 246]. From this point of view, Briggs moves on to review the architectural style of Italian churches and palaces, focusing on the masonry, the arches, and the role of the vaults, as presented below. Starting with Vitruvius (fl. 80–70?-23 BC) the Doric, Ionic and Corinthian styles were abandoned in favour of new buildings with purely decorative purposes, in which the scientific temperament of the work of certain more cultured architects (e.g., Brunelleschi) was more in the service of aesthetics than structural issues, in the strictest sense. In this regard, Briggs presents some examples of structural situations on the arches of the inner courtyards of the large palaces, and on the columns of separation in the naves of the churches that rest directly on the capitals of the Corinthian columns. For example, he reports the case of the Florentine palaces that dispensed with columns and instead mounted cornices of the Romanesque style. More precisely, those mentioned by Briggs are the Florentine cornices of *Palazzo Strozzi* (c. 1500) and Palazzo Riccardi (1430). The cornice of *Palazzo Riccardi* is 3 m high and has a projection of 2.54 m. [Cfr. 14, III, pp. 247–248]. The two buildings were finished with large stone blocks (*ashlar*) to produce an effect of majesty, although the stone facade was in fact thin, and the walls were masonry (bricks or debris). Like all Italian Renaissance palaces for the use of nobles and wealthy families, they were bright, spacious and usually had very high vaulted ceilings. Strozzi and Riccardi were also made in the precise Romanesque style, but contained variants such as the *ricasco* curve in each of the walls. The *ricasco* was made with woven matting and mortar, and the ceiling was composed of lattice beams made of wood and held together with rods or ropes. Regarding the structure of the roofs of Italian Renaissance churches – according to Briggs – several had roofs with a slightly steep slope and mostly made of wooden trusses, such as the Christian basilicas of Rome. In these cases, the thick wooden beams were an integral part of the structure. In addition to the palaces with stone cornices, other Italian buildings and churches of the Renaissance were characterized by very protruding eaves. For example, think of the *Cappella Pazzi* in Florence (1429) where the deep overhang reaches almost 2.45 m, made possible thanks to the care of the structural engineer of the time who doubled the thickness of the roof rafters right at the point where these could have given way, being more subject to flex under the load of the protruding eaves. *But what exactly was the source of inspiration through which the mixture of art and science, made possible in some architects, manifested itself with the realization of such works?* The rediscovery of Vitruvius' text in the early Renaissance aroused interest among Renaissance architects and immediately became the first authentic handbook-guide in both design and practical building: for example, baking lime, seasoning timber, painting on stucco [14, III, pp. 249–250]. It seems that even the Greek Theophrastus (ca. 372–287 BC) served as a guide, at least for Alberti who cited him in his *De re aedificatoria* (1485) *although Theophrastus' passages on building materials were more descriptive than scientific* [14, III, p. 250]. However, only a few men dedicated to the art of design balancing aesthetics and science were able to produce magnificent works. Nevertheless, e.g., "[...] was he [Leonardo da Vinci] able to contribute anything useful to the science of building, and were there any other pioneers in the same promising field?" [14, III, p. 250]. It is not clear to what extent Leonardo da Vinci would have carried out – as did Galilei [12, 15] and Evangelista Torricelli (1608–1647) [24, 25, 36] in his own way – specific research in mathematics or mechanics. However, there is no doubt that during his youth Leonardo learned something from a book on the mechanics of the early fifteenth century and from relations with the mathematician Pacioli (1445–1519) [21, 14, III, p. 250]. Now without going into too much detail, all the comments of Briggs on Leonardo, and the umpteenth modern synthesis from 1960s, when Briggs was writing, are already very well known, we think it useful to report some history of science considerations on Leonardo with respect to history of technology. This is to make more evident the historiographical methodology within the debate on the interplay between the history of technology and the history of science announced at the beginning. Leonardo, who was extremely interested in the deformability of beams in one of his drawings, created an instrument to measure the *deformability* [26] of a wire to breaking point. It must be noted that the elongations were imperceptible and therefore not always identifiable using the techniques of the time. Leonardo da Vinci, in a brief description, explains that the first thing to do is to look for the maximum load that the wire (ca. 2 *braccia* = ca. 1.16 m) can bear. Then it is necessary to find a convenient position to support the wire and then tie a certain container to the wire and pour fine sand through a small hole. A spring that closes the hole when the wire breaks determines the moment when the maximum load has occurred. Leonardo, as an excellent experimenter, does not limit himself only to the understanding of the phenomenon in its final act (the result), but also recommends incrementally recording the mass reached corresponding to the sand poured in, and also establishes one of the elementary rules of an experiment: its reproducibility. Briggs notes that Leonardo's experiment using sand for tensile tests would continue to be applied for a long time after the Renaissance. As for Leonardo's drawings concerning structures, the English historian dwells on beams and loaded struts. It was one of Leonardo's many studies attempting scientific solutions to the structural problems he faced. Resolution was often accomplished by first methods of decomposing a single force (or velocity) along the components of a Cartesian plane. (Although the theory on the decomposition of static forces as we understand it today was still not clear, at this time). One of the problems he faced was also the lack of adequate scientific (symbolic) language, as that would only appear in the middle of the seventeenth century [Cfr. 27, 28]. In spite of this, his studies on the existence of struts, which had no precedent, and on weight-bearing pillars were not overtaken at least for the whole of the eighteenth century. Leonardo is not exempt from criticism and errors. Although he did not obtain accurate results on the studies of loaded beams, he used a method [22] which allowed him to deal in detail with problems in statics. However, Leonardo does not seem to have completely overcome the structural problems of his time. On the basis of Briggs' reflections, it could also be assumed that the reason for this lack would be the fact that in that period there were no worthy interlocutors (on structural problems) with whom to compare and improve research results. Briggs concludes that "Despite Leonardo's pioneering efforts, there is no evidence that architects of his own or the next generation made use of the science of structural mechanics." [14, III, p. 252].

3 Concluding Remarks: The Role of Machines/Mechanics into Engineering/Architecture Designs During the Renaissance

The complexity of the historical and historiographical analyses are remarkable. Only thanks to a work on the history of science which is also extremely attentive to its foundations has it been possible to identify a certain influence of mechanical science, which when applied to architecture takes the name of building science. It remains to be understood in detail *how* and to what extent, author by author. We remark that an inquiry based solely in the history of technology could not reveal this aspect. For example, the use of geometry was particularly fruitful in the representation of architectural objects: buildings, fortifications [Cfr. 8, 9, 10, 11, 12], urban plans [Cfr. 3]. For example, think of the process of measurement, and the sensitivity of the gears of machines that required a correct numerical localization of the positions of objects in space. It should be also noted that the first scholars who systematically applied geometric design to architecture, with explicit reference to the supporting elements, were Leonardo da Vinci and Bernardino Baldi (1533–1617). The latter is well known to historians of science, also thanks to the contribution of Pierre Duhem who dedicated many pages to him [38, II, pp. 127–141] considering him a figure of importance although not very original. Thus, at least for the Italian Renaissance, those of Leonardo and Baldi represent perhaps the greatest successes in the application of mechanical science to design. The general understandings obtained by Galileo are used as empirical criteria for sizing in carpentry works: scaffolding, trusses, slabs (see Pisano's works). The way of doing architecture passes from the medieval reproduction of predefined modules to designing for objectives that can be justified by: a) the breaking strength of masonry and stone materials, dominant in construction, is so high that it does not in fact impact sizing; b) problems in static equilibrium that existed, such as the absorption of thrust in an arch, were solved by following rules of an empirical nature; the criteria of geometric sizing, however, concerned only major structures; c) secondary but fundamental elements, such as floors, roof trusses and stairs were always trusted to the experience of the master builder. Machines are designed and built according to the laws of mechanics, as regards the determination of the relationship between power and resistance, but they seem to be sized only on an empirical basis as regards robustness. In addition, when Gothic architecture flourished in the 13th century, with its bold spires, the mechanical studies of Jordanus de Nemore (fl. 13th century) [7] appeared, the most important until Galileo; and this could not have been by chance. So far, however, no decisive written evidence has been found that supports an historical hypothesis in favour of the use of (formalized) notions of mechanical science in architecture and engineering. In general, the solution of structural problems was entrusted to the use of small-scale models, verifying that these were sufficiently resistant to then move to the realization of the prototype in real scale. However, it soon became clear that the full-scale structure was not as strong as the model. In this regard, think of the account of Camillo Agrippa (1535–1589) on the movement of the obelisk from St Peter's Basilica to the center of the square in front. In the early seventeenth century, scientific and technological knowledge was potentially available which could have allowed a structural sizing, in the modern sense, of architectural works, that is, by means of calculations.

Coming back to *No theory—no machine*, our answer is functioning machines can be constructed without any theory. This is *historical* fact and *historiographical* interpretation (also linked to perpetual motion) [4, 5]. This is certainly true for machines constructed in the long period preceding the 16th–17th centuries: many machines were constructed independently of a theoretical support. This depends on two internal *historical–scientific* facts, which have to be taken into account: a) As to the lever and connected machines: weights can be used in a correct manner without knowing the difference between weight and mass, and without having a precise idea of what a force is; b) As to hydraulic machines: the concept of water flow, velocity and the crucial relationships between these quantities and their relative proportions run, without scientifically knowing the relations between them, e.g., volume, advantage and speed. In addition, the concepts of the science of mechanics are theoretically strictly connected to machines, but, from a historical standpoint, a relatively broad use of machines preceded the science of mechanics. In fact, we can *historiographically* admit that if we establish: 1) *Mechanics*, as a part of theoretical physics, is the branch of scientific analysis which deals with motion, time and force; 2) *Machines* as rigid bodies connected by joints in order to accomplish a desired force and/or motion transmission; 3) *Mechanics* (maths-physics & geometry) as the final result of the gathering between machinery and machines. So that mechanics includes the culture of machines–technique/technology, which includes machinery. This is the theoretical perspective which does not correspond to the historical development. Then we can add to the historiography above analysed that in the Renaissance there is also a Culture of Machines/Mechanics which was not theory; it was also not a simple practice. For, there was no continuity in the relations mechanics culture of machines–machinery. Thus, a new perspective in the studies concerning the relations mechanics–culture of machines–machinery arises. But, these aspects should be approached (including the interplay between them) by internal history of science (physics, mathematics etc.) cultivated by historiographical—and—intellectual (ideas) history of science investigations [Cfr. 4, 5, 8].

References

1. Koyré, A.: Newtonian studies. Harvard University Press, Cambridge–MA (1965)
2. Pisano, R., Agassi J., Drozdova D., (eds) Hypothesis and Perspective in History and Philosophy of Science. Homage to Alexandre Koyré 1892–1964. Springer, Dordrecht (2017)
3. Pisano, R., Bussotti, P.: On the Conceptualization of Force in Johannes Kepler's Corpus: an Interplay between Physics/Mathematics and Metaphysics. In: Pisano R., Agassi J., Drozdova D., (eds). Hypothesis and Perspective in History and Philosophy of Science. Homage to Alexandre Koyré 1892–1964. Springer, Dordrecht, pp. 295–346 (2017)
4. Pisano, R., Bussotti, P.: Historical and epistemological reflections on the culture of machines around the renaissance: machines. Machineries and Perpetual Motion Acta Baltica Historiae et Philosophiae Scientiarum **3**(1), 69–87 (2015)
5. Pisano, R., Bussotti, P.: Historical and epistemological reflections on the culture of machines around the renaissance: how science and technique work? Acta Baltica Historiae et Philosophiae Scientiarum **2**(2), 20–42 (2014)
6. Pisano, R., Bussotti, P.: Fibonacci and the Abacus Schools in Italy. Mathematical Conceptual Streams, Education and its Changing Relationship with Society. Almagest **6**(2), 126–165 (2015)

7. Pisano, R., Capecchi, D.: Tartaglia's Science Weights. Mechanics in Sixteen-Century. Selections from Books VII-VIII. Springer, Dordrecht (2016)
8. Pisano, R.: A Tale of Tartaglia's *Libro Sesto & Gionta* in *Quesiti et Inventioni diverse* (1546–1554). Exploring the Historical and Cultural Foundations. Foundations of Science **25**(2), 477–50 (2020)
9. Galilei, G.: Ms. B – Trattato delle fortificazioni, con disegni diversi. Biblioteca Ambrosiana. D: 296 Inf.– 38499. (fl. 16th)
10. Galilei, G.: Breve instruzione all'architettura militare. In: Le opere di Galileo Galilei: Edizione nazionale sotto gli auspici di sua maestà il re d'Italia. Favaro A (ed.), Barbera, Firenze, II, pp. 15–75 (1890–1909)
11. Galilei, G.: Trattato di fortificazione. In: Le opere di Galileo Galilei: Edizione nazionale sotto gli auspici di sua maestà il re d'Italia. Favaro A (ed.), Barbera, Firenze, II, pp. 77–146 (1890–1909)
12. Pisano, R., Bussotti, P.: Galileo in padua: architecture, fortifications, mathematics and « practical » science. Lettera Matematica Pristem International (Springer) **2**(4), 209–221 (2015)
13. Pisano, R., (ed), A Cultural History of Technology in the Age of Encounters (1450–1650), Vol. III. In: Jennifer Alexandre, J., Bix, A., Suzanne Moon, S., Storey, W., (eds). A Cultural History of Technology in the Age of Encounters, 6 Vols. Bloomsbury, London (expected 2023)
14. Singer, C.J., Holmyard E.J., Hall, A.R., Williams T.I. (ed), A History of Technology. VIII Vols. (1954–1958). [I.: From Early Times to fall of Ancient Empires; II: The Mediterranean Civilizations and the Middle Ages; III: From the Renaissance to the Industrial Revolution; IV: The Industrial Revolution; V: The Late Nineteenth Century; VI-VII: The Twentieth Century, 2 Parts; VIII: Consolidated Indexes]. Clarendon Press, Oxford
15. Pisano, R.: Galileo Galilei: The Two Le Mecaniche. Translations, Text and Commentaries. Springer, Dordrecht, forthcoming (2022)
16. Tartaglia, N.: La nuova edizione dell'opera Quesiti et inventioni diverse de Nicolo Tartaglia brisciano Riproduzione in facsimile dell'edizione del 1554. Masotti A (ed), Commentari dell'Ateneo di Brescia, Tipografia La Nuova cartografica, Brescia ([1554]1959)
17. Singer, C.J.: A short History of scientific ideas to 1900. OUP, Oxford (1959)
18. Briggs M.S.: The architect in history. Da Capo Press, New York ([1927]1974)
19. Derry, T.K., Williams, T.I.: A Short History of Technology. OUP Press, Oxford (1970)
20. Pisano, R., Gaudiello, I.: Continuity and discontinuity. An epistemological inquiry based on the use of categories in history of science. Organon **41**, 245–265 (2009)
21. Pisano, R.: Details on the mathematical interplay between Leonardo da Vinci and Luca Pacioli. Bulletin British Society for the History of Mathematics **31**(2), 104–111 (2016)
22. Pisano, R.: Continuity and discontinuity. On method in Leonardo da Vinci' mechanics. Organon **41**, 165–182 (2009)
23. Da Vinci, L.: Ms. B, ff. 36–39, 47r. Institut de France (c. 1487– c. 1490)
24. Pisano, R., Bussotti, P.: Historical and Methodological Details on the *De motu gravium naturaliter descendentium* in Torricelli's *Opera Gometrica* (1644) In: Pisano, R., Dhombres, J., Radelet de Grave, P., Bussotti, P. (eds). See 25 (2022)
25. Pisano, R., Dhombres, J., Radelet de Grave, P., Bussotti, P. (eds). 378th Anniversary of Torricelli's *Opera Geometrica* (1644) Text, Transcription, Commentaries and Selected Essays as New Historical Insights, *Forthcoming* (2022)
26. Pisano, R., Capecchi, D.: Leonardo da Vinci. Recenti riflessioni storico–epistemologiche sulla deformabilità dei corpi. Proceedings SISFA 2007. La fisica nella scuola XLI(3), 120–129 (2008)
27. Knobloch, E., Mathematical language and mathematical progress retrieved: https://halshs.archives-ouvertes.fr/halshs-00004274v3/file/08_Knobloch.pdf

28. Knobloch, E., Mathematical Methods in Preindustrial Technology and Machines. In: M. Lucertini M., Gasca A.M., Nicolò F. (eds). Technological Concepts and Mathematical Models, Springer–Verlag, Berlin (2004)
29. Da Vinci, L., Codex Atlanticus, f. 119v
30. Galilei, G.: Discorsi e dimostrazioni matematiche intorno a due nuove scienze attenenti alla meccanica e i movimenti locali, del signor Galileo Galilei linceo. Elsevirii, Leida. (1638). In: Galilei, G., 1890–1909, VIII, pp. 39–362
31. Galilei, G.: Le opere di Galileo Galilei: Edizione nazionale sotto gli auspici di sua maestà il re d'Italia. Favaro A (ed), Barbera, Firenze (1890–1909)
32. Pisano, R., Bussotti, P.: Conceptual Frameworks on the Relationship between Physics–Mathematics in the *Newton Principia Geneva Edition* (1822). Foundations of Sciences Springer (2022). https://doi.org/10.1007/s10699-021-09820-2
33. Pisano, R., Bussotti, P.: A newtonian tale details on notes and proofs in *Geneva Edition* of Newton's *Principia*. Bulletin British Society for the History of Mathematics **31**(3), 160–178 (2016)
34. Pisano, R.: Il ruolo della scienza archimedea nei lavori di meccanica di Galilei e di Torricelli. Da Archimede a Majorana: La fisica nel suo divenire. Guaraldi Editore, Rimini, pp. 65–74 (2009)
35. Pisano, R., Capecchi, D.: On Archimedean Roots in Torricelli Mechanics, pp. 17–28. The Genius of Archimedes. Springer, Dordrecht (2010)
36. Duhem, P.M.: Études sur Léonard de Vinci, Hermann, Paris (1906–1913)
37. Duhem, P.M.: Les origines de la Statique, 2 Vols. Hermann, Paris (1905–1906)

Historical Details of Coulomb's Torsional Forces in Metal Wires

Agamenon R. E. Oliveira[1(✉)] and Raffaele Pisano[2]

[1] Polytechnic School of UFRJ, Rio de Janeiro, Brazil
agamenon.oliveira@lwmail.com.br
[2] IEMN, Lille University, Villeneuve d'Ascq, France
raffaele.pisano@univ-lille.fr

Abstract. The theoretical and experimental research carried out by Charles Augustin Coulomb (1736–1806) to calculate torsional forces in metal wires had two objectives: 1) based on the Newtonian mechanical paradigm to determine the torsional tensile strength of iron and brass wires, relating this to the length of the wire, its diameter and traction; 2) calculate the imperfection of the elastic reaction of the metal wires, and describe what possible consequences are taken, related to the laws of elasticity of the bodies involved. Based on historical documents—&—method, for the sake of brevity, we mainly aim to describe a) the early experiments and his combined torsional vibrations modelling, b) to detail Coulomb's construction of the torsional balance. We integrated the methods outlined in *History of Physics* and *History of Science and Technology*.

Keywords: History of mechanical oscillations · Torsional vibrations in wires · Elasticity problems in torsion · History of the torsion balance

1 Introduction

Our work is an historical analysis of the text. It aims to describe step—by—step what (*history of ideas*), when (*history of facts*), and how *(history of physics/applied science and technology*) Coulomb produced his Torsional Forces in Metal Wires experiment within a combined history of science and history of science and technology. We preserved the style of formulas typical of Coulomb's manuscript.

In 1784, Coulomb read his research on torsion in metal wires [1], with the purpose of determining their strengths and relating them to their elastic properties. In 1785, he presented to the French Academy of Sciences his first study on electricity and magnetism [2], in which, together with other subsequent studies, he was led to the fundamental law of electrostatics. Some year later Alessandro Volta (1745–1827) presented his battery to the French court [5, 6]. The experimental results of this 1785 study were related to one of his best-known achievements: the torsion scale. With the help of this new mechanical device, it was possible to outline the law governing the attraction and repulsion of forces of electrical charges in points, in other words, directly proportional to the product of charges and inversely proportional to the square of the distance between them, mathematically

M. Ceccarelli and R. López-García (Eds.): HMM 2022, HMMS 40, pp. 241–254, 2022.
https://doi.org/10.1007/978-3-030-98499-1_20

analogous to the Newtonian law of gravitation [3]. The method for determining torsional force is to suspend a cylindrical weight by a metal wire, so that its shaft is vertical or in the direction of the suspended wire. As long as the wire is not twisted, it will be at rest. However, if we twist the weight around its axis, the wire will also twist, making an effort to restore movement to the equilibrium position. If the twisted wire is released, it will wobble for a while. By observing the duration of a certain number of these oscillations, through the formulas of oscillatory movement, the torsion reaction force will be determined. By varying the suspended weight, length of the wire and its diameter, Coulomb expected to determine the laws of the torsion reaction in relation to the traction, length, and diameter of the wire, depending on the wire material. If the metal wire was perfectly elastic, if the air resistance did not change the oscillation amplitude, the suspended wire would oscillate until it was stopped. The decrease in the amplitudes of the oscillations cannot be attributed to anything other than air resistance and imperfections inherent in the elastic properties of the material due to torsion. Coulomb's expectation was that by observing the successive decrease of oscillation amplitudes and subtracting the part of the change that could be attributed to air resistance, through the oscillatory movement formulas applied to the experiments performed, the torsional force could be determined altered by the deviations of the expected elastic behavior. Coulomb divided his work into two parts. In the first, he assumed that torsional force was proportional to the torsion angle, made according to the experiments, as long as the oscillations were small. Some practical applications are also shown, extracted from this theory. In the second part, Coulomb uses a series of experiments to show changes in elastic forces and try to figure out their forms of variation for large oscillations. Below we present Coulomb's 1784 mathematical model (Fig. 1):

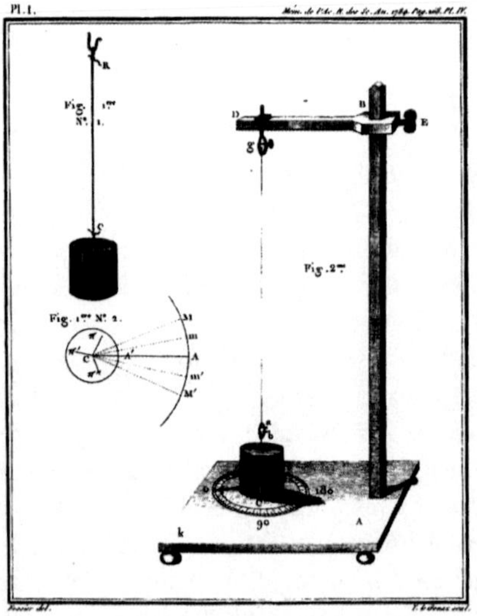

Fig. 1. 1784. Model used by Coulomb to study twisting in metal wires [1, p. 269]. Source: *Académie des science* Dossier Coulomb; National French Library BnF, Gallica.

In Fig. 1 above, a cylindrical body B is suspended by a wire in the direction RC so that the cylinder shaft remains vertical, aligned with the wire and upright. By rotating the cylinder around its axis, we try to determine the torsional force, assuming that it is proportional to the torsion angle and therefore also determine the formulas of the oscillatory moment. Looking at a horizontal section of the cylinder and projecting all its elements over to this circular section in the planes π, π′ and π″, and if assuming that a) the initial torsion angle is *ACM* = *A*, b) after a certain instant *t* the angle is *ACm*, we find that then the initial angle has been decreased from the angle MCm = S. Thus, the effective angular displacement is:

$$\mathrm{ACm} = (\mathrm{A} - \mathrm{S})$$

n is a constant that will depend on the nature of the wire, its length, and its diameter; if we call v the velocity of any point over the plane π, in time *t*, we can write the equation below for the impulse and the quantity of motion of the cylinder:

$$\mathrm{n(A - S)dt} = \int \pi \mathrm{rdv} \qquad (1)$$

where *r* is the distance Cπ from a point in the plane π to the axis of rotation *C*. Calling the radius of the cylinder *a*, the speed of an inner point of radius *r*, after a certain time t, is:

$$\mathrm{v} = (\mathrm{r/a})\mathrm{u}$$

where

$$\mathrm{n(A - S)dt} = \mathrm{du} \int \pi \mathrm{r}^2/\mathrm{a}.$$

Considering that: dt = (a/u) ds, and integrating Eq. (1), we obtain:

$$\mathrm{n}\left(2\mathrm{AS} - \mathrm{S}^2\right) = \mathrm{u}^2 \int \pi (\mathrm{r/a})^2 \qquad (2)$$

where

$$\mathrm{dt} = \mathrm{ds}/\left(2\mathrm{AS} - \mathrm{S}^2\right)^{1/2} \cdot \left(\int \pi \mathrm{r}^2/\mathrm{n}\right)^{1/2}$$

By applying the boundary condition when *S* = *0* and *S* = *π/2*, we can calculate the period of oscillation as follows:

$$\mathrm{T} = \left(\int \pi \mathrm{r}^2/\mathrm{n}\right)^{1/2} 180° \qquad (3)$$

To compare the torsional force with the force of gravity on the pendulum, if λ is the length of the pendulum, we obtain the following value for the period of oscillation:

$$\mathrm{T} = (\lambda/\mathrm{g})^{1/2} 180° \qquad (4)$$

where g is the gravity acceleration. For isochronism of oscillations, we obtain:

$$\int \pi r^2/n = \lambda/g. \tag{5}$$

From which we take the value of n, by using experimental results with the other known parameters. Coulomb considers how to calculate $\int \pi r^2$ that corresponds to the moment of inertia of a cylinder, and concludes [1, p. 233, VII] that, if M is the cylinder's mass and r its radius,

$$\int \pi r^2 = {}^1\!/_2 Ma^2$$

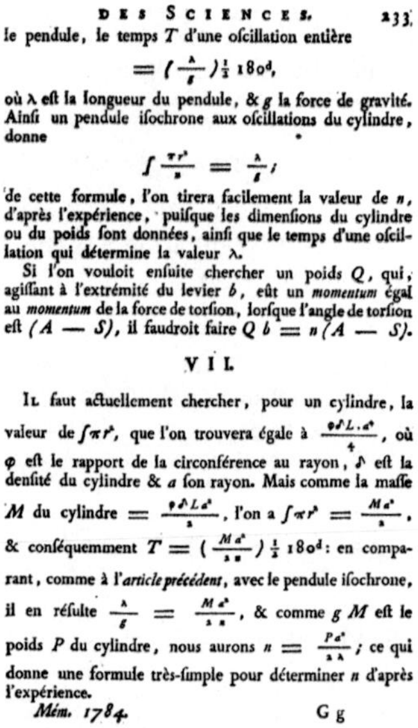
DES SCIENCES. 233

le pendule, le temps T d'une oscillation entière

$$= \left(\frac{\lambda}{g}\right)^{\frac{1}{2}} 180^d,$$

où λ est la longueur du pendule, & g la force de gravité. Ainsi un pendule isochrone aux oscillations du cylindre, donne

$$\int \frac{\pi r^2}{n} = \frac{\lambda}{g};$$

de cette formule, l'on tirera facilement la valeur de n, d'après l'expérience, puisque les dimensions du cylindre ou du poids sont données, ainsi que le temps d'une oscillation qui détermine la valeur λ.

Si l'on vouloit ensuite chercher un poids Q, qui, agissant à l'extrémité du levier b, eût un *momentum* égal au *momentum* de la force de torsion, lorsque l'angle de torsion est $(A - S)$, il faudroit faire $Q\,b = n\,(A - S)$.

VII.

Il faut actuellement chercher, pour un cylindre, la valeur de $\int \pi r^2$, que l'on trouvera égale à $\frac{\varphi \delta L . a^4}{4}$, où φ est le rapport de la circonférence au rayon, δ est la densité du cylindre & a son rayon. Mais comme la masse M du cylindre $= \frac{\varphi \delta L a^2}{2}$, l'on a $\int \pi r^2 = \frac{M a^2}{2}$, & conséquemment $T = \left(\frac{M a^2}{2 n}\right)^{\frac{1}{2}} 180^d$: en comparant, comme à l'*article précédent*, avec le pendule isochrone, il en résulte $\frac{\lambda}{g} = \frac{M a^2}{2 n}$, & comme $g\,M$ est le poids P du cylindre, nous aurons $n = \frac{P a^2}{2 \lambda}$; ce qui donne une formule très-simple pour déterminer n d'après l'expérience.

Mém. 1784. Gg

Fig. 2. 1784. Mathematical Model used by Coulomb to study twisting in metal wires [1, p. 233, VII]. Source: *Académie des science* Dossier Coulomb; National French Library BnF, Gallica.

Consequently, the period of oscillation is:

$$T = \left(Ma^2/2n\right)^{1/2}.180^\circ$$

Comparing this result with the period of oscillation of an isochronous pendulum, we will have:

$$\lambda/g = Ma^2/2n \tag{6}$$

and $Mg = P$, the cylinder's weight. Consequently:

$$n = Pa^2/2\lambda \tag{7}$$

The relation (7) will be very useful for the experimental calculation of n.

2 Coulomb's Mathematical Model

As we know, the pendulum isochronism is a first approximation for movements with small oscillations. If the angle of twisting exceeds a certain value, the oscillations are no longer isochronous. Coulomb then admits [1, p. 233, VII; Fig. 2] that it is possible to refine the previous model assuming that the torsional force can be changed by a certain amount *R*, to move away from the isochronism, as follows:

$$[\mathrm{n(A-S)-R]dt = du} \int \pi \mathrm{r}^2/\mathrm{a} \tag{8}$$

Using the same substitutions in (1), one obtains:

$$\mathrm{n}\left(2\mathrm{AS}-\mathrm{S}^2\right) - 2\int \mathrm{RdS} = \mathrm{u}^2 \int \pi \mathrm{r}^2/\mathrm{a}^2 \tag{9}$$

To solve the problem, Coulomb proposes to consider a complete two-part oscillation. The first from *M* to *A* where the torsional force accelerates velocity *u*, until the retarding force decreases it; the second part, after reaching angle *A* when all the forces that retard the movement compete. Considering the first part of the movement above, Coulomb assumes that $\mathrm{R} = \alpha(\mathrm{A}-\mathrm{S})^{\mathrm{m}}$.; that is, that the retarding portion of the torsional force has a polynomial shape in S. Thus, one obtains:

$$\mathrm{n}\left(2\mathrm{AS}-\mathrm{S}^2\right) + 2\alpha(\mathrm{A}-\mathrm{S})^{\mathrm{m}+1}/\mathrm{m}+1 - 2\alpha\mathrm{A}^{\mathrm{m}+1}/\mathrm{m}+1 = \mathrm{u}^2 \int \pi \mathrm{r}^2/\mathrm{a}^2 \tag{10}$$

For the boundary condition in which the angle vanishes, we have:

$$\mathrm{nA}^2 - \left(2\alpha\mathrm{A}^{\mathrm{m}+1}/\mathrm{m}+1\right) = \mathrm{U}^2 \int \pi \mathrm{r}^2/\mathrm{a}^2$$

where *U* is the new velocity at *A*. Substituting U^2 by its value given below:

$$\mathrm{U}^2 = \left(\mathrm{nA}^2 - 2\alpha\mathrm{A}^{\mathrm{m}+1}/\mathrm{m}+1\right)/\int \pi \mathrm{r}^2/\mathrm{a}^2 \tag{11}$$

We can then do full integration until the velocity becomes zero, or the oscillation is reached:

$$\left(\mathrm{A}-S'\right) = 2\alpha/\mathrm{n}(\mathrm{m}+1)\cdot\left(\mathrm{A}^{\mathrm{m}+1} + S'^{\mathrm{m}+1}\right)/\mathrm{A}+S' \tag{12}$$

In addition, if the retarding forces are such that at each oscillation, the amplitude is slightly decreased, we will have an approximate value $(\mathrm{A}-\mathrm{S}')$:

$$\left(\mathrm{A}-S'\right) = 2\alpha\mathrm{A}^{\mathrm{m}}/\mathrm{n}(\mathrm{m}+1)$$

If the above quantity is too small to be treated as an ordinary differential, then, for Z number of oscillations:

$$2\alpha/\mathrm{n}(\mathrm{m}+1)\cdot\mathrm{Z} = (1/\mathrm{m}-1)\cdot\left(1/\mathrm{S}^{\mathrm{m}-1} - 1/\mathrm{A}^{\mathrm{m}-1}\right) \tag{13}$$

Where S represents the value to which A tends after a number of Z oscillations.

$$\mathrm{S} = \left\{2\alpha(\mathrm{m}-1)/\mathrm{n}(\mathrm{m}+1)\cdot\mathrm{Z} + \left(\mathrm{A}^{\mathrm{m}-1}\right)^{-1}\right\}^{1/1-\mathrm{m}} \tag{14}$$

Which gives the value of S after any number Z of oscillations.

3 Coulomb's Physical Model

To understand the experimental apparatus used by Coulomb in his experiments with metallic wires, we can also make use of the Fig. 1 used to establish the mathematical model of the problem. In the figure mentioned, we can see how it is possible to replace wires and use different materials, diameters, and lengths. In addition, an angular displacement meter in degrees can be seen under the weight that pulls the wire and serves to measure the oscillations of the system. Coulomb uses three clove threads, due to the ease to find them in the market. They were sold in reels and numbered:

Iron wire:

- Wire no. 12; before breaking held 3 lb and 12 oz; has a length of 6 ft and weighs 324 mg.
- Wire no. 7; before breaking held a weight of 10 lb; it is 6 ft long and weighs 907.2 mg.
- Wire no. 1; breaks with a 33 lb. traction; is 6 ft long and weighs 3.6 g.

Brass wire:

- Wire no. 12 with a 2 lb and 3 oz burst load; 6 ft in length, weighs 324 mg.
- Wire no. 7 with a 14 lb rupture load; 6 ft in length, weighs 1.17 g.
- Wire no. 1, with a breaking load of 22 lb; 6 ft in length, weighs 4.28 g.

With the wires and their characteristics presented above, Coulomb performed a series of experiments, varying both the wires and the weights of the cylinders used. A 0.5 lb and 2 lb cylinder were both used, varying the traction on them, having to be careful not to divert them from their vertical position. Counting the number of oscillations and measuring the time elapsed it was found that the oscillations were of a low frequency and that they rarely exceeded 1 Hz. One of the objectives of these experiences was also to determine the reaction to the torsional force in the metallic wires, varying the length of the wire, its diameter, and the tension in them. One of the first conclusions that Coulomb drew [1, p. 241] from these experiments was that, if the torsion angle is not very large, the characteristics of these oscillations are that they are sensibly isochronous, which in other words means that the torsion force is proportional to the torsion angle. Consequently, the mathematical formulas for oscillatory movement, used previously in our models can be repeated here. Thus, as we saw earlier, the oscillation period is:

$$T = \left(Ma^2/2n\right)^{1/2} \cdot 180^\circ$$

where M = cylinder mass and a = cylinder radius, where n = torsional proportionality (elastic) constant. It follows that n is proportional to (M/T^2). Also, from the same experiments carried out, for the same metal wire, a weight (cylinder) of 2 lb produces oscillations in a time double that produced by a weight (cylinder) of 0.5 lb; therefore, the duration of the oscillations is proportional to the square root of the weights; moreover, more or less large pulls do not significantly influence the reaction of the torsional force. Evidently, the great tractive forces make the strands elongate, reducing their diameter, which reduces the duration of oscillations.

4 Variation of Torsion Force According to Length of Wire

In the previous section, we saw that the traction on the threads has only a slight influence on the torsional force. We will now look, within these same experiments, at how, for a given equal torsion angle, the length of the suspension wire increases or decreases this force. Of course, as we increase the length of the metal wire, we can increase the number of revolutions in the cylinder at the same proportion, without changing the torsion angle; thus, the reaction force to torsion must be for the same number of revolutions, in the inverse ratio of the wire length. Let us see if the reasoning is in line with the experiment. As we saw earlier:

$$T = \left(Ma^2/2n\right)^{1/2} \cdot 180°$$

or, for the same weight, T proportional to $1/n^{1/2}$. Thus, if n is in the inverse length ratio, as the theory says, T will be calculated like the roots of the lengths of the suspended wires. Experiments show the following results:

- Brass wire no. 7, 9 inches long, pulled by the weight of a 0.5 lb cylinder, produced 20 oscillations in 110 s, i.e., an oscillation whose frequency is 0.18 Hz.
- For the same brass wire no. 7, 36 inches in length, pulled by the weight of a 2 lb. cylinder, had 20 oscillations in 222 s, i.e., a frequency of 0.09 Hz.

In this way, the lengths of the threads are in a proportion of 1:4, while the times of oscillations of the threads are in a proportion of 1:2. Experiments prove that the times for the same number of oscillations are, for the same threads pulled by the same weights, the square root of the lengths of those threads, in perfect agreement with the theory. Coulomb emphasizes [1, p. 243] that many experiments similar to these have been carried out proving this law of variation.

5 Variation of the Torsion Force Using Different Metals

As previously seen, torsion force is equal to a constant coefficient, represented by n, multiplied by the torsion angle, for isochronous oscillations. This value of n depends on the material of the wire, its length, and its diameter. It was also seen that n can be expressed by $n = Pa^2/2\lambda$. With P being the weight of the cylinder that pulls the wire, a its radius and λ the length of the isochronous pendulum. In this way, we can, in the light of previous experiments, numerically calculate this stiffness. Using iron wire no. 12 which is pulled by a 2 lb. cylinder, with a radius of 0.85 inches, with 20 oscillations performed in 242 s, we will obtain the value of n = 1/715 lb. If we want to compare the values of torsional stiffness of two different metals, we can use iron wire no. 12 as an example to compare with brass wire no. 12. For iron wire no. 12 we have already calculated the value of n to be 1/715 lb. Doing the same calculation for the brass wire, also no. 12, we will have $n/n' = 3.34$, and thus the stiffness of the iron wire no. 12 compared to the brass wire no. 12, is in the proportion of 3½:1. This means that by twisting an iron wire in a circle, we will gain the same reaction as twisting a brass wire in 3½ circles. Next,

Coulomb compares [1, p. 250] the torsional stiffness of the threads with their structural cohesion. He notes that the aforementioned iron wire requires 60 oz while the brass wire breaks with 35 oz; if they had the same diameter their cohesive forces would be with each other as the ratio of 60:35 and their torsional forces in the proportion of 3½:1.

Structural cohesion studies were analyzed from Leonardo da Vinci (1452–1519 [7]) to Galileo Galilei (1564–1642 [10]) in his *Discorsi* (1638 [8, 9]). These studies only began to advance again in the 19th century. In the first decades of that century, Laplace (1749–1827) and some of his followers began to defend a point of view according to which phenomena such as optical refraction, the cohesion of solids, the capillarity effect, and chemical reactions were the result of the attraction of the forces exerted by the particles of matter, similarly to mathematical Newton's law of gravitational [3, 12–14]. A movement called the Laplacian project was created with the aim of universalizing Newton's ideas to other domains. However, the scholar who presented these ideas in a more developed way, as a theory of "mechanics-physics", was Denis Poisson (1781–1840). This project obviously had a clear intention of replacing Lagrange's analytical mechanics (1736–1813). The development of sciences in the 19th century was responsible for denying the validity of the ideas that supported the Laplacian project, despite the stimulus that this project gave to the progress of the mechanical sciences, even in the experimental field. Navier's studies (1785–1836 [11]) of structural analysis and the fact that he considered the modulus of elasticity as a property of materials, in addition to Cauchy's works (1789–1857) in stress theory, raised the level of knowledge about the cohesion and mechanical properties of solids, to a level completely unknown in the Coulomb era [5, 11].

6 Balance for Calculating Friction Between Fluids and Solids

When studying the effect caused by the relative movement between solid bodies and a given fluid, Coulomb does so by seeking to find a formula in which the portion of the fluid friction is broken down into several parts. Thus, he mentions a portion that depends on the shock of the fluid with the solid bodies and others that are attributed to fluid friction itself; among the terms due to friction, there is one that depends on adherence, which he believes to be constant. However, this term seems to be very small, being confused in the experiments with others that depend on the shock, therefore, it is difficult to calculate. Coulomb, in describing the experimental way to measure fluid friction, refers to mechanical paradigm at time period, carried out to discover this constant quantity, which are found in Book II of *Principia* [3, pp. 337–338; 12–14]. Coulomb then seeks with these experiments, developing a torsion balance, to find a faster way to determine this adherence (Fig. 3).

In Fig. 3 below an ADBE vessel, filled with a fluid from which it is desired to determine adherence in relation to solid bodies. A copper wire is attached to an abcd, a copper or lead cylinder is then suspended in the middle of the fluid; we then place a circle A′F B′ on the top of the vase, divided into degrees; this circle is at the level of the end d, attached to the cylinder. By making the cylinder rotate around a vertical axis, without deviating from its plumb, we can observe through the graduated cylinder how much each oscillation is changed; and as the torsional force of the wire that produces

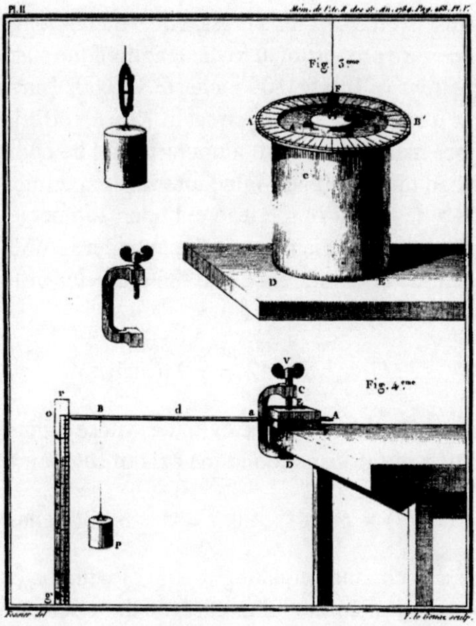

Fig. 3. 1784. Balance to measure fluid friction [1, p. 270]. Source: *Académie des science* Dossier Coulomb; National French Library BnF, Gallica.

the oscillations is known from previous experiences, we can then also know the changes due to imperfections of elasticity, causing the cylinder to oscillate in a vacuum or even in the air, and we can expect this medium to find constant amount due to adherence.

The first experiment conducted with the balance was made with a vessel filled with water, copper wire no. 12, 2.61 inches long, and a lead cylinder weighing 2 lb, which had already been used in previous experiments; the AB circle on which the oscillations were observed was 3.97 inches in diameter. Before starting the observations, we saw that the amplitudes of the oscillations decreased from 3 degrees and 55″ and observing the movement march with a magnifying glass, we noticed that distinctly 14 oscillations occurred before the movement was extinguished. If the successive decrease in each oscillation is assumed to be constant and that this decrease is attributed entirely to the fluid's adherence to the lead cylinder surface,

$$\left(A - S'\right) = (2\mu/n)$$

Where $(A - S')$ is the decrease of each oscillation; $n\,(A - S')$ is the moment of the torsion force and μ is the moment of the retarding force due to adherence.

However, since, after observing the oscillations, the arc traversed decreased from 0.13 inches in 14 oscillations, and the radius of the circle over which this decrease is observed was 1.98 inches, assuming that this decrease is constant, we have the angle $(A - S')$ and then the amplitude decreases with each oscillation of

$$(0.04\ \text{inches})/(2 \times 22 \times 0.09 \times 14)\text{inches} = 0.0024$$

As we had already discovered that for a brass wire 9 inches long, no. 12, n = 1/755 lb; and as the torsional forces are proportional to the length of the suspended wires, we will have for our 29″ long wire $\mu = 1$ lb × 0.09 inches/3.155.00. That is, the moment of the constant retarding force μ is approximately equal to 70% of 1 trillionth of a lb. Coulomb emphasizes that by other means this small amount would be completely negligible. In order to effectively obtain the adherence value after the experiment described above, it is necessary to note that the height of the lead cylinder, submerged by the water in the vessel, is 2.16 inches, and that the diameter of the cylinder is 1.58 inches. Thus, taking the relationship between the circumference and the diameter of the circle as 22/7 (π), the submerged cylinder surface is equal to

$$22/7 \times 1.58 \times 2 = 9.93(\text{inches})^2$$

But as the movement is made around the cylinder whose radius is 0.79 inches, if "d" is the grip, the momentum of the grip around the axis of rotation will be

$$d \times (22/7) \times (1.58)^2 \times 0.5 = d \times 3.92(\text{inches})^3$$

The above value of d is obtained equating d × 3.93 with the experimental findings. Coulomb adds another portion to the adherence of the circle that forms the base of the cylinder immersed in water. Coulomb does some arithmetic operations to obtain the adherence value for the ft^2 unit, providing even this unit with a very small value. This set of experiments showed, according to Coulomb, that for small speeds, the grip is proportional to the speed and for high speeds it is proportional to the square of the speed.

7 Elastic Force While Twisting the Wire

During the twisting of the stretched metallic wires, as was done in the previous experiments, we can observe two things: if the twisting angle is small, in relation to the length of the suspended wire, after oscillations it returns approximately to the position it had before the twisting; if the torsion angle is large, after the wire is twisted, it does not totally distort as in the previous case. After these two observations, the following experiments were conducted. The first of these experiments is done to determine the decrease in these oscillations, in other words, how much the elastic torsional force is altered in the oscillatory movement, as long as the oscillation center has not moved. The second, to determine the displacement of this reaction center, when the torsion angle is too large to imply the displacement of this center.

1st Experiment

A 6 inch long iron wire, loaded with a weight of 2 lb. Rotating the cylinder around its axis to determine how many degrees of amplitude each oscillation decreases.

- 90° torsion angle decreases 10° in 3½ oscillations
- 45° torsion angle decreases 10° in 10½ oscillations
- 22½° torsion angle decreases 10° in 23 oscillations
- 11¼° torsion angle decreases 10° in 46 oscillations

Coulomb observed a certain degree of uncertainty in the reduction of oscillations when the torsion angle is greater than 90° and that beyond this value the wire does not return to its original position anymore; he also noted that the center of torsion reaction is also shifted. Coulomb also measures this displacement, but for the sake of space we omit these results.
2nd Experiment
Iron wire no. 7 with 6.54 inches of length.

- 180° torsion angle loses 10° in 3½ oscillations
- 90° torsion angle loses 10° in 12 oscillations
- 45° torsion angle loses 10° in 27 oscillations
- Twist angle of 22½° loses 10° in 54 oscillations

Altogether Coulomb performs five of these experiments, of which we show only two above. The third experiment was made with iron wire no. 12, 6.54 inches in length, the fourth with brass wire no. 1 also 6.54 inches long and the last with brass wire no. 7 with the same length as the previous wires. In all experiments Coulomb measured for different degrees of torsion, in how many oscillations the movement lost 10°; it also measured the distortion caused by the position of the torsion center. Coulomb noted that the brass wires numbered 5 to 8 have less imperfection in their elasticity and have a much more extended range of elasticity than the iron wires. He again highlights the appearance of many irregularities in the results, which he attributes largely to the manufacturing process. Below we comment on the mathematical–physical role played in the Coulomb's argument:

1) Despite the uncertainties that appeared in these experiments, it is necessary to point out that within certain limits, these changes are approximately proportional to the oscillation amplitudes, as we saw and recorded in the previous observations. According to our observations, air resistance does not alter the experiences of measurements of oscillation amplitudes.
2) To build a torsion balance, it is always necessary to choose threads that have the least imperfect elasticity; brass wires are preferable to iron wires; the choice of wire diameter depends on the forces we want to measure. He reports that he built a magnetic balance which would be described in the studies to be sent to the Academy of Sciences, where he used alternatively a 3 ft long brass wire, i.e., no. 12 and no. 7; the torsional force is such that having twisted these eight-circle threads for 30 h, there was not even a degree of change or displacement of the center of torsion.
3) In all metal wires, the elastic reaction lasts only to a certain extent; the isochronism of oscillations indicates that the elastic force is almost perfect; but in addition to the torsion angle at which this occurs, the measure of the elastic force, the torsion reaction center moves almost entirely from any torsion angle that exceeds that of the elastic reaction. However, as we noted in previous experiments, the amplitude of the elastic reaction is no longer a constant amount, across all torsion angles, and it grows as the torsion increases.
4) From the experiments done, we will try to explain the elasticity and coherence of metals. The integral parts of the iron and brass wire, or any other metal have an

elasticity which we can define as perfect, i.e., that the forces necessary to compress or expand these integral parts and which are proportional to the expansions and compressions that they support; but they are not linked together only by consistency, constant quality and absolutely different from elasticity. In the first degree of torsion, the integral parts change figures, lengthen or compress without the points where they adhere to each other, change places, because the force necessary to produce these first degrees of torsion is less considerable than the force of adhesion, but, as long as the torsion angle becomes such, the force with which these parts is compressed or expanded, is equal to the coherence that unites these integral parts; therefore, they must separate or slide over each other. This sliding of the parts takes place throughout the ductile bodies; but if the body compresses by sliding the parts of one another, the extension of the contact points increases and the extension of the elasticity field becomes greater.

8 Concluding Remarks

Coulomb's study on twisting wire, in addition to being comprehensive and pioneering in many respects, contains an interesting theoretical approach and is anchored in a solid and meticulous experimental analysis. Despite the quality of his analysis, Coulomb found himself involved in a controversy over the experimental results obtained on his torsion balance until they were reproduced again in the laboratory [15–17]. Like other works of his, such as the theory of dry friction [4], he developed a mathematical model which guided his experiments in the sense of measuring some quantities, in careful and consistent experiments, so that in a second moment other parameters can be extracted through the analytical model [18].

For the specific case of twisting in metallic wires, Coulomb submits the hypotheses of elastic behavior of the wires to the sieve of the experimental analysis. Then, in a second phase of the study, he examines the difficulties that arise when the twist in the threads exceeds a certain limit [19].

The initial mathematical model, proposed by Coulomb, (in modern terms) uses an equation from Newton's second law, in which the *moment of the torsional force* is equal to the *variation in the angular quantity of motion* (*angular momentum*), with the advantage of falling into the solution of a differential equation of first order. In this way, he obtains a very useful formula that provides the period of oscillation of the wire according to geometric quantities and the constant of elastic stiffness of torsion. This formula will later test the approximation or departure from elastic conditions by the degree of isochronism of the oscillations.

In a second step, Coulomb refined the mathematical model admitting that the torsional force in the thread can be changed by a certain amount for which he proposes a polynomial shape. Again, the distortion suffered by the wire when moving away from linearity conditions can be obtained experimentally through the refined model. Finally, it is important to note that the study, divided into a mathematical model and a physical one, forms a whole that complements and leads to a series of interesting observations about this type of movement, especially in the construction of a balance to measure fluid friction.

References

1. Coulomb, C.A.: (1784) [Read] Recherches Théoriques et Expérimentales sur la force de torsion & sur l´élasticité de fils de métal: Application de cette théorie à l´emploi des métaux dans les Arts & dans différents balances de torsion, pour mesurer les plus petits degrés de force. Observations sur les lois de l´élasticité & de cohérence, pp. 229–269. Histoire de l'Académie Royale des Sciences, Imprimerie Royale, Paris (1787)
2. Coulomb, C.A.: Premier mémoire sur l´élasticité et le magnétisme. Construction et usage d'une balance électrique fondée sur la propriété qu´ont les fils de métal d´avoir une force de réaction de torsion proportionnelle a l ´angle de torsion. Détermination expérimentale de la loi suivant laquelle les éléments du même genre d´électricité se repoussent mutuellement, Collection de Mémoires relatifs á la Physique Publie par la Société Française de Physique, T.1, pp. 107–115. Mémoire de Coulomb, Gauthier-Villars, Paris ([1785]1884)
3. Newton, I.: The mathematical principles of natural philosophy. Transl. by Motte Andrew. London, Motte ([1726]1729)
4. Oliveira, A.R.E.: The Contribution of Coulomb to Applied Mechanics. In: Ceccarelli (ed). IFToMM International Symposium on History of Machines and Mechanisms, Cassino, pp. 217–226. Springer, Dordrecht (2004)
5. Gillispie, C.C., Pisano R.: Lazare and Sadi Carnot. A Scientific and Filial Relationship, 2nd edn. Springer, Dordrecht (2014)
6. Pisano, R.: On Lagrangian in Maxwell's electromagnetic theory. Scientiatum VI. História das Ciências e das Técnicas e Epistemologia, pp. 44–59. The University of Federate University of Rio de Janeiro Press, Brazil (2013)
7. Pisano, R.: Details on the mathematical interplay between Leonardo da Vinci and Luca Pacioli? J. Br. Soc. History Math. **31**(2), 104–111 (2016)
8. Galilei, G.: Discorsi e dimostrazioni matematiche intorno a due nuove scienze attenenti alla meccanica e i movimenti locali, del signor Galileo Galilei linceo. Elsevirii, Leida. (1638). In: Galilei, G., 1890–1909, VIII, pp. 39–362
9. Galilei, G.: Le opere di Galileo Galilei: Edizione nazionale sotto gli auspici di sua maestà il re d'Italia. In: Favaro, A. (ed.), Barbera, Firenze (1890–1909)
10. Pisano, R.: Galileo Galilei: The Two Le Mecaniche. Translations, Text and Commentaries. Dordrecht, Springer (2022, forthcoming)
11. Pisano, R., Capecchi, D.: Conceptual and mathematical structures of mechanical science between 18th and 19th centuries. Almagest **2**(4), 86–121 (2013)
12. Pisano, R., Bussotti, P.: Conceptual Frameworks on the Relationship between Physics–Mathematics in the Newton Principia Geneva Edition (1822). Foundations of Sciences, Springer (2022). https://doi.org/10.1007/s10699-021-09820-2
13. Pisano, R., Bussotti, P.: A Newtonian Tale Details on Notes and Proofs in Geneva Edition of Newton's Principia. Bulletin British Society for the History of Mathematics **31**(3), 160–178 (2016)
14. Bussotti, P., Pisano, R.: Newton's Philosophiae Naturalis Principia Mathematica "Jesuit" edition: the tenor of a huge work. Accad. Nazionale Lincei-Matematica Appl. **25**(4), 413–444 (2014)
15. Falconer, I.: Charles Augustin Coulomb and the fundamental law of electrostatics. Metrologia **41**, S107–S114 (2004)
16. Martinez, A.A.: Replication of Coulomb torsion balance experiment. Arch. Hist. Exact Sci. **60**, 517–565 (2006)

17. Oliveira, A.R.E:, [Read] Coulomb´s Memoir on Man-machine Interaction. In: 2nd International Conference on Mechatronics, Automation and Smart Materials, November 13–14, Paris, France (2017)
18. Heilbron, J.L.: On Coulomb´s Electrostatic Balance. Blondel-Dorries, pp. 151–161 (1994)
19. Gilmor, C.S.: Coulomb and the Evolution of Physics and Engineering in Eighteenth-Century France. Princeton University Press (1971)

Revisiting Inflection Circle via Level Luffing Crane for Academic Revival

Prashant B. Shiwalkar[1](✉) and Pramod M. Padole[2]

[1] Shri Ramdeobaba College of Engineering and Management, Nagpur, India
shiwalkarpb1@rknec.edu

[2] Visvesvaraya National Institute of Technology, Nagpur, India

Abstract. Repeated reference to a limited number of named straight line mechanisms in academics has introduced an element of dullness in the potentially challenging domain of synthesis and application. Young and vibrant domains like Compliant Mechanisms are likely to revive the interest in the synthesis of rigid body mechanisms as the starting topologies. Level Luffing Cranes are identified and described as one of the not-to-miss large size example of an approximate straight line mechanism. Historically the concept of Inflection Circle is more than three centuries old. Existence of the special point (Ball's point or Undulation point) on the Inflection Circle is known for around one and half century. Involved theoretical nature of such concepts of the Path Curvature Theory has resulted into the marked avoidance at the introduction level of the courses on mechanisms. This paper highlights the possible association between synthesis of these mechanisms and the concepts related to Inflection Circle. Only the easy to explain traits of Inflection Circle and the Ball's point are highlighted and their possible application in the synthesis of the double rocker kind of approximate straight line mechanisms is discussed. Three cases comprising of textbook and real life field examples are explained. Apparent limitations and introduction of higher level concepts of the path curvature theory are also discussed in the end.

Keywords: Level luffing · Straight line path · Inflection circle · Path curvature

1 Introduction

Necessity for converting the available rotational motion into a straight line existed even before appearance of Watt's linkage. Historically successful setups like tractor mounted post-hole-digger, cutting tool guidance linkages for machining etc. [1, 2] were in place at the time of transition to Industry 2.0 revolution. One of the concepts of the Path Curvature theory- Inflection Circle- was discovered almost eight decades before the Watt's creation [3]. For the domain of synthesis of straight line mechanisms, observations by Mruthyunjay "… the subject has never completely disappeared from the research horizon and indeed, has often bounced back with greater vigor" are relevant. The reasons cited are intellectual stimulation and genesis of innovation on one side and creativity along with aesthetic delight on the other [4]. The related technical literature published in last couple of decades has devoted less coverage to these related concepts having rich

M. Ceccarelli and R. López-García (Eds.): HMM 2022, HMMS 40, pp. 255–265, 2022.
https://doi.org/10.1007/978-3-030-98499-1_21

historical lineage [5, 6]. The enriching work and life stories of towering personalities like Burmester and Alleivi are almost forgotten today [7].

The intention of this exercise is to revive the interest of the entry level students of the course on Mechanisms in the historical concepts by analyzing their successful application. Of the possible list of successful applications, Level Luffing Crane (LLC) is chosen for further detailing.

2 Level Luffing Crane

At the turn of the nineteenth century cranes were introduced in major European seaports. Most of the European nations had high volume of trade with their colonies and other countries.

On the skyline of ship building ports like Cadiz (Spain), Ferrol (Spain), Goteborg (Sweden) or Hamburg (Germany), giant cranes cannot be missed. Among these cranes LLCs are examples of Approximate Straight-line Mechanisms. Movement of the hook or the grab of these cranes is similar to movement of the beak of a goose, almost parallel to the ground, achieved by folding of the neck. These cranes are also called Gooseneck cranes. Alternately LLCs are also identified as Double Jib cranes. Some of these cranes are nearing hundred years of service [8].

Picking up the clue from the sketches in the 'Lecture Notes on Mechanisms' [9] and very elaborate treatment in [10], this paper tries to explore the possible association between these two historical developments, the mechanism and the corresponding theoretical concept. After describing the Level Luffing Crane (LLC) and Inflection Circle (IC), their possible association is explored with graphical treatment.

2.1 Construction and Working

In 1926 remarkably advanced crane was installed in the port of Hamburg. This electric motor driven crane had a luffing jib [11]. With the luffing motion of the Jib the load can be moved inward or outward, along a trajectory in the vertical plane. The next remarkable development was introduction of second jib and a boom to mount the load pulley resulting in almost horizontal straight line traverse of the load (Fig. 1). This resulted in drastic saving in the energy cost per ton of cargo handled and increased the accuracy of location drastically [11]. During WW-II years the technology for the LLCs was well established. Post WW-II major replacement of devastated port cranes was with modified LLCs.

The LLC is a four bar approximate straight line mechanism of double rocker type. As shown in Fig. 1, Tension Boom and Compression Boom are the two rockers. Fly Boom is the coupler with pulley at the coupler point. This not-to-miss sight at the shipbuilding ports is not an experience for the inland populace who has no opportunity to visit such places. However net surfing can be the substitute [12]. For the students of mechanisms, a good number of possible LLC arrangements are included in [13]. For demonstration of the path generated by the coupler point, four mechanisms are used. First LLC, ME1, is selected from this source. ME2 is sourced from the chapter on Linkwork for Material Handling of [14].

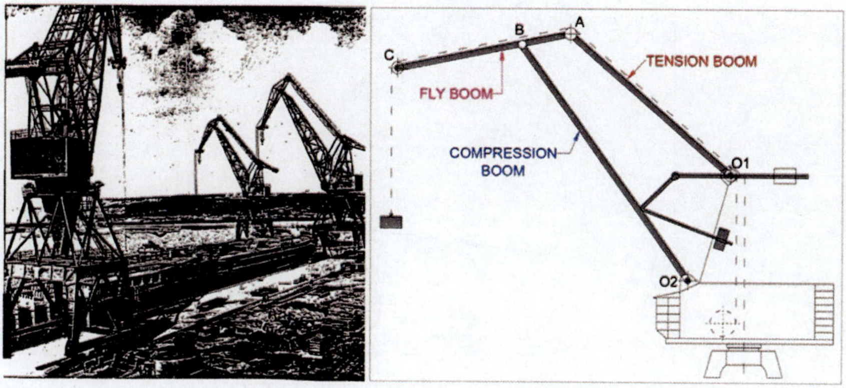

Fig. 1. Level Luffing Cranes and the terminology [1]

For ME3 actual link dimensions of a far eastern crane are used. It is in operation for over twenty five years [15]. This LLC has a minimum operating radius of 14 m and maximum of 38 m. This ensures a level travel of load through twenty four meters. All the required link lengths are included in Table1. Identified straight line segments and related parameters, *L* and Δ (Fig. 2 to Fig. 6) are also listed.

The possible paths, traversed by hooks of various cranes, between the two extreme limits of rockers, are shown in Fig. 2, 3, 4. All the figures are created in AutoCAD 2018 environment. Various softwares like MechAnalyzer [16] and SAM8.0 [17] are available to create the coupler point paths.

Table 1. Link dimensions and straight line parameters (in mm) for the selected LLC mechanisms

	O1A	O2B	O1O2	AB	BC	L	Δ
ME1	100	125	60	30	45	94	4.71
ME2	100	118	64	27	83	133	2.904
ME3	2300	27400	14200	6300	12913	134	5.5
ME4	100	120	69	48	88	70	3.29

2.2 Approximate Nature of the Path

To quantify the amplitude of departure from the exact straight line path different methodologies are suggested. The reported deviation in the path of crane in [15] is 1.2 m in a travel of 24 m. This amounts to 5% of the path length. This is indicated as (dy/dx). Authors have used alternate graphic approach suggested in [18]. The results of visual identification of the approximate straight line portion in the paths of selected cranes are shown in Fig. 2, 3, 4 and tabulated in Table1.

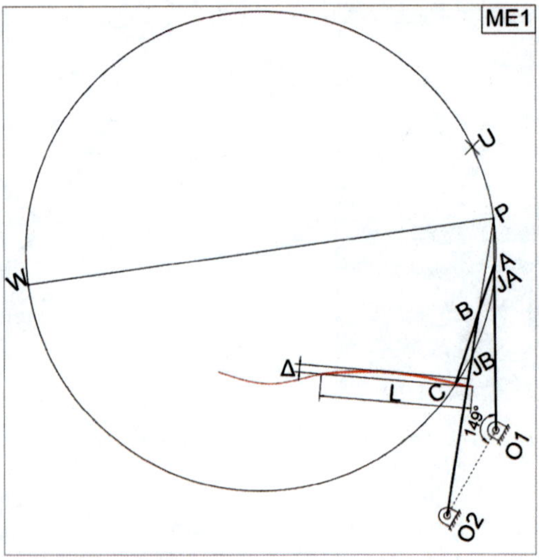

Fig. 2. Details of the crane mechanism ME1

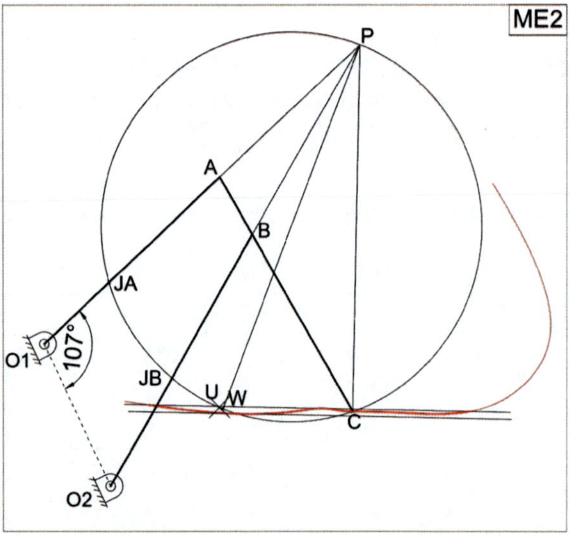

Fig. 3. Details of the crane mechanism ME2

3 Inflection Circle

In 1706 de Lahire submitted a paper to the French Academy of Science in which he has identified existence of the set of points in the moving plane which if made the coupler points will individually trace approximate straight line paths of considerable length [3].

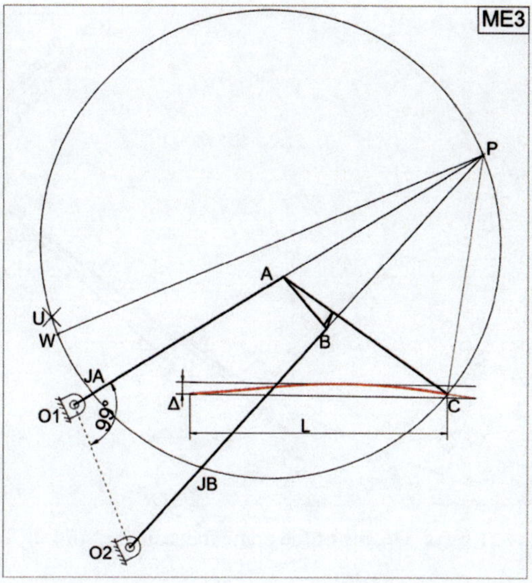

Fig. 4. Details of the crane mechanism ME3

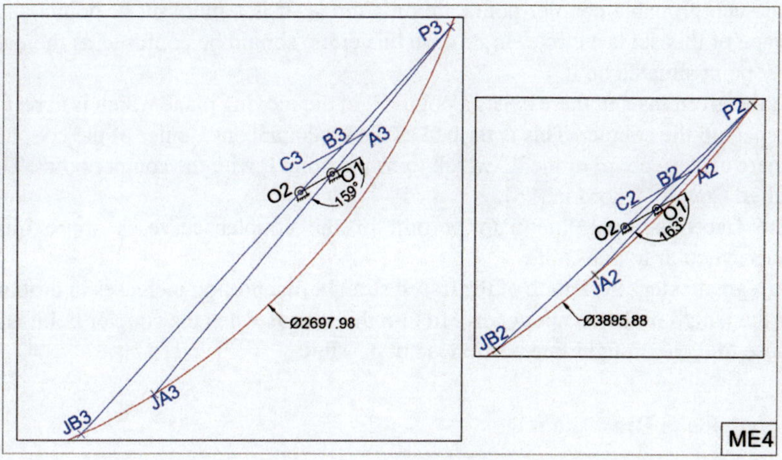

Fig. 5. Inflection Circles at two different crank poses for ME4

3.1 Description

Given below is the summary of notable aspects of the IC as understood from few of the easy to access texts [13, 19, 20]. Further discussion can be better understood by referring Fig. 5 and Fig. 6.

a. Existence of the set of such probable coupler points is instantaneous. Even with a minor input motion the IC can change dramatically. For the crane ME4, three different

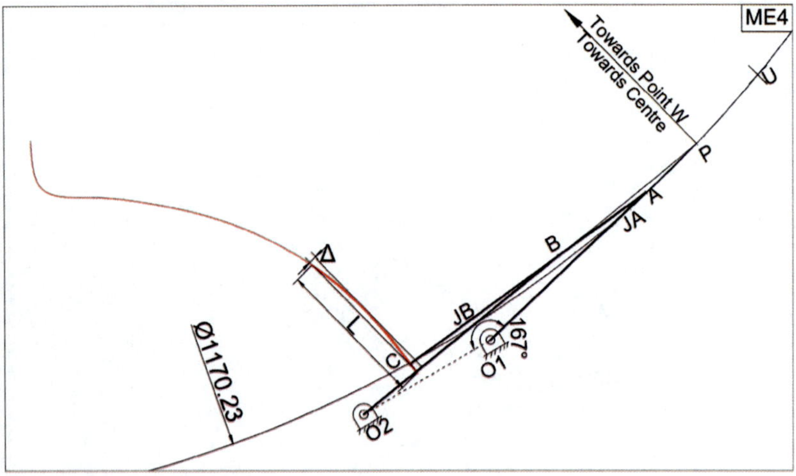

Fig. 6. Details of the crane mechanism ME4

poses of the mechanism and corresponding three different ICs with considerably different diameters are shown in Fig. 5 and Fig. 6.

b. All such probable coupler points lie on a circle. It is a mere set of points and the shape of this set is a circle. In no case this circle should be confused as the path of any point situated on it.
c. At the given instant, there exists a point 'P' in the moving plane which is at rest with respect to the coupler. This is termed as the Instantaneous Center of the coupler.
d. There exists a chord of the IC which joins the point P with the coupler point. This is called Coupler Chord in [25].
e. The traced straight line path portion of the coupler curve is approximately perpendicular to this chord.
f. To a great extent the length of the traced straight line portion increases in proportion to the length of the coupler chord. In turn this means when the coupler point is very close to P, the straight line portion is not possible.

3.2 Methods to Draw the ICs

In 1765 Leonhard Euler presented a paper to the Academy of Sciences at St. Petersburg. In this paper the first version of today's Euler-Savary equation is derived in relation to gear profiles. Subsequently, in 1845 Felix Savary derived the most general form of the equation which helps in plotting a point where the radius of curvature is infinite, or which is the most likely point to trace a straight line path. This usable equation is detailed out in the referred texts already cited above. It quantifies the distance of such inflection point from the pole P, when the distances of both the end points of the link under consideration are known. In Fig. 5 for the Tension Boom O1A the inflection point is at JA3 and for the Compression Boom O2B, it is at JB3. The third point which is always in the inflection state is P. The circle passing through these three inflection points P, JA3 and JB3 is the intended IC. When O1A reaches the new pose by travelling 4° more the new inflection points are given by JA4 and JB4. This gives a very different set of inflection points which are located on the circle of different diameter. This method is used in this study but the second method, Bobillier construction is equally helpful.

Etienne Bobillier while teaching a course in 1830s arrived at the said method of plotting the Inflection Circle for the mechanisms. This is a complete geometric construction and is detailed out in detail predominantly in [13, 20].

3.3 Ball's Point on the ICs

Historically the coupler points for generating straight line path were identified early. Equally important set of coupler points generating the circular path in the coupler curves were identified much later. Conveniently these points were called Circling Points and equivalent to the IC, this set of circling points were called Circling Point Curves. They help in creating dwell mechanisms [1]. Few important properties of these curves, today better known as Cubic of Stationary Curvature (CSC), are listed below [13]. Simultaneously refer Fig. 7, where the curve is superimposed on the instantaneous IC.

1. CSC is unique to the orientation of the mechanism.
2. Pole tangent and pole normal are intersecting at the Pole P.
3. Excepting the points very close to the pole P, points on the curve generate coupler curves with circular arcs.

As shown in Fig. 7, the point of intersection of the IC and the CSC is of significance. This point is called Ball's point after the British mathematician Robert Ball. Similar discussion can be studied in [5, 22].

Elaborate discussion on location of the Ball's point U (also called Undulation point) w.r.t. Inflection pole (point J in Fig. 7) and its' effect on the straight line path can be found in [10]. The best suited situation for the LLC is also discussed therein. Relevant discussion for the general cases can be had from [23].

In this graphical exploration of the concepts in relation to the LLCs, method suggested in [24] is adapted to plot the Ball's point U for every studied crane mechanism. These are included in Fig. 2 through Fig. 6. The graphical method used to plot the Undulation point U for the mechanism ME2 is included in Fig. 8.

4 Syntheses with Inflection Circle

In this study the curve traced by the double rocker crane mechanism has been analyzed in depth to highlight the fact that many times reorientation of the fixed link, O1-O2 was the accepted compromise to arrive at the required path of the output point. In case of ME4, reorienting the O1-O2 link vertically will provide the horizontal path at the output point C. Being a four bar mechanism, the output path is of approximate nature. Historically a more accurate tool path was obtained by employing 'Double Kite' mechanisms in such situations [2]. However, in IC based synthesis reorientation of the fixed link can be avoided and still the required orientation of the output path can be achieved [25].

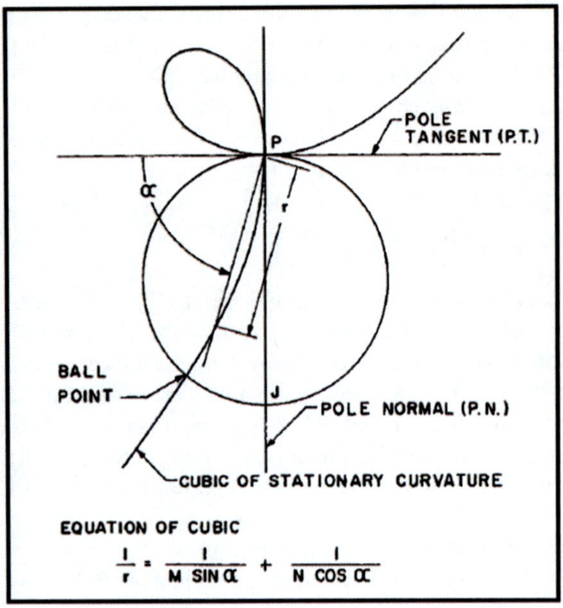

Fig. 7. Inflection circle, cubic of stationary curvature and the Ball's point [21]

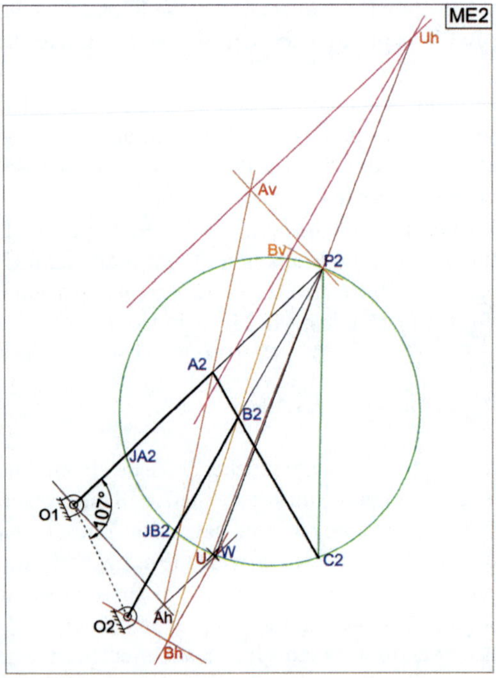

Fig. 8. Graphical construction for determination of Ball's point U for ME2

In Fig. 2 and Fig. 3 complete traverse of point C for ME1 and ME2 is not a straight line path. In synthesis of approximate straight line mechanisms, right from the time of James Watt, selection of the desired portion of total path is very common. The acceptable portion of the total path is analyzed in more detail in three chosen mechanisms. By trial and error method the exact pose of the mechanisms where obtained IC contains the coupler point C are included in Figs. 2, 3, 4.

Survey of the LLCs in general and elaborate designs of the Fly Boom in particular leads to an important observation. Successful designs of Fly Boom can lead to a maximum usable portion in the available path of the coupler point as in ME3. This is a very strong point in favor of ME3 and is apparent from Fig. 4. The location of Ball's point U w.r.t. the Inflection Pole w is expected to play a major role in design of LLC. However in all the chosen cranes no such strong relation can be pointed out. Practical difficulties in handling the oversized beyond the drawing board kind of ICs are discussed in [26].

5 Observations and Conclusion

Multiple approaches like analytical or Evolutionary Algorithms based numeric methods are being proposed for creating a usable synthesis methodology. Many often than not they are validated with examples of traditional named straight line mechanisms. Even the young domain of Compliant Mechanisms is using these named rigid body straight line mechanisms as the starting point [27–29]. Initial steps for application of numerical methods to the synthesis using intersecting ICs are included in [26].

The present graphical analysis of a successful approximate straight line mechanism, in the sequence given, which is the sequence in which the theory has evolved, suggests one of the recommended methods for academic revival of these concepts. Hence it is concluded that instilling desire for learning the concepts of Path Curvature Theory is possible.

6 Discussions

In the literature for synthesis of mechanisms the 'Ball-Burmester Point' appears quite often [30–32]. Historically, the concept of Cubic of stationary curvature appears much later than the Inflection Circle concept [33]. Ball's point, which is the intersection of these two curves, surely leads to a better straight line coupler curve. To rejuvenate the interest in the concepts of Path Curvature theory, it is advisable to create the required base with help of easy to understand and possible to trace Inflection Circle as the starting point and then to move towards more involved concepts of higher order curves like the CSC. The concept of 'Tricentric Sextant' coupler curves can follow after this. Ideally this can be included in the courses for Masters in Mechanisms.

References

1. Nolle, H.: Linkage coupler curve synthesis: a historical review-II. Developments after 1875. Mech. Mach. Theory **9**, 325–349 (1974)

2. Kempe, A.B.: How to Draw A Straight Line; A Lecture on Linkages, Project Gutenberg EBook [EBook #25155] (1877)
3. Koetsier, T.: From kinematically generated curves to instantaneous invariants: episodes in the history of instantaneous planar kinematics. Mech. Mach. Theory **21**(6), 489–498 (1986)
4. Mruthyunjaya, T.S.: Kinematic structure of mechanisms revisited. Mech. Mach. Theory **38**, 279–320 (2003)
5. Torres-Moreno, J., Giménez-Fernández, A., Carbone, G., Ceccarelli, M.: Kinematic and dynamic analysis of old mechanism by modern means. History of Machines and Machine Science. Springer (2015)
6. Pennestri, E.: Application of path curvature theory as a tool in the design of mechanical systems. Industrial Problems on Machines and Mechanisms IPRoMM. BITS Pilani (2020)
7. Ceccarelli, M., Koetsier, T.: Burmeter and Allievi: a theory and its application for mechanism design at the end of 19th century. J. Mech. Des. **130**, 07230–07231 (2008)
8. Weinreich, D.: Slew Cranes in Shipyards-A Study. Emscor Inc., Houston, Texas (1985)
9. Breteler, A.K.J.: Lecture notes on Mechanisms (wb3303) Chap. 3. TUDelft (2004)
10. Dijksman, E.A.: Approximate straight-line mechanisms through four-bar linkages. Rev. Roum. Sci. Techn.-Méc. Appl., 319–372 (1972)
11. Hamark, J.: Technology and productivity in the port of Gothenburg c. 1850–19651. Int. J. Maritime Hist. **26**(2), 265–287 (2014)
12. Pham, D., Ranganathan, S., Weng, C., Yu, G.: ENSC3001: Luffing Crane Project. https://www.youtube.com/watch?v=LmpEgpXD7tc. Visited on 9 Dec. 2021
13. Hartenberg, R.S., Denavit, J.: Kinematic Synthesis of Linkages. McGraw-Hill Book Company, NY (1964)
14. Artobolevsky, I.I.: Mechanisms in Modern Engineering Design, vol. 1. Mir Publishers, Moscow (1975)
15. Kim, D.S., Lee, J.: Structural design of a level-luffing crane through trajectory optimization and strength-based size optimization. Struct. Multidisc Optim. **51**, 515–531(2015). https://doi.org/10.1007/s00158-014-1139-2
16. Pandey, D., Kushwah, J.S., Chittawadigi, R.G., Rama Krishna. K., Saha. S.K.: Effective teaching of mechanism synthesis using mechanalyzer software. History of Machines and Machine Science. Springer (2019). https://doi.org/10.1007/978-3-030-20131-9_319
17. SAM 8.0: https://www.artas.nl/en/downloads. Last access 22 Dec. 2020
18. Hain, K.: Applied Kinematics. McGraw Hill, NY (1967)
19. Hall, A.: Kinematics and Linkage Design. Prentice Hall Inc., Englewood Cliffs, NJ (1964)
20. Tao, D.C.: Applied Linkage Synthesis. Addisun-Wesley Publishing Company Inc., London (1964)
21. Tesar, D.: The analytical theory of coplanar motion applied to approximate four-bar straight line mechanisms. Ph.D. Thesis, Georgia Institute of Technology (1964)
22. Goessner, S.: The planar Euler Savary Equation in vectorial Notation (2017). https://www.researchgate.net/publication/318661368
23. Ting, K.L., Wang, S.C.: Fourth and fifth order double Burmester points and the highest attainable order of straight lines. J. Mech. Des. **113**, 213–219 (1991)
24. Gössner, S.: Ball's Point Construction Revisited. In: Pisla, D., Corves, B., Vaida, C. (eds.) EuCoMeS 2020. MMS, vol. 89, pp. 28–34. Springer, Cham (2020). https://doi.org/10.1007/978-3-030-55061-5_4
25. Shiwalkar, P.B., Moghe, S.D., Modak, J.P.: Novel methodology for inflection circle based synthesis of straight line crank rocker mechanism. J. Mech. Robot. (2021). https://doi.org/10.1115/1.4053373
26. Shiwalkar, P.B., Moghe, S.D., Shiwalkar, J.P., Modak, J.P.: Inflection circle based approach to synthesis of approximate straight line mechanisms. History of Machines and Machine Science. Springer (2019). https://doi.org/10.1007/978-3-030-20131-9_154

27. Hricko, J., Havlík, Š: Compliant Mechanisms for Motion/Force Amplifiers for Robotics. In: Berns, K., Görges, D. (eds.) RAAD 2019. AISC, vol. 980, pp. 26–33. Springer, Cham (2020). https://doi.org/10.1007/978-3-030-19648-6_4
28. Hricko, J.: Straight-line mechanisms as one building element of small precise robotic devices. In: Hajduk, M., Kaukolova, L. (eds.) Industrial and Service Robotics, pp. 96–101. Trans Tech Publications Ltd., Switzerland (2014). https://doi.org/10.4028/www.scientific.net/AMM.613.96
29. Zhao, K., Schmiedeler, J.P.: Using rigid-body mechanism topologies to design path generating compliant mechanisms. J. Mech. Rob., **8**, 014506-1-8 (2016). https://doi.org/10.1115/1.4030623
30. Ting, K.L., Schmiedeler, J.P.: Using rigid body mechanism topologies to design path generating compliant mechanisms. J. Mech. Robot. **8**, 014506-1-7 (1991)
31. Yin, L., Huang, L., Huang, J., Tian, L., Li, F.: Solution-region-based synthesis approach for selecting optimal four-bar linkages with the Ball-Burmester point. Mech. Sci. **10**, 25–33 (2019)
32. Yin, L., Huang, L., Huang, J., Xu, P., Peng, X., Zhang, P.: Synthesis theory and optimum design of four-bar linkage with given angle parameters. Mech. Sci. **10**, 545–552 (2019)
33. A Century of Research on Kinematics (1870–1970): A Bibliographic Compilation. www.iri.upc.edu/people/thomas/repository.html. 8 August 2021

Developing a Virtual Museum of the Escuela Técnica Superior de Ingenieros Industriales of Madrid

Oscar Monjas Hebrero, Luisa Martínez Muneta(✉), Antonio Carretero Díaz, and Juan de Dios Sanz Bobi

Mechanical Engineering Department, Escuela Técnica Superior de Ingenieros Industriales, C/ Jose Gutiérrez Abascal, 2, Universidad Politécnica de Madrid, 28006 Madrid, Spain
luisa.mtzmuneta@upm.es

Abstract. The objective of this paper is to present the development and selection of artefacts, materials, and objects related to the industrial engineering career that are part of the collection of the Escuela Técnica Superior de Ingenieros Industriales (ETSII) of Madrid (Spain) and their adaptation to new technologies. Due to the characteristics of the building and the location of the pieces of the col-lection, developing an organized visit with its corresponding narrative is very complex, and an adaptation has been made to new technologies, proceeding to the creation of a virtual museum that allows the discovery, conservation and dissemination of the heritage of this engineering educational center.

Keywords: Industrial heritage · Virtual museum · Industrial museum · Photogrammetry · Cataloguing

1 Introduction

The advances, inventions and testimonies that the 170 years that the career of Industrial Engineer has been established in Spain are enormous, originating a historical, technological, social, architectural and scientific heritage throughout the national territory.

Before its teaching at the so-called Royal Industrial Institute, there were institutions that promoted the development of Spanish Industry. One of them was the Royal Cabinet of Machines, founded in 1791 by Agustín de Betancourt [1]. The first years it functioned as an industrial museum, being the first in Spain. In 1824, the Institution was given a boost, and the Conservatory of Arts was founded with a double function: museum of industrial machines and center of teaching of industrial grades. Finally, with the foundation of the Royal Industrial Institute, all the pieces that had been collected from 1791 to 1850 were transferred to the Convent of the Trinidad Calzada [2] and the Industrial Museum was created [3].

It was found with a set of geometric models, of which 45 corresponded to descriptive geometry and geometry of space. There were 312 models and models of machinery,

M. Ceccarelli and R. López-García (Eds.): HMM 2022, HMMS 40, pp. 266–278, 2022.
https://doi.org/10.1007/978-3-030-98499-1_22

and specific parts of machines, including also various models of architectural elements. Motion and transformation systems, 26 Betancourt differential motion devices, stoves, fans, pumps, mills, and various types of gears. It is possible that some of these machines belonged to the Roy-al Machine Cabinet (for example, the machines described as dissertated by Betancourt or cranes). The rest of the museum was made up of an abundant amount of natural and industrial products. This material was inventoried since December 1864 and was used when the Royal Institute was closed [3].

It is unknown if some of these elements, which were part of the Museum, would arrive after the closure of the Royal Industrial Institute (1867–1901) at the Central School of Industrial Engineers. What can be said is that the different institutions that have developed a teaching activity of Industrial Engineering, have combined it with the development of museums for the dissemination of industrial activities to the general audience.

1.1 Historical-mining Museum Don Felipe de Borbón y Grecia

The Don Felipe de Borbón y Grecia Historical-Mining Museum is a "Museum with a marked university character and open to all audiences with a wide collection of minerals from all over the world classified according to their composition and origin" [4]. It is one of the greatest references to university museums in the community of Madrid (Spain). It shares a location with the Escuela Técnica Superior de Minas y de la Energía of the Universidad Politécnica de Madrid (UPM), with the School managing and developing the Museum.

Its collection is one of the oldest in Spain, since the first copies date back to 1831. Due to its antiquity it has extensive exhibitions (minerals, fossils, measuring and topographic instruments, models…). In addition, it has a public and accessible database of all copies with included photographs. It also carries out a work of research, dissemination and publication of books and works, complying with the definition of the International Council of Museums (ICOM) of a museum: "preserves, investigates, communicates, exhibits and exhibits, for purposes of study and education, art col-lections, scientific, among others, always with a cultural value".

1.2 Museum "Joaquín Serna"

The Telecommunications Engineer Joaquín Serna Sturla, member of the 1962 class of the Escuela Técnica Superior de Ingenieros de Telecomunicaciones of the UPM, was responsible for rescuing the materials used in the School of Telegraphy (1913), its restoration and creation of a first exhibition in 1984 [5]. Reconstruction and dissemination work continued until the creation of the Joaquín Serna Museum after his death in 2002.

The Museum has 915 pieces that show the technological evolution in the following fields [6]: Image, Instrumentation, Radio, Sound, Telephony and Telegraphy.

The museum's catalogue is digitized in the Polytechnic Digital Collection, being free and open access.

1.3 Polytechnic Digital Collection

In 2007, the Library of the UPM began to digitize, through texts, images, videos and audios, those documents and pieces belonging to the University that had a scientific, historical and / or teaching value. Thus creating the Polytechnic Digital Collection. The Collection had in 2018 16,900 digitalized objects, composed of more than 140,000 files [7].

Currently, the collection section on ETSII collects 493 elements [8]. One of the departments with which there was a greater collaboration was with the Teaching Unit of Applied Chemistry, since 128 elements were digitized and many of them have files with additional information [9]. The Polytechnic Digital Collection performs a very important task when it comes to preserving the heritage of the UPM. Despite this, the Collection does not have an intentionality, such as that which a Museum could have when creating an exhibition or selection of pieces that share similar characteristics, author, or era, creating a more general vision on the subject in question.

The fact that many Colleges of the UPM have created their own museums to not only preserve, but also disseminate their historical heritage, made clear the need for a similar project in the ETSII. Machines and other artifacts are ex-posed in the corridors of the School and in each of the departments without a unified cataloging and, in some cases, in a state of poor conservation. This situation should change with the aim of remedying those pieces, some that date even from the nineteenth century, fall into oblivion or deteriorate.

2 Virtual Museums

In the early 1990s, with the introduction of new interconnection facilities, simple graphic tools and the access of more people to a computer, the boom of Internet began. This massive growth brought with it the emergence of a new profile of users, mostly ordinary people not linked to the academic, scientific, and government sectors [10].

With the advent of the Internet, some museums saw the possibility of reaching more audiences through online experiences, and others opted for the use of CD-ROM. These museums can be considered as the first virtual museums, being the Museum of Computer Art (MOCA), in 1993, the first virtual museum not to have a physical headquarters [11].

With the creation of virtual museums, initiatives were born to collect all the collections in a single portal such as CERES [12], HISPANA [13] (Spanish platforms) and EUROPEANA (at European level) [14]. The latter being the most complete, and updated daily.

Today, virtually all museums with physical headquarters have a virtual part. According to the virtual experience they offer, different types can be distinguished [15]:

- Those that are little more than advertising platforms for a physical museum. They take advantage of the Internet as a "bulletin board" but do not offer access to any type of virtual content.
- Those who accompany the physical museum with a virtual sample of its collections, through images and information. They also provide access to some typical services

of traditional museums such as the shop or the library. This type would be similar to the approach of the Polytechnic's Digital Collection.

- Those that incorporate, in addition to all the above, new functions only existing in the virtual space: exhibition, educational, social, cultural, recreational, technological, etc., and where the physical museum is completed by an equally important virtual component, and where the presence of users and communication with the Museum is constant.
- Those that lack physical headquarters and exist only on the Internet. A full virtual museum.

These last two are divided in turn, into three types according to the communication environment between museum and visitor:

- **Environments based on 360° panoramas.** It is the simulation of a space, usually the physical route of the museum, through photographs that make up 360° panoramas, on which you can interact (approaching some works, obtaining information from them by clicking on a certain point, etc.). Some examples of museums of this type are: the exhibition of Impressionism carried out by the Mapfre Foundation or the Natural Sciences Museum of Madrid (Fig. 1). Although the photographic quality is better than in other types of interactive environments, the movement of the camera is quite unreal and the freedom of interaction with the environment is much reduced, giving rise to a less immersive experience.

Fig. 1. 360° visit to the Museum of Natural Sciences. [16].

- **Environment based on 3D modeling.** It consists of the creation of a three-dimensional space, using modeling software, in which the visitor can move virtually thanks to computer gaming engines. It allows a greater immersion within the museum by the visitor, being also the interaction much greater. As a disadvantage, the photorealism that can be achieved in these environments are lower than those of 360° panora-mas. Some examples of this type of environment are the Visit to the ancient site of Gizeh (Fig. 2) or the city of Paris through the centuries (Fig. 3) developed by the company Dassault Systèmes [17].

Fig. 2. Giza Pyramids 3DEXPERIENCE – Source: Dassault Systèmes.

Fig. 3. Paris 3D: Through the Ages – Source: Dassault Systèmes.

•**Environments Based on Photogrammetry.** They are a middle ground between those based on 3D modeling and 360° panoramic photos. Through the realization of multiple photographs, in different positions, it is possible to reconstruct virtually in a 3D model a real piece or physical model with great accuracy. This technique can also be applied to reconstruct an archaeological site or interiors of historical places (Fig. 4). It is also applicable, with good results, to pieces of small or medium size, as perceived in Fig. 5. The advantage of this type of environment is that it preserves the realism of the photos, and at the same time, creates a more immersive experience for the viewer. As a disadvantage, the initial investment in equipment and personnel for optimal results can be very high for a museum.

Due to the health situation that the world has been facing since March 2019, caused by COVID-19, many of today's museums have sought alternatives for the public to enjoy their collections from their homes. The process of digitizing museums, such as the Prado, has accelerated [18]. Therefore, if we intend to create a virtual museum today, it will also be necessary to create a virtual experience.

The main objective of this project is the cataloguing, conservation and dissemination of the historical heritage of the ETSII of the UPM. To do this, the following stages were developed:

1. Cataloguing of items with historical value. A unified catalog was created that includes the items collected by the Digital Collection of the Polytechnic and some that were not collected at the time.

Fig. 4. Reconstruction of the interior of St. Sophia's Church in Ohrid with photogrammetry by Miguel Bandera. Source: Sketchfab.

Fig. 5. 3D reconstruction through photogrammetry of the Lady of Elche by Francisco Javier Luengo Gutiérrez. Source: Arqus3D.

2. Cataloguing of personalities of the School: relevant men and women coming from the ETSII.
3. Cataloguing of historical events. A timeline of all relevant events that affected to our institution.
4. 3D reconstruction of parts through photogrammetry. Some items due to their historical value will be reconstructed for conservation with this technique, this will also allow their better dissemination.
5. Creation of the School Museum. The bases, objectives and methodologies of the Museum will be established. A tour of the exhibited pieces of the School will be designed and their feasibility will be studied.
6. Virtual museum. A route with 360-degree panoramas will be created based on the physically created one.
7. Preparation of future collections
8. Website. The site is the support of the physical and virtual parts of the Museum. On the one hand, the physical route will have posters at each point of interest with QR codes to link to the information on the web pages. This will be in audio and text format for greater accessibility. And on the other, the virtual tour and 3D reconstructions will also be available. In addition, it will be the platform where all the research on the heritage of the School is collected.

9. Social networks. Social networks such as Twitter and Instagram will be used to give more publicity and divulgation to the Museum.

3 Methodology

One of the main objectives of this work has been to create a simple and exportable, but complete cataloguing. Based on the fields of the Polytechnic Digital Collection.

- The name of the item.
- Photo: inserted into Excel file.
- Label No.: Differentiating number. #0000, 4-digit number.
- Type of collection:
- Exposed. It is part of the physical tour of the Museum.
- Part of the School building.
- Topography devices.
- Thermal engines.
- Location in the ETSII.
- Year of manufacture.
- Materials of the object.
- Dimensions.
- Description of the item.
- Use/function of the item.
- Extra information/ external links: Link to the ETSII Heritage website for more information.

In addition to a collection of historical pieces, it has been chosen to create a list of historical personalities related to the School. This ranges from the Royal Industrial Institute to one of the current emeritus professors of the ETSII.

Finally, a timeline of historical events has been established from the inauguration of the Royal Industrial Institute (1850) to the present day.

The methodology followed to complete the different catalogs has been to consult the different sources available, which are:

- Polytechnic Digital Collection. Many of the pieces that have been included in the catalog of historical pieces, especially those belonging to the group of Topographic Devices and Thermal Engines, had prior information in the Polytechnic Digital Collection. This project has complemented this information and has corrected, some very specific cases, in which it was incorrect.
- Faculty and staff of the ETSII who have collaborated in the process of identification and research of some of the pieces, and suggested personalities to be included in the Catalog.
- Research. In addition to reviews in different historical texts and magazine reviews, it stands out as one of the main sources of information for the catalog of historical events and personalities has been the final project of Pablo Valbuena: "History of the Higher Technical School of Industrial Engineers of Madrid from 1901 to 1972" [19].

As a result of all this, this information has been included in tables that can be consulted on the website (Fig. 6).

3.1 Virtual Tour

In 2015, Google launched a free application called Google Expeditions aimed at the educational community. Its objective was to create virtual tours in remote places such as Machu Pichu, Antarctica… visited with virtual reality. Years later, in 2018, Google made Tour Creator available to everyone, so anyone can now create his or her own tour in VR.

ITEM	ID	TYPE	STATUS	YEAR	MATERIALS	DESCRIPTION	USE/FUNCTION	LINK
PALACE OF ARTS AND INDUSTRY	#0000	Building	Exposed	1887	Iron and brick columns	Central nave crowned by a dome with two side pavilions on the Paseo de la Castellana	Hosts the ETSII and the MNCN	http://patrimonio.industriales.upm.es/index.php/museo-etsii/palacio-de-las-artes-y-la-industria/
THE MACHINE	#0001	Exhibited collection	Exposed	1832	Steel	Watt type steam engine	Minting of coins in the Fábrica Nacional de Moneda y Timbre in Madrid until 1900	http://patrimonio.industriales.upm.es/index.php/museo-etsii/la-maquina/
MODEL OF THE MOUNTAIN	#0002	Exhibited collection	Exposed	1948	Wood	Model (1:3) of the section of a locomotive of the MZA 1700 series	Decorative	http://patrimonio.industriales.upm.es/index.php/museo-etsii/la-montana/
RELIEFS	#0003	Building	Exposed	1952	Marble	Plaques with motifs of the different specialties of the Career of Industrial Engineer in the stage of the Royal Industrial Institute	Decorative	http://patrimonio.industriales.upm.es/index.php/museo-etsii/relieves/

Fig. 6. Image of the cataloguing tables

Given the good results obtained in other museums, such as the National Museum of Natural Sciences [20], with which the ETSII shares a building, it was decided to opt for Tour Creator to transfer the physical tour of the Museum to the virtual tour.

To create the tour, Tour Creator offered two options: use photos taken with cameras capable of capturing 360° images or use one of the functionalities of the Google Street View application to create panoramic views from several normal images taken with a smartphone. The methodology is quite simple; it is enough to follow the orange dot on the mobile camera until the panorama is completed.

Since it is not possible for this project to have a camera to capture 360-degree panoramas, it has been decided to use the Xiaomi Redmi Note 5 Pro smartphone to take the virtual tour.

In addition, it is possible to link the panoramas with the GPS location of the phone and upload the panoramas to Google Maps, to be able to see them from the perspective of Street View (Fig. 7).

Once the panoramic views had been created at each point of the physical tour of the Museum, they were uploaded to Tour Creator and the virtual tour was generated. In a complementary way, information was added to the virtual tour through texts, images and audios, so that there are no differences with the physical tour.

The tour can be visited through the Google Poly platform, and if desired, with virtual reality headsets such as Google Cardboard or Daydream View on Google Expeditions.

Fig. 7. Virtual tour of the main hall

Since June 2021, Google has stopped supporting Google Poly and the models must be transferred to other virtual navigation systems.

4 Creation of the Website

With the aim of creating a platform, accessible to all, that would support the ETSII Museum and for the dissemination of the heritage of the Higher Technical School of Industrial Engineers of the Polytechnic University of Madrid, a website was designed. This is how "*Patrimonio ETSII*" (http://patrimonio.industriales.upm.es/) emerges (Fig. 8).

The website was created through the WordPress platform. It is a content management system, that is, a computer program that allows you to create a work environment for the creation and administration of content (text, multimedia…), mainly used in web pages.

In the blog section, written entries about the history of the institution or characters related to it are included. Also we have added voice interviews to some relevant engineer biographies thanks to Prof. Miguel Angel Sebastian from UNED (Universidad Nacional de Educación a Distancia of Spain).

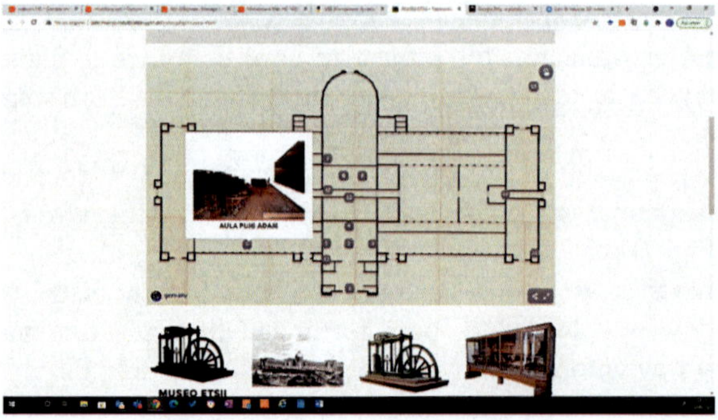

Fig. 8. Detail of the content of the ETSII heritage website.

To spread the activities carried out on the "Patrimonio ETSII" web site, the pieces that make up the museum's collections and the history of the School of Industrial Engineers, two accounts were created on the Instagram and Twitter platforms with the same nickname or name (@museo_etsii).

The approach is to create interest in the aforementioned issues with a direct language, understandable by anyone without prior technical knowledge and interesting from a historical, scientific, or engineering point of view.

4.1 3D Reconstruction of Parts Through Photogrammetry

Twenty pieces has been 3D reconstructed with photogrammetry to include in the website.

Photogrammetry is the technique that studies and seeks to precisely define the shape, dimensions and position in space of any object, essentially using measurements made on one or more photographs of that object.

To achieve an optimal level of accuracy when reconstructing a piece in 3D, more than one photograph is required. This is because, if two images are available from different perspectives, the position of a 3D point can be found as the intersection of the two projection rays, this is known as triangulation.

Triangulation is a process similar to that performed by a human being unconsciously. When observing an object, the eyes send two independent images to the brain, and this is responsible for creating a three-dimensional image through the superposition of both [21].

Now, 3D reconstruction is a more complex problem, since it is intended to make a reconstruction from all perspectives of the object. For this, it is not enough with two images, but a minimum of 100 images from multiple perspectives have been necessary to obtain optimal results to develop the mesh (Fig. 9) for each reconstructed object. To gain realism it is necessary to apply on the mesh colors and textures giving rise to a continuous surface. The results, as seen in Fig. 9b, are close to reality or, rather, the reality captured by the camera.

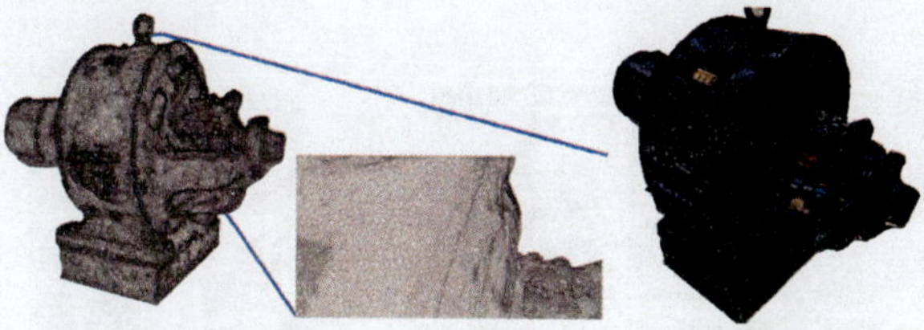

(a) (b)

Fig. 9. (a) Mesh and detail of the polygonal structure of an electric motor. (b) Final result including textures and colors

The reconstruction of industrial parts presents the following drawbacks or characteristics to take into account:

1. Materials: some materials used in industry (some paints, brass, aluminum, crystals, polished steels…) cause undesirable reflections.

2. Shapes: the pieces usually have a large number of complicated surfaces, holes, grooves (motors, gears…) that promote the appearance of shadows and prevent a good reconstruction (see Fig. 10a).
3. Size: Some industrial parts can be large in size. Its displacement (some may be anchored to the ground) and its good lighting are practically impossible.
4. Shadows: In some cases it is necessary to employ natural light if possible, Light bulbs generates shadows that could impair the photogrammetric process.

ReCap © (Reality Capture), like other Autodesk products, is developed for compatibility with other Autodesk tools, such as those that work with CAD and BIM has been selected to reconstruct the pieces virtually. The program has two basic modes: one for aerial photogrammetry and the other for short-range photogrammetry (objects and closed spaces). In addition to photos, ReCap can also process laser scans and align them with photo-based projects to generate 3D models. Although ReCap can automatically register images, it also allows manual selections to improve the overall result. The most outstanding functionality is the automatic cleaning or Auto Cleanup, which automatically identifies and eliminates unwanted points, such as people and moving objects. Another useful option is the ability to convert scan data into meshes, which can be exported for processing in other applications.

Each point of the ETSII museum tour has a poster. This will be the way to access all the information of the object, which is available on the website of 'Patrimonio ETSII' (http://patrimonio.industriales.upm.es). Access is made through a QR code.

Each of the pieces or stops of the tour of the School Museum have a historical and scientific investigation behind. This has been reflected in a series of online cards accessible with a QR code, printed on the corresponding poster.

The concepts explained in each file try to introduce the visitor not only to the historical context where they were used, but also to their scientific basis. Therefore, when writing the files of each piece of the Museum, a clear and simple language has been chosen, as brief as possible and accessible to an audience without technical knowledge and of different ages.

(a) (b)

Fig. 10. (a) Photo set of the typewriter for photogrammetric reconstruction. (b) Typewriter reconstruction with ReCap Photo (image corrections: + 40% brightness 40% contrast)

In addition, to improve the experience of the visit and make it more interesting, the online files have spoken audios, so that the visitor can use his own phone as an audio guide while appreciating the piece.

5 Conclusions

The number of items to be catalogued is still very extensive. Many of the Departments of the ETSII hide true historical jewels that must be cataloged and exhibited. It is also important to emphasize the role of women in the history of ETSII in future research. Of all the personalities collected in the catalog, there are only three women.

Some of the lines of future work that should be explored are as follows:

- Physical design of exhibitions at the School and their export to three-dimensional virtual tours.
- Creation of a narrative that reflects the history of the School from 1975 to the present. Interviews, testimonies, photographs, videos, documents, etc.
- Cataloguing of pieces from the rest of the departments that have not been able to participate in this initial project.
- Study of the impact of our School on the industrial development that Spain experienced from the 1950s. Projects carried out and influence of the graduated engineers from the School in the Spanish and international Industry.
- Study of the impact of the creation of the Royal Industrial Institute. Recovery of innovations in Spanish technology.

In conclusion, it is necessary and important that the management of the ETSII continues to bet on projects of this type. It is part of its task to work on its conservation and dissemination of both the pieces that are there, and the legacy of the people who have given their lives for the School. Knowing the past of industrial engineers is essential to decide what steps they should take in the future.

The digitization process of this museum has been accelerated since the Covid-19 pandemic. Virtualization gives us the opportunity to divulge and virtually operate with our collection and remain open 24 h a day to all people.

Acknowledgments. The Authors wants to acknowledge the support received from the Management Board, the Library and Faculty of the ETSII.

References

1. Egorova, O.V.: (2009) The First Steam Machine in Cuba: Little-Known Pages of Agustin de Betancourt's Work and Life. In: Yan HS., Ceccarelli M. (eds) International Symposium on History of Machines and Mechanisms. Springer, Dordrecht. https://doi.org/10.1007/978-1-4020-9485-9_12
2. Convento de la Trinidad Calzada (Madrid). Wikipedia. Retrieved August 25, 2021, from https://es.wikipedia.org/wiki/Convento_de_la_Trinidad_Calzada_(Madrid)

3. Pavón, JMC.: (1998). The Royal Industrial Institute of Madrid (1850–1867). An historical overwiev Quaderns d'història de l'enginyeria, 2002, https://core.ac.uk/download/pdf/41788859.pdf
4. Museo Histórico-Minero Don Felipe de Borbón y Grecia. Retrieved August 25, 2021. http://www.minasyenergia.upm.es/museo-historico-minero.html
5. ETSIT. Catálogo del Museo. Retrieved August 25, 2021. https://www.etsit.upm.es/escuela/unidades-administrativas-y-servicios/biblioteca/museo-de-telecomunicacion-profesor-joaquin-serna/catalogo-del-museo.html
6. ETSIT. Blog del Museo. Retrieved August 25, 2021 https://www.etsit.upm.es/escuela/unidades-administrativas-y-servicios/biblioteca/museo-de-telecomunicacion-profesor-joaquin-serna/blog-del-museo.html
7. BIBLIOTECA UPM. (2020a). Colección Digital Politécnica http://cdp.upm.es/exlibris/dtl/u3_1/dtle/www_r_spa/icon/presentacionCDP.html
8. BIBLIOTECA UPM. (2020b). Colección Digital Politécnica ETSII. ETSI Industriales. http://cdp.upm.es/R/U2X6AI4HK55634AHX27TQQQD9QHDIN3BLKQ1418AG2CPDA9GL3-02460?func=results-brief
9. BIBLIOTECA UPM. (2020c). Colección Digital Politécnica ETSII. ETSI Industriales. Unidad Docente de Química Aplicada. http://cdp.upm.es/R/U2X6AI4HK55634AHX27TQQQD9QHDIN3BLKQ1418AG2CPDA9GL3-02614
10. Leiner, B.M., et al.: A brief history of the internet. SIGCOMM Comput. Commun. Rev. **39**(5), 22–31 (October 2009). DOI:https://doi.org/10.1145/1629607.1629613
11. Dodson, A.D.B.: Introducing Ourselves. http://moca.virtual.museum/about.htm (2002)
12. Red digital de colecciones de Museos de España http://ceres.mcu.es/pages/SimpleSearch?index=true
13. Hispana https://hispana.mcu.es/es/inicio/inicio.do
14. Purday, J.: Think culture: Europeana.eu from concept to construction **33**(2), 170–180 (2009). https://doi.org/10.1515/bfup.2009.018
15. Jahn, V., Gonzalo Vélez Jahn, A.: Museos virtuales-Presente y Futuro. Primera Conferencia Venezolana Sobre Aplicacion de Computadoras en Arquitectura (1999). https://itc.scix.net/pdfs/6132.content.pdf
16. Museos de Madrid 360° https://www.livingmadrid.com/museos-de-madrid-visita-virtual-360/
17. DassaultSystèmes. Giza 3D; Paris 3D. https://www.3ds.com/stories/
18. Museo Nacional del Prado.: Actividades del Museo Nacional del Prado con el covid-19. MUSEO DEL PRADO. https://www.museodelprado.es/busqueda-actividades/tag/covid-19 (2020)
19. Valbuena Vázquez, P.S.D., Scala Estalella, J.J.: Historia de la Escuela Técnica Superior de Ingenieros Industriales de Madrid desde 1901 hasta 1972 Proyecto Fin de Carrera. http://oa.upm.es/5331/1/PFC_PABLO_VALBUENA_VAZQUEZ.pdf (1996)
20. Virtual Tour Museo de Ciencias Naturales de Madrid. Retrieved September 25, 2021 https://artsandculture.google.com/exhibit/yAISzR1n3bdnLw
21. Grussenmeyer, P., Hanke, K., Streilein, A.: Architectural photogrammetry. In: Kasser M., Egels Y. (eds.) Chapter in « Digital Photogrammetry ». Taylor & Francis, pp. 300–339 (2002)

Kinematic Modelling in Greek Astronomy: The Design of Heavenly Mechanisms

Teun Koetsier(✉)

Department of Mathematics, Faculty of Science, VU-University, Amsterdam, The Netherlands
t.koetsier@vu.nl

Abstract. The Babylonian astronomers kept a close track of the phenomena in the sky in a tradition based on arithmetic. The Greeks introduced a completely new approach. They developed kinematical models to explain the phenomena. These models are idealized mechanisms. In memory of Stephanos Paipetis and Carlos López-Cajún we give in this paper a brief survey of the development of the heavenly mechanisms the Greeks designed to explain the motion of the stars, the Sun, the Moon and the planets. Carlos and Stephanos were my colleagues and my friends. They passed away in 2020. I shared with them the admiration for the mathematicians and engineers of classical antiquity.

Keywords: Greek astronomy · History of kinematics · Autolykos · Plato · Eudoxus · Hipparchus · Ptolemy

1 Geometrical Modeling in Greek Astronomy

Diogenes Laërtius wrote "Further we are told that Pythagoras was the first to call the heaven the universe and the earth round (i.e. spherical), though according to Theophrastus it was Parmenides, and according to Zeno it was Hesiod" ([5], p. 11). Aëtius wrote: "The Pythagoreans held the Sun to be spherical" ([5], p. 35). These quotations show the tendency that existed in early Greek thinking to approach the universe, and in particular the heavenly bodies, in a geometrical way. We do not know who came up with the idea to explain the motion of the stars by assuming that they are fixed to a spherical surface rotating uniformly about a fixed axis. However, since Parmenides the existence of such a sphere of the fixed stars was generally accepted in Greek astronomy. Its center coincided with the center of the spherical earth.

As for the mathematics corresponding to this model one of the oldest extant works in Greek astronomy is *On the moving sphere*, written by Autolykos of Pitane around 320 BCE [1]. Autolykos wrote in the second half of the 4th century BCE, but the contents of his books are undoubtedly much older. The book starts with the following text: "It is said that points move along uniformly, when in equal time (intervals) they trace equal or similar lengths (of arc). If then on a certain line a certain point, moving uniformly, goes through two line (segments), the ratio of the time intervals in which the point has gone through either of the lines will equal the ratio of the line segments. Axis of a sphere is

M. Ceccarelli and R. López-García (Eds.): HMM 2022, HMMS 40, pp. 279–291, 2022.
https://doi.org/10.1007/978-3-030-98499-1_23

the diameter of the sphere which remains fixed while the sphere rotates about it. Poles of the sphere are the ends of the axis" ([1], p. 1).

The second sentence is identical with Proposition 1 of Archimedes' *On Spirals*. Autolykus' book deals with a sphere rotating uniformly about an axis. In Propositions 5 and 6 of the book a fixed large circle of the sphere (obviously fixed in the sense that it does not share the rotational motion of the sphere) is introduced. This circle defines the visible and the invisible part of the sphere. This circle is the horizon. Let us consider Proposition 6 of *On the moving sphere*:" If on the sphere the large fixed circle defining the visible and the invisible (part) of the sphere is inclined to the axis, it will be tangential to two circles which are equal and parallel to another. Of these the one towards the visible pole is always visible, and the one towards the invisible (pole) always invisible" ([1], p. 10). See Fig. 1.

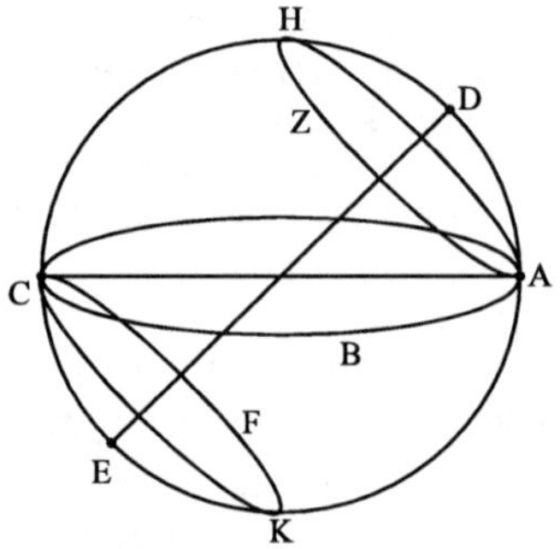

Fig. 1. Autolykos' moving universe

Autolykos' proof runs as follows: Let the large circle ABC define the visible and the invisible. Let the visible pole be D. Draw the large circle ADC. Let point E on circle ADC be such that arc CE is equal to arc AD. E is obviously the invisible pole. It is then clear that the circle AZH drawn around the pole D and the circle CFK drawn around pole E are both tangent to the horizon and they represent the areas on the rotating sphere that are respectively always visible and invisible.

2 The Model in Plato's Timaeus

Autolykos lived c. 360–c. 290 BCE. Plato wrote his *Timaeus* roughly at the time Autolykos was born. In it he gives a model of the universe. The Sun, the Moon and the planets move in a plane which is not perpendicular to the axis of the sphere of the fixed stars; it intersects this sphere in the circle called Ecliptic which is the center of the belt of constellations called the Zodiac. Possibly the Greeks became acquainted with the signs of the Zodiac, defined earlier by the Babylonians, through Cleostratus of Tenedos about 550 BCE ([11], p. 84). At the time of Plato the Zodiac was known.

Look at this part of Plato's description of how God created the world (Fig. 2.): "This entire compound he divided lengthwise into two parts which he joined to one another

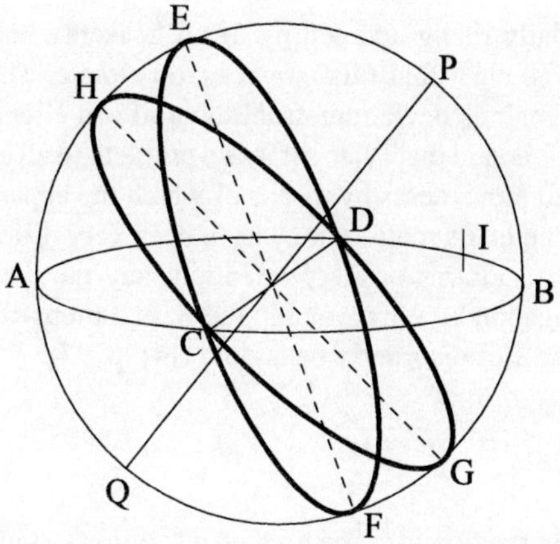

Fig. 2. The universe with its equator HCGD and the Ecliptic ECFD

at the center like the letter X, and bent them into a circular form, connecting them with themselves and each other at the point opposite to their original meeting point, and, comprehending them in a uniform revolution upon the same axis, he made the one the outer and the other the inner circle. Now the motion of the outer circle he called the motion of the same, and the motion of the inner circle the motion of the other or diverse. The motion of the same he carried round by the side to the right, and the motion of the diverse diagonally to the left. And he gave dominion to the motion of the same and like, for that he left single and undivided but the inner motion he divided in six places and made seven unequal circles having their intervals in ratios of two and three, three of each, and bade the orbits proceed in a direction opposite to one another. And three [Sun, Mercury, Venus] he made to move with equal swiftness, and the remaining four [Moon, Saturn. Mars, Jupiter] to move with unequal swiftness to the three and to one another, but in due proportion" ([9] p. 1166).

The letter X in the quotation corresponds to the intersection of the ecliptic, the circle that carries the Zodiac, and the equator of the sphere of the fixed stars. Those two circles are created by connecting the end points of the legs of the letter X in such a way that those legs become the two circles.

The two legs of the letter X in Plato's Timaeus are the 'outer circle' HCGD, the celestial equator, and the 'inner circle' ECFD, the Ecliptic. The Sun, the Moon and the planets move in the plane of the inner circle, the Ecliptic. The celestial sphere rotates about the axis PQ perpendicular to the plane of the celestial equator. The celestial equator as part of the whole celestial sphere rotates daily from East to West. That is the dominating motion. The Sun, the Moon and the planets have an extra annual motion. They move in the plane of the Ecliptic in the opposite direction, from West to East. Of course, next to these two circles we have the horizon ABI. Elements of the geometry of this more sophisticated model of the universe are also treated in Autolykos' *On the moving sphere.*

The model of the universe that we find in the *Timaeus* is based on the recognition that the plane of the annual motions of the Sun, the Moon and the planets must be inclined to that of the daily rotation of the sphere of the fixed stars. The latter rotation is dominant and accounts for the daily rising and setting of all heavenly bodies. However, in the course of time it became clear that this model is too simple. The Sun, the Moon and the planets exhibit anomalous deviations from the uniform circular movement in both longitude and latitude. It is said that Plato set it as a problem to all earnest students to find the uniform and ordered movements by means of which the apparent movements of the planets could be explained. His motives may have been very different from the motives of modern astronomers.[1] Plato's primary intention may have been to save his astral theology from these vagabonds. Why worship the stars if these divine beings cannot do better than set an example of disorderly behavior? ([4], p. 97).

3 Eudoxus' Models

Eudoxus may have been the first to come up with a solution to the problem set by Plato. Eudoxus studied geometry with Archytas. At the age of 23 he traveled to Athens where he listened to the lectures given by sophists. He was a pupil of Plato for some time.

The results given by Autolykos were certainly known to Eudoxus. We may even owe them to Eudoxus, but they may be older. Eudoxus' complex system has been reconstructed in the 19th century by Schiaparelli on the basis of very modest evidence. Others have elaborately commented on Schiaparelli's reconstruction (See, for example, [12, 13]). The reconstruction is too lovely not to describe its basic structure. Eudoxus' system is an attempt to refine the simple model described in the *Timaeus* based on the following idea, already implicit in that model. The movement of the fixed stars is modeled by means of one sphere, the celestial sphere, rotating about its axis. In one day it completes a full rotation. As we have seen, if one considers a longer period of time it becomes clear that the Sun moves with respect to the celestial sphere. This anomaly is taken care of by Eudoxus by adding another sphere coinciding with the celestial sphere, which rotates about an axis fixed inside the celestial sphere. This second sphere takes a year to complete a full rotation. Clearly the 'equator' of this second sphere, to which the Sun is attached, coincides with the Ecliptic in the celestial sphere. The angle between the axes of the two spheres would have to be some 24°, which is the obliquity of the Ecliptic. The same idea is applied to the Moon and the planets. In a first approximation of their movement spheres are introduced that rotate about coinciding axes fixed inside the celestial sphere perpendicular to the ecliptic.

Unfortunately for none of the heavenly bodies the two-sphere model was sufficiently in accordance with the phenomena. In order to correct the anomalies further rotating spheres were added by Eudoxus. In the course of time Eudoxus' system was unable to satisfy astronomers. In his model of concentric spheres the distance of the Sun, the Moon and the planets to the earth is constant, which is not in accordance with the fact that the Sun and the Moon have a variable apparent diameter. Simplicius wrote: "[...] the theories of Eudoxus and his followers fail to save the phenomena, and not only those which were first noticed at a later date, but even those which were before known and actually accepted by the authors themselves" ([5], p. 68). Simplicius mentions a pupil of Eudoxos, Polemarchos of Cyzocos (c. 340 B. C.) who knew the discrepancy between the

[1] For a balanced view on Plato's attitude with respect to astronomy see: ([2], pp. 107–112).

model and the phenomena as for the apparent size of the heavenly bodies, but minimized its importance. Others like Calippus (about 370–300 BCE) and Aristotle, added spheres to Eudoxus' model. Calippus did this to account for the unequal motion of the Sun in longitude (the velocity of Sun against the backdrop of the Ecliptic is not constant in the course of the year) and Aristotle found it necessary to add spheres when he attempted to envisage a working material mechanical system that could replace the set of seven separate abstract systems that Eudoxus had defined ([5], pp. xlvi–xlix).

4 Apollonius' Epicycle Model

The Greeks did not possess the notion of instantaneous velocity and a general treatment of non-uniform motion was for them impossible. Yet the Greeks found a way to handle non-uniform motion in astronomy in a quantitative way. The motion of the Sun, the Moon and the planets is non-uniform. By combining uniform circular motions Eudoxus succeeded in creating precisely defined non-uniform motions, although the result turned out to be only roughly in accordance with the observed phenomena.

The next ideas along these lines led to the final Greek solution with respect to the motion of the heavenly bodies, the stars, the planets the Sun and the Moon. This solution has come to us in a book written by Claudius Ptolemaeus or Ptolemy in English (circa CE 100–circa CE 175), an astronomer working in Alexandria, the main city in Greco-Roman Egypt at the time. The original title of the text was 'Mathematical Systematic Treatise'. The modern title is *Almagest* derived from the Arabic 'Almagisti' which in its turn comes from the Greek word 'greatest' (μεγίστη). At the end of Antiquity the *Almagest* had become the standard textbook on astronomy, which it would remain for more than a millennium. The success of the book contributed much to the loss of earlier works on astronomy.

In *Almagest* III.3 Ptolemy writes with respect to the irregularities or anomalies in the motions of some of the heavenly bodies the following: "The reason for the appearance of irregularity can be explained by two hypotheses, which are the most basic and simple. When their motion is viewed with respect to a circle imagined to be in the plane of the Ecliptic, the center of which coincides with the center of the universe (thus its center can be considered to coincide with our point of view), then we can suppose, either that the uniform motion of each [body] takes place on a circle which is not concentric with the universe, or that they have such a concentric circle, but that their uniform motion takes place, not actually on that circle, but on another circle, which is carried by the first circle, and [hence] is known as the 'epicycle'. It will be shown that either of these hypotheses will enable [the planets] to appear to our eyes, to traverse unequal arcs of the Ecliptic (which is concentric to the universe) in equal times" [10].

We do not know who invented these hypotheses or models. It is, however, quite possible that Apollonius invented the second hypothesis, the *epicycle model* (Fig. 3). As we will see he studied this model. The Eudoxian model to explain the retrograde motion of the planets is brilliant but complex. Moreover, the motion of the Sun, the Moon and the planets takes place in the plane of the ecliptic, if we abstract from deviations in longitude. The idea to try to explain the retrograde motion by means of a plane model is very natural. Moreover, this is what the epicycle model can easily accomplish.

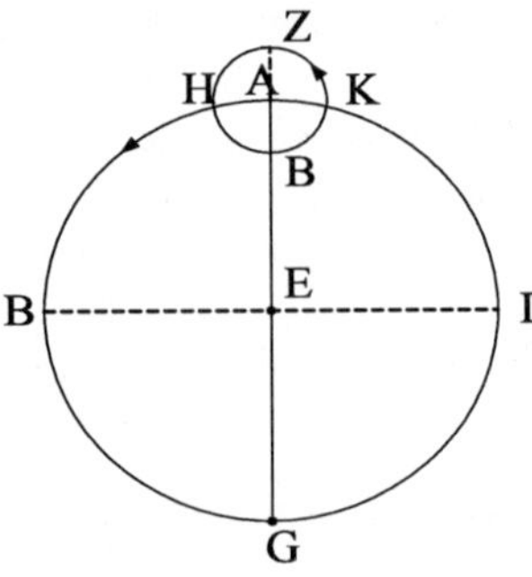

Fig. 3. The epicycle model

Two circles are involved: the *epicycle* and the *deferent*. The *epicycle* ZHBK, or rather its center, moves on the circle ABGD with center E. This circle ABGD is called the *deferent*. The body moves on the epicycle ZHBK. It is quite obvious that seen from the point of view of the terrestrial observer the motion is non-uniform. Suppose the epicycle moves in the direction from A to B. If the body on the epicycle is in the position K it appears to lag behind, if it is in H it appears to move faster than the epicycle.

5 Hipparchus' Theory of the Motion of the Sun (About 150 BCE)

As we have seen Ptolemy also mentions another hypothesis: the eccentric hypothesis or the *eccentric model*. We do not know who invented the model, but it may have been Hipparchus. From the period between 350 BCE and the year 0 many Babylonian astronomical cuneiform tablets are extant. They contain calculations and predictions concerning the lunar phases, positions of planets etc. This was an arithmetic tradition in astronomy which kept a close track of the phenomena and differed from the more speculative tradition in which Eudoxus and Apollonius were working. The two great Greek mathematicians were developing kinematical models for astronomical phenomena, but they did not go beyond a qualitative comparison between the model and the phenomena.

In this respect Hipparchus (about 150 BCE) represents a revolution. Hipparchus was familiar with Greek geometry and with Babylonian observations. As far as we know he was the first to try to really adapt the models to the actual observed data. He is famous for his investigation of the motion of the Sun. It led to the first sine-table. Consider a circle and suppose that a cord is seen under an angle α from the center of the circle. Hipparchus' table differed from the modern sine-tables in the sense that it gave the lengths of the cords corresponding to a sequence of angles α. The lengths were given in terms of the Babylonian sexagesimal system. They were denoted using the Greek alphabet. Many of the entries in the table are approximations, inevitably so because the cord and any 'unit of length' are often incommensurable.

The Babylonians had divided the yearly circular path of the Sun against the background of the Zodiac, in twelve parts of 30°. According to Neugebauer the Sumerians had a unit of distance, the *danna*, equal to approximately 7 English miles. The danna was subdivided in 30 UŠ or 30 "lengths". These units were also used to measure time. A period of one danna covered the time needed to travel a distance of one danna. One day or one revolution of the sky took twelve danna. This is where, according to Neugebauer, the subdivision of the ecliptic in 12 times 30° comes from. Later all other circles were divided in 360°. Cf. [3, 6, 7].

The twelve parts got the names of twelve constellations of stars with which the twelve parts more or less coincided (The modern signs in unicode are: Aries ♈, Taurus ♉, Gemini ♊, Cancer ♋, Leo ♌, Virgo ♍, Libra ♎, Scorpio ♏, Sagittarius ♐, Capricorn ♑, Aquarius ♒and Pisces ♓.).

At a certain moment the Babylonians realized that the velocity of the Sun in the course of a year is not constant. The number of days elapsing between the vernal solstice and the equinoxes is not the same as the number of days between the equinoxes and the winter solstice. The equinoxes, autumnal and vernal, and the two solstices divide the year in unequal parts. We know that earlier than 200 BCE a system was created in which the Sun moved in two parts of the year with a different constant velocity. Later another system was used in which the velocity of the Sun increased linearly for half a year and afterwards decreased linearly for half a year. It is quite possible that Hipparchus discovered the eccentric model while pondering these problems. The Sun never moves backward with respect to the Ecliptic, so this particular aspect of the epicycle model is not needed her. The basic idea of the eccentric model is that if we observe a uniform circular motion from an eccentric point we get a non-uniform motion in which for half a year the speed decreases and then increases for half a year. This is how Ptolemy described the motion. Consider Fig. 4. The body travels with uniform motion on the circle ABGD with center E. Let the point Z on the diameter AED represent the observer. Let the angle BEA be equal to the angle GED. Ptolemy writes "the body will traverse the arcs AB and GD in equal times, but will [in doing so] appear to have traversed unequal arcs of a circle drawn on center Z". In this eccentric motion A is the *apogee*, by definition the point farthest from the observer, and D is the *perigee*, by definition the point nearest to the observer. Clearly, to the observer in Z the least speed occurs at the apogee A and the greatest at the perigee D.

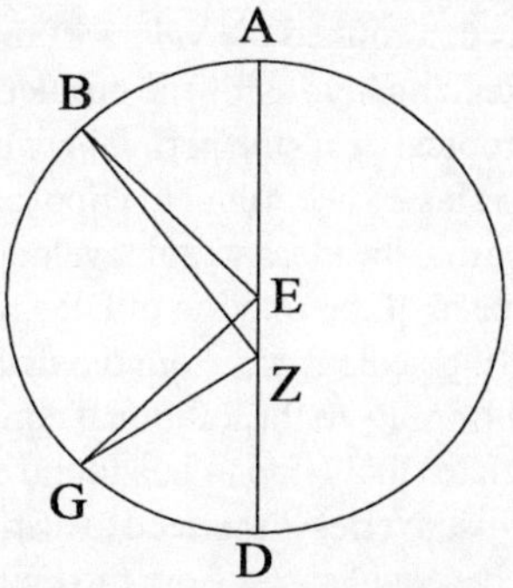

Fig. 4. The eccentric model

If in an epicycle model the two angular velocities are equal but have an opposite direction, it is equivalent to an eccentric model. We do not know who discovered this property. It may have been Hipparchus. Consider in Fig. 5 the two parallel line segments DE and AB. When DE and AB rotate with equal velocity, say counterclockwise, in the same direction about, respectively, D and A, the distance BE remains equal to AD and ADEB is in every position a parallelogram (Excluding the two special positions in which AB and DE coincide with the same straight line.). Clearly one and the same motion of point E can be realized by means of an eccentric motion about D and an epicyclic motion composed of a motion of B about A and a motion of E about B ([10], Book 1, chapter 5).

Let us briefly discuss Hipparchus' theory of the Sun. Hipparchus' model describes the motion of the Sun with respect to the Zodiac. See Fig. 6. The earth T is in the center of the universe. D is a fixed point at a distance *e* from the center. The Sun rotates uniformly about D on a circle with radius *R*. P_1 is the position of the Sun at the vernal equinox in the sign Aries. The unknown parameters in the model are e, R and the angle $DTP_1 = \lambda_a$.

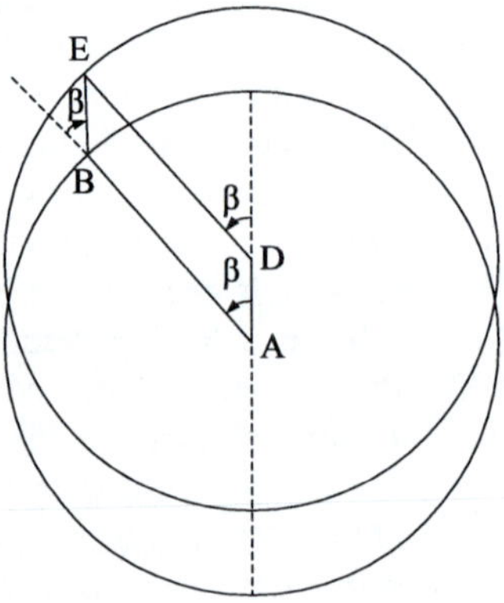

Fig. 5. The equivalence of the eccentric and the epicycle models

Let us see how Hipparchus determined the values of these parameters. As for e and R he determined the ratio e/R leaving here aside the problem of the distance between the Sun and the earth. A solar or tropical year (the period between two succeeding positions of the Sun at the vernal equinox) lasted according to Hipparchus 365¼–1/300 days ([10], Book 3, chapter 1). This implies that the mean angular velocity of the Sun $\omega = 360°/365¼ = 59' 8''$ per day. This is the speed of the rotation of DP about D. Moreover Hipparchus used the lengths of two seasons: he calculated from the data that he possessed that from P_1 to P_2 it takes 94½ days and from P_2 to the autumnal equinox it takes 92½ days ([10], Book 3, chapter 4). Ptolemy relates that Hipparchus found $e/R = 1/24$ and $\lambda_a = 65° 30'$.

One of Hipparchus' major discoveries is the need to distinguish the Zodiac, the yearly path of the Sun divided in twelve equal signs, from the Ecliptic, the circle on the sphere of the fixed stars. In this way the Zodiac is fixed with respect to the seasons. Whenever spring starts the Sun is in the vernal equinox and by definition in the sign Aries ♈. This means that signs of the Zodiac move with respect to the constellations of stars that they originally represented: the solar year moves slowly with respect to the ecliptic. Every

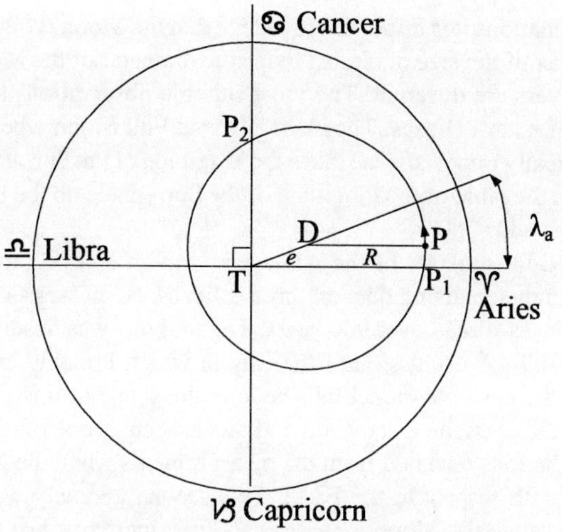

Fig. 6. Hipparchus' theory of the motion of the Sun

hundred years the signs of the Zodiac rotate one degree backwards with respect to the ecliptic. This phenomenon is called the *precession.*

6 Ptolemy's Contributions

Ptolemy followed in the footsteps of Hipparchus and other Greek astronomers who attempted to come up with models that were fully in accordance with the observations. This asked for a lot of ingenuity. By definition the orbit of the Sun is in the plane of the Ecliptic: it only possesses a rotation in longitude. Next to the anomalies in the rotation in longitude the orbits of the Moon and the planets possess deviations in latitude that must be taken into account as well.

Ptolemy essentially follows Hipparchus as for the model of the motion of the Sun. Hipparchus had given an epicyclic model for the motion of the Moon. In the model the epicycle takes care of the (first) anomaly in the motion of the Moon: the deviations from the uniform rotation. Ptolemy doesn't follow Hipparchus because he discovered that the motion of the Moon deviates from the predictions that result from the epicyclic model: there exists a second anomaly. The model he comes up with also handles this second anomaly. It is highly original.

In the case of the Moon we have a very complicated situation. The tropical month, the period in which the Moon returns to the same point on the Ecliptic, is not constant. The synodic month, the period between two successive conjunctions of the Sun and the Moon or the complete cycle of the Moon's phases, is not constant either. The angular velocity of the Moon varies from 10° to 14° per day. This leads to a new "month": the anomalistic month which is the period in which the Moon returns to the same velocity. This period in principle corresponds to the period of the rotation of the epicycle. Moreover the Moon has varying latitude to the North or South of the Ecliptic with a maximum of 5°.

As for the calculations it is important to realize that the Moon is big and has a considerable daily parallax of the size of almost twice the diameter of the Moon: observations by different observers are different. The most suitable observations to be used as data are observations of lunar eclipses. They take place at Full Moon when the Sun and the Moon are diametrically opposed. One takes the longitude of the Sun at the exact moment of the eclipse from the table of the longitude of the Sun (made on the basis of the theory of the Sun) and one adds 180°.

Ptolemy started in Book IV of the Almagest with an epicycle model of which he determined the parameters using data on three eclipses. He possessed a list of observations covering eclipses spread over 900 years. The first one was observed in Babylon in 721 BCE ([8], p. 169). A discussion of the way in which Ptolemy exactly determined the parameters of the epicycle model falls beyond the scope of this paper Once he had determined the parameters, he discovered and this is seen as one of his major contributions, that the predictions deviated from the measurements when the Moon and the Sun are in quadrature with respect to the Earth. This second anomaly was later called the *evection.* It occurs when the Moon is in its exact first quarter or last quarter phases. In such positions the angle between the Sun and the Moon as seen from the Earth is 90°. In these positions the apparent size of the epicycle is larger.

Consider the epicycle model for the Moon in Fig. 7. The arrow pointing at the Mean Sun doesn't point at the actual position of the Sun; it points at the position that the Sun would have if it were executing a uniform rotation about the earth. However in order to couple the motion of other heavenly bodies to the motion of the Sun, the Mean Sun is used. The position of the Mean Sun is determined by means of the angle ΩS, starting from a fixed point on the ecliptic, for example, Aries. The position of the epicycle of the Moon is determined by means of the angle ΩS + μ, also measured from Aries. Obviously ΩS, μ and ΩS + μ are linear functions of time.

The epicycle model for the Moon was fine except for the fact that Ptolemy needed a mechanism to pull the Moon somewhat towards the Earth in positions of quadrature in order to explain the evection.

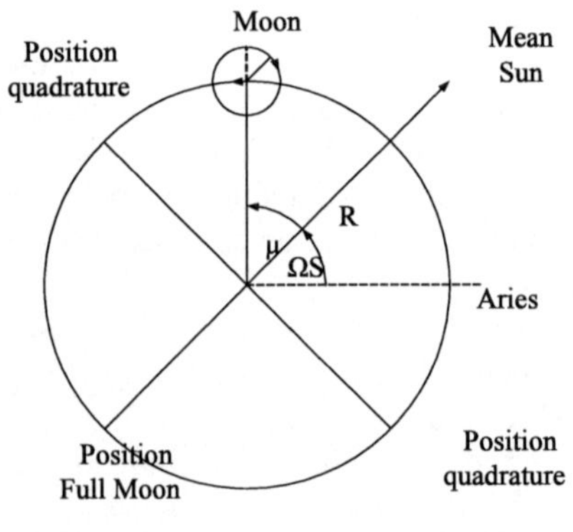

Fig. 7. The epicycle model for the Moon.

Ptolemy did this by adding what is from a modern point of view an abstract crank and slider mechanism to the simple epicycle model ([6], p. 85 and Fig. 78). See Fig. 8. The connections A, B and C are hinges. Point A is a fixed and so is the line AC along which C can slide during the rotating input of crank AB.

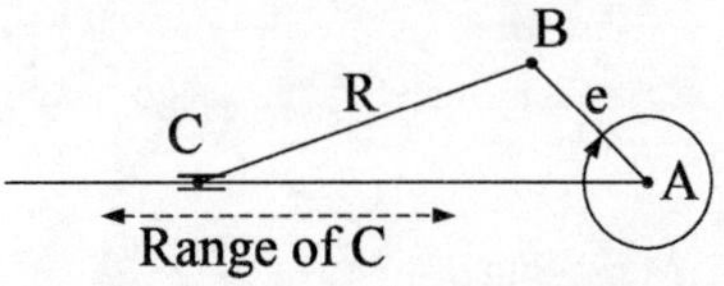

Fig. 8. Crank and slider mechanism

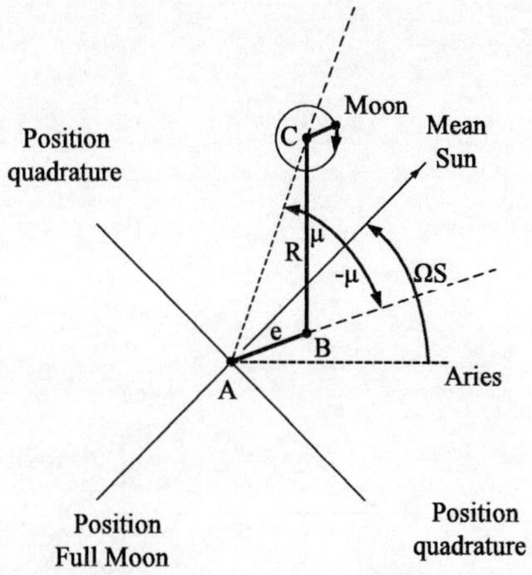

Fig. 9. Ptolemy's model for the motion of the Moon

The basic idea is that we start from the crank slider motion. A is the position of a terrestrial observer and C is the center of the epicycle of the Moon. A is fixed but AC rotates about A with the corresponding angular velocity. We also maintain the angular velocity of the Moon about the center of the epicycle from the first model. The distance of the epicycle from the earth is regulated by means of the crank AB; the crank moves in such a way that its direction, measured from Aries, is given by the angle $\Omega S + \mu$. The result is visible in Fig. 9.

Suppose AB has a length e and the length of BC equals R-e. The value of μ is a linear function of time. When $\mu = 0$ or 180° we have either Full Moon or New Moon and the center of the epicycle is at a distance R from the earth. The model behaves exactly as the epicycle model at these moments. However, when $\mu = 90°$ or $\mu = 270°$ the 'slider' of the crank-slider mechanism is its other opposite position, we have positions of quadrature and the center of the epicycle is at a distance R-2e from the earth.

7 Ptolemy's Contributions Continued

In his theory of the outer planets, Saturn, Mars and Jupiter, Ptolemy in a sense applies the crank-slider mechanism in a different way. See Fig. 10. The point D, E and C are hinges. C slides on the line DC. Now the link ED is fixed. A rotation of EC about E (in the case of Ptolemy a uniform rotation) yields a non-uniform rotation of C about D. The mechanism is applied by Ptolemy as follows (See Fig. 11).

T is the earth. Point E is fixed as well. It is the center of uniform motion (This point is later called the 'equant'), because EC rotates uniformly about E. The position of C on

EC is determined by the constant length R of CD. C is in Ptolemy's model the center of an epicycle on which the planet P rotates with a constant angular velocity relative to the direction of EC.

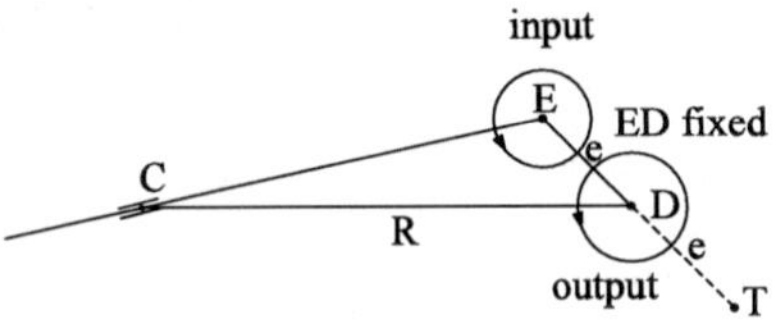

Fig. 10. Crank and slider mechanism

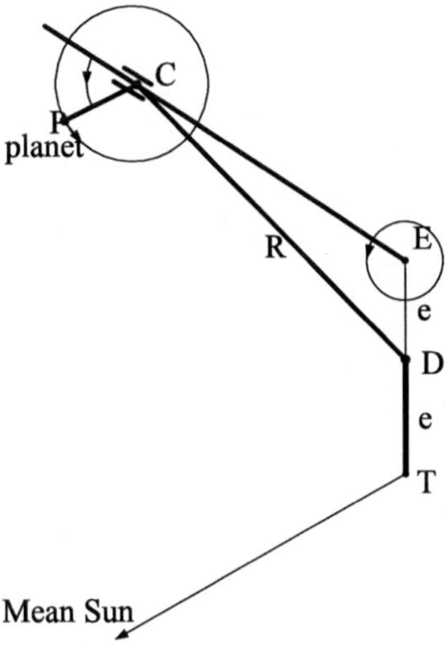

Fig. 11. Ptolemy's model for the outer planets

8 Conclusion

We will not discuss Ptolemy's models for the planets Venus and Mercury. In the models the idea of the slider crank mechanism is used also; a point moves on a circle, but its rotation is uniform with respect to another point which is not the center of the circle. This particular point has been the source of much debate. Aristotle had defended a very strict position with respect to the uniform circular motions that were allowed in mathematical astronomy: i) They had to be uniform as seen from their own center, and ii) they had to be concentric with the whole universe ([8], pp. 34–35). Eudoxus' theory was in accordance with these requirements, but not able to account for the varying distances of the planets. Apollonius and Hipparchus were forced to abandon the second requirement. Ptolemy abandoned the second requirement as well. Technically there were no objections; the models that he uses are well defined. Yet later astronomers have criticized him for applying a uniform rotation with respect to a point that is not the center of the circle: for example, Ibn al-Haitham and Copernicus.

For Aristotle there was an essential difference between the natural motion of the heavenly bodies and natural motion on earth. The natural motion of the stars, the planets and the Sun and the Moon was circular in the sense indicated above. Natural motion on earth was rectilinear. Yet, the design of the heavenly mechanisms remains strikingly similar to the design of earthly mechanisms. Mechanical engineering design is always subjected to external conditions. In that sense the Aristotelian restriction to circular motion in the case of the heavenly mechanisms is remarkable but nothing more.

References

1. Bruin, F., Vondjijdis, A. (trans., ed.): The Books of Autolykos: On the moving sphere and on Risings and Settings. American University of Beirut, Beirut (1971)
2. Bulmer-Thomas, I.: Plato's astronomy. The Class. Quart. New Ser. **34**(1) (1984)
3. Clagett, M.: Greek Science in Antiquity. Reprinted in 1971 by arrangement with Abelard-Schumann, Ltd. (1955)
4. Farrington, B.: Greek Science, Its Meaning for Us. Penguin Books 1944, References to the 3rd edition by Spokesman (1980)
5. Heath, T.L.: Greek Astronomy. In: Baker, E. (ed.) J.M. Dent & Sons Ltd., London and Toronto. E. P. Dutton & Co. Inc., New York (1932)
6. Neugebauer, O.: A History of Ancient Mathematical Astronomy, Part I. Springer (1975)
7. Neugebauer, O.: Astronomy and History, Selected Essays. Springer, New York, etc. (1983)
8. Petersen, O.: A Survey of the Almagest. Odense University Press (1974)
9. Plato: The Collected Dialogues including the letters, edited by E. Hamilton and H. Cairns with and introduction and prefatory notes, Bollingen series LXXI, Princeton (1973)
10. Ptolemy's Almagest, translated and annotated by G. J. Toomer, Duckworth, The Old Piano Factory, London (1984)
11. van der Waerden, B.L.: Science Awakening. P. Noordhoff Ltd, Groningen (1954)
12. Yavetz, I.: On the homocentric spheres of Eudoxus. Arch. Hist. Exact Sci. **52**, 221–278 (1998)
13. Yavetz, I.: A new role for the hippopede of Eudoxus. Arch. Hist. Exact Sci. **56**(1), 69–93 (2001)

Representation, Encoding and Geometry of Machines and Mechanisms

The Contribution of Jesuit Johann Schreck to Development of Chinese Scientific Culture in 17th Century

Michela Cigola(✉)

University of Cassino and Southern Lazio, Cassino, Italy
michela.cigola@unicas.it

Abstract. This article aims to investigate the role played by Johann Schreck (1576–1630), known by his Latin monicker Terrentius. He was a keen scholar, an academician of the Lincei, a Jesuit, a missionar in China, a tireless traveller and insatiable and passionate collector of books and scientific instruments. Schreck was one of the most interesting figure both in European and Chinese cultural world of the first half of 17th century. He contributed to the spread of European scientific and mechanical knowledge in China.

Keywords: Johann Schreck · Jesuits history · Society of Jesus missionaries · Chinese history of science · Mechanical engineering history

1 Society of Jesus

The order of Society of Jesus, founded by Spanish Ignatius of Loyola (1491–1556) is a male religious institution. Some of the order's cultural centres were promoted by St Ignatius in Rome: the Collegium Romanum founded in 1551, and the Collegium Germanicum founded in 1552.

The history of Jesuit missions is integral to the history of relations between China and the West, infact Jesuit missionaries brought Western science and technology in China and achieved significant results in the intercultural dialogue with their Chinese collaborators [1, 2].

The first was the Italian Matteo Ricci (1552–1610), he began to collaborate with several of his disciples who spread Western scientific knowledge and writing books on science and technology based on European knowledge. Some of his disciples became Christians, the most important among them was Xu Guangqi (1562–1633), with whom he translated the six volumes of Euclid's Elements and wrote many other works [3]. Another important Jesuit missionary was Sabatino de Ursis (1575–1620). He arrived in China from Italy in 1603. In 1607, Matteo Ricci invited him to Beijing because of his knowledge on astronomy, hydraulics, and mechanics. He began collaborating with Ricci and Xu on the Chinese translation of Euclid's Elements. After Ricci's death, his collaboration with Xu strengthened; in 1612, Sabatino de Ursis orally translated some European texts on Western hydraulics knowledge [4] which was later drafted in Chinese by Xu from the notes while studying with the Jesuit with the title "Taixi Shuifa" (On Western Hydraulics).

M. Ceccarelli and R. López-García (Eds.): HMM 2022, HMMS 40, pp. 295–304, 2022.
https://doi.org/10.1007/978-3-030-98499-1_24

2 Johann Schreck Biographical Notes

Johann Schreck (1576–1630), known by his Latin monicker Terrentius, was a keen scholar, an academician of the Lincei, a Jesuit, a tireless traveller and insatiable and passionate collector of books and scientific instruments. He has a dual and inseparable nature as a man of faith and a man of science [5].

Johann Schreck, was born in 1576 in Belgium in Bingen, and at the age of 14 he began his studies at the University of Freiburg. In 1596 he became magister in medical faculty.

After Freiburg Schreck continues his medical accademic curriculum in Basel. In 1600–1603 he studied in Paris where he collaborated with the famous mathematician Francois Viète (1540–1603) counsellor and member of the court of Henry IV of France. In this period he choosed his monicker Terrentius as a sign of his admiration for Marcus Terrentius Varro (116 BC–27 BC), one of the Ancient Rome's greatest scholar.

After university Johann Schreck completed his training with 10 years of travelling in Europe, from 1600 until his entry into the Society of Jesus. Schreck made his first trip to Rome in 1600, on the occasion of the Jubilee, but then returned there several times: in 1603 and 1604 and then in 1610.

He matriculated on 1 July 1603 in the Faculty of Arts of Padua university where he followed the courses of the academic year 1603–1604 and probably the course of Galielo Galiei (1564–1642), who in these year was professor in Padua.

Probably on one of his trips to Rome, Schreck met Prince Federico Cesi (1585–1630), scientist and naturalist from a noble Italian family and founder of the Accademia dei Lincei (Lincean Academy). The aim of the Academy was to promote free and autonomous scientific research and many of the best minds in Italy and Germany joined it.

Prince Cesi asked Schreck to become a member of the Lincean Academy; Galileo Galilei became a member of the academy in April 1611, followed immediately by Johann Schreck in May of the same year.

Being a Lincean led Sckreck to enter the most fervent and interesting environment of Italian science. For some time he worked on researching the botany of Central America, rearranging the work "Nova plantarum, animalium et mineralium mexicanorum historia" [6], an encyclopedia on plants, animals and minerals by Francisco Hernández de Toledo (1514–1587) a naturalist and court physician to the King of Spain.

In these same years Schreck was mulling over his decision whether to become a Jesuit. Between 1596/98, he told his friends the decision he had come to and on 1/11/1611 he entered the novitiate of Sant'Andrea in Monte Cavallo and left the Lincean Academy. His abandonment of the Lincei caught other academics by surprise, to the extent that Galileo wrote a letter to Federico Cesi in which called the fact that Shreck had become Jesuit "a great loss…" [7].

During his theology studies, Schreck met the French Jesuit Nicolas Trigault (1577–1628). Trigaul arrived in China in 1610, and in 1612 he received orders to return to Europe to recruit new missionaries, to seek financial sponsorship and to buy and accumulate books and tools for what would become the Jesuit library in Beijin.

Trigault persuaded Schreck to become a missionary in China and chose him as his companion for his European tour (1616–1618) and his return journey to China. On

16 April 1619 the Jesuits sailed fron Lisbon with a huge collection of volumes and instruments.

Schreck arrived in October 1618 in Goa, on 22 July 1619 he reached Macau. In 1621 he arrived in Hangzhou, and in late 1623 Beijin where he died in 1630.

3 Conversion of European Scientific Knowledge into Chinese by Jesuit in XVIIth Century

In China Jesuit missionaries began working on the systematic convertion of European scientific knowledge into Chinese.

After the work of Matteo Ricci, Xu Guangqi and Sabatino de Ursis one of the most important of these works is the "Qiqi tushuo" (Diagrams and explanations of the wonderful machines of the Far West) by Johann Schreck and Wang Zheng in 1627, a kind of encyclopaedia of European mechanical knowledge in Chinese] [8]. This was the first volume to present this kind of device to the Chinese public, since, despite being written in 1612, the work of Sabatino de Ursis and Xu Guangqi was not published until 1629.

The volume contained commented reproductions taken from the works of the Frenchman Besson [9], Italians Ramelli [10] and Zonca [11], Dalmatian Faustus Verantius [12] e from German Heinrich Zeising [13].

Jacques Besson (1540–1576) professor of mathematics in Grenoble then in Lyon and Orleans, was the first to introduce a text in France with purely mechanical representations, drawing on Renaissance treatises on the subject. In 1569, King Charles IX appointed him "King's master of machines"; following the appointment, he moved to Paris where he began his "Théâtre des instruments et machines" written and published in Latin and followed by other editions in French, German, Italian and Spanish (Fig. 1). The volume consists of 60 plates representing various devices. All the machines are represented in a three-dimensional perspective.

Agostino Ramelli (1531–1600ca.) studied mathematics and architecture. After being appointed engineer for the Duke of Anjou, the future King Henry III, he moved to France. His "Le diverse ed artificiose macchine" ("The Diverse and Artifactitious Machines") was published in Paris in 1588 (Fig. 2). It consists of 195 chapters, each of which contains the illustration and description in French and Italian of a different machine. In the volume, machines are placed in natural environments and are represented in pseudo perspective with refined lighting effects.

Fig. 1. Frontpage of Besson 1579 edition

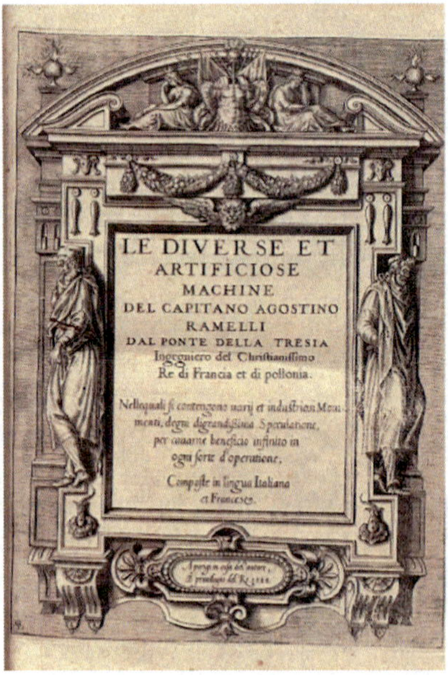

Fig. 2. Frontpage of Ramelli 1588 edition

Vittorio Zonca (1568–1602) architect of the city of Padua, is best known for publishing the "Novo Teatro di machine et edificii per varie et sicure operazioni" ("New Theatre of Machines for Various and Safe Operations") a text that features 42 mechanical and hydraulic devices (Fig. 3). The work presents machines in natural or architectural environments, and great care is taken with three-dimensional perspectives and treatment of chiaroscuro.

Faustus Verantius (1551–1617) born in Dalmatia but a citizen of the Republic of Venice, is remembered for his book "Machinae Novae" which contains 49 illustrations that illustrate various devices, whose settings are much more measured and concise than in previous treatises accompanied by a description in five languages (Italian, Latin, German, Spanish and French) (Fig. 4).

Heinrich Zeising, out whom very little is known, in 1613 in Leipzig published his "Theatrum Machinarum" describing 140 assorted devices (Fig. 5). The work synthesises the treatises of Besson, Ramelli and Zonca. In fact, Zeising redesigns the machines described in these treatises and places them in very complex environments with strong chiaroscuro, populated by many characters.

Fig. 3. Frontpage of Zonca 1607 edition

Fig. 4. Frontpage of Veranzio 1615 edition

Fig. 5. Frontpage of Zeising 1613–66 edition

4 "Qiqi Tushuo"

Johann Schreck drew on the European treatises for the drafting of his "Qiqi tushuo". It is quite clear that he made use of the science and technology library that he and Nicolas Trigault had patiently put together during their travels to the courts of Europe and then taken as far as China.

It should not be forgotten that for some years, first as a disciple of Galileo and then as a member of the Accademia dei Lincei, Schreck had frequented Europe's most vibrant cultural environment and had certainly acquired the knowledge that he was to re-use in his works from the Chinese period.

Certainly Schreck's collaboration with Wang Zheng in compiling the volume had a strong influence especially on the illustrations which could be by Zheng creating a sort of translation from European machine's representation to Chinese ones [14].

"Qiqi tushuo" includes 81 illustrations. From the graphic organisation of the pages it seems that it was closely inspired by the work of Zonca, whose layout is faithfully copied, with a large box and the upper part dedicated to a description of the device.

Excluding the first nine drawings, the other 64 more or less faithfully refer to the originals taken from the treatises referred to above. In fact, 12 are based on Besson, 13 are reworkings of Ramelli's drawings, and four recapture Zonca, five Veranzio and six Zeising. One of the machines can be found in the treatises of Zonca and Zeitung, three in those of Ramelli and Zeising, and five in those of Zonca and Zeising. Percentagewise, therefore, the work of Ramelli is present in 25% of Schreck's work, the work of Zonca in 5%, that of Zonca in 11% and Zeising in 44%, while the remainder draws on illustrations by more than one author. The preponderance and the large numbers of references to Zeitung's treatise as an example to borrow from comes as no surprise; in fact, Schreck must have been aware that this treatise was itself a sum of the others and therefore the easiest source of inspiration and reference.

Figures 6–7 compare tables taken from the works of Ramelli, Besson and Zeitung with tables from "Qiqi tushuo". A faithful reproduction of the original drawings in every detail, although the graphic style is completely Chinese.

In Fig. 6 there is a comparison between a well represented in "Qiqi tushuo" (Fig. 6a) and a well from the work of Ramelli (Fig. 6b). It is evident that the general layout of the drawing is identical in the choice of the depth of perspective and in every small detail, even if it completely lacks graphic connotations of materials (such as the wood of the uprights or the ground on which the well stands). In Ramelli's drawing, the representation is mature and therefore more correct from the point of view of three-dimensional treatment. In Fig. 6c there is a well's figure from Zeisung's treatise (Fig. 6c); the authors, who worked 25 years after the Italian, copied the drawing identically but, perhaps in an attempt to make his plagiarism less evident, added an architectural background and an onlooker in half- darkness beyond the well.

In Fig. 7 there is a comparison between a well represented in "Qiqi tushuo" (Fig. 7a) and a well from the work of Besson (Fig. 7b). The device is a lift for weights to be used on building sites. The original drawing shows a group of workers intent on their work, both on the ground and on scaffolding. Schreck, however, reduces the setting to a minimum, leaving only a small portion of crenelated masonry at the top. He also almost completely eliminates the human presence, which is reduced to a single machine

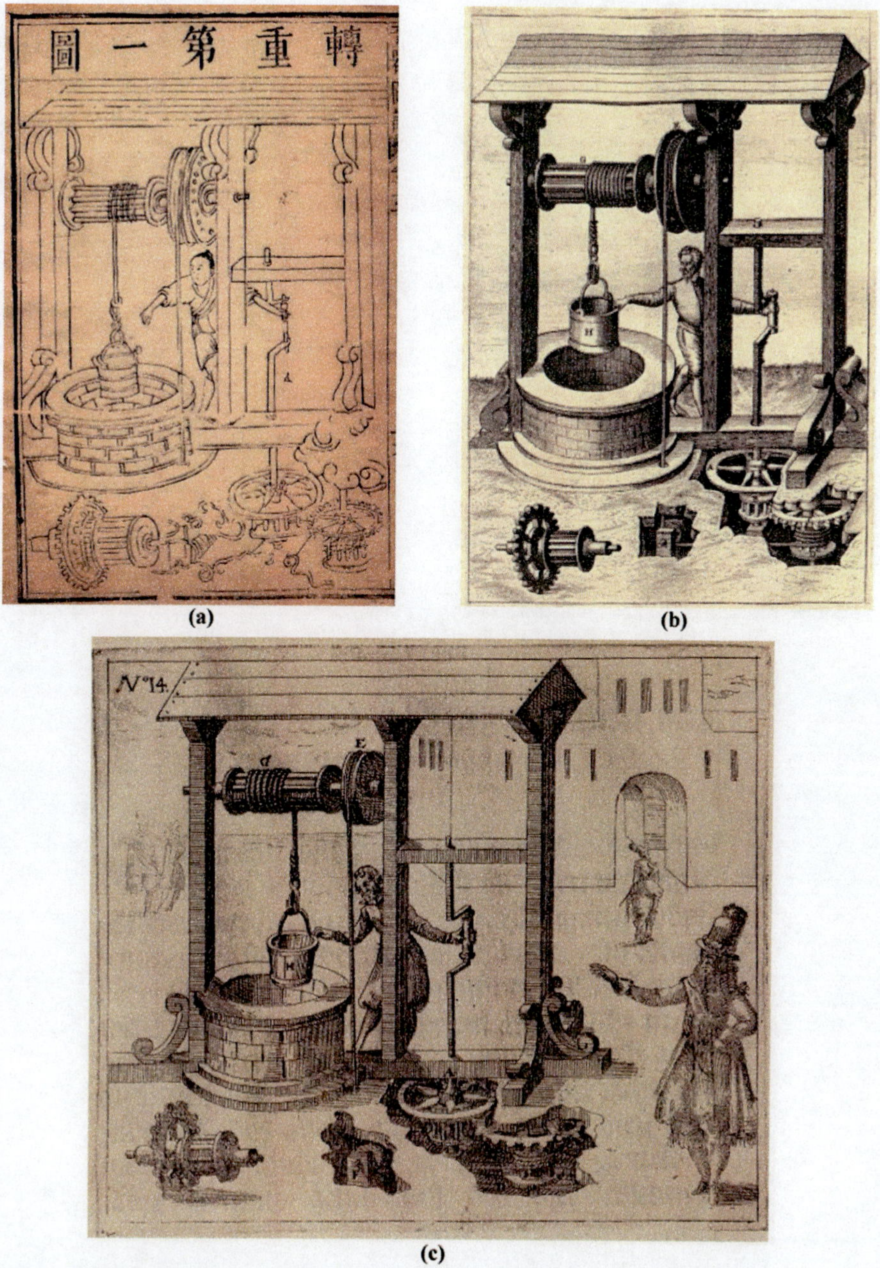

Fig. 6. (**a**) Well in "Qiqi tushuo", (**b**) Well in Ramelli's volume, (**c**) Well in Zeising's volume

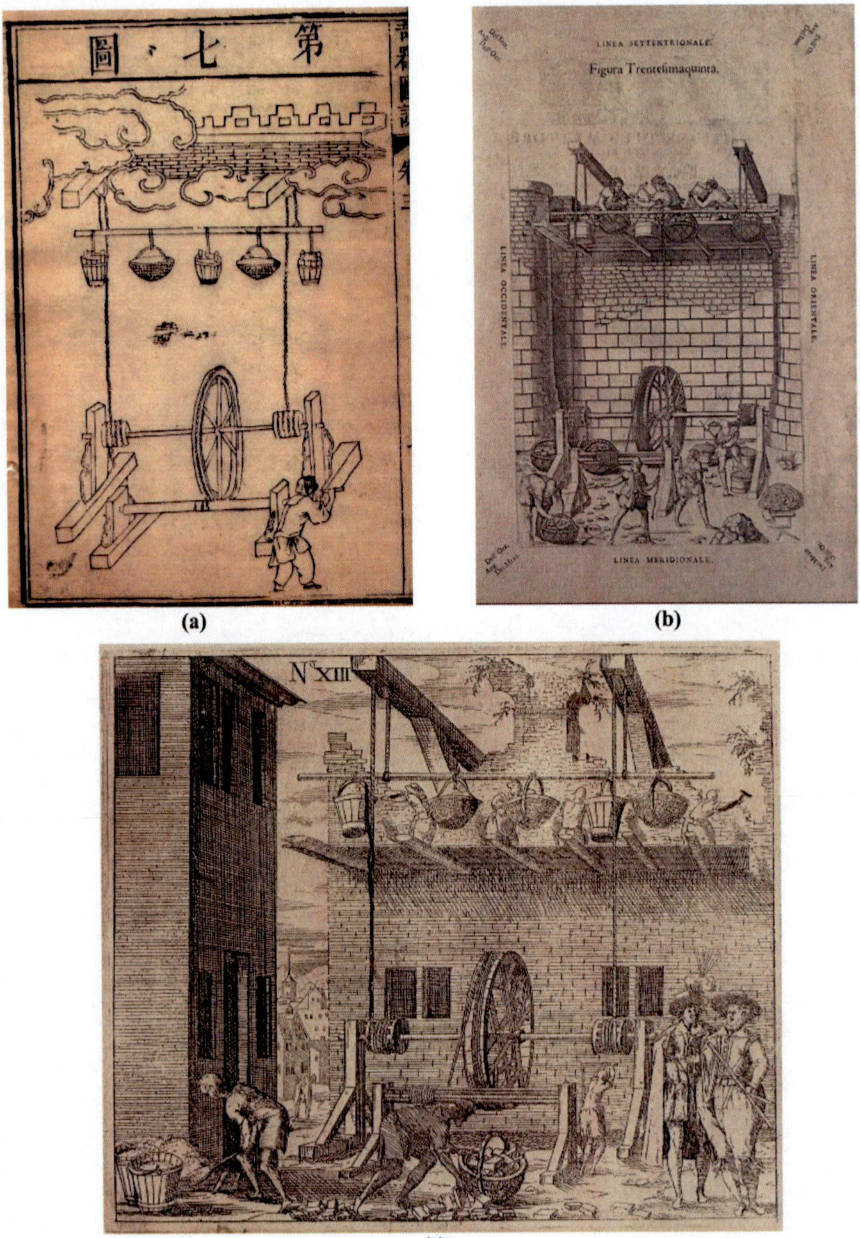

(a) (b)

(c)

Fig. 7. (**a**) Winch in Qiqi tushuo", (**b**) Winch in Besson's Volume, (**c**) Winch in Zeising's volume operator indispensable for clarifying the functioning. In Fig. 7c there is a figure from Zeisung's treatise (Fig. 7c) that is identical to that of Besson, except for the inclusion of an architectural backdrop to the right and the treatment of the upper part of the wall. This last detail leads us to believe that Schreck did not take the drawing from the French original but from the German copy.

5 Conclusions

From 1600–1618, during his period of academic studies and during his tour in the courts of Europe collecting books and scientific instruments, Johann Schreck met some scientists, artists, writers, scholars of the sparkling world of intelligence and culture that laid the foundations of what is today the scientific culture not only of Europe and also the Far East. These included the Italians Galileo Galiei, Federico Cesi, and Cardinal Federico Borromeo; the Germans Lucas Brunn, Johann Faber, Gaspar Schoppe and perhaps Christopher Clavius; the Flemish Pieter Paul Rubens who in 1617 in Antwerp portraied him in Chinese-style clothes.

Thanks to these experiences and thanks to the huge amount of technical and scientific European volumes and instrument that he collected for the Jesuit library in Beijing Johannes Schreck contributed with his confreres Jesuit missionaries to the transmission and continuance of scientific knowledge between China and Europe.

In conclusion Johann Schreck Terrentiuns was one of the main protagonist not only of the European cultural world of the first quarter of the 17th century cultural but also of Chinese. He was a great traveller, a careful scholar of botany, a doctor who annotated the books on which he refined his training.

Most importantly Johann Schreck was a man in whom the nature of a scholar went hand in hand with that of a man of faith.

References

1. Masini, F.: Western Humanistic Culture Presented to China by Jesuit Missionaries (XVII to XVIII Centuries). Institutum Historicum, Rome (1996)
2. Cigola, M.: The influence of the Society of Jesus on the spread of European mechanical knowledge in China in the XVIth and XVIIth centuries. In: Explorations in the History of Machines and Mechanisms. Proceedings of HMM2012, Springer, Dordrecht, pp. 69–79 (2012)
3. Sun, S., Tang, Y.: Li Madou yu Xu Guangqi. Xinhua Press, Beijing (1993)
4. Cigola, M.: Sabatino de Ursis e la diffusione della cultura meccanica europea nella Cina del XVII secolo. L'Idomeneo 30, 219–226 (2020)
5. Golvers, N., Johann Schreck Terrentius, S.J.: His European Network and the Origins of the Jesuit Library in Peking. (Series: De diversis artibus, tome 107). Brepols (2020)
6. Hernández, F.: Rerum medicarum Novae Hispaniae Thesaurus, seu Plantarum, Animalium, Mineralium Mexicanorum Historia cum notis Joannis Terentii Lineæi. Roma (1648)
7. D'Elia, P.: Galileo in Cina: relazioni attraverso il Collegio romano tra Galileo e i gesuiti scienziati missionari in Cina, 1610–1640, Roma (1947)

8. Cigola, M.: The Qiqi tushuo by the Jesuit Johann Schreck: European "Theatra Machinarum" in China in the XVIIth century. In: Proceedings of the 32th International Congress of the Italian Society of Historians of Physics and Astronomy. The Scientia Socialis Press, Siauliai (Lithuania), pp. 77–86 (2013)
9. Besson, J.: Theatre des Instruments Mathematiques & Mechaniques de Jacques Besson Dauphinois, docte mathematichien. Avec l'interpretation des Figures d'iceluy, par Françoise Beroald, A Lyon, Par Barthelemy Vincent, Avec Prililege du Roy (1579)
10. Ramelli, A.: Le Diverse et artificiose machine del capitano Agostino Ramelli dal Ponte della Tresia, Ingegniero del Christianissimo Re di Francia et di pollonia. Nelle quali si contengono varij et industriosi muvimenti, degni di grandissima speculatione, per cavarne beneficio infinito in ogni sorte d'operazione composte in lingua italiana e francese, Paris (1588)
11. Zonca, V.: Novo Teatro di machine et edificii per varie et sicure operazioni. Cò le loro figure tagliate in Rame é la dichiaratione, e dimostratione di ciascuna. Opera necesaria ad Architetti, et a quelli, che di tale studio si dilettano. Di Vittorio Zonca Architetto della Magnifica Comunità di Padova.In Padova appresso Pietro Bertelli (1607)
12. Verantius, F.: Machinae novae Fausti Verantii Siceni: cum declarationem latina, italica, hispanica, gallica et germanica. Venetiis (1615)
13. Zeising, H.: Theatrum machinarum. Henning Gross, Leipzig (1613)
14. Zhang, B., Tian, M.: Wang Zheng (1571–1644). In: Ceccarelli, M. (eds.) Distinguished Figures in Mechanism and Machine Science. History of Mechanism and Machine Science, vol. 7. Springer, Dordrecht, pp. 247–259 (2009)

Drawing the Vibrations of Sound

Arturo Gallozzi[1](✉) and Rodolfo Maria Strollo[2]

[1] University of Cassino and Southern Lazio, Cassino, Italy
gallozzi@unicas.it
[2] University of Rome Tor Vergata, Rome, Italy
strollo@ing.uniroma2.it

Abstract. Through the construction of a simple apparatus, consisting of two pendulums -which oscillate at the same time-it is possible to graphically reproduce Lissajous' experiments on the commuting vibrations of the diapasons. This equipment was first built by Samuel Charles Tisley (London 1829-Unknown), under the name of harmonographer. Similar apparatuses were produced in the nineteenth century and subsequently, a list -limited to the first nineteenth-century phase- is proposed in the contribution that illustrates the essential characteristics of the harmonograph. Born as a means of analysing vibrations- but which, by its principle belongs to mechanics, for its applications to acoustic science and geometry for the ability to draw a wide variety of "harmonious" curves, which in the paper are referred to as *Lissajous curves "damped"*.

Keywords: Harmonograph · Tisley · Lissajous curves · Bowditch curves · Spiller · Pendulum · Tuning fork

1 Introduction

The construction of machines to draw curves has found, over the centuries, wide spread. In particular, eighteenth-century literature is rich in examples; it is enough to remember the work of Giovanni Battista Suardi (1711–1767) [5] or Giovanni Poleni (1683–1761) [4], to make only brief references. The treatises do not lay a boundary between geometry and mechanics, between mathematics and technique, not often neglecting purely playful aspects. In the 18th and 19th centuries multiple experimental equipment will be produced, with a considerable variety of mechanical devices. All aimed not only for productive purposes, but also with demonstrative value. Often the mechanisms were identified as "mathematical games", i.e. applications of mathematics "delicious and curious". In fact, these devices, built on precise scientific observations, are reflected in physics, geometry and mechanics [1]. The instruments are often diversified with ornamental techniques and variants by various builders, in a process of progressive improvement. These "machines for drawing", with their extraordinary illustrations, often find a first application in aesthetic contemplation and entertainment of "mathematical games". The mechanical device is thus also transformed into a simulation of natural phenomena.

Around the second half of the nineteenth century, following this line of thought, an instrument is created that takes the name of Harmonograph. By some builders it will

M. Ceccarelli and R. López-García (Eds.): HMM 2022, HMMS 40, pp. 305–313, 2022.
https://doi.org/10.1007/978-3-030-98499-1_25

also be called Pendulograph. The apparatus, in its simplest conformation, consisted of pendulums attached to a pen. The movement of the pendulum determines the coordinates of the pen, drawing the resultant of two or three harmonic motions. In fact, the final drawing can be assimilated to the representation of a frequency interval, as in the experiment and to the curves codified by Jules Antoine Lissajous (1822–1880). In addition, the pendulums will also have the characteristic of simulating the damping of a musical note, due to the dissipation of the mechanical energy that develops in the movement of the pendulum.

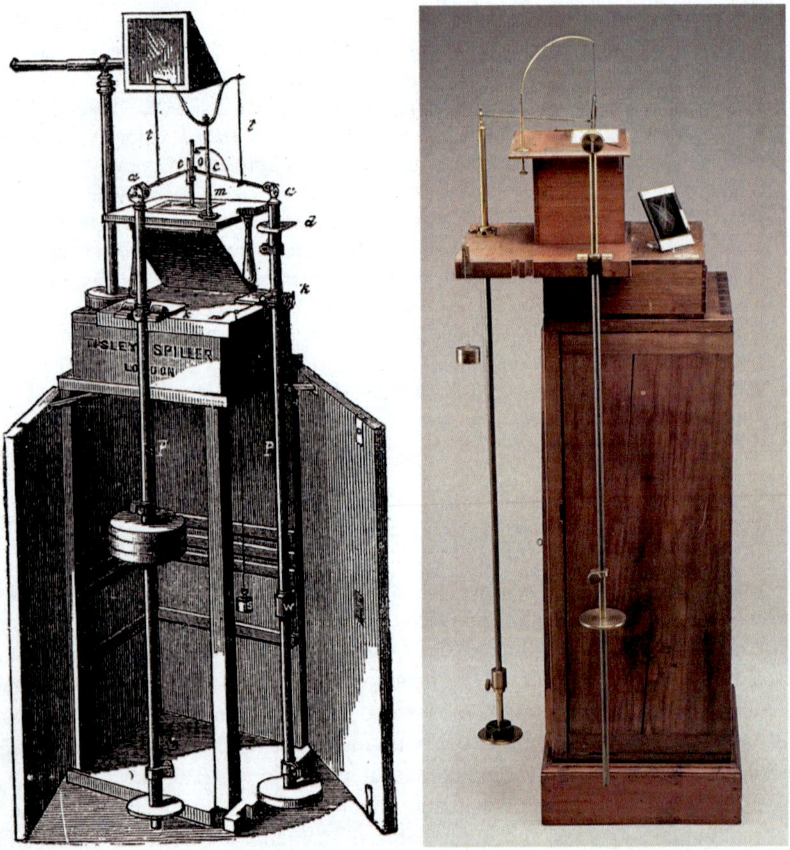

Fig. 1. Tisley's compound pendulum, 1873. "Tisley's compound pendulum" Engineering 17, 101–102 (6 February 1874)

2 The Harmonograph

Samuel Charles Tisley in 1873 invented a compound pendulum, Fig. 1, and later, in 1877, presented and demonstrated his instrument, called "harmonographer", at the Royal Society Meeting in London Fig. 2.

The mechanical apparatus, using the movement of the pendulums, was able to create a well-defined geometric image. In the 1870s and 1880s, the firm Tisley & Spiller (later "S.C. Tisley and Co.") produced numerous scientific instruments in London, encouraged by the expansion of science teaching in England after 1871, Fig. 3.

Fig. 2. Tisley's Harmonograph, 1877. Left: Tissandier G., Le Ricreazioni Scientifiche, ovvero l'insegnamento coi giuochi, Treves, Milano (1897), p. 174; Right: The Collection of Historical Scientific Instruments, Harvard University – Inventory Number:1997-1-0451

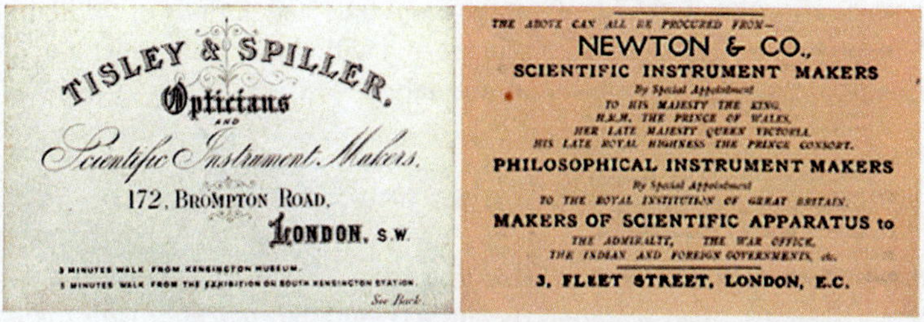

Fig. 3. Tisley & Spiller, Advertising tickets

Tisley's harmonograph is therefore a mechanical apparatus that uses pendulums to create a geometric image. In the description of his instrument Tisley specifies that "... *it was designed to record the figures shown in Lissajous' experiments with the Diapason*" [6].

The drawings created can therefore be traced back to the curves of Lissajous. The Armonographs spread from the mid-19th century and reached the peak of popularity in 1890. However, such devices cannot be definitively attributed to a single person. In particular, after "Y-Suspended Pendulum" produced in 1815, presented and studied by James Dean and Nathaniel Bowditch, the official inventor of Pendulum is considered Hugh Blackburn (1823–1909), professor of mathematics at the University of Glasgow. The latter is known for studying, in 1844, a pendulum hanging from a V-shaped rope, known as the "Pendolo Blackburn". However, in 1893 H. Irwine Whitty stated in his book on the harmonograph that the device was first built by Mr. Tisley, by Samuel Tisley and George Spiller, "the famous opticians", who opened a shop in 1872 on Brompton Road [9]. Tisley's first instrument was a simple "lateral" harmonographer, which used two pendulums to control the movement of a pen over a drawing surface. One pendulum moves the pen back and forth along one axis and the other pendulum moves the drawing surface back and forth along a perpendicular axis. The motion of the point expressing the curve of the figure, produced by the sum of the two motions mentioned, is analytically described by the parametric equation of the curve in parameter t:

$$y = A_1\, e^{--k1\omega 1t} \cos(\omega 1t + \varphi 1)$$

$$x = A_2\, e^{--k2\omega 2t} \cos(\omega 2t + \varphi 2) + A_3\, e^{--k3\omega 3t} \cos(\omega 3t + \varphi 3)$$

(with the centre of the Cartesian reference placed at the quiet point of the pen).

By varying the frequency and phase of the pendulums in relation to each other, different designs are created and, subject to varying conditions, unrepeatable. More complex harmonographs consist of three or more pendulums or pendulums connected to each other. These can return a rotary movement, when one or more pendulums are mounted on hinges to allow movement in any direction [8]. In particular, the shapes relative to the graphs obtained with the Tisley harmonograph, while responding to the laws of the motions generated by the oscillations of the two pendulums that make up the apparatus, can assume a virtually infinite number of shapes dependent on the "phase" type (function of oscillation that imposes on the pen by the relative positioning of the oscillating masses of them, which affect the amplitude and periods of oscillation) and the friction conditions (between paper and pen and with the air), all difficult to quantify and codify and in any case practically unrepeatable.

3 Tisley's Harmonograph and Lissajous Curves

The apparatus is widely described, with a rich iconographic apparatus, in the volume of Gastone Tissandier "*The scientific recreations—that is, teaching with the games*", for the types of the Brothers Treves publishers in Milan, 1882. Among the various chapters, all with play-didactic purposes, Tisley's harmonographer is presented as part of the one dedicated to "*The Physics Without Appearances*", to deepen the elementary concepts of acoustics proposed by the treatment [7]. A paragraph outlines the physical-acoustic aspects of the curves that the instrument produces. The treatment completes precise indications for the practical construction of the mechanism, summarised by comprehensive explanatory drawings, Fig. 4.

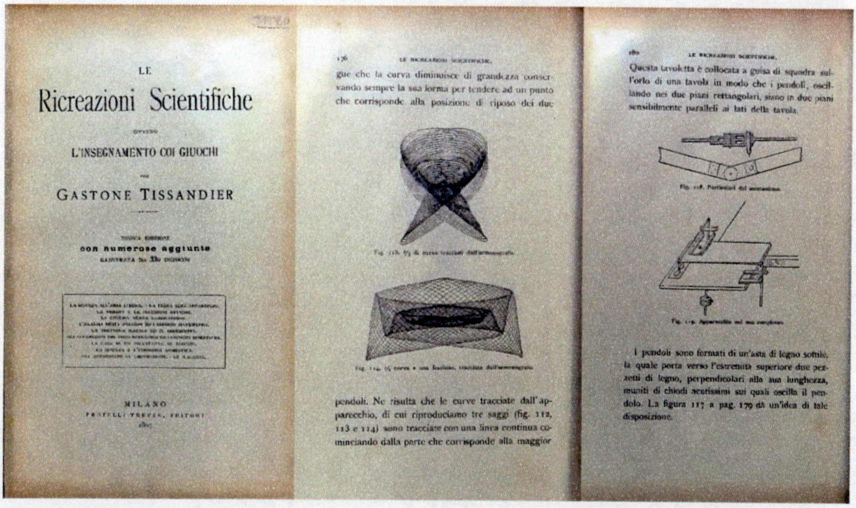
LE

Ricreazioni Scientifiche

L'INSEGNAMENTO COI GIUOCHI

GASTONE TISSANDIER

con numerose aggiunte

MILANO

gue che la curva diminuisce di grandezza conservando sempre la sua forma per tendere ad un punto che corrisponde alla posizione di riposo dei due

pendoli. Ne risulta che le curve tracciate dall'apparecchio, di cui riproduciamo tre saggi (fig. 112, 113 e 114) sono tracciate con una linea continua cominciando dalla parte che corrisponde alla maggior

Questa tavoletta è collocata a guisa di squadra sull'orlo di una tavola in modo che i pendoli, oscillando nei due piani rettangolari, siano in due piani sensibilmente paralleli ai lati della tavola.

I pendoli sono formati di un'asta di legno sottile, la quale porta verso l'estremità superiore due pezzetti di legno, perpendicolari alla sua lunghezza, muniti di chiodi acutissimi sui quali oscilla il pendolo. La figura 117 a pag. 179 dà un'idea di tale disposizione.

Fig. 4. Tissandier G., Le Ricreazioni Scientifiche, ovvero l'insegnamento coi giuochi, Treves, Milano (1897). Cover page, pp. 176, 183

The device created, is fundamentally faithful to the design principles expressed in the text but structurally different from it, is essentially formed by two pendulums. One of the pendulums is connected to the frame by means of a gimbal joint, a mechanism that makes possible all the oscillations within the infinite vertical planes passing through the axis of the pendulum at rest. The second pendulum is connected to the main frame by means of a pin that binds it in the oscillations according to a single plane. On the upper end of the shaft of the first pendulum, a tablet shall be fixed at right angles to the shaft on which the sheet of paper intended to accommodate the drawing of the curve is placed. By analysing the motion of the tip that describes the figure numerous variables that, for practical – operational conditioning, generate an unrepeatable graph with respect to its first definition. However, there are several scientific fields in which there can be considerable formal analogies with the figures obtained by the harmonographer. The first is certainly that of mechanics and acoustics that, with the figures of Lissajous finds a certain affinity, not only formal but also analytical, suggesting the hypothetical definition, for the curves of the harmonograph, of "*Lissajous curves damped*".

The figure of Lissajous is a curve formed by the trajectory of a point that simultaneously participates in two mutually perpendicular harmonic vibrations. The curves of Lissajous are named after the French physicist Jules Antoine Lissajous (1822–1880). The figures are obtained by reflecting a ray of light subsequently by two mirrors connected to two diapasons placed at right angles between them, Figs. 5 and 6.

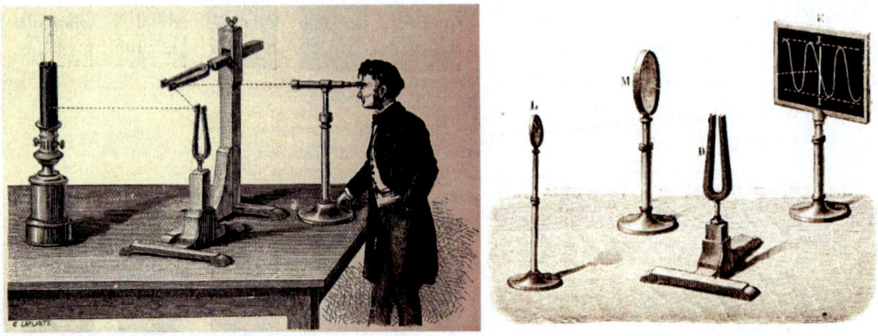

Fig. 5. Experiment of Lissajous. The light-figures produced by the compound vibrations of two forks vibrating at different pitches. In: Privat-Deschanel A. and de Ochoa y Ronna E., Tratado elemental de física, Hachette, Paris, pp. 696–697 (1872) [3]

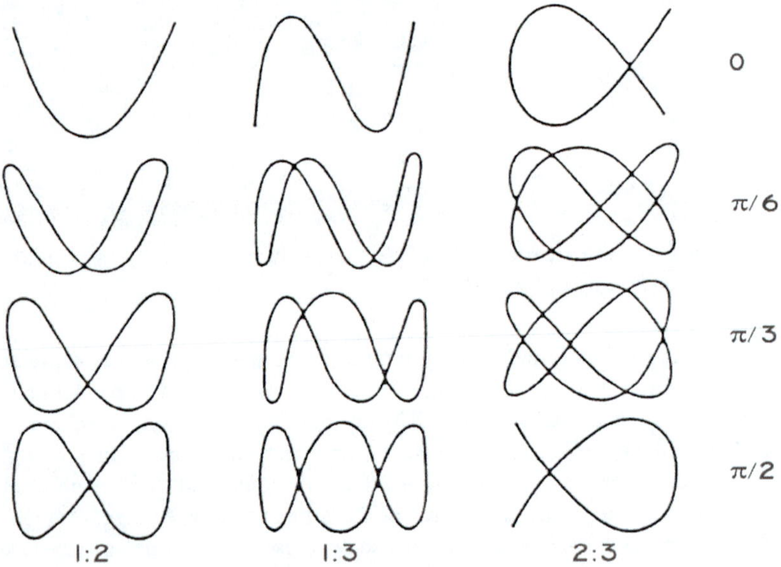

Fig. 6. Some simple Lissajous curves

The final drawing of the curves represents a frequency range. In fact, as in Lissajous' experiment, pendulums have the characteristic of simulating the damping of a musical note. The slow decay to silence, thanks to the mechanical energy that dissipates within the movement. The images produced by the harmonograph are to be understood as drawing from the outer edges (maximum volume) in the center (silence) – representing the path of decay.

The design produced by the harmonograph can therefore be assimilated to the display of particular sounds.

The frequency of the pendulum is the same as the sound, only slowed down by about 1000 times, allows it to be displayed. It is therefore the frequency that determines the image you get. By altering the frequency of the pendulums (changing the length, shifting the weights upwards or downwards) you can create images that represent a particular sound frequency Fig. 7.

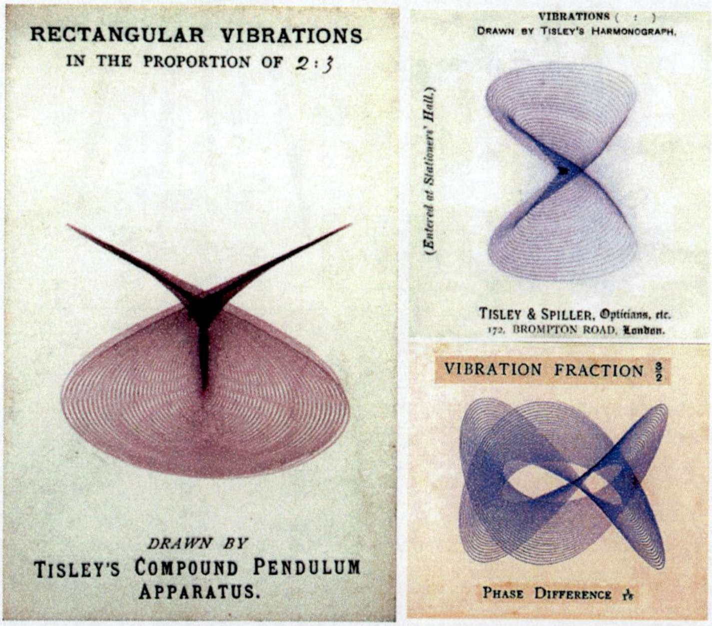

Fig. 7. Tisley & Spiller, Drawn by Tisley's Compound Pendulum and Tisley's Harmonograph. Rectangular Vibrations in the Proportion of 2:3 and 3:2

Moreover, the interest of the harmonographer lies in the fact that the periods of the pendulums can be agreed among themselves. So, for example, if one pendulum is set to swing twice, while the other swings three times, a new series results. In addition, the variety of results can be increased by changing the respective oscillation and phase amplitudes of the pendulums.

Lissajous' experiment and its connection to the harmonic movements of the pendulum, through the harmonograph, have literally made it possible to "see the sound".

The numerous variations on the apparatus, produced by various manufacturers, testify to the interest and ingenuity in the mechanical representation of curves Fig. 8 and Table 1.

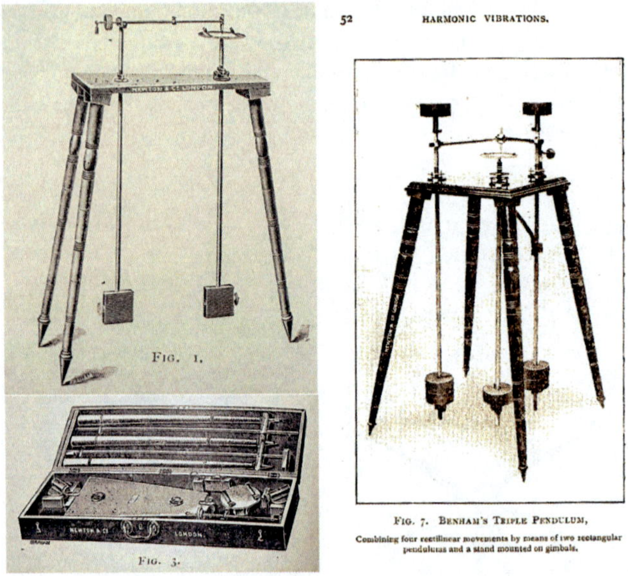

Fig. 8. Other Harmonograph models, inspired by Tisley's design – Left: Newton's Harmonograph, Right: Benhan's Triple Pendulum. In: Goold J and al, Harmonic Vibrations, Newton & Co., London (1909). pp. 5, 20, 52 [2]

Table 1. Instrument designers Harmonographs and their derivatives in the 19th century

Instrument	Designer	Year
Y-suspended pendulum	James Dean; Nathaniel Bowditch	1815
Blackburn pendulum	Hugh Blackburn	1844
Tisley's compound pendulum	Samuel Charles Tisley	1873
Tisley's harmonograph	Tisley & Spiller	1877
Sympalmograph	John Browning; Charles Edwin Benham	1877
Dobson duplex pendulum	John Dobson	1877
Double pendulum	George Hartley Bryan	1894
Pendulograph	John Andrew	1895
Quadruple harmonic motion pendulum	S. J. Michael J. Hoferer	1899
Twin-elliptic pendulum	Charles Edwin Benham	1906
Benham's triple pendulum	Charles Edwin Benham	1909

4 Conclusions

The construction of the Armonographs and its derivatives was widespread in the mid-19th century. Rich is in fact literature and interest in these instruments. The involvement of scholars, in addition to the demonstration of simple physical phenomena, has also involved the design of sophisticated mechanical equipment made of precious materials. However, perhaps, the most intriguing beacon concerns the production of the design of the curves. Therefore, the construction of the Haronographs has found its most fortunate application as demonstration devices. Used to illustrate the "vibrations" in both mechanics and physics of sound and harmonies of music. Today, with the advent of computer science, the production of "Lissajous curves" is extremely facilitated, with extraordinary results. However, watching the progressive progress of curves, designed by a mechanical device in an infinite variety of forms, still has an extraordinary aesthetic appeal.

References

1. Gay, F.: Intorno agli omolografi, strumenti e modelli per la geometria descrittiva. In: Quaderni IUAV.13.2000, DPA, Venezia (2000)
2. Goold, J., Benham, C.E., Kerr, R., Wilberforce, L.R.: Harmonic Vibrations and Vibration Figures. Herbert C. Newton – Newton & Co. Scientific Instrument Makers, London (1909)
3. Privat-Deschanel, A., de Ochoa y Ronna, E.: Tratado elemental de física, pp. 696–697. Hachette, Paris (1872)
4. Poleni, G.: Epistolarum mathematicarum fasciculus. Giovanni Manfrè, Padova (1729)
5. Suardi, G.B.: Nuovi istromenti per la descrizione di diverse curve antiche e moderne, e di molte altre, che servir possono alla speculazione de' Geometri, ed all'uso de' Pratici. Col progetto di due nuove macchine per la nautica ed una per la meccanica, e con alcune Osservazioni sopra de' Poligoni rettilinei regolari. G.M. Rizzardi, Brescia (1752)
6. Tisley, S.C.: On a Compound-Pendulum Apparatus. Report of the British Association for the Advacement of Science (1873)
7. Tissandier, G.: Le Ricreazioni Scientifiche, ovvero l'insegnamento coi giuochi, Treves, Milano (1897)
8. Whitaker, R.J.: Harmonographs. I. Pendulum design. Am. J. Phys. **69**(2), 162–173 (2001)
9. Whitty, H.I.: The Harmonograph. Jarrold & Sons, London (1893)

Check for updates

Nomograms: An Old Tool with New Applications

Giovanni Mottola and M. Cocconcelli(✉)

University of Modena and Reggio Emilia,
Via G. Amendola 2, 42122 Reggio Emilia, Italy
{giovanni.mottola,marco.cocconcelli}@unimore.it

Abstract. In this paper, we consider the history of nomograms as a computational tool in mechanical engineering, together with their potential applications for teaching purposes, and summarize the mathematical methods used to derive them. Nomograms are graphical descriptions of a mathematical problem, such that the desired solution may be derived through a simple geometric construction, which usually requires nothing more than a straightedge. This way, a reasonably accurate solution to a complex problem can be quickly obtained even in adverse environmental conditions by low-skilled users; moreover, a nomogram can provide immediate insight on the relationship between the variables. Nomograms date back to the 1800s and have been used by engineers for decades, due to their convenience over manual computation, before computers became widespread. While nomograms have now been largely superseded as engineering tools, our analysis shows that they can still have some applications in workshops and for teaching purposes.

Keywords: Nomograms · Graphical · Vibrations · History of engineering · Mechanisms

1 Introduction

In their work on a given problem, engineers usually develop an analysis which is ultimately condensed in one or more equations to be solved. These could be, for instance, the *ordinary differential equations* (ODEs) that describe the time-behavior of a system, or an algebraic equation that defines the optimal design parameters of a component to be realized. In the following, we will mostly focus on engineering equations having only one or finitely many solutions; conversely, we disregard equations that define a *constraint* (which correspond to a boundary in a vector space, with infinitely many points).

Once the right equation (whether algebraic or differential) for the model at hand has been found, finding the solution is usually a mechanical procedure. Indeed, after the mass advent of personal computers in the '60s and the corresponding development of prepackaged numerical algorithms that are not domain-specific (and can thus be used as "black boxes"), equation solving has become little more than a minor step in most cases.

M. Ceccarelli and R. López-García (Eds.): HMM 2022, HMMS 40, pp. 314–329, 2022.
https://doi.org/10.1007/978-3-030-98499-1_26

This, however, was not the case up until a few decades ago; indeed, entire teams of human "computers" were sometimes hired by research centers to manually develop tedious arithmetical steps of numerical methods. Calculating by hand is however a long and risk-prone procedure, where a minor mistake can invalidate all the following work and which provides little to no insight that can be used to check the intermediate results.

Graphical methods have been used by engineers since before our profession was formally established; while their accuracy is physically limited by the tools used, such as pencil and straightedge, they offer interesting advantages. For example, they are quite immediate to understand and employ, such that they may be used by low-skilled operators; furthermore, they can provide reasonably precise results in a fraction of the time required for analytical procedures. Their strongest asset, however, is the intrinsic visualization of the data: this way, one has immediate insight over the results and especially over the relationships between variables, which is useful when exploring engineering alternatives.

Nomograms [5] are graphical methods developed for solving equations. Essentially, they are graphs drawn such that the mathematical relationship between the variables has a simple graphical equivalent (such as three points being collinear). Thus, they can be used to find a numerical result by graphical means: one "enters" the graph with given input parameters (corresponding, in general, to points on lines) and finds a solution by a geometrical construction, which usually requires no more than pencil and straightedge. A thin thread can also be used, to avoid drawing lines, so the nomogram can be reused.

Nomograms have been used extensively by engineers up until the '60s due to their practical convenience in a number of applications, including those in the field of mechanical engineering. Here, we will offer a bibliographic retrospective on such methods, especially from the point of view of Machine Mechanics.

While no longer in widespread use, nomograms still offer interesting advantages and have some niche applications, for instance in shop-floor and open-field work, where digital tools would be less practical due to their lack of ruggedness. We also believe nomograms can have interesting advantages as teaching tools, again owing to their immediate visualization: thus, we present nomograms that we created specifically for such applications, as examples of how a nomogram can be developed and used. These nomograms have been designed through the open-source Python package "pyNomo" [4].

The goal of this work is to discuss the role of nomograms in the history of mechanical engineering, with a special attention to the field of Mechanisms and Machine Science. An application of nomograms for educational purposes is also presented, both for clarifying complex equations and for furthering interest in classical methods among students.

The rest of this paper is structured as follows. In Sect. 2, we present a history of the applications of nomograms in engineering, particularly in machine and mechanism design. In Sect. 3, we present the basic methods used to create nomograms and how they may be combined to solve more complex equations; we also present nomograms targeted for classical problems in Mechanisms and

Machine Science. Moreover, we discuss potential applications of nomograms, for example in industry and for educational purposes. Finally, in Sect. 4, we draw conclusions and offer directions for future work.

2 Nomograms in History of Machines and Mechanisms

Nomograms are a part of the larger history of *graphical methods* for computation: these were the most used techniques for numerical analysis in science and engineering, before computers became widespread. For more than a century, graphical methods were the object of research interest and were commonly a part of technical education programs; in this paper, we can only provide a very brief description of the landmarks in the development of nomograms, with a focus on their use for mechanical engineering. For a broader historical perspective, we refer the reader to [6,13,14,18,30]; we are not aware of any historical analyses of nomography specifically for our field.

To restrict the scope of our review, we only consider purely graphic approaches that can (at least in principle) be applied with nothing more than pen and paper. Thus, we disregard more general approaches using mobile elements, such as slide rules, another instrument (conceptually related to nomograms) that was in use until the '60s.

We also remark that the terminology is not unequivocal: several works on "nomograms" use in fact graphical methods that are quite different (and generally more cumbersome to use) than standard nomograms [17], for instance by using graphs with multiple lookup curves on squared paper; here, we only consider methods that require some form of geometrical construction. Moreover, especially in older sources, alternative terms such as "nomographs", "alignment charts" and "abacs" are used interchangeably [6]. To avoid confusion, in this paper we only use the term "nomogram", first introduced by French mathematician and engineer Maurice d'Ocagne (1862–1938) in the late 19th century [5] to distinguish his work from previous contributions: our bibliographic search on scholarly research engines shows that this definition remains the prevalent one.

2.1 Invention and Diffusion: 1800–1960

In the history of mathematics, nomogram (from Greek words νόμος, meaning "law", and γραμμή, which means "line") is a relatively recent term, which should strictly not be applied to calculating devices published before the seminal work by d'Ocagne [13]; at the same time, d'Ocagne was influenced in his research by several previous works, which are useful to understand the main concepts in nomography as he developed it.

Several mechanical devices have been developed since ancient times to ease complex computations; these mostly used movable components that allowed the user to keep track of numbers and of their respective relationships. An example is the abacus, using beads moving on wires. Devices that have been identified as precursors to nomograms include:

1. Astronomical calculating instruments [13,30], such as *astrolabes* and *equatoria*, used to compute the position of the Sun, the Moon and the planets; examples are the *Albion* by Richard of Wallingford (dating to the early 1300s) and the *jovilabe* 1600s) used by Galileo Galilei to study the orbits of the moons of Jupiter.
2. *Sundials* [13,16] and *moondials* [30], which, while not strictly computing instruments, are based on calculations and present graphically their results.
3. *Volvelles* and *slide charts* [13], related to the slide rules mentioned previously; composed of movable pieces of paper, they are designed for specialized computations.

The first necessary element was the development (in the 17th century) of coordinate geometry by Descartes, allowing the drawing of graphs of a function F. In the following, we consider mostly three-variable nomograms, which provide solutions for the equation

$$F(v_1, v_2, v_3) = 0 \tag{1}$$

We then define auxiliary functions F_i in the variables of interest v_i and in the coordinates x and y of a Cartesian plane: the F_i's are such that the set of equations

$$\begin{cases} F_1(x, y, v_1) = 0 \\ F_2(x, y, v_2) = 0 \\ F_3(x, y, v_3) = 0 \end{cases} \tag{2}$$

holds if and only if Eq. (1) is satisfied. Then, three families of curves in the x-y plane can be graphed, one for each of the conditions in Eq. (2). Let us denote each curve by the corresponding value of the fixed parameter v_i: a point at the intersection of three curves is then a solution to Eq. (1). Most of the time, no exact intersection of three curves is available, as the values of v_i's have been discretized (to avoid cluttering in the graph); the closest values are then interpolated by the user from simple graphical approximation. Some of the main features of the nomogram are already present in this simple concept:

- from two values v_i, the third can be derived graphically, even if the relationship in Eq. (1) cannot be inverted analytically; solving direct and inverse problems (depending on which variables are defined as input) is thus equivalently easy.
- the approximation of a given solution is immediately visible from the distance to the closest curve of the output variable, thus providing a visual feedback of the error;
- the precision is limited by the resolution of the graph and by the user's capabilities, especially in approximating a quasi-exact solution; however, large errors due to misreading are unlikely (and out-of-scale errors are impossible, since one only plots the curves corresponding to meaningful ranges for the v_i's);
- graphical methods are targeted for a specific equation, while other calculating devices (such as slide rules) are more general; however, once a graph has been produced, its usage is immediate and less prone to user errors.

With reference to Eq. (2), suppose that $F_1 = v_1 - x$ and $F_2 = v_2 - y$; then, we can set $F_3 = F$. The nomogram is obtained by drawing the curves (called *isopleths*) on plane x-y corresponding to constant F_3. See for example the Pouchet nomogram in Fig. 1a.

The first known example of such a graph (also called *intersection nomograms*) was presented by French manufacturer Louis-Ézéchiel Pouchet (1748–1809), who generalized the *discrete* multiplication tables (known since Pythagoras) to *continuous curves*, namely hyperbolas; this way, multiplying between non-integer numbers is also possible through interpolation (see Fig. 1a). Pouchet's work is also historically significant as he was the first to introduce the expression "graphical calculus" to describe his approach [29].

An issue of the method above was that it required graphing complex curves on paper, which was a time-consuming process to do by hand. A further improvement was offered by French engineer Léon-Louis Lalanne (1811–1892), who suggested to use nonlinear scales for graphs. Consider again the multiplication graph introduced by Pouchet: the isopleths in this case are given by $xy = C$, with C being a constant value. If the x and y values are reported on logarithmic axes, however, the graph can be simplified: indeed, taking the logarithm of both sides, one has $\log(x) + \log(y) = \log(C)$, meaning that on a log-log plot the isopleths become straight lines. While drawing nonlinear scales is somewhat more complex, this is more than compensated by the advantage of drawing straight lines instead of curves, as the former can be defined by only two points and are thus much easier to plot. See Fig. 1b: here, the graph in Fig. 1a has been redrawn using a nonlinear scale for V. Lalanne described his concept as *anamorphosis*, in reference to a painting technique using exaggerated projective distortions. The use of nonlinear scales later became an essential tool in devising simple nomograms for complex equations.

The final element in this development was presented by d'Ocagne in several works, which in 1899 culminated in the first book on nomography [5]. There, the author put to use the tools of projective geometry, a branch of geometry that had recently been introduced by Poncelet and others, and in particular of the principle of *duality* between points and lines. Through this concept, each straight line in Lalanne's anamorphic graph was represented by a point; the condition of having three curves through a point then corresponds to having three points aligned on the same line. This also simplifies the procedure to read the nomogram, which is much less cluttered: one only needs to draw a straight line through two points, corresponding to the two input variables, to find the third one, which corresponds to the output. The other idea advanced by d'Ocagne was to substitute the (orthogonal) coordinate axes with three curves C_i $(i = 1, 2, 3)$ in the plane, thus introducing *alignment nomograms*. Each variable v_i then defines a point on the corresponding curve, which is graded along its length; the scale may be linear or nonlinear (Fig. 1c and 1d), depending on which approach makes the resulting nomogram simpler to draw and to use. If each curve is parametrically defined by $C_i = \left(f_{ix}(v_i), f_{iy}(v_i)\right)$ in the Cartesian plane, the condition for alignment of three points is then given by

$$\begin{vmatrix} f_{1x}(v_1) & f_{1y}(v_1) & 1 \\ f_{2x}(v_2) & f_{2y}(v_2) & 1 \\ f_{3x}(v_3) & f_{3y}(v_3) & 1 \end{vmatrix} = 0 \tag{3}$$

If a function $F(v_1, v_2, v_3)$ can be rewritten in the form from Eq. (3), then a nomogram can be developed for solving Eq. 1. A further advantage of alignment nomograms is that they allow to easily solve equations in more than three variables; for instance, to solve $F(v_1, v_2, v_3, v_4, v_5)$ one may first derive v_3 from v_1 and v_2 and then obtain v_5 from v_3 thus obtained and v_4; the two nomograms (for the two steps of the solution) can then be combined, by having a curve C_3 that is common to both. While writing an equation in this form is a complex problem in general, the resulting nomogram is significantly easier to use and understand than an intersection nomogram. Another aspect to observe is that the nomogram for a given equation is not unique: with a different choice of scaling (for instance, logarithmic instead of linear) the scales are defined by different curves. Creating easy-to-use nomograms thus required skill and experience, a limitation that has now been largely superseded by software [4]. One can also modify a nomogram by multiplying the matrix in Eq. (3) by another (constant) matrix and then taking the determinant: geometrically, this is equivalent to applying a linear transformation in the plane (such as scaling, rototranslations, stretching and shearing), which results in a valid nomogram that still solves the original equation, but may be easier to read.

D'Ocagne greatly helped to popularize his invention through a number of publications, in which he presented example nomograms for several applications, such as physics, hydraulics, topography, navigation, aviation and accounting [30]. By the 1920s, nomograms had become the main research topic in graphic computation [29, p. 142, Fig. 3]; moreover, they were of interest also in terms of their mathematical analysis. For example, German mathematician David Hilbert, in his famed list of 23 unsolved problems presented in 1901 [20], posed the 13th one in terms of nomographic analysis, by asking whether a 7th-degree polynomial equation (which cannot be solved in closed form through algebraic functions) can be solved instead through nomograms. A solution was found by d'Ocagne himself, which however relied on a movable element in addition to a nomogram and was thus considered as unsatisfactory.

A practical problem in nomography was to determine whether a given equation of three variables can in fact be solved through nomograms, and if so, to write it in determinant form as in Eq. (3): this problem was finally solved in a practical form in 1959 by Polish mathematician Mieczyslaw Warmus [32], who also presented a useful classification of the functions that can be written in nomographic form.

By the '50s, research in nomography, at first mostly published in French [29, p. 141, Fig. 2] after the pioneering works by Pouchet, Lalanne and d'Ocagne, had become of interest at the global level; we observe, for example, a significant amount of works by authors from the former Soviet Union [13], possibly due to delayed access to computers.

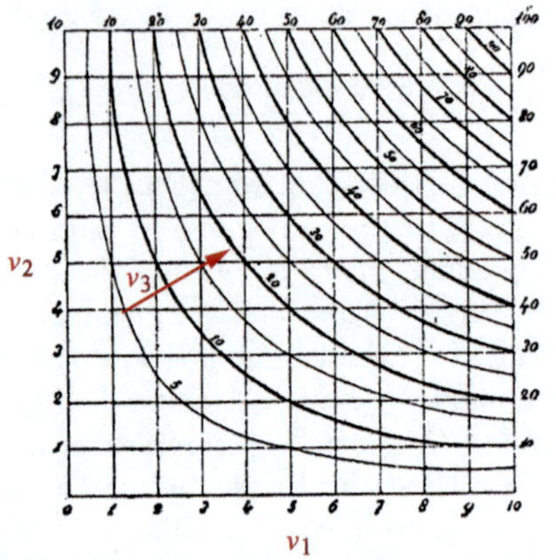

(a) Intersection nomogram.

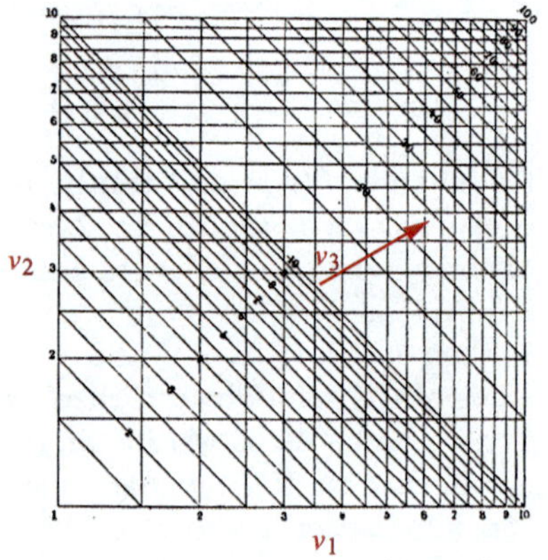

(b) Intersection nomogram with nonlinear scales.

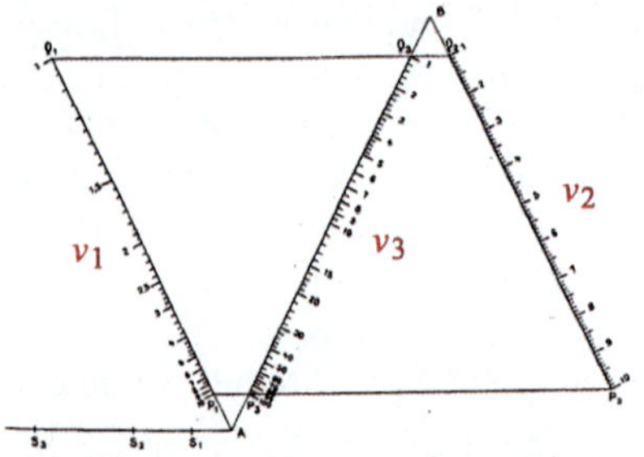

(c) Alignment nomogram, first version.

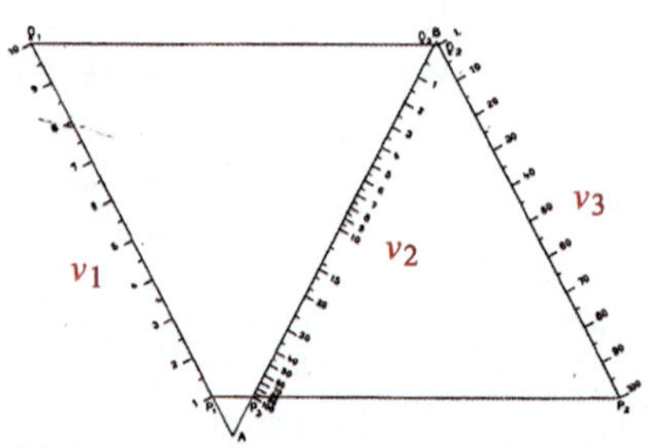

(d) Alignment nomogram, second version.

Fig. 1. For the same task (calculating the product of two variables, that is, $F = v_3 - v_2 v_1 = 0$) different charts can be provided, with the alignment nomogram offering an elegant solution. Notice that this can be drawn in different ways; the one in Fig. 1d has two linear scales (which are easier to draw). Charts reproduced from [5], with some changes for clarity.

2.2 Decline: 1960–1990

The research on nomography began to decline in the 1960s [29, p. 140, Fig. 1], as computers became commonly used in academia and industry. Nevertheless, nomograms were still used by engineers; published papers frequently developed calculations in nomographic form for practical applications. Topics that saw a significant use of nomograms until very recently were the analysis of internal combustion engines, manufacturing technology, design of hydraulic systems and civil engineering (especially for the study of soil mechanical properties): in all these cases, one needs to work with complex relationships that are often approximated from numerical data, thus the reduction in accuracy is not an issue. On the other hand, the possibility of obtaining a quick result that can be used in the field makes nomograms appealing for these applications.

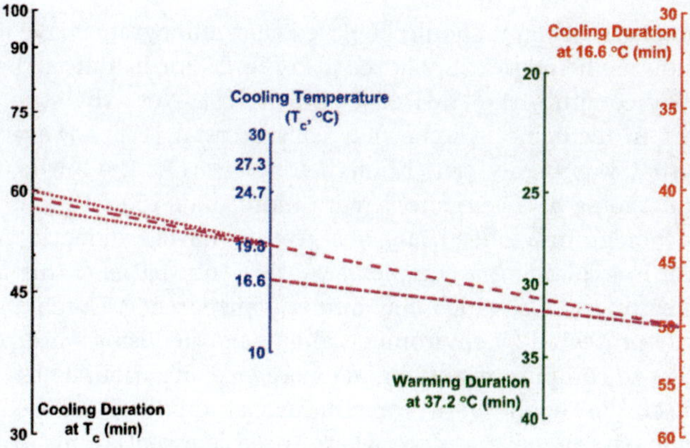

Fig. 2. Nomogram for thermal therapy of tissue injuries. Reprinted with permission from [24].

Regarding applications in Machine Mechanics, we observe that until the '70s nomographic solutions for mechanism design problems were appreciated for their practicality [1]: indeed, mechanism design generally leads to complex equations in which it is convenient to explore different design solutions, by designating different parameters as outputs. In particular, [1] showed a nomogram to optimize a four-bar linkage for function generation and observed how, after some practice, one can arrive in a relatively short time at a suitable solution with acceptably small errors. As it was common, the nomogram was also offered in large format on graph paper [1] for practical use.

Other nomograms were developed during this period specifically for mechanism design; for instance, planar, four-bar mechanisms were studied in [3,34], while [26] considered their spherical equivalents. A planar mechanism with higher (contact) pairs, namely a cam-follower linkage, was studied in [8] with the goal of preventing undercut. We also cite [7], on the vibration analysis of a spatial (Hooke's) joint.

Nomograms proved useful also in the design of gears [28,33], to present large amounts of experimental data in a form that is more convenient for designers (with respect to numerical tables). Some applications were also proposed for vibration analysis [21,27].

2.3 Resurgence? 1990–Today

After the '90s, nomograms have largely disappeared from engineering research, having been replaced by code listings that can be easily distributed. Nevertheless, nomograms are still present in published literature: for instance, they are still used in design standards and manufacturers' catalogs, to present results in a compact way. A common application is the Smith chart [16], used in electrical engineering for the design of transmission lines.

Outside of engineering, it should be noted that nomograms have had a significant impact on medicine [17,23,24]; see also Fig. 2. For instance, they are often used to quickly compute the BMI of a patient [16]. Note that the application of nomograms in medicine is somewhat controversial [17]; however, even in a 2008 review that was highly critical of their usage [17], the author noted that they were undergoing a "resurgence", with more than 150 citations in the year 2007 alone. Nomograms, indeed, are observed to have a didactic value, which can be helpful in explaining a complex situation to a patient; this is especially true in comparison to PCs, which may hinder transparency. Nomograms are also low-cost, easily protected by environmental damage (by using waterproof paper) and allow users to compute results in an emergency situation. These advantages can also be useful for engineers in some industrial applications.

Finally, nomograms have a place where speed is essential: for example, they are used to predict the behavior of forest fires or the lift of hot-air balloons [16]. Artillery was another relevant application, in which nomograms were in use up to World War I [18].

Considering in particular the field of Machine Mechanics over the last 30 years, we found several works on nomograms for the design of gear trains [9–12]; they are also still used in some cases for mechanism design, for planar [19] or spherical [22,25] linkages, for the optimization of motion laws in cams [15], for the analysis of efficiency in kinematic chain [2] or for studying tire dynamics in a vehicle [35].

3 Nomograms: Methods and Applications

In this Section, we present a very brief description of how nomograms can be produced, using modern tools, for a few example problems in Machine Mechanics. While a complete guide on drawing nomograms [6] is beyond the scope of this work, we believe some examples of nomograms can effectively show their advantages and limitations.

3.1 Modern Nomographic Tools: PyNomo

Traditionally, a notable limit of nomograms is the fact that they require skill in their design, not to mention the time needed to manually draw them in high resolution. Recently, however, some software tools and packages have been presented that allow us to automate nomogram design, which can help reintroduce nomography in practical usage. Indeed, as observed in [23], it is desirable to have a tool that *"combines computerization with classical nomograms"*. At the time of writing, the most complete tool for this task (that we are aware of) is pyNomo [4], a library for the Python programming language written to automate nomogram design. We therefore have created a few nomograms with pyNomo to illustrate some basic concepts in Machine Mechanics. The output of a pyNomo script is a PDF file containing the nomogram in high-resolution vector graphics; this is achieved by automatic compilation of a LaTeX program generated in Python.

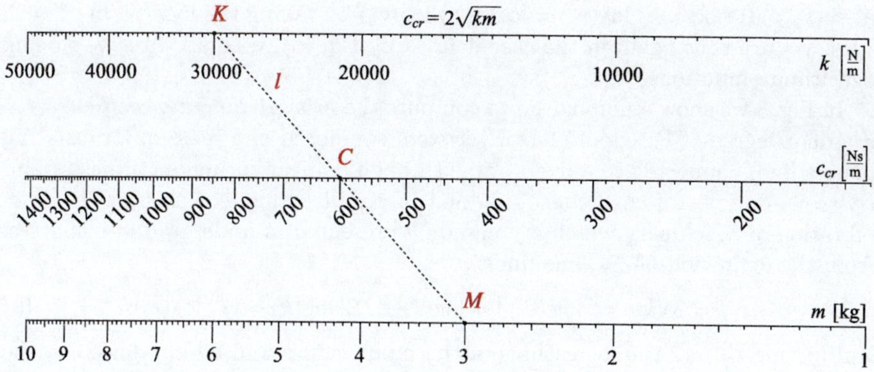

Fig. 3. Nomogram for calculating the critical damping coefficient c_{cr} of a 1-DoF damped harmonic oscillator; points K, C and M are defined by the values k, c and m of the stiffness and damping coefficients and of the system mass, respectively (each on the corresponding scale). For a critically-damped system, points K, C and M are aligned on a line l.

In the following, we refer to the classification in the pyNomo documentation, which identifies 10 types of nomograms. Traditionally, the most typical nomogram is the one defined by three parallel[1] straight lines (Type 1 in pyNomo), as shown in Fig. 3. The most general alignment nomogram determinant of Eq. (3) becomes simpler: if the j-th scale (along the line $x = x_j$) starts at $y = y_{jS}$ and ends at $y = y_{jE}$ $(j = 1, 2, 3)$, one has

$$\begin{matrix} f_{jx}(v_j) = x_j \quad \text{(constant)} \\ f_{jy}(v_1) = y_{jS} + \Delta y_j g_j(v_j) \end{matrix} \quad \text{with} \quad \begin{cases} g_j(v_{jS}) = 0 \\ g_j(v_{jE}) = 1 \end{cases} \quad \text{and} \quad \Delta y_j = y_{jE} - y_{jS} \tag{4}$$

where $v_j \in \left[v_{jS}, v_{jE}\right]$ is a variable corresponding to a point on the scale and g_j is a general monotonic continuous function; for linear scales, for example, it holds $g_j(v_j) = \frac{v_j - v_{jS}}{v_{jE} - v_{jS}}$. Substituting in the determinant from Eq. (3) and expanding, one has

$$(x_3 - x_2)\,\Delta y_1 g_1(v_1) + (x_1 - x_3)\,\Delta y_2 g_2(v_2) + (x_2 - x_1)\,\Delta y_3 g_2(v_3) + C_0 = 0 \tag{5}$$

where C_0 is a constant (once the nomogram has been defined): thus, any weighted *sum* of functions g_j (with weights $(x_i - x_k)\,\Delta y_j$ that can be set by a proper choice of the start and end points) can be achieved with a nomogram such as the one shown in Fig. 3.

A *product* of functions can also be computed by this type of nomogram: indeed, the equation $v_3 = v_1 v_2$ (for which Eq. (1) can be written as $F(v_1, v_2, v_3) =$

[1] In most nomograms of this type, the lines are vertical, for clarity. Here the lines are horizontal, due to lack of space; the y axis in the drawing plane is parallel to the scales.

$v_3 - v_1 v_2 = 0$) becomes $\log(v_3) = \log(v_1) + \log(v_2)$ by taking the logarithm of both sides. We are thus again in the case defined by Eq. (5), with all three g_j's being logarithmic functions.

In Fig. 3 we show a nomogram to compute the *critical damping coefficient* c_{cr} of a one-Degree-of-Freedom (1-DoF) system composed of a mass m (translating along a line) connected to a fixed frame through a linear spring of stiffness k and a viscous damper of coefficient c, acting in parallel; this is a classical topic in Vibration of Machinery which is commonly presented in undergraduate courses. From the definition of c_{cr}, one finds

$$c_{cr} = 2\sqrt{km} \Rightarrow \log(c_{cr}) = \log(2) + 1/2\log(k) + 1/2\log(m) \tag{6}$$

Multiplying by 1/2 the logarithms of the input values and adding log(2) to the result is "embedded" in the way the scales have been drawn (which is possible since 1/2 and log(2) are constant values) and does not complicate the nomogram. An isopleth has been added to Fig. 3 to show how a computation is performed: for $k = 30000\ \frac{\mathrm{N}}{\mathrm{m}}$ and $m = 3$ kg, one has $c_{cr} = 600\ \frac{\mathrm{Ns}}{\mathrm{m}}$, which is found by drawing a straight line l over the points K and M in Fig. 3 and finding the intersection C of l with the scale for c_{cr}.

A given equation can be represented in nomographic form in different ways. For example, another possible representation of an equation containing a product of functions $v_3 = v_1 v_2$ can be achieved with a "N" (also called "Z") type nomogram (Type 2 in pyNomo), corresponding to Figs. 1c and 1d; this nomogram is a graphical embedding of a multiplication between functions, but (unlike Fig. 3) requires using only one logarithmic scale. Consider three scales for the variables v_i, each starting at point $P_{jS} = (x_{jS}, y_{jS})$, corresponding to $g_j(v_j) = 0$. The direction of each scale is defined by vector $(\Delta x_j, \Delta y_j)$, which we assume to have unit magnitude, for simplicity. Equation (3) then becomes

$$\begin{vmatrix} x_{1S} + \Delta x_1 g_1(v_1) & y_{1S} + \Delta y_1 g_1(v_1) & 1 \\ x_{2S} + \Delta x_2 g_2(v_2) & y_{2S} + \Delta y_2 g_2(v_2) & 1 \\ x_{3S} + \Delta x_3 g_3(v_3) & y_{3S} + \Delta y_3 g_3(v_3) & 1 \end{vmatrix} = 0 \tag{7}$$

If the scale for variable v_2 passes through P_{1S} and P_{3S}, and if the scales for v_1 and v_3 are parallel, it can be shown that linear scales can be set for v_1 and v_3, such that $g_1(v_1)$ and $g_3(v_3)$ are linear functions and that it holds $v_2 = v_3/v_1$ (and thus again $v_3 = v_1 v_2$). Since the only nonlinear scale is the one for v_2, the resulting nomogram is simpler to draw and, in some cases, more compact than the equivalent Type 1 nomogram for the same equation (obtained by using logarithmic scales, as explained above).

As another example application, we present a "concurrent scale" (or "angle") nomogram in Fig. 4. This kind of nomogram (Type 7 in pyNomo) is defined by having three scales passing through a common point, which for simplicity is taken to be also the origin $P_{1S} = P_{2S} = P_{3S}$ of all three scales; the origin O of the coordinate plane is also set in this point. Defining the angles α_j formed by the j-th scale with the x-axis, one has

$$\frac{\sin(\alpha_3 - \alpha_2)}{g_1(v_1)} + \frac{\sin(\alpha_1 - \alpha_3)}{g_2(v_2)} + \frac{\sin(\alpha_2 - \alpha_1)}{g_3(v_3)} = 0 \tag{8}$$

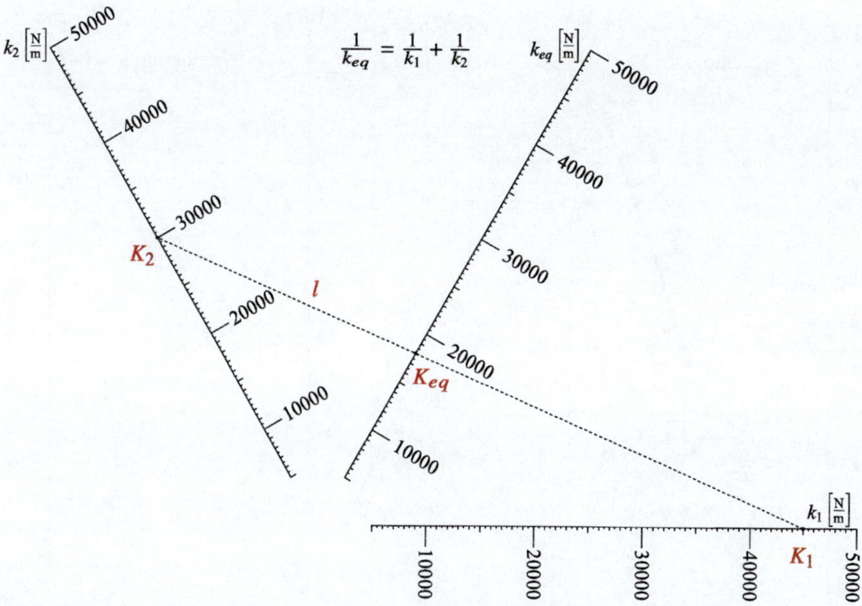

Fig. 4. Nomogram for the stiffness k_{eq} of a spring, equivalent to two springs in series having stiffnesses k_1 and k_2, respectively. In this case, all three scales have the same units of measurement.

A possible application for this nomogram that is of interest for Machine Mechanics is computing the combined stiffness of a system of two linear springs (having stiffnesses k_1 and k_2, respectively) attached in series. It can be shown that the resulting system is statically equivalent to a single spring of stiffness k_{eq}, having

$$\frac{1}{k_{eq}} = \frac{1}{k_1} + \frac{1}{k_2} \tag{9}$$

which can be directly implemented in a Type 7 nomogram. See Fig. 4, where the isopleth l has been drawn for an example problem with $k_1 = 45000\ \frac{\mathrm{N}}{\mathrm{m}}$ and $k_2 = 30000\ \frac{\mathrm{N}}{\mathrm{m}}$ (corresponding to points K_1 and K_2, respectively); the stiffness $k_{eq} = 18000\ \frac{\mathrm{N}}{\mathrm{m}}$ is found at K_{eq}. Notice that unrealistic cases having $k_j = 0$ are automatically excluded by the graphic procedure, making it impossible to obtain incorrect results from a mistake in the input. The angles α_j can be chosen according to convenience; having $\alpha_3 - \alpha_2 = \alpha_2 - \alpha_1 = 60°$ leads to the same unit length being used for all three scales, as in Fig. 4.

3.2 Applications

To illustrate concrete applications of nomography in machines' and mechanisms' analysis, we present an example design of a more complex nomogram, devised to solve a practical problem in the kinematic analysis of a spatial mechanism. This

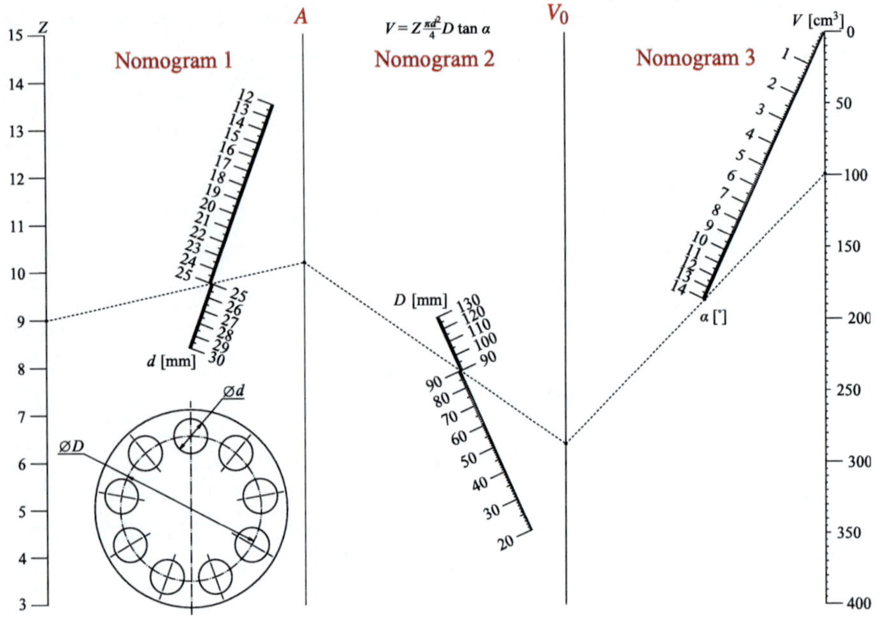

Fig. 5. Combined nomogram (composed of three "intermediate" Type 2 nomograms, denoted as 1, 2 and 3) for calculating the variable displacement of a volumetric axial piston pump.

design has been inspired by a collaboration with a company working on variable-displacement axial piston pumps (with a swashplate to control the displacement). One minor need reported by the company was to compute the pump displacement from measurements on its main parameters: this is useful when working on a shop floor, as the pump data (such as the serial number and product specification) is not directly available. The displacement V can then be computed from easily measured quantities, as

$$V = \frac{\pi d^2}{4} Z D \tan(\alpha) \tag{10}$$

where d is the diameter of the pistons, Z their total number, D the diameter of the cylinder through which the piston axes pass and α the swashplate inclination.

Clearly, Eq. 10 can be implemented in any computation software, for instance as a spreadsheet. However, operating a digital device when disassembling a pump (which is generally filled with lubricant) for maintenance operations is clearly inconvenient. Moreover, there is a significant risk of operator errors due to incorrect data entry.

Nomograms provide an interesting possibility for this task. However, Eq. (10) contains five variables; thus, it cannot be directly expressed as a nomogram of one of the types presented previously. We then divide the corresponding nomogram in parts, as follows.

1. We create a nomogram to compute $A = \frac{\pi d^2}{4} Z$; this is the total cross-sectional area of the pistons. This is achieved through a simple Type 2 nomogram with three variables, as described in Subsect. 3.1, where $v_1 = d^2$, $v_2 = Z$ and $v_3 = A\frac{4}{\pi}$.
2. The result from the previous nomogram is used for a second step, by joining two nomograms at the scale for A; the corresponding expression is $V_0 = DA$. Again, being a product of two variables, this result can be computed with a Type 2 nomogram.
3. Finally, a third nomogram is introduced, joined at the previous one at the scale for V_0. This computes the final result $V = V_0 (\tan(\alpha))$.

The final result is shown in Fig. 5; here, the two vertical scales in the middle correspond to variables A and V_0, respectively. The values of these intermediate quantities are not required in the final result; thus, no ticks are displayed, to avoid cluttering the graph. The isopleth on the graph shows an example calculation for a realistic pump design; on one corner of the nomogram, a sectional view of the central cylinder block of the pump is shown for reference, to clarify the definition of diameters D and d.

This nomogram has been printed on laminated paper, to protect it from oil spills, and introduced in shop-floor practice for quick computations. The isopleths for each step can be found with any rigid element having a straight-line segment of sufficient length.

We also suggest an application in education. As noted in [6,23,30], nomograms allow us to easily understand the behavior of equations with complex algebraic expressions; also, nomography is a graphical approach that makes it unlikely to misinterpret inputs or results. Besides nomograms, other graphical methods, such as those used for planar mechanism kinematics and statics, are currently taught in mechanical engineering courses. While these methods could be replaced by purely analytical approaches, they are still used due to their pedagogical value. We thus expect nomograms to easily fit in existing courses; in particular, nomography extends graphical approaches to any equation that can be written in determinant form as Eq. (3). An experience on nomograms for math education in high schools was reported in [31]; a test on their usefulness for university-level education would then be worthwile to understand their strengths and limitations. We also believe that mentioning examples of nomograms, together with their uses and applications, within a course in Machine Mechanics, could also enhance students' interest in historical methods for engineering analysis and in the broader historical development of our field.

4 Conclusion and Outlook

We have presented a brief history of nomograms, a graphical method once commonly used to solve equations. We focused on their applications in mechanical engineering, with the goal of performing a didactic experiment for a course in Machine Mechanics, to appreciate their pedagogical potential. We also presented some original nomograms developed with pyNomo, a nomographic software package. We expect that computers, which once made nomograms obsolete, can indeed help in reviving interest in these tools.

References

1. Adams, D.P.: Nomographic synthesis of generator linkages. J. Eng. Ind. **82**(1), 29–38 (1960). https://doi.org/10.1115/1.3662986
2. Aleksandrov, I.K.: Determining the limiting efficiency of a kinematic chain. Russ. Eng. Res. **31**, 539–540 (2011). https://doi.org/10.3103/S1068798X11060037
3. Antuma, H.J.: Triangular nomograms for symmetrical coupler curves. Mech. Mach. Theory **13**(3), 251–268 (1978). https://doi.org/10.1016/0094-114X(78)90049-6
4. Boulet, D., Doerfler, R., Marasco, J., Roschier, L.: pyNomo documentation (2020). http://lefakkomies.github.io/pynomo-doc/index.html
5. d'Ocagne, M.: Traité de nomographie. Gauthier-Villars, Paris (1899)
6. Doerfler, R.: On jargon – the lost art of nomography. UMAP J. **30**(4), 457–493 (2009). https://www.comap.com/product/?idx=1048
7. Éidinov, M.S., Nyrko, V.A., Éidinov, R.M., Gashukov, V.S.: Torsional vibrations of a system with Hooke's joint. Sov. Appl. Mech. **12**, 291–298 (1976). https://doi.org/10.1007/BF00884975
8. El-Shakery, S.A., Terauchi, Y.: A computer-aided method for optimum design of plate cam-size avoiding undercutting and separation phenomena–II: design nomograms. Mech. Mach. Theory **19**(2), 235–241 (1984). https://doi.org/10.1016/0094-114X(84)90046-6
9. Esmail, E.L.: Nomographs for synthesis of epicyclic-type automatic transmissions. Meccanica **48**, 2037–2049 (2013). https://doi.org/10.1007/s11012-013-9721-z
10. Esmail, E.L.: Configuration design of ten-speed automatic transmissions with twelve-link three-DOF Lepelletier gear mechanism. J. Mech. Sci. Technol. **30**(1), 211–220 (2016). https://doi.org/10.1007/s12206-015-1225-4
11. Esmail, E.L., Hussen, H.A.: Nomographs for kinematics, statics and power flow analysis of epicyclic gear trains. In: Proceedings of the ASME 2009 International Mechanical Engineering Congress and Exposition, vol. 13, pp. 631–640. ASME, Lake Buena Vista (2010). https://doi.org/10.1115/IMECE2009-10789
12. Esmail, E.L., Pennestrì, E., Juber, A.H.: Power losses in two-degrees-of-freedom planetary gear trains: a critical analysis of Radzimovsky's formulas. Mech. Mach. Theory **128**, 191–204 (2018). https://doi.org/10.1016/j.mechmachtheory.2018.05.015
13. Evesham, H.A.: Origins and development of nomography. IEEE Ann. Hist. Comput. **8**(4), 324–333 (1986). https://doi.org/10.1109/MAHC.1986.10059
14. Evesham, H.A.: The History and Development of Nomography. Docent Press, Mountain View (2010)
15. de Freitas Avelar, A.H., Roschier, L., Fernandes Soares, L., Oliveira Ávila, P.H.S.: Analytical solutions and computational nomograms for maximum pressure angle for cam mechanisms for full and half cycloidal and harmonic motion curves. J. Mech. Eng. Sci. **235**(15), 2725–2736 (2021). https://doi.org/10.1177/0954406220962823
16. Glasser, L., Doerfler, R.: A brief introduction to nomography: graphical representation of mathematical relationships. Int. J. Math. Educ. Sci. Technol. **50**(8), 1273–1284 (2019). https://doi.org/10.1080/0020739X.2018.1527406

17. Grimes, D.A.: The nomogram epidemic: resurgence of a medical relic. Ann. Intern. Med. **149**(4), 273–275 (2008). https://doi.org/10.7326/0003-4819-149-4-200808190-00010
18. Hankins, T.L.: Blood, dirt, and nomograms: a particular history of graphs. Isis **90**(1), 50–80 (1999). https://doi.org/10.1086/384241
19. Hassaan, G.A.: Nomogram-based synthesis of complex planar mechanisms, part I: 6 bar-2 sliders mechanism. Int. J. Eng. Tech. **1**(6), 29–35 (2015)
20. Hilbert, D.: Mathematische probleme. Archive für Mathematik und Physik **1**(1), 44–63 (1901)
21. Hohenberg, R.: Detection and study of compressor-blade vibration. Exp. Mech. **7**, 19A-24A (1967). https://doi.org/10.1007/BF02327002
22. Hwang, W.M., Chen, K.H.: Triangular nomograms for symmetrical spherical non-Grashof double-rockers generating symmetrical coupler curves. Mech. Mach. Theory **42**(7), 871–888 (2007). https://doi.org/10.1016/j.mechmachtheory.2006.05.008
23. Kattan, M.W., Marasco, J.: What is a real nomogram? Semin. Oncol. **37**(1), 23–26 (2010). https://doi.org/10.1053/j.seminoncol.2009.12.003
24. Khoshnevis, S., Brothers, R.M., Diller, K.R.: Level of cutaneous blood flow depression during cryotherapy depends on applied temperature: criteria for protocol design. ASME J. Med. Diagn. **1**(4), 041007 (2018). https://doi.org/10.1115/1.4041463
25. Lu, D.M.: A triangular nomogram for spherical symmetric coupler curves and its application to mechanism design. J. Mech. Des. **121**(2), 323–326 (1999). https://doi.org/10.1115/1.2829463
26. Meyer zur Capellen, W.: Nomogramme für die Krümmung sphärischer und ebener Bahnkurven. Mech. Mach. Theory **18**(3), 249–254 (1983). https://doi.org/10.1016/0094-114X(83)90098-8
27. Miconi, D.: Vibration control in industrial plant: a methodological approach. J. Vib. Acoust. Stress Reliab. **109**(4), 335–342 (1987). https://doi.org/10.1115/1.3269450
28. Seireg, A.A., Houser, D.R.: Evaluation of dynamic factors for spur and helical gears. J. Eng. Ind. **92**(2), 504–514 (1970). https://doi.org/10.1115/1.3427790
29. Tournès, D.: Notes & debats—Pour une histoire du calcul graphique. Rev. d'Histoire des Math. **6**(1), 127–161 (2000). http://www.numdam.org/item/RHM_2000__6_1_127_0/
30. Tournès, D.: Du compas aux intégraphes: les instruments du calcul graphique. Repères-IREM **50**, 63–84 (2003). https://publimath.univ-irem.fr/biblio/IWR03005.htm
31. Tournès, D.: Calculating with hyperbolas and parabolas. In: Barbin, É., et al. (eds.) Let History into the Mathematics Classroom. History of Mathematics Education, pp. 101–114. Springer, Heidelberg (2018). https://doi.org/10.1007/978-3-319-57150-8_8
32. Warmus, M.: Nomographic Functions. Państwowe Wydawnictwo Naukowe, Warsaw (1959)
33. Wellauer, E.J., Holloway, G.A.: Application of EHD oil film theory to industrial gear drives. J. Eng. Ind. **98**(2), 626–631 (1976). https://doi.org/10.1115/1.3438951
34. Wunderlich, W.: Nomogramme für die Wattsche Geradführung. Mech. Mach. Theory **15**(1), 5–8 (1980). https://doi.org/10.1016/0094-114X(80)90028-2
35. Zotov, N.M., Balakina, E.V.: Using the $\varphi - s_x$ nomogram in calculating the dynamics of a braked wheel. J. Mach. Manuf. Reliab. **36**, 193–198 (2007). https://doi.org/10.3103/S1052618807020161

The Armillary Sphere: A Representation of the Celestial Sphere and Knowledge Symbol

Barbara Aterini(✉)

University of Florence, Florence, Italy
barbara.aterini@unifi.it

Abstract. The armillary sphere is a model of the celestial sphere born in ancient Greece. It is made up of graduated metal circles that connect the poles and represent the equator, ecliptic, meridians and parallels; at its center a sphere represents the Earth. It is used to solve various astronomical problems and to explain the movements of stars in the sky. This scientific tool is based on projective geometric principles. Since the fifteenth century it has found a place in paintings and frescoes, where it is used as a symbol of knowledge and wisdom, so much so that it is represented alongside famous people.

Keywords: Armillary sphere · Stereographic projection · Measuring instruments · Projective geometry · Astronomy

1 Introduction

The armillary sphere is a model of the celestial sphere, that is, of that imaginary sphere of arbitrary radius on whose surface all the stars are projected. This sphere is said to be geocentric if it has the center of the earth as its center, local if it has the observer's eye as its center, heliocentric if it has the Sun as its center.

Its name derives from the Latin *armilla* (circle, bracelet) - from armus '*homerus*', since it is composed of graduated metal circles that connect the poles and represent the equator, ecliptic, meridians and parallels; at its center is a sphere which, at least initially, represented the Earth. It is an instrument that was used to solve various astronomical problems as well as to explain the movements of the stars in the sky. In fact it can be used to trace the path of the sun for a given day of the year or to determine the coordinates of a star. The armillary spheres are divided into two categories: 'observational' such as those used by Ptolemy and the Danish astronomer Tycho Brahe (1546–1601), and 'demonstrative'. The latter are used to illustrate the celestial mechanics of stars and planets moving around the Earth (Fig. 1).

2 Brief History of the Armillary Sphere

It is not clear exactly when the armillary sphere was invented, it probably originated in ancient Greece. Already known to Chinese astronomers since the fourth century BC,

M. Ceccarelli and R. López-García (Eds.): HMM 2022, HMMS 40, pp. 330–342, 2022.
https://doi.org/10.1007/978-3-030-98499-1_27

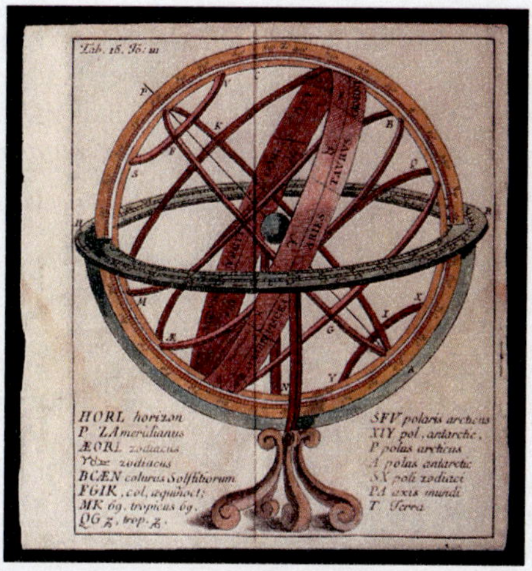

Fig. 1. Print representing an armillary sphere from the 18th century. The earth is placed in the center and surrounded by the projection in the vault of all its notable points, listed and explained in the references placed on the sides of the pedestal of the armillary sphere. The constellations and related zodiac signs are indicated in the ecliptic belt.

it seems to have been independently invented in the West by the Greek-Alexandrian astronomer Eratosthenes (276–194 BC) around 255 BC. The Arabs, who collected the legacy of the Greco-Roman culture, further perfected it by producing wonderful specimens and introduced it to the West through Andalusia around the year 1000.

Used above all as a didactic tool, its operating principle certainly inspired various measuring instruments even if we only have the armillary astrolabe described by the astronomer and cartographer Claudius Ptolemy (100 AD -?) In the Almagest (140 AD).

The Ptolemaic geocentric model positions the spherical and immobile Earth at the center of the cosmos and makes stars, sun, moon and planets move, offering the great advantage of adhering perfectly to the point of view of us terrestrial observers who have the illusion of being stationary at the center of the cosmos.

As the Copernican model began to prevail, the central sphere came to represent the Sun. Often, the armillary spheres were built in pairs: one representing one model, the second the other; this is to teach the differences between the two.

In the Middle Ages this device became a didactic tool used to demonstrate the principles of astronomy. Later the Danish astronomer Tycho Brahe (1546–1601) built a whole series of different types of armilles, the most important of which had only one complete circle (a graduated declination circle of wood and brass with a diameter of 2.7 m) and was used for direct observation of the position of celestial bodies. In Europe, armillary spheres were common tools in the late medieval and early modern era.

Islamic and Renaissance astronomers built large armills of various types, equipped with detailed scales and alidade with sights, used to measure different coordinates of the celestial bodies. Smaller models were also built, useful for demonstrating the principles of spherical astronomy and the motions of celestial bodies. The latter were enriched with additional rings representing the Sun, the Moon and the planets.

This instrument assumed such importance that many surviving armillary spheres from 1500 and later centuries were made of precious metals for collectors. In the 18th century the armillary spheres were also made of wood and cardboard. They were used until the 19th century, mainly as educational tools to teach the difference between the Ptolemaic and the Copernican model of the universe.

3 How It is Made

The armillary spheres, or armilles, were among the first complex mechanical apparatuses and constitute a group of ancient astronomical instruments that differ from each other, even considerably, both in form and in the use for which they were intended.

The instrument described for the first time by Claudius Ptolemy around 140 AD, consisted of a series of large circles on which target could be set, and was used to determine the position of the Sun, the Moon and the planets in celestial coordinates (ecliptic). Starting from the assumption that man on Earth sees the stars moving around him according to "apparent" movements. Let us imagine that the observer is the center of a sphere of arbitrary radius; on this sphere the stars and their apparent movements are centrally projected, as the observer sees them. In other words, the visible cosmos is transferred to that apparent sphere called the "celestial sphere" (Aterini 2019a, b).

The center of this sphere is the Earth, given its small size; the geographic poles are projected onto the north (PN) and south (PS) celestial poles; the meridians, the parallels and the equator look the same. The Zenith is on the vertical of the observer, at the opposite extreme we find the Nadir. The observer can see everything that is above the horizon, that is, in the hemisphere on the side of Zenith.

The apparent motion of the Sun manifests itself on the celestial sphere as a maximum inclined circle of 23° 27′ above the level of the equator. This circle is called ecliptic; the Sun moves counterclockwise on it, completing one revolution in 365.2422 days (Fig. 2).

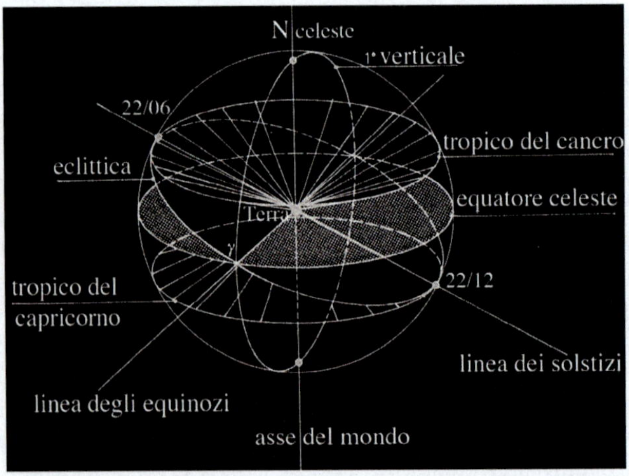

Fig. 2. The celestial sphere according to the geocentric model.

The armillary sphere is made up of rings with graduated scales, some of which are able to rotate relative to others. These rings represent the circles on the celestial sphere, such as the meridians, the parallels, the equator, the ecliptic horizon, the tropics and the coluri, i.e. the two celestial meridians, intersecting the ecliptic respectively at the equinox and solstice points. In particular, the rings that define the armilla, the coluri (equinoctial and solstitial) and the equator represent the firmament, that is, the sphere on which the fixed stars lie. The outermost, and usually larger, band surrounding the sphere represents the constellations of the zodiac. The ecliptic is drawn on it divided into 12 sectors of 30° each, on which the symbol of each constellation is engraved to represent the entire zodiacal band. The line that crosses that band is the ecliptic, the path the sun follows across the sky. The stand can be decorative, but it also allows you to place the sun in its astrological home for a certain date and to demonstrate the times of sunrise and sunset.

The central globe represents the spherical earth, immobile and placed at the center of the cosmos. The sphericity of the earth was an already established fact and its immobility and centrality rested on theoretical arguments based on Aristotelian physics and discussed among others by Claudius Ptolemy in the Almagest.

The web of parallels and meridians constitutes the sphere of ether on which the stars are seen; the mechanism allowed the clockwise rotation of this around the axis passing through the poles in a time interval of 24 h, fully accounting for the motion of the stars that we can see during the night. This first motion in turn set in motion the seven internal spheres on which the five planets then known (Saturn, Jupiter, Mars, Venus, Mercury) the Sun and the Moon were fixed. These two elements, receiving their main movement from the sphere of the stars, are seen to rise and set in the time interval that constitutes day and night. In the Ptolemaic model, the seven internal spheres also have slow proper motions: thus, the counter-rotation of the lunar sphere determines the different phases of the moon offering that intermediate division of time called the month as well as the four lunar phases form the basis for the further division in weeks.

Finally, the very slow counter-rotation of the solar sphere determines the movement of the sun with respect to the stars along the zodiac. This movement determines the slow daily variation of the position from which the sun rises and sets as well as the different height in the sky of its arch for all observers placed on earth. This results in the cycle of the seasons over the course of that period called the year. Thus, Ptolemaic astronomy anchored the flow of time - released in days, months and years - to a solid geometric vision of the cosmos, an almost perfect synthesis that contained the knowledge accumulated by Egyptians, Babylonians and Greeks, and which will last for almost 1400 years. Until the late Renaissance of our time.

The armillary spheres were used by setting them to the appropriate latitude by setting the outer meridian rings in a position perpendicular to the horizon and parallel to a line drawn from north to south. Their orientation was established by sighting a celestial object (star, sun, moon or planet) whose position on the ecliptic was known, using a divided ecliptic ring and a ring corresponding to latitude. It was possible to find the position of a celestial body on the ecliptic using an internal latitude ring, which in turn contained a ring that could be rotated without interrupting the latitude ring.

4 The Fundamental Projective Principles of the Armillary Sphere

The armillary sphere, whose plane projection is the astrolabe, is based on precise geometric projective principles (Aterini 2019a). In particular, we can say that it is based on the 'stereographic projection' that is the operation by which the points on the surface of the sphere are projected from a fixed point P, called the pole, on a plane π parallel to the plane tangent to the sphere in the pole. In this way a biunique correspondence (homology) is obtained between the points of the sphere and those of the plane π which preserves the shape and therefore is said to be compliant. The angles are maintained, so the projection of a circle μ of the spherical surface from a pole V is still a circle μ′. In stereographic projections, the point of view is on the earth's surface and in the site opposite to that of the plane's tangency. The distances between the parallels are not proportional to the real ones. In orthogonal projections the point of view is located at infinity and the projection rays are parallel. The parallels are as close as possible the farther you are from the point of tangency (Fig. 3).

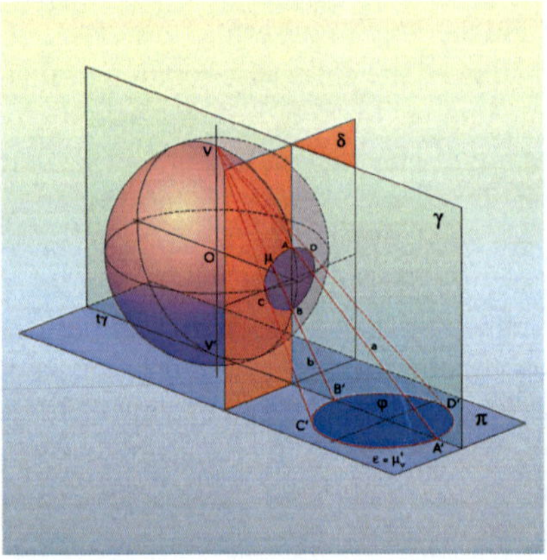

Fig. 3. The Stereographic projection of the sphere.

The stereographic projection of the circumferences of the Almucantarath succession (ie the parallels of the celestial sphere) from the Pn on the plane ζ ne defines the shadow on the same plane, whose true magnitude represents a sequence of arcs of circumference also present in the astrolabe. These allow the observer to define the height of the stars and, consequently, the time (Aterini 2019b).

Through a simple representation in parallel perspective (Aterini 1995) it is possible to represent the equator and the celestial tropics (Figs. 4 and 5).

5 The Armillary Sphere in Iconography: A Symbol of Knowledge

Due to the complexity of its mechanisms and the ability to solve important astronomical problems, the armillary sphere became the very image of knowledge, so much so that

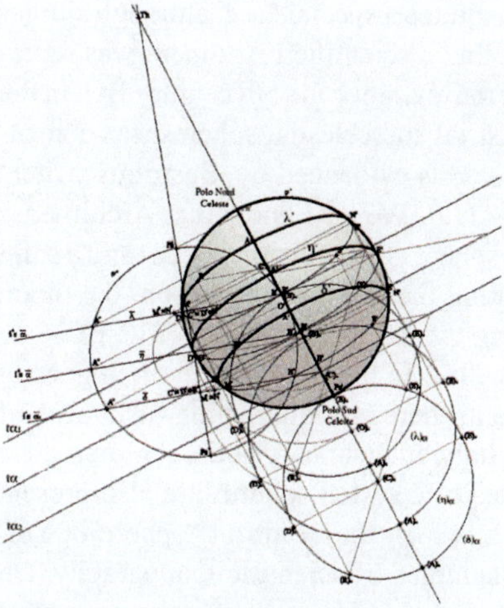

Fig. 4. Given the plane α (celestial equator), consider two planes α1 and α2 parallel to it, for the two points of intersection between the sphere and a straight line r passing through the center of the sphere and inclined by 23° 44′ with respect to the plane α, we find two circles λ e δ of the two planes with the sphere represent the celestial tropics.

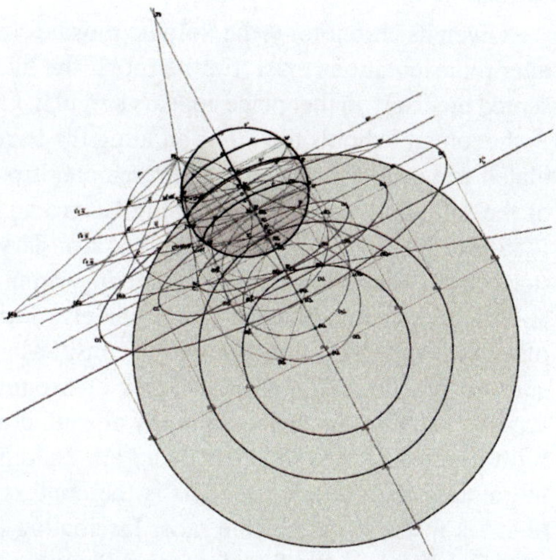

Fig. 5. Using the same type of representation we consider the plane (ζ) parallel to α e passing through Ps. The stereographic projection of the fundamental circumferences (η, λ, δ) from the Pn on the plane ζ can be considered their shadow on the same plane. These in true size represent the fundamental entities of the astrolabe.

we find it in murals and in many paintings. One of the oldest representations is found in a floor mosaic of the ancient city of Solunto, near Palermo, which dates back to the Hellenistic Roman period (De Rose 2016). It is a mosaic in *opus vermiculatum* - found in the 'House of Leda' and dating back to the second century. B.C. -which depicts an armillary sphere with the earth globe in the center, according to the Ptolemaic conception, surrounded by movable rings while, in a corner, the figure of a wind can be glimpsed, represented as a youthful head in profile in the act to blow (Fig. 6).

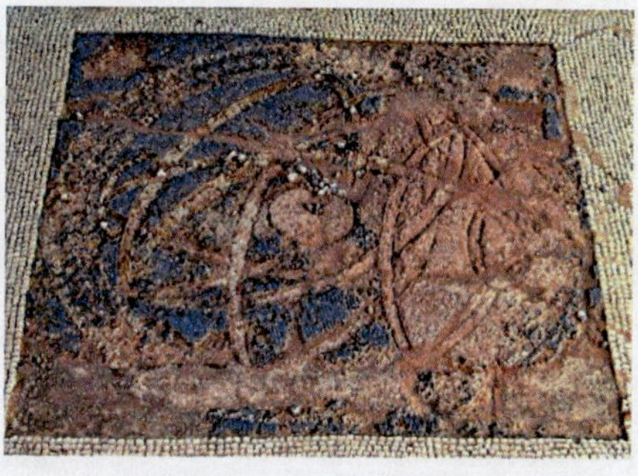

Fig. 6. The armillary sphere in the mosaic of the "Casa di Leda" in Solunto.

Given its chronology, the Solunto mosaic constitutes a special case, although similar later representations exist. Furthermore, the fact that a scientific instrument was represented precisely in that place appears singular. Certainly, since the 5th century BC. In the Pythagorean Schools the art of building the terrestrial and celestial spheres was consolidated and used for teaching and research purposes, as evidenced by illustrious writers of the following centuries including Cicero and Ovid. We also know that Archimedes (287–212 BC), had built two models of armillary sphere - an empty (celestial) and a solid (terrestrial) very showy one - apparently simulating the motions of the sun, the moon and the five visible planets. Unfortunately, during the Sack of Syracuse (212 BC) one of the two spheres was stolen by the consul Marcellus. In the Greek tradition in general and Archimedean in a Magno-Greek environment, there is a flourishing of images of armillary spheres in the iconography of southern Italy. In fact not only did sculpture give birth to mixed forms of science and art; examples of celestial spheres are also present in painting and mosaic art. This is the context in which the beautiful "sphere of Solunto" is inserted, one of the most fascinating examples of scientific iconography. On the other hand, in the first century AD we find examples of armillary spheres in wall painting: Apollo Cosmocrator Pompeian wall painting of the 1st century A.D. found in the House of Apolline in Pompeii and today preserved in the National Museum of Naples, it depicts the Sun god, Apollo-Helios with the celestial sphere to symbolize his sovereignty over the movement of the stars and the Cosmos. On the right we recognize the sphere represented with two circles.

From the first century onwards, the scientific meaning overlaps to the symbolic meaning that will become prevalent in iconographic representations. Thus also in another fresco, recently restored, also from Pompeii, we see Urania, the Muse of astronomy, who holds a celestial sphere in her hand, her attribute together with the rod. The sphere is similar to the previous one, where two armilles are distinguished, the painting is part of the cycle of the Muses, found in Pompeii in the house of Iulia Felix (62–79 AD) (Figs. 7 and 8).

Fig. 7. Apollo Cosmocrator, 1st cent. A.D. wall painting found in the House of Apolline in Pompeii.

Fig. 8. Urania, the Muse of Astronomy, portrayed in a fresco in the House of Julia Felix in Pompeii (62–79 AD)

Fig. 9. The planisphere of Stabia (Napoli), fresco from the middle of the 1st century. A.D. Detail of the ceiling of the portico, Villa San Marco.

Even more detailed is the armillary sphere visible in the fresco from the triclinium of Moregine, a complex located about 600 m south of the walls of Pompeii, which emerged in 1959 during the works on the Salerno-Reggio Calabria motorway. Against the Pompeian red background, the muse Urania holds a globe in her hands that reproduces the same model as Solunto, five celestial spheres, visible only in one hemisphere with the equator in the center and the other circles, tropical and polar, intersected by the coluri and from the zodiacal belt, with the earth inside.

Another fresco from the Neronian era, known as the "planisphere of Stabia", shows the iconographic value of art inspired by astronomy (Fig. 9). The painting represents an allegory of "everything", in which the seasons are included. It is a transparent globe, where drawn circles, a sort of flat ribbons, have scientific value. Two bands are vertical and divide at the poles of the figure, top right, perhaps a deformation of the perspective. The third band is articulated around an immaterial axis, placed on the plane of the horizon and perpendicular to the imaginary axis of the other two circles. The poles are visible on the external vertical circle, this circle, seen from the poles is oblique and forms an angle of about 45° with the plane of the equator. Only the upper part is visible, but the rest is presumed, hidden by the allegories of the seasons. Despite the excessive angle, more than double the amount due, the ecliptic was recognized in this oblique circle, and in the other two circles the colures, as the outer circle marks the points of the equinox, the other those of the solstice. The fragment was found on the ceiling of the portico of Villa San Marco and is kept in the archaeological museum of Castellammare di Stabia (De Rose 2016).

In the medieval period we still find the representation of armillary spheres in miniatures, especially Arabic (Fig. 10).

Fig. 10. Miniatures where the armillary sphere is represented.

On the other hand, the Arabs, in addition to producing astronomical instruments, were also authors of important theoretical treatments on the subject. In particular, with regard to the armillary spheres, the eighth-century treatise of Al-Fazārī is important. These astronomical instruments used as didactic tools starting from the third century BC. - As well as instruments of observation - they became extremely popular in Europe only in the late Middle Ages. However, it was in the Renaissance, and up to the Romantic period, that they assumed the role of a true symbol of wisdom and knowledge, so much so that they appeared in numerous artistic representations usually in the hands of scientists. In fact, from the fifteenth century the armillary sphere comes out of the illuminated manuscripts and begins to be painted on paintings and frescoes, since in addition to being considered an important mechanical instrument, it appears as a sort of synthesis of knowledge in the field of astronomy and therefore such as to represent the wisdom achieved. Thus also in pictorial representations it is placed in the hand or in any case close to illustrious characters to enhance their wisdom and knowledge. Among the most important characters we find Saint Augustine portrayed in his studio by the Venetian painter Vittore Carpaccio (1465–1520) (Fig. 11).

Fig. 11. Vision of Saint Augustine, detail of the painting; at the top right you can see the armillary sphere, a symbol of knowledge.

Fig. 12. Gentile da Fabriano, Personification of Astronomy (1411-12), Palazzo Trinci, Foligno.

Another famous example is that of the personification of Astronomy (1411-12), a late Gothic work by Gentile da Fabriano (1411-12), in Palazzo Trinci, Foligno (Pulvirenti 2014) (Fig. 12).

The Studiolo by Federico da Montefeltro in the Ducal Palace of Urbino (1473-76) is entirely Renaissance. It is a small room whose walls, covered with wooden inlays, reproduce with studied perspective effects shelving, with half-open doors, containing numerous books and scientific objects including an armillary sphere. In the same studio, the Duke of Montefeltro had commissioned the Flemish painter Joos van Ghent, also known as Justus of Ghent (circa 1430-circa 1480), to paint 28 portraits of illustrious men of the past. The paintings show the various characters in a loggia seen from below (since they should have been above the band covered with wooden inlays). Among these is Ptolemy (1476), whom the painter depicts with the inevitable armillary sphere in his hand (Fig. 13).

Fig. 13. Examples of the representation of the armilla in the *Studiolo* of Urbino

In Guercino's 'Personification of Astrology' (1591–1666), the armillary sphere once again becomes an incisive symbol of wisdom and knowledge (Fig. 14).

Fig. 14. Giovanni Francesco Barbieri known as Guercino, Astrology (1650)

In the seventeenth century the painters Peter Paul Rubens (1577–1640) and Jan Brueghel the Elder (1568–1625) in one of the paintings of the Allegory of the five senses - The sight - insert the armillary sphere as a symbol of knowledge (Fig. 15).

In this detail taken from an Allegory of the five senses painted during the 17th century by the Flemish painter Jan Brueghel the Younger (1601–1678) in collaboration with Hendrick van Balen (1575–1632), the armillary sphere stands at the foot of the allegorical figure representing sight, along with other astronomical instruments. Once again he gives importance to the painted scene (Fig. 16).

Fig. 15. P. P. Rubens and J. Brueghel the Elder. La Vista (1617), oil on panel, Museo del Prado, Madrid.

Fig. 16. J. Brueghel the Younger and H. van Balen, Allegory of Sight.

At the beginning of the nineteenth century the armillary sphere is still present in the allegories of astronomy, but then the interest of the painters towards it decreases, not only for the fact that it is an object that is now outdated from a technical point of view, but also because we are witnessing a distancing between scientific and humanistic knowledge. On the other hand, this does not mean that is stop building these tools that are made of wood or metal to be used, as happened in previous centuries. Just think of the armilla made by the mathematician, cosmographer and architect Egnazio Danti (1536–1586) installed on the facade of the church of Santa Maria Novella in Florence in the second half of the sixteenth century and still admirable today (Danti 1573). It is a very simplified instrument, as it only reports the equinox line and serves, in fact, to determine the days in which the equinoxes fall (Fig. 17).

Just as we can think of the monumental armillary spheres with numerous rings and precious decorations, first of all the one created under the direction of the astronomer and cosmographer Antonio Santucci (1550–1613) between 1588 and 1543 at the behest of

Fig. 17. E. Danti, equinoctial armilla on the facade of the church of Santa Maria Novella in Florence.

Fig. 18. Armillary sphere made by Antonio Santucci and currently preserved in the Galileo Museum in Florence.

Ferdinando I de Medici (Fig. 18). The geocentric sphere, according to the Aristotelian-Ptolemaic conception, is built in wood and metal; the diameter measures 2 m, is 3.70 m high and wide 2, 450 m. The sphere, whose wooden parts are richly painted and covered with pure gold leaves, rests on a base with four sirens. This example is similar to another, smaller in size, built by Santucci in 1582 for the King of Spain Philip II and now kept in the library of the Escorial.

The armillary sphere certainly represented in art, as in life, a royalty sprung from knowledge, so much so that the flag of Portugal sees it inserted as part of the national coat of arms since 1816 but, since the fifteenth century, King Manuel I had adopted it as a part of its coat of arms as a symbol of Portuguese explorations and conquests at the beginning of the Modern Age. For the same reason in the treasury of the Portuguese Crown we find the *Scepter of the Armillare*, also known as the Scepter of the United Kingdom of Portugal, Brazil and Algarves. Made of gold, it is part of the Crown Jewels and was created in 1817 for the proclamation of King John VI. Currently, the armillary sphere is back in the limelight as a decorative object in squares or gardens, on the one hand as a symbol of the past, on the other for the unusual geometric shape that intrigues, so much so that today it is offered as a ring that can also be purchased online.

6 Conclusions

The armillary spheres, born in ancient Greece as an astronomical instrument, soon became so important that they were represented in mosaics and paintings over the centuries until the early nineteenth century. In the Renaissance period they were very popular, especially among kings and nobles, so much so that they became not only an object of study, but also real works of art to be collected. We can say that the armillary sphere has never been forgotten, even if in the nineteenth century it was set aside by painters, it continued to be built for study or curiosity. This astronomical instrument, as complex as it is ingenious, has undoubtedly played in the course of history the dual role of mechanical instrument and symbol of knowledge.

References

Aterini, B.: The astrolabe: a mechanism for reading the stars. In: Zhang, B., Ceccarelli, M. (eds.) Explorations in the History and Heritage of Machines and Mechanisms. HMMS, vol. 37, pp. 227–242. Springer, Cham (2019a). https://doi.org/10.1007/978-3-030-03538-9_19

Aterini, B.: Geometric instruments for the orientation and measurement: the astrolabes. In: Advances in Historical Studies (AHS), **8**, 58–78. Springer (2019b). http://www.scirp.org/journal/ahs. https://doi.org/10.4236/ahs.2019.81004

Danti, E.: Trattato dell'uso della Sfera, Giunti, Firenze (1573)

Pulvirenti, E.: Astronomia e arte: la sfera armillare (2014)

De Rose, L.: La sfera armillare di Solunto, in XVIII Seminario di Archeoastronomia, 19–20 marzo, Sestri Ponente, Genova (2016)

(2021). http://www.museocieloeterra.org/qrcode/005

(2021). https://catalogo.museogalileo.it/oggetto/SferaArmillare_n04.html

Aterini, B.: La Prospettiva Parallela. Alinea, Firenze (1995)

Developments of Mechanical Design and Automation

"A Great Body of Machinery": From Robots to Mecha and Back

Nadia Ambrosetti(✉)

University of Milan, Milan, Italy
nadia.ambrosetti@unimi.it

Abstract. The paper reconstructs some episodes of the diffusion of robots in Japan both in literature and in manga and anime production until the Seventies; the popular cultural phenomenon of mecha is put in a relationship with some salient moments of the introduction of industrial robotics in the country.

Keywords: Japan · Engineering · Mecha · Robot · Anime · Manga · Atom Boy · Tetsujin 28 · Grendizer · Mazinger Z · Go Nagai · Gakutensoku · Unimate · Wabot · SCARA

1 At the Beginning

In 1898 the English writer H. G. Wells published probably his most famous novel, "The War of the Worlds" [1], in which he tells the story of a Martian attack empowered by the use of enormous[1] ("higher than many houses") war machines with three legs, called tripods. They not only can walk, but also smash or grasp everything on their passage with their "long, flexible, glittering" tentacles; they can even produce a ringing metallic paced sound. Such perfect and autonomous weapons come from an advanced technology, that is about to annihilate the defenseless humans. The final defeat of the Martians is only due to their immunodeficiency to the bacteria of the Earth.

The tripods can represent the first example of a fictional character which will become very popular in 20th-century Japan: the warrior robot.

2 The First Imagined and Built Robots

The word robot comes from the Czech language and its use spread due to the avantgarde play *R.U.R.* (first set in 1921) by Karel Čapek (1890–1938). Robots were humanlike beings built in a factory as a low-cost human workforce alternative, in order to free humanity from tiring and repetitive activities, as they are programmed to perform tasks

[1] Their impressive dimensions can be compared to a classical ancestor, the gigantic Talos, set by Zeus to protect the island of Crete, by preventing potential attackers, and pirates from landing. Its defeat and destruction are due to the magic of Medea, who succeeds in draining its vital fluid.

M. Ceccarelli and R. López-García (Eds.): HMM 2022, HMMS 40, pp. 345–359, 2022.
https://doi.org/10.1007/978-3-030-98499-1_28

and routines. At the end, this utopian project fails when robots are given the abilities to think and feel emotions for themselves, because they rebel against the weak and lazy human race.

After its translation in 1923, such work had a great impact in Japan until the Thirties and it was a great inspiration for the diffusion of a new genre of robot stories, which go under the name of "Early Showa Robot Literature": just to quote the most famous ones, the short story "Robot" (Jinzo Ningen) by Hatsunosuke Hirabayashi (1892–1932) and "The Robot and the Weight of The Bed" by Sunjugo Naoki (1891–1934)[2]. These are two perfect examples of the general attitude and the growing fears of Japanese people for the introduction of emerging technologies in the landscape of the rapid industrial development Japan was experiencing in those years: robots were represented as latent, but actual threats to life and culture [2], as it also appears in "Android Era", a 1929 drawing by Nihofu Mizushima (1884–1958), showing a traditionally dressed Japanese couple being threatened by a number of robots. As a matter of fact, Japan had already experienced the half-century of industrialization and technological innovation in the Meiji period (1868–1912) [3] and some local companies were already working in the automation field, like the Yaskawa Electronic Distributor Association (now Yaskawa Electric Corporation), founded on July 16, 1915, by Keiichiro Yasukawa (1849–1934).

Another relevant stimulus to discuss on the robot theme came from the silent movie "Metropolis" by the German director Fritz Lang (1927), premiered in Tokyo at the Syoitiku Kinema, Yurakucho on 3 April 1929, just some months before Japan experienced the deepest economic downturn in modern history.

Metropolis is a moral fiction about the threats of the rampant technological transformation of the Twenties and the failure of capitalism to have a social conscience.

The movie proved a wild success but paradoxically encouraged the modernistic Japanese fascination with robotics.

"The beginning of an exceptional time for robots [took place] in Japan" [4]: in 1926 the marine biologist Makoto Nishimura (1883–1956) resigned from Hokkaido Imperial University, and started building a robot in Osaka, without any specific engineering skill; in 1929, his Gakutensoku (learning from the laws of nature), actually the first Japanese robot, was exhibited in the main cities of Japan, China, Korea and Germany [5] in a sort of triumphal tour. The robot was similar to the mechanisms of classic automata, that is partly mechanical (the pegs on the drum allow a basic programming of the sequence of motions), partly pneumatic. Nishimura's concept was far more human-like and therefore more social-acceptable reply to the Westinghouse's Televox, which was a bulky, loosely humanlike, utilitarian device designed by the American engineer Roy J. Wensley to switch telephone calls [6] and known through Japanese press reports on foreign automata [7].

[2] This last story, written in 1931, is set in a far future, and tells of a lead engineer in the Robotics Department at KK Electric Housewares Manufacturing Works, whose plan is to leave a lifelike robot designed to control the faithfulness of his widow, after his death, and to possibly punish the woman. A very similar episode is also present in Homer's Odyssey (VIII, 272–284): Hephaestus creates a trap to weigh whoever lies down in the nuptial bed in his absence, and to trigger a net if the load exceeds the threshold of Aphrodite's weight.

The robot success continued in the Thirties, when Yasutaro Mitsui built his steel humanoid, known mainly from a photograph: it is not clear what actions he could perform, but it seems probable, from the signs of wear on his legs, that he could bow. The electric part of the equipment appears to be a fake.

The literary magazine Shinseinen (New Youth) represented the cultural trend of that time, devoted to contemporary Western narrative.

In 1934 Gajo Sakamoto's Tank Tankuro manga strip (Tanku tankurō) was first published on Chugai Shougyou Shimbun. The protagonist has a spheric body with three holes and a head. Hidden in the holes are objects of all kinds that the reader discovers from time to time, such as weapons or a propeller that allows him to take off like a helicopter. He is also capable of mutating in different forms to fight against his enemy, Kuro Kabuto (Black Helmet).

The first robot warrior had come to (fictional) life.

In 1935 detective story "Jinzoningen jigen" (The Robot Case) by the science fiction novelist Unno Juza (1897–1949), a perfect murder is set: the robot scientist Takeda is killed by its robot, controlled by a love rival, who is given an alibi by the machine [8]. In 1939 a more famous novel by Unno Juza displays another image of robot in "Jinzo ningen efu shi" (Mr. F. the robot): this time the robot (Mr. F), built by the Russian scientist Dr. Ivanov, is taught Japanese by two children of a Japanese merchant, living in Vladivostok. The robot takes the form of one of the children, heads to Japan, and, under control of Dr. Ivanov, destroys a Russian tank hold by the Japanese army. Later it runs berserk and brings destruction everywhere.

The robot not only looks human but can even replace a human being to carry out war actions with superhuman skills. A failure in control system however makes an unpredictable weapon out of it.

A robot to fear had appeared, just when the robot boom was fading because of the war Japan was fighting. The warrior and the threating robots will play a relevant role in the following years.

3 As Far as Your Jets Can Take You

After the long parenthesis (1937–1945) due to the tragic wars (Second Sino-Japanese War and Pacific War), and to cultural censorship during the American occupation from September 1945 to October 1949, manga came to new life and would play a major role in the spread of this type of character.

The terrifying consequences of the atomic bomb also left enduring effects in this genre, as it can be shown by the postwar icon Astro Boy[3], whose original name explicitly refers to the power of the nuclear weapon: Tetsuwan Atomu (Atom Boy or Mighty Atom, literally Iron-Fist Atom), by Tezuka Osamu (1928–1989).

This robot character first appeared in manga at the end of the Fifties, followed by a TV series (1963–1966): a mechanical boy, born on April 7th, 2003, who, despite the negative connotation of its name, is eager to support the world with his considerable

[3] Astro Boy is the most famous Western name of the character; the reason why its lacks references to atomic power can be easily imagined. For its spread in the US, see [9].

skills[4] and superhuman abilities, such as super-sight and super-hearing, because it has been created with cutting edge (for the time) technologies. Its jet-propelled boots allow Astro Boy to fly through the skies at ultrasonic speed. It has a 100,000 HP strength and can fire missiles. It can also have (good) feelings: it goes to school and has the appearance of a nice boy. Atom's nuclear heart would bring help, not death [10].

Atom became rapidly a TV star and an icon of Japan's techno-scientific progress. It was linked to "postwar optimism fueled by an industrial shift from military to consumer technologies for everyday use" [8]. As a matter of fact, from the Sixties, the most famous consumer electronics, home appliances and car manufacturing companies[5], set before the war, had returned to production, supported by the government, in terms of cooperation and protectionist policies. In this landscape, industry was contributing to strengthening national identity. In the Sixties, Astro Boy was inspirational for young roboticists, at the turning point from traditional robots to humanlike ones, designed to help people [8, 11–13].

In 2003, to celebrate Astro Boy birthday, the character returned to TV screens in a 50 episodes comeback series produced by artists influenced by Tezuka; at Robodex 2003 in Yokohama it was also protagonist of a famous handshake with Tezuka's son, Makoto[6].

4 Tetsujin 28

Openly inspired by Astro Boy and under the influence of war experience, another manga artist, Mitsutero Yokohama (1934–2004), published his Tetsujin 28 (Iron Man #28, later Gigantor in the US) in 1956 in the magazine Shōnen [4, 12].

While Astro Boy was apparently like a human boy as for aspect and size, in Yokohama's story, Tetsujin has a huge body[7] and was built as the 28th trial to make a powerful weapon for the Pacific war by the Japanese army. After the end of the war, it comes under the control of twelve-year-old private detective, Shotaro Kaneda, son of the robot developer. Tetsujin is equipped with rockets on its back to fly. It is not the product of an advanced technology, as it uses brute-force to submit its enemies (both humans and robots), and it's not AI-based, rather all its movements are exclusively remote controlled, as his operator guided it from the rear. Therefore, Tetsujin is presented as a mere, value-neutral instrument, that can be used for good or evil reasons, like industrial robots and differently from Atom Boy.

In 1963 the anime series of T-28 first introduced enormous warrior robots and contributed to create an actual genre. Its success continued for decades, at least until the 21st century: in addition to videogames, a live-action film was based upon the series in 2005, and on March 31, 2007, the movie, "Tetsujin 28-go: Hakuchu no Zangetsu" (Tetsujin #28: The Daytime Moon) was released in Japanese cinemas.

[4] Obviously for the time: fluency in fifty languages, and calculation speed of one arbitrarily complex operation per second.

[5] Just to quote the most famous ones and their date of foundation: Mitsubishi (1870), Toshiba (1875), Nissan (1933), Fujitsu (1935), and Toyota (1937).

[6] http://astroboyworld.blogspot.com/2008/04/robodex-2003.html.

[7] A super-robot 20 m tall and weighing 25.08 iron tons, inspired by the US B-29 bombers that flew over the Kobe port, where the artist lived as a boy.

Today people can admire a statue of Tetsujin 28 in all its bulk at Wakamatsu Park, in Kobe (Fig. 1).

Fig. 1. The Tetsujin Statue in Kobe

One manga/anime generation later, a human character would pilot the robot from within a cockpit: mecha were on the go.

5 Mecha

The term mecha is based upon the Japanese abbreviation of the word mechanical; originally it not only refers to machines, such as vehicles (mainly cars and spaceships), weapons, and computers, but also to cyborgs, with both organic and biomechatronic body parts.

The presence of mecha in manga/anime culture can be dated back to 1972, when the manga artist Go Nagai (1945-) launched his Majingā Zetto (Mazinger Z) in the comics magazine Shonen Jump and as an animated series on Fuji TV [14].

The idea of robots controlled by the inside, a kind of exoskeleton or empowering suit, was not new[8], but Nagai brought it to a worldwide success.

The plot pivots again on a lonely schoolboy, Koji Kabuto, a human who will fight against enemies; during an earthquake, he finds in a secret underground laboratory, the 20-meter-tall robot Mazinger Z, forged by his grandfather in the Z-alloy, based upon Japanium, a new element extracted from Mount Fuji mines; before dying, Juzo Kabuto can barely reveal to Koji the use of the robot.

The Earth is immediately in need of his help, as a mad scientist, Dr. Hell, is attacking with his remote-controlled monsters, found during an archaeological expedition in Rhodes. Each episode has a fixed structure: as a new robot threatens the Earth, Mazinger Z comes into action with a precise ritual: it is kept at the Photonic Research Institute near Tokyo, and it is activated by Koji by docking inside the robot's head a special aircraft, called the Pilder. The fight begins, Dr. Hell's robots are destroyed, and the hero brings the robot back at the institute for repair.

The word majin (both "demon" and "magical" person) gives origin to the name Mazinger, meaning that it is neither good nor evil by itself; the pilot, Koji, is in the robot head, and he acts like its brain, by forming a cybernetic link with the machine, and is on the front line exactly as the robot. Such a complementarity is also evident from the fact that, even if Koji, in the cockpit, operates the robot with typical airplane instruments (joystick, buttons, dials, knobs, radar)[9], the pilot suffers physical pain from the damage undergone by the machine when hit by missiles, rays, or shots, thus becoming the technological evolution of the samurai, and humanizing the machine, while still maintaining its mechanical character [16].

Koji is not left alone in his war against Dr. Hell: Mazinger Z is helped by other, less powerful, robots, such as Afurodai Ēsu (Aphrodites A), operated by Sayaka Yumi, the daughter of the director of the Institute, Dr. Yumi, who was an apprentice of Juzo Kabuto, and the comic counterpart, the ramshackle Boss Borot, built out of junks found in an abandoned warehouse, only useful for distracting enemies. Such a character will be present also in Gurēto Majingā (The Great Mazinger), the famous sequel of the Mazinger Z series, begun in 1974[10].

---

[8] Both sci-fi literature (I. Asimov, Runaround, 1942; Robert Heinlein, Starship Troopers, 1959) and manga (Atom Boy, 1954) had told of people fighting on or inside robots or wearing robotic powered armor. The last feature will be present also a later live-action superhero TV series, the Mighty Morphin Power Rangers (1993–1996). The fighters can become powerful superheroes wearing different suits and helmets with visors. In addition to enhanced strength, durability, agility and combat ability, the rangers have Zords (animal-like robots) they can combine into a bigger Megazord.

[9] or using voice commands to activate weapons, like the samurai did in their duels [15].

[10] Kenzo Kabuto, the once thought dead father of Koji Kabuto, raised an orphan, Tetsuya Tsurugi, and created for him a stronger version of Mazinger Z to fight against humanity's new enemy, the Mycenae Empire, led by the Great General of Darkness and his army of Battle Beasts. Tetsuja will also have helpers: Jun Hono, a half Japanese, half African American girl, driving Venus A (Venus Ace), and the aforementioned Boss. Tetsuya arrives just in time to Koji's aid as the Mycenae Battle Beasts overwhelm Mazinger Z. Kenzo then left Japan's defence in the hands of Tetsuya and the Fortress of Science.

UFO Robo Gurendaizā (UFO Robot Grendizer in the US, Goldrake/Goldorak in Europe) appears in 1975 and it is linked to the Mazinger saga by the presence of Koji Kabuto as a minor character, here becomes a helper of the alien protagonist.

A bellicose alien empire, Vega, has enslaved and ravaged the far away, peaceful, advanced planet of Euphor. The prince of Euphor (Duke Fleed), however was able to escape the massacre by subtracting the combat machine. Grendizer can travel through space stowed in its carrier aircraft. His pilot finds refuge on Earth two years before the start of the plot, is then discovered, treated, and adopted by a humanist scientist: Professor Genzō Umon, director of a space research centre located at the foot of the Yatsugatake Mountains. He gives Duke Fleed the identity of Actarus, his son, and conceals Goldorak in an underground base under the center. Externally human in all aspects, Actarus leads an ordinary life and works as a stable boy at the neighboring Shirakaba ranch, co-managed by Rigel and Procyon. When the Earth is attacked by Vega forces, he decides to fight against them, even if he hates war.

There are some notable differences from Nagai's previous robots: the protagonist transforms his clothing into that of a pilot, by means of a kind of nanotechnology operated suit; curiously, the robot is also, at the beginning of the fighting, transported on a spaceship, operated by Actarus, while, at the beginning of the battle, the pilot seat moves into the robot's head, and the robot is released, while the aircraft moves autonomously away from the combat area. While the second difference can be explained as a variant, to make the narration more compelling, the first one links Grendizer with the contemporary manga series Kōtetsu Jīgu (Steel Jeeg).

Its plot is not very innovative: Professor Shiba, a well-known Japanese scientist, during archaeological research in Kyushu, discovers an old bronze bell. Having deciphered the ancient inscriptions on it, he discovers the existence of a technologically advanced people: the ancient Yamatai[11] kingdom, ruled by the evil queen Himika. The Yamatai kingdom had hibernated in the rock waiting to come back to life: only the magical bronze bell would allow to conquer the world. For this reason, the professor decides to hide the bell; during a serious laboratory accident, his 3 years-old son Hiroshi is unfortunately involved. Professor Shiba saves his life by transforming him into a cyborg and miniaturizing the bell in his chest, making him invulnerable. Hiroshi grows up without being aware of all this and becomes a great Formula 1 champion.

When the queen and her helpers, the monsters Haniwa, awaken, Professor Shiba is killed, but thanks to a computer where he had poured all his knowledge, the scientist can explain his son how to transform himself into Jeeg, the steel robot, to defend humankind. Hiroshi, joining the metamorphical gloves given by his father and wearing a radio transmitter pendant, exploits the energy derived from the bronze bell and transforms his body into Jeeg's head; then to complete the robot, he needs the collaboration of the helper Miwa, who throws the rest of the components from the Big Shooter to him.

Two relevant themes, linked to technology, are present. First, the computer where Professor Shiba's knowledge is kept, is not a simple repository: the characters can talk with the professor's avatar displayed on the screen and they receive orders and advice on what to do in real time; it appears as an actual AI implementation.

[11] All names of the evil characters are taken from Japanese tradition [17].

The second theme is the cyborg hero: this time, technology is brought into the human being, not for research purposes, but to save him from a certain death. Therefore, the metamorphosis is very useful to enhance Jeeg's skills and strengthen the link between man and machine.

Still in the context of references to technological innovations, we find another famous mecha: Muteki kōjin Daitān 3 (Invincible Steel Man Daitarn 3). It was first broadcast on Japanese TV in 1978, directed by Yoshiyuki Tomino (1941).

The Meganoids, cyborgs created on Mars by Professor Haran Sozo, escape his control. At the head of them, there is the evil Don Zauser (Don Zauker), a robot with a primitive physical structure and a humanoid brain, which speaks in an unintelligible way; and Koros, his priestess and interpreter, a creepy female cyborg with a glacial charm. The Meganoids want to enslave humanity and turn the best specimens into Meganoids. They have developed a technology which makes the Meganoid Commanders able to transform into Megaborgs, huge combat robots. Haran Banjo, son of Professor Sozo, pilots the gigantic (120 m) solar-powered[12] robot Daitarn 3, whose face is expressive. Like Grendizer, having arrived on Earth from space, it has an external module used to bring it to the battle site.

A systematic analysis of all manga/anime mecha would be too long, but it is worth adding some further examples.

In 1979, Yoshiyuki Tomino and his staff created another epic mecha character: Kidō Senshi Gundam (Mobile Suit Gundam), inspired by the powered armor from the novel Starship Troopers. Even in this case the story beginning is very similar to other animes'. In Universal Century calendar year 0079, a group of rebel colonies of Earth forms the Principality of Zeon and attacks Earth Federation with monstrous Zakus, a giant humanoid weapon. In the chaos, the teen-ager Amuro-Ray finds the RX-78-2 Gundam, the experimental battle robot designed by his scientist father along with its instruction book: so, he decides to use Gundam and defeats the enemy. Amuro, thanks to his mechanical intuition and the need for able bodies of all ages, becomes a pilot and the worst opponent of Zeon [18].

Like Go Nagai's mecha, Gundam is a very tall humanoid robot with samurai-style armor, but it is in fact far more realistic. It is only a mobile suit, cast in a regular metal alloy, without transformation skills, or superhuman power, and can be easily broken by weapons of current technology. As Hornyak [4] writes, the armors are "products of industry, with prototypes and mass production-manufactured commodities that can be repaired or re-placed when damaged"[13] (Fig. 2).

Between 1974 and 1981, new mecha aired on TV screens: some of them share the feature of being driven by three pilots, all with their own skills. Such a characteristic is the result of a different design: the mecha is ready for the battle, only when its parts are properly combined.

The first one, aired in 1974, is Getter Robo (Getter Robot), which introduced the idea of separate Get machines, bond to wild animals, combining to form a more powerful

[12] During the Seventies, a global oil crisis was on the go, culminating in two shocks, in 1973 and 1979.

[13] Their names feature coded alpha-numeric model numbers. "RX-78–2", for example, means that it is a second-generation prototype, rolled out in the year Universal Century 0078.

Fig. 2. The 18-m tall statue of Gundam outside the Tokyo Museum, in Odaiba.

robot, in three different versions, appropriate for distinct conditions and situations. While studying the peculiarities of the mysterious Getter Rays to use them as a source of energy, Professor Saotome realizes that the Earth is in danger. The people of the Dinosaur Kingdom, which have been in hibernation since the Mesozoic era, are awakening and preparing to recapture the planet. Saotome then builds a robot, the Getter Robot, and sets out to find three pilots capable of fully using the 3 different configurations. In the end, Ryoma Nagare (Eagle), Hayato Jin (Jaguar) and Musashi Tomoe (Bear) are chosen and, with them at the helm, the Getter Robot will face the threat from the past.

Mirai Robo Daltanious (Future Robot Daltanious) appeared in 1979. The plot is like the previous ones: in 1995, the Earth is suffering the raids of the Saar Interstellar Empire. However, in a squatter area of the ruined Japan, some orphans leaded by the brave Tate Kento, survive with theft or charity. Once they find the underground base of the hibernated Āru hakase (Doctor Earl), come from the planet Helios in 1945. Kento, Hiiragi Danji and their companions awaken him and unearth his base, causing an immediate attack by the Akron. So, Doctor Earl forces Kento and Danji to drive the Atorausu (Antares) robot and the Ganpa (Gunper) plane respectively, and the two, surprisingly, defeat the enemy on the battlefield. During the Akron's retaliation, Kento and Danji are in serious trouble, but, just as they are about to be defeated, the robotic lion Verarios (Beralios) shows up. At this point, Doctor Earl informs the two boys that the robot and the lion can unite to form the invincible Daltanious, which is 56 m tall and weighs 678 tons, the first combining mecha with an animal component (later present also in the Transformers and in the Power Rangers).[14]

In 1980 Uchū Taitei Goddo Shiguma (Space Emperor God Sigma) is aired in Japan. In 2050, humanity has made tremendous progress in space technology, but is suddenly attacked by a mysterious enemy, the Elder's army, emerging from the future. In fact, in 2300, Elders, being at an overwhelming disadvantage in front of Earth's weapons based upon Trinity energy, will be invaded by mankind. The Elders' purpose in 2050 is to steal this energy power before it can be used against them: in a few words, they want to change history. Earth's defence is entrusted to three robots, originally designed for peaceful purposes, based on Trinity energy: Kuraiou (Sky Thunder King, red coloured), Kaimeiou (Sea King, blue), Rikushinou (Land Earthquake King, yellow). God Sigma is completed by combining the three. At the signal of "Sigma Formation!", a triangular formation is taken, and the Big Wing filled with Trinity energy is summoned and united by the shouts of the three pilots "Trinity Charge!". The total height was initially set at 265 m, but due to problems such as depiction in the play, it was reset to 65 m after the broadcast. Its weight is 1,200t. Although it can be operated by only one main pilot, Toshiya, the output drops if the other two generators cannot be controlled. Its autonomy allows it a round trip from Earth to Jupiter. Each robot is equipped with a system to assess the skills of the driving pilot: Sky Thunder King, while controlling God Sigma, requires advanced manoeuvring skills, and it does not work unless a qualified pilot is operating in the cockpit.

Sengoku majin Goshogun (GoShogun or Macron 1 or Gotriniton) appears in 1981 and the landscape depicted is still the one of a humanity trying to expand its control over the space. Professor Sanada, analysing the fragments of a meteorite that crashed in Siberia, discovers a new form of energy (the Bimler) which makes the teleportation of matter feasible up to 10000 km, and implements its use in the spaceship Goodsander (Good Thunder), and in the large robot GoShogun, equipped inside with three aircrafts: King Arrow, Jack Knight, and Queen Rose. The powerful Dokuga Crime Syndicate intends to steal this technology but finds the courageous opposition of Professor Sanada. Before dying, he transfers his conscience to a large computer called Father, equipped with an autonomous artificial intelligence. The control of the base where GoShogun is housed passes to Captain Sabarath, while the three aircrafts and the robot are piloted by three young people, respectively Shingo Hojo, Killy Gagley, and Remy Shimada.[15] Meanwhile, Kenta Sanada, a 10-year-old boy, son of the dead professor, unites to the team: highly inventive, but initially a slacker and troublemaker, is tutored by OVA, an educational robot designed for him. Over time he learns how to communicate with machines and robots, and with spirits of the earth's ecosystems as well, and finally becomes the living embodiment of Bimler energy, merging with GoShogun.

[14] The vocal command for union is "Cross in" and the cross-shaped emblem represents the coalescence. Initially, Daltanious is powered with sigma energy, but later with hyperspace energy to enhance output for attacks. By releasing the energy, it is possible to convert and store energy such as enemy attacks and explosions, and the capacity is said to be about one star.

[15] The robot becomes ready for battle after the landing of the three aircrafts inside its body, and then it is controlled by Shingo. Thanks to the "Three for one" mode, King, Jack, and Queen can combine and form a smaller multipart robot, Tri, piloted by Remy.

In this excursus, it appears evident the influence of Western literature and cinema, which introduces the idea of the robot as something more than a toy, however complex and mechanically evolved, like in Japanese tradition.

Japanese culture creatively reworks these ideas: in fact, it is possible to glimpse a sort of evolutionary line. Astro Boy is a friendlier robot, with the resemblance of a nice child, powered by nuclear energy for peaceful purposes. A simple and naive idea of robot underlies. Tetsujin 28 is completely different: inexpressive, enormous, remote controlled, it reflects early approaches to industrial automation. Mecha appear as a mix of the two: dimensions remain huge, but the pilot drives the robot from the inside and shares danger and pain with it; in Jeeg's case, the connection between the two becomes even closer, in the form of the cyborg. Then, in Gundam, the mecha, though reduced to an exoskeleton, loses its uniqueness, and becomes an industrial product, with a serial number, that overcomes the uncanny valley. Finally, we meet robots that allow and enhance the exploitation of extremely powerful energy sources (another key theme of the early 1980s technology), along with the use of multipurpose parts that can recombine to form different robots, suitable to meet different needs. Later, during the Eighties, in the anime scenario, we assist to a sort of modeling: mecha evolve to the classical "deus ex machina" idea they literally embody, or, we can also say, to the archetypical epic hero, who duels with all his strength (and technologies) against the invader champion, to defend his own homeland.

6 In the Meanwhile, the Japanese Industry

Japanese culture has undoubtedly played a significant role in fostering a positive attitude towards robots, which helps to explain how Japan has always been fond of robotics[16] and is nowadays among the countries with the highest number of robots employed in industry and in service tasks. Some well-known aspects could be quoted. First, the automata called karakuri (literally mechanism), originally made between the 17th and 19th centuries, which performed scenes taken from mythology or ordinary life in festivals across Japan and were extremely appreciated by the audience. Automata, and later robots, became therefore very popular and familiar to the Japanese people, but they remained largely useless. Moreover, the animistic religious component leads to admitting the possibility that objects, even of common use, can perform actions. Again, the superiority of spirit over matter, as in some of the mentioned mecha, takes the form of the survival of a scientist in a computer or of the bond between human and machine, or in the absence of characterization of the evil robots, which are simply called (and are) monsters, lacking human guide.

Anime is a mass cultural phenomenon, the representation of the imaginary of millions of Japanese. One might wonder if the mecha are mere science fiction or if they are somehow linked to actual technology. As a matter of fact, Japanese robotics was just taking her first steps and coping with problems similar to mecha's ones.

In the period between the end of WW2 and the Cold War, the rate of Japan's economic growth was so high that the era is also known as Japanese economic miracle [19]. In

[16] In 1971 the first national robot association was established in Japan.

1961, the industrial robotic pioneers Joseph Engelberger (1925–2015) and George Devol (1912–2011) introduced the first industrial robot, the Unimate[17] 1900, in Chicago. The robot consisted of a robotic arm designed for typical factory activities (transporting objects from assembly line, painting and arc welding) and was first installed in a General Motors plant in the same year [19, 20]. This could considerably enhance facility safety and productivity by shifting skilled workforce to more critical operations[18] and would have a great influence on Japanese industry.

In 1968, while Japan was managing to achieve a double-digit growth and to become the second country in the world GDP ranking, the manufacturing sector (30% of that GDP) experienced heavy labor shortages: "automation and streamlining processes in automobile manufacturing sites where production capacity was increased in order to catch up with the effects of rapid motorization, had become urgent" [21].

So, the following year, Kawasaki Heavy Industries signed a technical agreement with Engelberger and Devol, to introduce the robot in Japanese industry, as it happened with the first industrial robot of Japan, the Kawasaki-Unimate 2000 [21, 22], and in 1970, when Hitachi developed the HIVIP Mk.1, the world's first vision-based fully automatic intelligent robot assembling objects from plan drawings [23]; such an achievement makes automation of mecha storage, assembling and repair systems not only possible but even realistic.

In the same year, Professor Masahiro Mori (1927) of the University of Tokyo, published "The Uncanny Valley" [24], a robotics milestone. He analyzes the results of his experiment and speculates that sensation of familiarity and pleasantness, generated by human-like robots and automata, can increase as their anthropomorphism rises; this growth is however limited to a point (the uncanny valley), where the extreme realism arouses repulsion and disturbance. Industrial robots are very well accepted, due to their lack of similarity to humans: they can perform manufacturing tasks with only one arm, and are based on functionality, displaying no affinity with humans. Humanoid robots' familiarity to humans increases as well as the resemblance to people: at some point, an excess of likeness arouses fear and robots are seen as corpses or zombies. This is particularly true for robots with a whole body, but not for parts, such as prosthetic hands, which appear to be less disturbing. This study paved the way for the creation of humanoid robots, and its findings can also explain the roughly humanlike but not anthropomorphic bodies of mecha.

Professor Ichiro Kato (1925–1994), a pioneer in the field of autonomous robots, and his mechanical-engineering students constructed the world's first life-sized humanoid robot in 1973 at Waseda University in Tokyo. It was built in two versions: Wabot 1 (WAseda roBOT), a heavy metal robot equipped with basic vision and speech synthesis in Japanese; the 10-fingered robot Wabot 2, created between 1980 and 1984, became famous by reading a musical score, playing tunes on an organ with hands and feet, and accompanying a singer at the Tsukuba Science Exposition of 1985 [12].

In 1978, Professor Hiroshi Makino supported by five Japanese companies developed the robot prototype SCARA (Selective Compliance Assembly Robot Arm), which allowed to solve the peg-in-hole alignment problem, a very typical problem of the

[17] Literally "Universal Mate"; Patent Number 2,988,237.

[18] Displaying the same separation between control activities and task execution as in Tetsujin.

assembly process [25], as shown in some anime, when, for instance, Jeeg succeeds (or fails) in joining the robot components. In the following year, Nachi developed the first electromotor-driven robots, replacing the previous era of hydraulic drive: like in mecha stories, energy was a relevant theme.

During the 1970s, major industrial manufacturers, such as Fanuc[19] [26] or Yaskawa, decided to take advantage of microelectronics technology, power electronic technology and computer technology, enhancing their CNC systems. Starting from the 1980s, robots appeared as 'machines of the future', and engineers supported industrial progress to foster manufacturing competitiveness. The advent of powerful and cheaper technologies produced a step change in industrial robotic: the use of precision force sensors and lasers allowed robots to check manufacturing components along assembly lines; they were provided with a human-like sight and touch and could have new interactions with the environment [27, 28].

As a result, robots evolved from plain mechanical devices designed, and programmed to perform repetitive tasks (possibly for mere entertainment), built in a small number of pieces, to mass-market sophisticated machines, able to perform (or avoid) multistep actions on demand, to support humans.

7 Conclusions

It is quite trivial to state that anime/manga and robotics are immediately identified as expressions of the Japanese culture on a global scale [29, 30]. They both could be studied from the point of view of history, tradition, religion, psychology, economy, industry, technology, and politics, or as a celebration of national pride, but a complete covering of the topic is beyond the scope of this paper.

After this review of the history of Japanese mecha and robots, it appears that the design of mecha has been repeatedly influenced (or, better, inspired) by technological and above all robotic evolution, albeit in a non-systematic way.

We can however point out some remarks. First, both mecha and robots are tools built to take care of humans: mecha defend mankind against threats, by increasing, under different respects, the strength, and skills of the protagonists, while industrial robots lighten human work in factories and make it safer.

Then, in some cases, the stories of the mecha anticipate some technologies or features that will be fully developed and implemented only later (artificial intelligence, vocal recognition, use of solar energy, nanotechnologies, computer vision, modularity).[20]

Finally, the scientist and the engineer, supported by their lab staff, occupy a prominent place, as they are the expression of science and technology in a system: it would be

[19] It is suggestive that the geography of many mecha stories, the slopes of Mount Fuji, is the place where the headquarters of FANUC, one of the main industrial automation and robotics industries, are located.

[20] It is also worth mentioning here the anime series Doraemon, a cat robot come from the 22nd century to help a clumsy and bullied boy, Nobita, in his daily life: a first example of robot companion, first aired in 1973. Still in the last decade, Japanese robotics companies, supported by the Japan's Robot Strategy of the government (former Abe administration), planned to create humanoid and "animaloid" robots to provide company and perform domestic tasks.

impossible to have Mazinger without a Professor Kabuto and a Fortress of the Science, and we wouldn't have a Wabot, without a Professor Kato and the School of Science & Engineering of the Waseda University.

References

1. Wells, H.G., The War of the Worlds. William Heinemann (1898)
2. Nakamura, M.: Marking Bodily Differences: Mechanized Bodies in Hirabayashi Hatsunosuke's 'Robot'and Early Showa Robot Literature. Japan Forum. Taylor & Francis (2007)
3. Jacobowitz, S.: Between men androids, and robots: assaying mechanical man in meiji literature and visual culture. Mechademia: Second Arc **9**, 44–60 (2014)
4. Hornyak, T.N.: Loving the machine: the art and science of Japanese robots. Kodansha International (2006)
5. Frumer, Y.: The short, strange life of the first friendly robot. IEEE Spectrum (2020)
6. Schaut, S.: Robots of Westinghouse, 1924-today. Mansfield Memorial Museum (2006)
7. Inoue, H.: Nippon Robotto Sōseiki, 1920–1938 (The Genesis of Japanese Robots, 1920–1938). NTT Publishing (1993)
8. Ito, K.: Robots, A-bombs, and war: cultural meanings of science and technology in japan around World War II. In: Filling the Hole in the Nuclear Future: Art and Popular Culture Respond to the Bomb, pp. 63–97. Lexington Books, Plymouth (2010)
9. Ladd, F., Deneroff, H.: Astro Boy and Anime Come to the Americas: An Insider's View of the Birth of a Pop Culture Phenomenon. Jefferson (2008)
10. Power, N.O.: God of Comics: Osamu Tezuka and the Creation of Post-World War II Manga. University Press of Mississippi (2009)
11. Schodt, F.L.: Manga! Manga! The World of Japanese Comics. Kodansha (1983)
12. Schodt, F.L.: Inside the Robot Kingdom: Japan, Mechatronics, and the Coming Robotopia. Kodansha (1990)
13. Schodt, F.: Astro Boy Essays: Osamu Tezuka, Mighty Atom, and the Manga/Anime Revolution. Stone Bridge Press (2007)
14. Patten, F.: Watching Anime, Reading Manga; 25 Years of Essays and Reviews. Stone Bridge Press (2007)
15. AA.VV. Mangamania. 20 anni di Giappone in Italia. IF - Immagini & fumetti, vol. 5 (1983)
16. Aumont, J.: Un'estetica industriale. (A proposito di Goldrake). In: L'immagine al plurale: serialità e ripetizione nel cinema e nella televisione, pp. 103–126. Marsilio Editori (1984)
17. Pellitteri, M.: Mazinga nostalgia. Storia, valori e linguaggi della Goldrake-generation. Coniglio Editore (2008)
18. Heldreth, L.G.: In search of the ultimate weapon: the fighting machine in science fiction novels and films. In: The Mechanical God; Machines in Science Fiction. Greenwood Press (1982)
19. Beckley, M., Horiuchi, Y., Miller, J.M.: America's role in the making of Japan's economic miracle. J. East Asian Stud. **18**, 1–21 (2018)
20. Menzel, P., D'Aluisio, F.: Robo Sapiens: Evolution of a New Species. MIT Press (2000)
21. Hashimoto, Y.: Half a century of Kawasaki Robotics: The Kawasaki Robot Story 1968–2018. Kawasaki Robotics (2018)
22. Wallén, J.: The history of the industrial robot. Linköpings universitet (2008)
23. Koga, M., Hosoda, Y., Moriya, T.: Humanoid robots. Hitachi Rev. **58**(4), 151–156 (2009)
24. Mori, M.: The uncanny valley. Energy **7**(4), 33–35 (1970)
25. Makino, H.: Development of the SCARA. J. Robot. Mechatron. **26**(1), 5–8 (2014)

26. Shibata, T., Yano, M., Kodama, F.: Empirical analysis of evolution of product architecture: Fanuc numerical controllers from 1962 to 1997. Res. Policy **34**(1), 13–31 (2005)
27. Behnke, S.: Humanoid robots-from fiction to reality? Künstl. Intell. **22**(4), 5–9 (2008)
28. Funk, M., Irrgang, B.: Robotics in Germany and Japan. Peter Lang (2014)
29. Kovacic, M.: The making of national robot history in Japan: monozukuri, enculturation and cultural lineage of robots. Crit. Asian Stud. **50**(4), 572–590 (2018)
30. Sabanović, S.: Inventing Japan's 'robotics culture': the repeated assembly of science, technology, and culture in social robotics. Soc. Stud. Sci. **44**(3), 342–367 (2014)

A Historical Development of LARM Finger Design Shape

Emanuela Chiavoni[1(✉)], Michela Cigola[2], Alekos Diacodimitri[1], and Marco Ceccarelli[3]

[1] University of Rome Sapienza, Rome, Italy
{emanuela.chiavoni,alekos.diacodimitri}@uniroma1.it
[2] University of Cassino and Southern Lazio, Cassino, Italy
cigola@unicas.it
[3] University of Rome Tor Vergata, Rome, Italy
marco.ceccarelli@uniroma2.it

Abstract. The paper presents a historical evolution of the design of the LARM hand finger in terms of aesthetics and protective cover using shape design considerations and procedures. The development of the mechanical design of the fingers is outlined up to reaching a specific attention to the design of the envelope of artificial fingers with aesthetic and functional requirements, leading to an investigated and proposed solution also in the form of a shape patent. The historical development of this approach is based on a multidisciplinary collaboration integrating the design of the mechanisms and functionality of the grasp mechanics with aspects of shape design and graphics design to reach an anthropological aesthetic solution that has elaborated with a gradual development.

Keywords: LARM hand · Drawing · Functional design · 3D modelling · CAD drawing · Colours test

1 Introduction

In Robotics, in the last decade, aspects of man-machine interactions have increasingly been taken into consideration also for design and functional solutions. An important aspect that is increasingly of interest and importance especially in service robots is the aspect of acceptance by potential users, configuring in problems for ergonomics, aesthetics and understanding of the functioning of the robotic structure for an adequate awareness of the operation and benefits, with satisfaction also on a psychological level.

In particular, especially in anthropomorphic robotic structures with the possibility of prosthetic applications and/or support for the human body, ergonomics and aesthetics are of particular importance and need, requiring specific analyzes and solutions even from an aesthetic point of view.

Once the functional and structural characteristics of the basic model of the prosthesis had been established it was possible to move on to the redesign phase when the key structure of the previously defined object gradually takes on its final form during

M. Ceccarelli and R. López-García (Eds.): HMM 2022, HMMS 40, pp. 360–371, 2022.
https://doi.org/10.1007/978-3-030-98499-1_29

steps controlled by drawing instruments. This operation is necessary not only for purely aesthetical reasons, in order to formally characterize the object and define it based on a unitary visual 'style', but also functionally, so as to develop forms and lines that can fluidly accompany the movements of the prosthesis. The enhanced aesthetics of the prosthesis, not necessarily based on the total mimesis of forms and colors, can actively help the user to psychologically accept it.

The first step in the redesign of the prosthesis involved studying the visual references that could be used to define both the formal and aesthetic characteristics of the mechanical finger and the principles governing its design and style. A decision was taken to look for these references in the world of visual imagery associated with science-fiction entertainment products. Over the years productions such as films, TV series, comics, illustrations, and animated films have inputted into the creation and stratification of what is, to all intents and purposes, a codified and structured visual language. "It is not heretical to say that the present produces the future, and that we are directly involved in this process. The idea of a future we have imagined and helped to produce has contributed to the birth of the field of future studies, dedicated amongst other things to the study of the socio-technical relationship between 'action, knowledge and responsibility' in future societies" [12]. Fiction and reality pursue and influence each other, pre-empting trends, imagining possibilities, and materializing visions. Science fiction doesn't necessarily have to be realistic, but simply plausible: to narrate a reality in which the spectator can become immersed without losing his incredulity.

2 A Short History of LARM Hand

The LARM hand was developed starting from experiences in grasp mechanics with grippers and human manipulations with solutions progressively obtained for the fingers operated by a specially designed mechanism (Ceccarelli 2012).

The development of the LARM hand is part of an activity that is widespread both in research laboratories and in practical commercial applications for the development of gripping systems that can be implemented and adapted on robotic systems as well as humanoids. The vastness of the sector is demonstrated by a vastness of designs and prototypes that adopt the most varied solutions both in terms of mechanisms and aesthetic forms as reported in a vast literature of which the volume (Carbone 2013) can be considered an emblematic example. In particular, several solutions similar to the LARM hand can be recognized in the aspects both in the mechanical design and in the external aesthetics of the structure of each finger with a body that encloses the actuating systems of the phalanges as well as in the mechanisms that allow the implementation and function of the finger in grasping actions. The peculiarity and the innovative aspects of the LARM hand compared to those similar solutions are recognizable in the anthropomorphic functionality with a component structure that lends itself well to an aesthetic adaptation as proposed in this paper.

Figure 1 summarizes the evolution of the design solutions of the four versions produced by Professor Ceccarelli's team in the years 1998–2006 (Carbone et al. 2002, Carbone and Ceccarelli 2008) with a last version represented in Fig. 1a, having the characteristics of a three-fingered hand with a user-oriented functionality thanks to the structure with industrial PLC and low-cost components.

In particular, the body of the finger is structured in such a way that the mechanism that activates the three phalanges with an anthropomorphic motion, that is, the three phalanges move in a coordinated manner with an anthropomorphic motion. The mechanism always remains inside the body of the finger with an anthropomorphic size and shape with different scales depending on the version of the prototype built over the years.

Figure 1b shows the first functional prototype of the finger from 1998 in which the external structure of the finger is highlighted mainly for the function of containing the actuation mechanism and supporting structure for the gripping actions that are obtained by the finger on an object against a palm with a circular structure equipped with a compliance by means of a spring (Carbone et al. 2002).

Figure 1c shows the prototype of version 2 developed in 2003 in which the size of the finger is related to a human solution, while still having a mechanical structure aimed above all at containing the actuation mechanism and at a mechanical compactness without particular attention to the anthropomorphic shape of the phalanges as indicated by the fact that the connecting screws and the axles of the relative movements have been designed with adjustment heads external to the surfaces of the finger.

Finally, the last finger version of the 2006 prototype represented in Fig. 1d represents the finger with an external structure of the phalanges with a first attempt to represent the roundness of the finger while continuing to use metal materials without particular attention to the aesthetics of the operation.

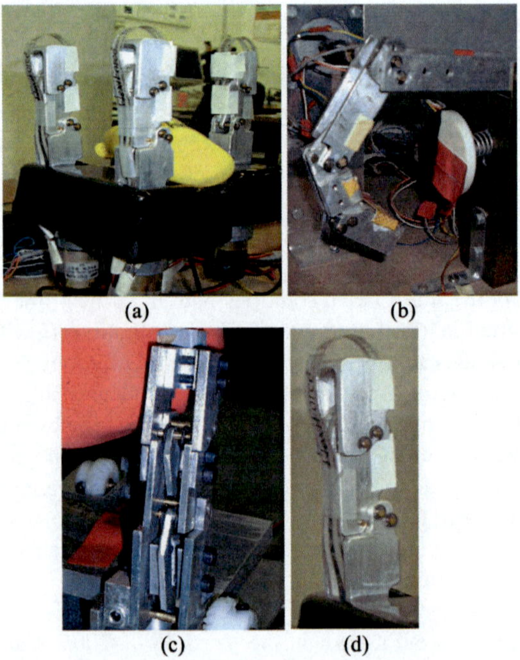

Fig. 1. LARM hand: a) the full LARM hand 4 in 2006, b) first finger design in 2000, c) Human sized prototype in 2003, d) low-cost user-oriented prototype in 2006

The several versions of the LARM Hand have been developed with the fingers, all characterized by the fact that they have been designed with an external compact mechanical structure to contain the actuation mechanism in its anthropomorphic motion functionality and to produce the necessary grasping force upon contact of one or more phalanxes with an object in contact.

Furthermore, the possibility of using different materials was also subsequently investigated by using 3D printing manufacturing technology with plastic-type materials and also with colors that can be attractive for the use and implementation of the finger in various applications, not only practical but also ludic ones. Figure 2 represents a last example of this constructive evolution, showing a prototype built by a 3D printer with PLA material using the red color with the aim of enhancing more than anything else the relative movement of the phalanges rather than the shape itself.

Fig. 2. Finger design for LARM hand in red PLA by 3D printing in 2019

3 Multidisciplinary Interactions

In the 1990s at the University of Cassino started the multidisciplinary interaction between DART (Laboratory of Documentation, Analysis, Survey of Architecture and Territory) and LARM (Laboratory of Robotics and Mechatronics). Two lines of research have been developed: the history of the representation of machines and mechanisms and robotics applied to study characteristics and issues concerning the analysis and conservation of built heritage.

The very official beginning was in 1993 with an article by M. Ceccarelli and M. Cigola for an international Congress on "Design representation from Antiquity to XVIII century" in Sapienza University of Rome [1].

After this contribution, the study on history of the representation of machines and mechanisms developed in some others researches on Marcus Vitruvius Pollio [2, 3], Gaspard Monge [4], and Descriptive Geometry and its applications for Mechanism Science [5].

In 2010, Cassino team (LARM and DART) was part of the "ThinkMotion Project. Digital Mechanism and Gear Library goes European" Grant Agreement No 250485 with Technische Universitaet Ilmenau (IUT) Germany, Universidad del País Vasco (UBC) Spain, Universitatea Politehnica DIN Timisoara (UPT) Romania, Rheinisch-Westfälische Technische Hochschule Aachen (RWTH) Germany, Institut Français de Mécanique Avancee (IFMA) France. The ThinkMOTION project as a part of European leads to the world's largest, free access digital library of content in the field of motion systems, which is essential for all technical domains.

The second research line was on Robotics for Cultural heritage. Our first case was the survey of the pavement of Montecassino Abbey [6, 7]. The interaction between the staff at DART, experts in architectural surveying, and the team at LARM, experts(ìxpets) in robotized systems developed a hexapod robot with a large platform for carrying survey devices. So, Cassino team planned and built some prototypes of robotics devices for Cultural heritage during years [8].

In 2015 Cassino won a grant of 250.00 Euros for the research project "Heritage. Robotic services for BBCC", PI prof. M. Ceccarelli. HeritageBot was funded by FILAS Project – Regione Lazio carried out by the Department of Economics of the University of Cassino and Southern Lazio and involving researchers from DART laboratories, LARM, IMPRENDILAB and FINLAB. Aim of the project is to realize a prototype of remote-controlled mobile platform called HeritageBot able to constitute itself as an element of innovation for its ability to acquire digital data independently and in places otherwise inaccessible [9].

The collaboration between DART and LARM Laboratories, despite their apparent differences in terms of fields of research, for years have found time and motivations for opinion exchanges and cultural contacts, as well as research topics for genuine and fertile collaborations and multidisciplinary interactions.

4 Shape Design Evolution

4.1 Requirements

During their careers many artists have found themselves having to deal with technological prefiguration. Some of them have even made it the main topic of their artistic research; over the years the forms, lines and surfaces developed by these artists have contributed to the development of the visual imagery used in several sci-fi subgenres. Several major authors have been selected for the project presented here. The first is Syd Mead (1933–2019), probably one of the most important and best-known concept designers in the world of science fiction. He has a long history of involvement in this genre and has signed off on many of the more famous sci-fi films: from Blade Runner (Ridley Scott,

1982), to Aliens (James Cameron, 1986), and Elysium (Neill Blomkamp, 2016). "By now there are millions, maybe even tens of millions of Earth's residents, who have formed an idea of the future as visualized by Syd Mead. They program computers and fix cars, teach school and write stories, go to movies and eat hamburgers. What they have in common is a view of the world, and man's place in it, in which stylized objects loaded with advanced technology are sought-after harbingers of things to come. They have come to understand, through Syd's imagination, and the films based on it, that among the endless possibilities of a technically mature culture, that man and machine can, and perhaps should, share a seamless, symbiotic environment" [14].

Although most of Mead's creations involved the design of transportation vehicles, stage sets and props, rather than robotics design, he has been chosen as the visual reference for the redesign of prostheses not only due to the easily recognizable forms that characterize his drawings (Fig. 3), but above all for the traditional approach he used in his work. The idea was to adopt his method of focusing on drawing by hand, the use of graphic signs, and line and color as elements of a research aimed at enhancing his intuitive approach to spatially understand the object, without using 3D modelling and CAD drawing. Mead's work has a unique trait thanks to his use of ink, colored alcohol-based markers, and the pencil and ruler. So, a decision was taken to maintain his work philosophy, i.e., the use of drawing by hand as the main design tool.

A review of possible references regarding mechanical limbs in the film industry led to comparisons with other classic sci-fi films, including the mechanical skeleton used in Terminator (James Cameron, 1984); this structure is a perfect imitation of the shape of the bones of a finger using simple joints that allow the phalanges to rotate. The Terminator model is a good example of the concept of mimesis of the forms and functions of the mechanisms of the human body: the forearm displays all the basic characteristics of a real human skeleton: the radius/ulna articulation simplified by a spherical metal joint

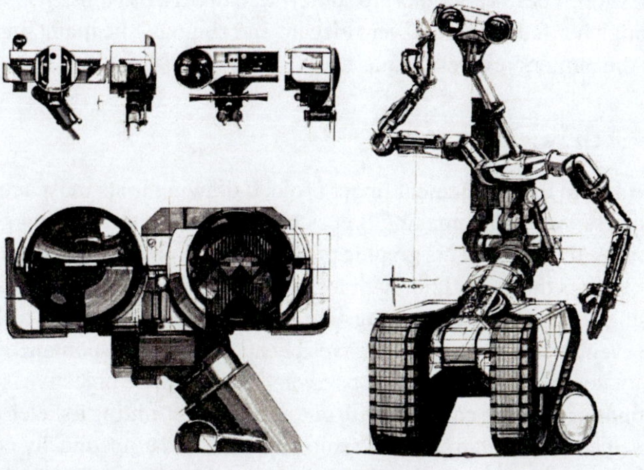

Fig. 3. Syd Mead, studies for the design of the robot protagonist of the film Short Circuit (John Badham, 1986) "Extract from Mead S., and Hodgetts C.: The Movie Art of Syd Mead: Visual Futurist. Titan Books, London (2017)".

that allows pronation; the metal wires that pass precisely in the center of the wrist like tendons; and the pistons that coincide with the internal bones of the hand. And yet the object, albeit in its mimesis, has obvious visual characteristics, making it different to a real human skeleton; this is undoubtedly due to the story's narrative requirements, but it also functions to affirm its identity and state that its existence is something unique and independent; it uses the skeleton's salient characteristics as a reference and reprocesses them using a visual identity mode.

Similar statements can apply to the other references that were analyzed, for example Luke Skywalker's robotic hand in Star Wars: The Force Awakens (2015, J. J. Abrams) or the robotic hand with a reflecting sphere by the Japanese comics author Katsuhiro Otomo, which even if it has only four fingers, exhibits the same proportional ratios between big "metal bones" and joints. In all the above cases, the joint immediately displays its presence, function, and the allowed degree of movement.

Another reference, chosen in relation to the problem of how to display the artificial robotic nature of the prosthesis, is provided by the concept designer Jock [15], author of studies for the robotic protagonist of Ex-Machina (2014, Alex Garland). The procedure Jock used to create the character called Ava involved endowing this nevertheless graceful and organically shaped female figure with visible mechanical features. Clearly the artificial being, although based on a human form, is searching for a very recognizable visual identity. The designer explained his intentions during an interview "when he shared six of his concept images for 'Ex Machina' with Business Insider, taking us through his process for the creation of Ava. To get Ava right, Garland and Jock spent a lot of time talking about what the movie's AI should not look like. "We went through so many variations in the early stages of designing Ava. I started out with a figure much closer to human, with internal lights and a few subtle oddities in the joints, but Alex really pushed for a far more robotic look; and of course, his instincts were right. To present something that is entirely mechanical, and then ask how the viewer feels about it, that's a really interesting question. These variations are somewhere in between those two initial ideas" [16]. So, although her female forms were elegant and sinuous, she maintains her robotic nature, as per the authors' express request.

4.2 Analogical Drawing and Color

Drawing is the key to the mechanical finger project; drawing in its most authentic form, as a tool to express every idea, making it possible to establish the ensuing graphic steps required to finalize the object. This graphic process is more creative compared to the next phases, and is always the most fluid, with less rigidities, thus allowing – at any moment - rethinks, changes, or even backtracking. The communicative power of a sketch is still very valuable even in a digital world, perhaps because in a more spontaneous manner it suggests, stimulates, and influences the observer. Obviously, the objective is not to make the different kinds of drawing compete with one another, e.g., pitting a sketch against a 3D model of the same object, but to adopt a rapid approach and empathically convey ideas. At the start of any design process, drawing by hand enables the transcription of thoughts and ideas in the quickest, most direct, and conscious manner. Drawings are precious tools that explicitly render a trace, always linking it to what one wishes to achieve. The designer uses sketches that are often very concise to enter into contact with the object, but he has to add intangible values such as imagination and creative and emotional aspects to its form, type, and function. A sketch is a schematic drawing of an object which is usually executed freehand; it leads to the creation of a series of images which, defined by conventions and precise graphic symbols, recall the object to be produced, thus making it possible to perform checks between the design process and their material implementation. To all intents and purposes, it is a methodologically ordered ensemble required to clearly foresee what these future objects will look like. The unbridled freedom of materializing one's own personal visions on a sheet of paper using simple tools is unparalleled as regards the author's involvement with the object he or she is creating, above all the control and verification of its parts and characteristics. The roots of these principles lie in the Italian tradition of drawing freehand, but they persist and resist in time due to their expressive ease that permits immediate comprehension. Since the digital revolution has led to changes in the way we produce images and 'classically' transmit them, the act of drawing instead sparks curiosity and raises questions in the minds of those who practice it; it enables analyses and critical interpretations, maintaining the power incorporated in the faithful representation of the object. The procedure that is triggered makes it possible to envisage the results in advance; it is extremely economical and sustainable and allows drawers to control the level and quality of the object they wish to produce. During the initial phase numerous chromatic tests were performed after defining the form of the mechanical finger: several representations were made using bright colors in order to instill greater visual-perceptive wellbeing. Other images were produced in chromed colors, satin-chrome colors, and with metallic effects so as to give them a more modern and futuristic appeal. No test was performed using skin-colored tones because the goal was to make the limb standout. In every design process color is an extremely important element due to the ways in which it can be used: color is not a characteristic of the material, but when light is reflected off the material, we see it as color. In addition, color changes depending on the intensity of natural and artificial light. Psychic wellbeing always interacts with the design of artificial limbs, so its color must be studied and assessed on a case-by-case basis for each object.

We supplement the text with a list of specific technical terms which are the requisites that have been considered for the design of robots.

- Manageability
- Waterproofing
- Capacity of tactile recognition
- Easy to grip
- Remote control
- Technical reliability
- Easy to operate even by non-experts

The analyzes concerning the colorimetric values (data verification, tables, Ral codes, RGB codes, colorimetric tests, color sensors, etc.…) are still being processed.

5 Form and Study of the Prototype

The 3D prototype acted as the basic reference for the entire redesign of the prosthesis. The prototype, printed in 3D, has four simple forms: parallelepipeds joined together. During this phase their form depends only on the function and movement that the joints have to allow. Drawing was used to analyze the two main conformations (open finger and closed finger) in order to understand: the reciprocal position between the four elements of the finger; the action they perform during movement; and the position of the screws, which were left in place. Drawing cannot modify these parts of the object; they are the rigid base used during the task at hand. They are fixed points or red lines that act as reference for the drawing (Fig. 4).

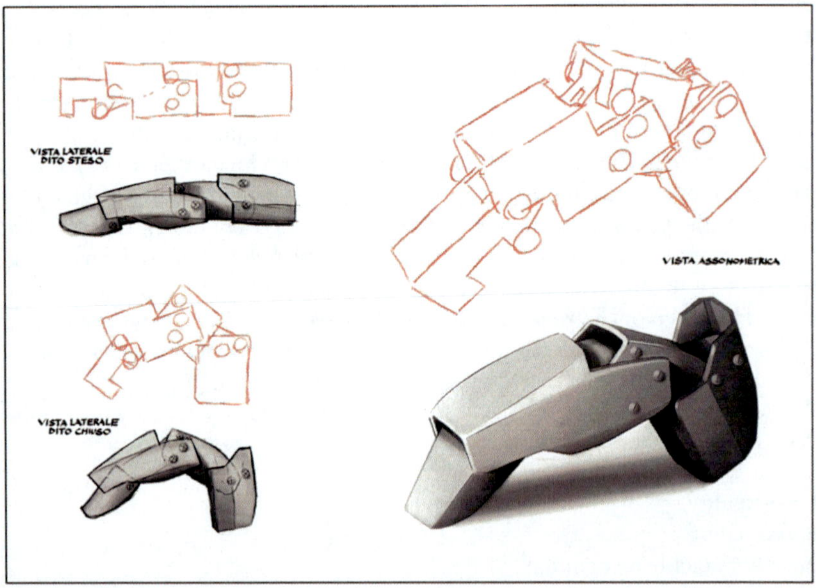

Fig. 4. Redesign of the prosthesis, comparison between the prototype and the final model in its open and closed conformation and in an axonometric view that helps to better understand the spatial conformation of the surfaces (Drawing: A. Diacodimitri)

The final design tends to be more fluid and organic, rounding off all the form's sharp corners. Fluid forms always enhance movement, pinpointing the lines of force on the surface. These organic forms are obviously inspired by the real forms of a human finger, but they still let their artificial nature standout. In the case of the chosen references this is particularly obvious close to the joints between the individual elements, where the prosthesis reveals its internal mechanisms. Two-dimensional drawing is the only work tool used during this phase. This enhances the overall impact of the form rather than the effective three-dimensional development of the object, which is not openly clarified (Fig. 5).

The pictorial chiaroscuro achieved using digital drawing techniques suggests the movement of the surfaces and the depth of the volumes; however, it does not go into the details of its real conformation. The next step is to convert these ideas into a real three-dimensional model: the drawings were imported into a three-dimensional modeler (Fig. 6) and used as an initial reference with which to create the generating curves of each element. During this phase several discrepancies due to the less precise freehand drawing are eliminated: when the drawing appears ambiguous, the designer corrects the model and proceeds to clean up the forms. The ensuing model can thus be used to project the two-dimensional final drawings in double orthogonal projection as front and side views and longitudinal and transversal sections.

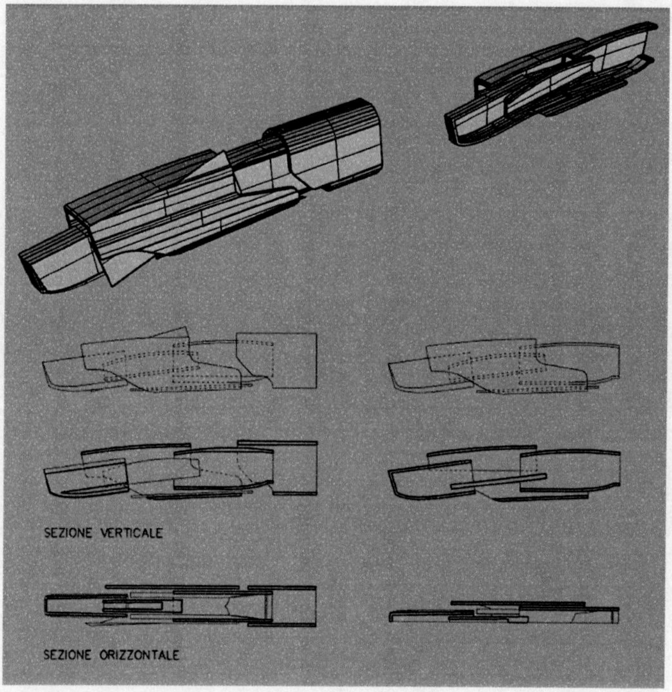

Fig. 5. Redesign of the prosthesis, executive phase. Elaboration of the three-dimensional model in Rhino environment and two-dimensional executive drawings (lateral views and sections) (Drawing: A. Diacodimitri)

Fig. 6. Colours test of shell for artificial fingers with aesthetic purposes with specific solution each phalanx (Drawing: A. Diacodimitri)

6 Conclusions

The shell to protect the structure of anthropomorphic artificial fingers can be used in numerous ways. This is also true regarding the production of an aesthetics suited to enhance the anthropomorphic functions and acceptance by human users in man-machine interaction applications. The shell model can also be employed in the use of artificial hands as prostheses and in medical systems. This model is described with reference to a predefined form of production. Other forms of production are obviously possible based on the same inventive idea. (Ceccarelli/notes to the patent).

In conclusion this mechanical design of fingers is the ultimate development of the LARM experience in interaction between human and robotic devices. This experience started several years ago with projects by professor Marco Ceccarelli and professor Giuseppe Carbone. These projects aimed to develop support for people with walking problems and evolved in the direction of replacing human body parts with robotic devices, opening up new and bold horizons not only in technology and mechanics.

References

1. Carbone, G., Civitillo, R., Ceccarelli, M.: Design and test of an articulated mechanism for a 1-DOF anthropomorphic finger. In: ASME Design Engineering Technical Conferences, DECT 2002, Montreal, CD Proceedings, Paper MECH-34300 (2002)
2. Carbone, G., Ceccarelli, M.: Design of LARM hand: problems and solutions. J. Control Eng. Appl. Inform. **10**(2), 39–46 (2008)
3. Carbone, G. (ed.): Grasping in Robotics, Series: Mechanisms and Machine Science, vol. 10. Springer, Dordrecht (2013)
4. Ceccarelli, M., Cigola, M.: Evoluzione della rappresentazione grafica nel progetto dei meccanismi. In: Il disegno di progetto dalle origini a tutto il XVIII secolo. Storia e sviluppi, Roma, Sapienza, p. 25 (1993)

5. Cigola, M., Ceccarelli, M.: Marcus Vitruvius Pollio. In: Ceccarelli, M. (ed.) Distinguished Figures in Mechanism and Machine Science. HMMS, vol. 26, pp. 309–346. Springer, Dordrecht (2014). https://doi.org/10.1007/978-94-017-8947-9_15
6. Cigola, M., Ceccarelli, M.: Machine designs and drawings in Renaissance Editions of de Architectura by Marcus Vitruvius Pollio. In: Sorge, F., Genchi, G. (eds.) Essays on the History of Mechanical Engineering. HMMS, vol. 31, pp. 291–307. Springer, Cham (2016). https://doi.org/10.1007/978-3-319-22680-4_17
7. Cigola, M., Ceccarelli, M.: Influencia de la obra de Gaspard Monge en el desarrollo de la ingenieria mecanica italiana en el siglo XVI. In: Actas de IV Congreso Internacional de Patrimonio e Historia de la Ingenieria, Las Palmas de Gran Canaria, pp. 143–156 (2009)
8. Cigola, M. (ed.): Distinguished figures in Descriptive Geometry and its applications for Mechanism Science. HMMS, vol. 30. Springer, Cham (2016). https://doi.org/10.1007/978-3-319-20197-9
9. Cigola, M., Ceccarelli, M.: A robot application for analysis, survey and conservation of historical architectures. In: Balaguer, C., Abderrahin, M. (eds.) Robotics and Automation in Construction, pp. 328–354. I-Tech Educational and Publishing, Wien (2008)
10. Ceccarelli, M., Cigola, M.: Service robots for restoration of goods of cultural heritage. In: Ceccarelli, M. (ed.) Service Robots and Robotics: Design and Application, pp. 213–228. IGI Global, Hershey (2012)
11. Ceccarelli, M., Blanco-Moreno, F., Roig, P., Carbone, G., Cigola, M., Regidor, J.L.: A robotic solution for restoration of fresco paintings. Int. J. Adv. Rob. Syst. **12**(160), 1–14 (2015)
12. Ceccarelli, M., Cigola, M., Recinto, G. (eds.): New Activities For Cultural Heritage. Springer, Cham (2017). https://doi.org/10.1007/978-3-319-67026-3
13. Spaziante, L.: Immaginare il futuro prossimo: costruire mondi attraverso la fantascienza audiovisiva. Rivista di estetica **71** (2019). https://doi.org/10.4000/estetica.5506. http://journals.openedition.org/estetica/5506. Accessed 12 Nov 2021
14. Schwarz, J.O., Liebl, F.: Cultural products and their implications for business models: why science fiction needs socio-cultural fiction. Futures (2013). https://doi.org/10.1016/j.futures.2013.03.006
15. Mead, S., Hodgetts, C.: The Movie Art of Syd Mead: Visual Futurist. Titan Books, London (2017)
16. Jock personal page. http://www.4twenty.co.uk. Accessed 03 Nov 2021
17. Here's the concept art that inspired the robot from the year's best sci-fi movie. https://www.businessinsider.com.au/ex-machina-ava-concept-art-2015-4. Accessed 03 Nov 2021

Evolution History of Non-circular Gears

Olga V. Egorova[1](✉), Nikolay N. Barbashov[2], and Roman M. Kiselev[2]

[1] V.I. Vernadsky Crimean Federal University, Simferopol, Russia
tmm-olgaegorova@yandex.ru
[2] Bauman Moscow State Technical University, Moscow, Russia
kiselevrmbmstu@gmail.com

Abstract. Recently variable-ratio gears are increasingly being used to drive the modern machines. For this purpose cams and non-circular gears are often used. In the past, the production of non-circular gears was a very complicated process, which explains their relatively limited use in practice. In the 20th century, Professor F. Litvin played a great role in the development of theory and manufacture of non-circular gears. Modern technologies of the 21st century make the manufacturing relatively cheap and easy, using high-precision machine-tools, 3D prototyping and spraying. The paper considers the history of non-circular gears evolution and application.

Keywords: Non-circular gear · History of machine and mechanism science · Variable gear ratio · Gear-cutting tool

1 Introduction

Non-circular gears (NCG) are designed for certain applications. While a regular (circle shaped) gear is designed to transmit torque to another engaged member of transmission with the least amount of noise and wear and the highest efficiency, a non-circular gear's primary goal could be to get a variable gear ratio, axle displacement oscillations, and other important properties and characteristics.

They were not used widely in the past because of the high prices for their production. Today, in the 21st century, due to the modern technologies and use of computer-controlled machine-tools as well as new mathematical models for calculations of the non-circular shape, the NCG have started actively to replace mechanisms with cams. Specially designed equipment, rapid prototyping and electron discharge machining allowed to establish non-circular gear manufacturing by duplicating or rolling. The enhanced potential in non-circular gear design and production explains the growing interest in this type of gears.

The original version of this chapter was revised: References and figure caption have been updated.The correction to this chapter is available at https://doi.org/10.1007/978-3-030-98499-1_40

M. Ceccarelli and R. López-García (Eds.): HMM 2022, HMMS 40, pp. 372–381, 2022.
https://doi.org/10.1007/978-3-030-98499-1_30

2 The Earlier Application of Non-circular Gears

2.1 Dondi's Astrarium

Giovanni de Dondi (1330–1388) was one of the first, who used NCG while building his astronomical device, which simulated the solar system and is known as the Astrarium (Fig. 1). Also, he authored the manuscript «Tractatus Astrarii» documenting, with text and drawings, the design and manufacturing phases of a very sophisticated clock. In particular, the manuscript contains one of the earliest documented procedures for the design of non-circular gears" [1].

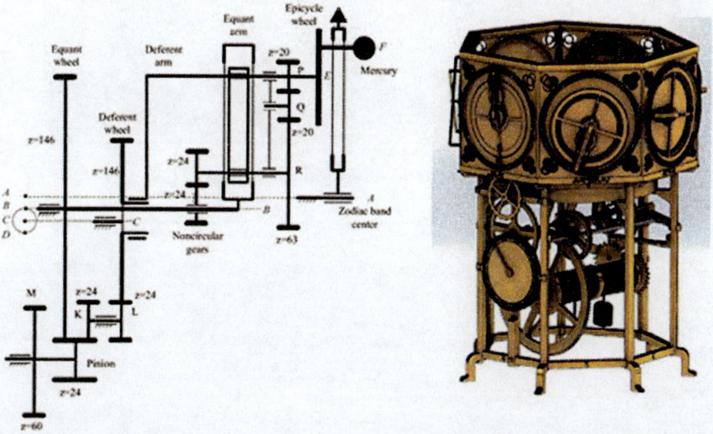

Fig. 1. Schematic diagram [23] and a general view of the Astrarium (drawing courtesy of Laura Guida [24])

2.2 Leonardo da Vinci and Non-circular Gears

Reference to non-circular gears could be found in the works and sketches of Leonardo da Vinci (1452–1519) [2], who supposed to create a clockwork mechanism and elliptical shaped gears with a variable transmission ratio were ideal for this purpose (Fig. 2).

2.3 Redtenbacher's Non-circular Gear Mechanisms

Evolution of NCG in the 19th century was connected with Ferdinand Jakob Redtenbacher (1809–1863), one of the most prominent mechanics of the 19th century. He manufactured several models of non-circular gear mechanism to demonstrate them during the "Mechanics" courses he taught (Fig. 3) [7].

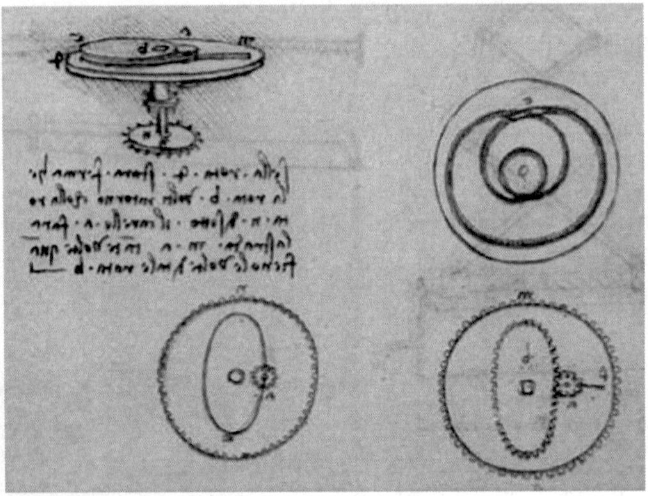

Fig. 2. Leonardo da Vinci's drawings [2]

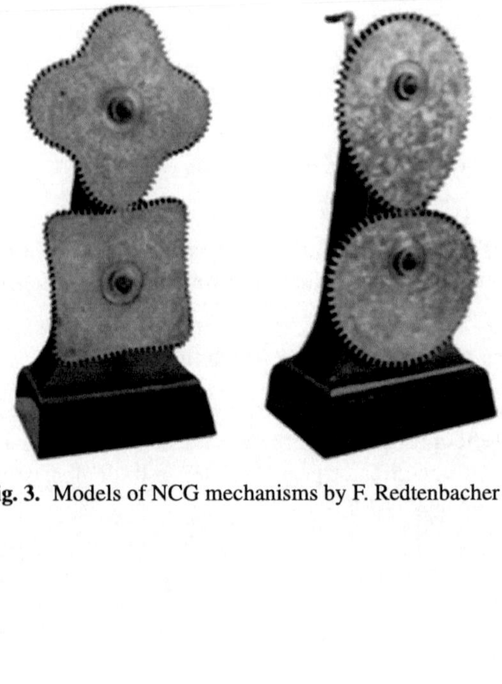

Fig. 3. Models of NCG mechanisms by F. Redtenbacher [3]

2.4 Reuleaux's Models

Later on, Franz Reuleaux (1829–1905) developed his teacher's (Redtenbacher) ideas and found some new applications for non-circular gears (Fig. 4) in industry. Thus, he proposed to use them in textile machinery, mechanical presses and high- torque hydraulic motors as an alternative to cam mechanisms [4].

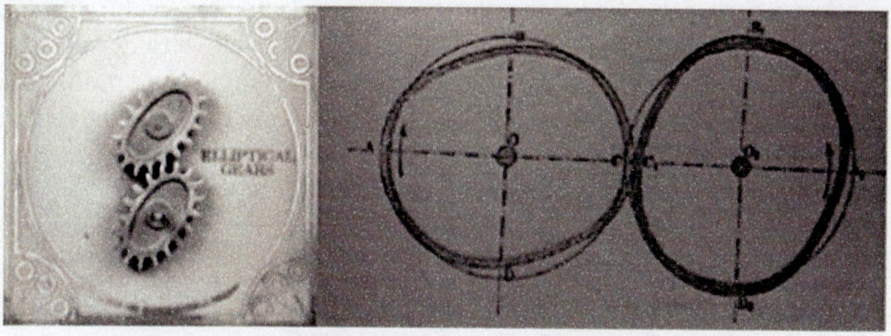

Fig. 4. Reuleaux's models of NCG mechanism [4]

2.5 "Principles of Mechanism" by S.W. Robinson

The book "Principles of mechanism" by S.W. Robinson, published in 1903 (Fig. 5), shows in detail the main principles of NCG mechanism design and operation [5]. An entire chapter of the book is devoted to the profile and ratio calculations as well as application, with examples and photographs of various shaped non-circular gears used in Mechanical Engineering.

(a) (b)

Fig. 5. A hardcover a) and a page b) of "Principles of mechanism" book by S.W. Robinson [5]

3 Non-circular Gears in the 20–21st Centuries

3.1 Ivan Ivanovich Artobolevsky (1905–1977)

Descriptions of several mechanisms, composed of non-circular gears, can be found in the works of the Soviet (Russia) Professor Ivan I. Artobolevsky [9]. Basically, he showed the principle of operation of these mechanisms, giving a detailed description of their structure and properties, without touching on the problems associated with their use and manufacture. Artobolevsky considered NCG as an alternative to gears with a non-circular profile in order to show in the training course that the involute profile is not the only one, but is only the most successful for the current level of technology, but can be displaced in the future.

3.2 Faydor Lvovich Litvin (1914–2017)

In the 20th century, Professor Faydor Litvin played a great role in the development of NCG theory and manufacture [6, 7]. In his book "Non-circular Gears: Design and Generation" [8] he represented the extension of the modern theory of gearing applied to the design and manufacture of the main types of non-circular gears: conventional and modified elliptical gears, eccentric gears, oval gears, gears with lobes, and twisted gears as well as updated theoretical descriptions of the methods of generation of non-circular gears by enveloping methods similar to those applied to the generation of circular gears. He also offers new developments intended to extend the application of non-circular gears for output speed variation and generation of functions.

However, Faydor Litvin understood that the main problem of developing and using NCG arose when cutting teeth on non-circular wheels. When using the run-in method, it is required to provide nonlinear dependencies between the movements of the tool and the gear, which led to the need to create a special gear-cutting machine, and a significant increase in the cost of manufacturing. Nevertheless, similar work was carried out. In the works of F.L. Litvin [6, 8] showed several machines that allow to obtain non-circular wheels, but the author noted that the existing methods of cutting gears need improvement for serial or mass production.

When using the run-in method, a gear rack was usually used. The task was to carry out such a movement of the tool, in which it was rolled with its centroid along the centroid of the gear being cut. For this purpose, an annular or rack-and-pinion milling cutter was used, which is a series of trapezoidal disk cutters, the axial distance between which should have been equal to the pitch between the teeth. However, the production of ring cutters was associated with great difficulties, as the exact value of the pitch of the teeth of the cutter was difficult to withstand. The cutting process turned out to be quite productive only in the manufacture of wheels of a small module. If the cutting of the teeth required several passes along the depth and length of the teeth, the cutting process was periodically interrupted to return the workpiece to its initial position. Also, this machine did not allow cutting oblique NCG.

Faydor Litvin recommended abandoning ring cutters and using conventional worm cutters, as well as developing machines that differ from existing mechanisms that carry out the required connection between the movement of the tool and the gear being cut.

In his works, he suggested using a device that allows you to engage an exemplary NCG with a rail. When cutting the teeth, the model and the cut gears, located on a common mandrel, took part in rotation around their axis, as well as in longitudinal and transverse movements. The tool rail moved only in the direction of the axis of rotation, performing a slotting motion. Thus, it was possible to realize continuous contact of the rail and the NCG.

Also, Faydor Litvin described a machine for cutting NCG with a chisel. Such a machine most effectively performed NCG with small centroid sizes (Fig. 6). The point O1 of the pantograph coincides with the center of rotation of the worm sector, which is the driver. The point O2 is the center of rotation of the wheel being cut. The point O3 is the center of rotation of the copier outlined by a curve, which is a transformed curve with respect to the centroid of the wheel being cut. The transformed curve is obtained by multiplying the radius of the centroid by the value m. When the driver rotates, the copier runs around a fixed circle, the diameter of which is m times larger than the diameter of the initial circle of the used chisel. In the process of cutting, the chisel makes a chiseling movement without having a rotational movement.

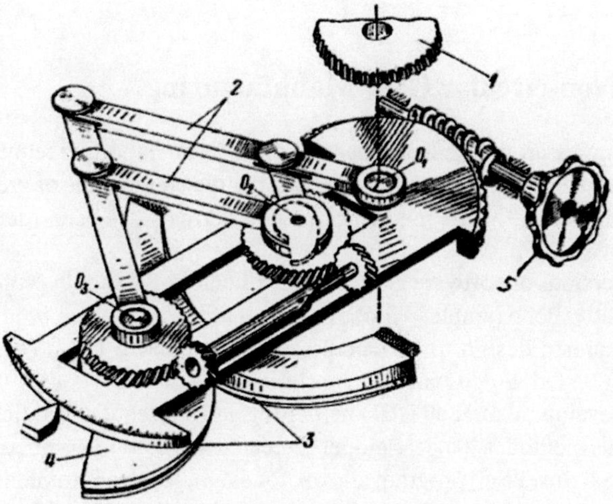

Fig. 6. NCG cutting device [8]

4 Collection of NCG Mechanisms in Moscow

In Russia in Moscow at the Bauman Moscow State Technical University the unique collection [10] of five models of NCG mechanisms is stored (Fig. 7). Two of them were made in the workshop of Alexander Clair and purchased in the 19th century. One model was developed on the results of the dissertation work of the Soviet (Russia) Professor Karl Sergeevich Tarkhanov and mainly devoted to the methods of NCG manufacturing. The models are used during "Theory of Mechanisms and Machines" course to demonstrate students the possibilities of no-circular gears. However, all five models were detailed described in the paper "Non-circular Gears: Innovative Manufacturing Technologies" [11].

Fig. 7. Mechanisms from the BMSTU collection [5]

5 Modern Non-circular Gear Manufacturing

For non-circular gear profiles, ease of manufacturing is an important requirement. In this connection, a more efficient method of determining the coordinates of profile points has been proposed in [12, 13]. The method of profile normals has been used to determine the profiles of gear teeth.

The latest versions of software allow the coordinates of the tooth profile to be calculated automatically. For example, «Globas Group» offers a software product that allows the entire mechanism design to be calculated and simulated [14]. The results of the calculations can be fed directly into the machine tool software.

One more developing method is 3D modelling and subsequent printing of gears on a 3D printer. In this method, a 3D gear model is created in a computer-aided design (CAD) system. "Autodesk Inventor" program allows, for example, an approximate 3D model to be created (Fig. 8), by taking a profile from an existing gear [11]. "MathLab" program allows to describe the gear profile mathematically and create an accurate 3D model of a non-circular gear with any profile shape [15]. The advantages of both methods are their high accuracy and ease of manufacturing. In addition, modern software packages allow not only to create an accurate 3D model and to calculate the strength of even the most complex structures, taking into account their features by the FEM or other methods [16, 17].

6 Modern Designs with Non-circular Gears

Some designs with non-circular gears have been used in mechanical engineering, but they have not been widely used, primarily due to the complexity of profile calculating and manufacturing [18–21].

Fig. 8. 3D model with non-circular gears.

One application of NCG was the conveyor belt drive. The conveyor belt was driven by a Maltese cross and made an intermittent movement. The use of non- cylindrical gears can significantly reduce angular acceleration without reducing the angular velocity of the main shaft. This reduces the load on structural elements, reduces dynamic effects that lead to fatigue stresses and reduction of carrying capacity. This solution increases the reliability of the conveyor and increases the service life of its components.

Many crank-slide mechanisms need to move the slider at a constant speed, so the crank must be designed to rotate at a variable angular velocity. This can be achieved effectively by the use of NCG placed in the crank drive. These can be agricultural machine cutting units, piston pumps, oscillating conveyor drives, and etc. Application of NCG can improve both: the reliability of the crank-slide mechanism and its performance.

In some cases, non-circular gears are used in automatic lathes for slow rotation of the camshaft during working operations and fast rotation during idle strokes. Sometimes they are used in printing machines, in textile machines for obtaining fabrics with a certain pattern, in silk-winding machines for obtaining a barrel- shaped bobbin, and etc.

Also, a machine-tool for milling keys has been developed in which the rotation of the crank mechanism was communicated from NCG in order to feed at an approximately constant speed.

NCG are widely used in hydraulic machines. Rotary machines with improved characteristics based on hydraulic machines are gaining great popularity. A distinctive feature of such hydraulic machines is the absence of support reactions with certain combinations of geometric parameters, which makes it possible to significantly simplify the design of support units, mechanical seals and the system as a whole, and the geometric shapes of the parts and their mutual arrangement make it possible to create compact devices. In addition, the design of the new hydraulic machine such that it can easily be connected in series based on a single shaft. Switching of high and low pressure mains is carried out without the use of special devices [22]. Often, the use of NCG in the design of hydraulic systems and pumps makes it possible to reduce the dimensions of the system and increase its quality.

But the most important advantage of NCG is the ability to synthesize variators with any law of changing the gear ratio. When designing, as a rule, they rely on already existing variators using gear wheels of an evolvent profile, cam mechanisms and other mechanisms, however, the use of NCG allows you to synthesize a car according to any given law of motion of input and output links, while NCG often make the car more compact, allow you to work at high speeds, in addition, in many works it has been shown that NCG, due to a larger contact area, have a greater fatigue life than other analogues. For example, in contrast to cam-type mechanisms, NCG allow high loads to be transmitted and are highly reliable. They also do not require special fixtures or springs to maintain contact continuity in the upper pair, resulting in a simplified design and increased manufacturability.

7 Conclusions

1. The main problem, namely the complexity of NCG manufacturing, which has existed for centuries, has been solved. Modern technologies make it relatively cheap and easy using high-precision machine-tools, 3D prototyping and spraying, while modern laser technology provides the necessary conditions for controlling the wear.
2. Non-circular gears have a longer service life and can handle heavier loads than comparable mechanisms. They will replace cams and could also reduce the risk of accidents, as jamming is much less common with NCG than with cams.
3. Non-circular gears make it possible to reproduce any, even the most complex law of motion. In addition, the NCG can operate at much higher speeds than their counterparts, thus increasing machine productivity.
4. NCG models give students an insight into a little-known area of Mechanics, allowing them to observe in person the movement of a non-regular gear, which greatly expands the horizons of future engineers and increases their interest in the study of engineering disciplines.
5. For some applications, planetary gearboxes with NCG are more efficient than planetary gearboxes with spur gears and could even provide a solution that cannot be achieved with spur gears.
6. NCG enable the development of more compact and efficient variable-speed drive, whose gear ratio can be derived naturally from the NCG geometry. Thus, the gear ratio change will be smoother and more shockless, which is a significant advantage compared to current variable-speed drives without NCG.
7. Studies have shown that NCG allow for the greatest range of gear ratio change, which also facilitates the development of new devices with large amplitudes of output link motion.

Acknowledgments. The authors express their gratitude to the leadership of the Bauman Moscow State Technical University and TMM Department in particular for their support.

References

1. Zhytomyr, S.V.: Ancient astronomy and Orphism, Janus-K (2001)

2. Moon, F.C.: Curves of constant breadth. In: The Machines of Leonardo Da Vinci and Franz Reuleaux: Kinematics of Machines from the Renaissance to the 20th Century. Springer, Dordrecht (2007)
3. Wauer, J., Moon, F.C., Mauersberger, K.: Ferdinand Redtenbacher (1809–1863): pioneer in scientific machine engineering. Mech. Mach. Theor. **44**, 1607–1626 (2009)
4. Pickover, C.A.: Reuleaux triangle. In: The Math Book: From Pythagoras to the 57th Dimension, 250 Milestones in the History of Mathematics. Sterling, New York/London (2009)
5. Robinson, S.W.: Principles of Mechanism a Treatise on the Modification of Motion by Means of the Elementary Combinations of Mechanism, or of the Parts of MacHines 1838–1910, BiblioBazaar (2010)
6. Litvin, F.L.: The main issues of designing non-circular wheels, collection Lonitomash №6 «Theory and calculation of gears» Mashgiz (1949)
7. Litvin, F.L.: Non-circular wheels. Mashgiz (1956)
8. Litvin, F.L., Fuentes-Aznar, A., Gonzalez-Perez, I., Hayasaka, K.: Noncircular Gears: Design and Generation. University Press, Cambridge (2009)
9. Artobolevsky, I.I.: Mechanisms in modern technology. Lenand (2019)
10. Tarabarin, V.B.: Collection of models of BMSTU mechanisms, Moscow (2019)
11. Egorova, O.V., Barbashov, N.N., Abdullina, L.R., Kiselev, R.M.: Non-circular gears: innovative manufacturing technologies. In: Quaglia, G., Gasparetto, A., Petuya, V., Carbone, G. (eds.) Proceedings of I4SDG Workshop 2021, vol. 108, pp. 586–594. Springer, Cham (2022). https://doi.org/10.1007/978-3-030-87383-7_6
12. An, I.K., Belyaev, A.E.: Cutting non-circular gears. In: Proceedings of the International Conference on Theory and Practice of Gears (1998)
13. An, I.K., Belyaev A.E.: Involute tooth profiles of non-circular wheels. In: Proceedings of the II Interuniversity Branch Scientific and Technical Conference (1999)
14. Zhytomyr, M.R.: Specialized machines and automatic machines of the "Linotype" Mashgiz (1949)
15. Popov, V.V.: Mathematical model of the dynamics of a straight-toothed involute gear (2019)
16. Popov, V.V., Sorokin, F.D.: Modelling of gears for nonlinear dynamics analysis. In: IOP Conference Series: Materials Science and Engineering (2020)
17. Sorokin, F., Vakhlyarsky, D., Gouskov, A.: High rise of ring resonator frequency split due to combination of two harmonics of density defect. Appl. Math. Model. **103**, 376–387 (2021)
18. Padalko, A.P., Padalko, N.A., Gear drive with noncircular gear. J. Theor. Mech. Mach. (2013)
19. Kireev, S.O., Ershov, Yu.V., Padalko, N.A.: Calculation of the parameters of the teeth for the manufacture of oval gears, Technical sciences (2010)
20. Sobolev, A.N., Nekrasov, A.Ya., Arbuzov, M.O.: Modeling mechanical transmissions with non-circular gears. J. Bull. BMSTU "Stankin" (2017)
21. Prikhodko, A.A., Smelyagin, A.I.: Creation and research of planetary mechanisms of intermittent motion with elliptical gear wheels. J. Bull. BMSTU Ser. Mech. Eng. (2019)
22. An, I.K.: Synthesis, geometric and strength calculations of planetary mechanisms with non-circular gears of rotary hydraulic machines (2001)
23. Addomine, M., Figliolini, G., Pennestrì, E.: A landmark in the history of non-circular gears design: The mechanical masterpiece of Dondi's astrarium. Mech. Mach. Theory **122** (2018). https://doi.org/10.1016/j.mechmachtheory.2017.12.027
24. Guida, L.: Virtual prototyping of Dondi's astrarium (in Italian), Bachelor Thesis, University of Rome Tor Vergata, Italy (2017)

Dynamic Analysis of a Semi-automatic Telegraph Key

Cosimo Fonte and Marco Cocconcelli(✉)

Department of Sciences and Methods for Engineering,
University of Modena and Reggio Emilia,
Via G. Amendola 2, Reggio Emilia, Italy
{Cosimo.fonte,Marco.cocconcelli}@unimore.it

Abstract. The first telegraph key model was invented in 1846 by Alfred Vai, a close collaborator of Samuel F.B. Morse. Since then the telegraph key has changed several designs from the mechanical semi-automatic or *bug* to automatic systems, representing for years the main means of communication as well as the first means of electrical communication invented. Today telegraphy replaced by the telephone and the Internet, however, these devices are still used by about 3 million Amateur radio operators. In this paper, a dynamic analysis of a precise telegraph key model has done. The model is the *Vibroplex Original Deluxe* which is a semi-automatic model made by Vibroplex Company Inc., a historical company manufacturer of these devices, founded by Horace Greeley Martin (1873–1937). The feature of this device is to automatically generate points (in Morse code sense) by means of an oscillating mass. The validated model will be useful to test different setting parameters and how to manage them to get comfort for the operator.

Keywords: Multi-body system · Dynamic · Morse-key · Semi-automatic key

1 Introduction

In the 19-th century the national and international communications have experienced a period of enormous expansion thanks to the born of the Morse code and the patent of specific devices called *keys*, for the generation and transmission of messages. At the beginning, telegraphers used the traditional Morse telegraph key invented by Samuel Morse and Alfred Vail in the 1840s. The standard Morse key (also known as *straight key*) had a spring-loaded movable lever, suspended between two pivot points, with a hard rubber knob which the operator grasped between the thumb and first two fingers to operate. A small up-and-down motion was used to open and close a circuit. The output is a sequence of binary signal, point (or dits) and lines (or dashes), that is transmitted through a cable or wireless (after the invention of the radio). The mechanical semi-automatic key (Fig. 1), also known as *bug*, was born in the United States in 1902 by the

M. Ceccarelli and R. López-García (Eds.): HMM 2022, HMMS 40, pp. 382–394, 2022.
https://doi.org/10.1007/978-3-030-98499-1_31

inventor Horace Greeley Martin (1873–1937) as a solution to a common problem among telegraphists that used the straight key, the so-called "glass-arm" (the same Martin suffered this illness): epicondylitis notoriously widespread among the telegraphists due to the long hours of work, the semi-automatic key allowed to automatically generate the points thus limiting the operator's effort. The first model developed by Martin took the name of Autoplex [1] and was an electromechanical system powered by dry batteries. This model did not find much diffusion since the telegraph companies considered its cost excessive. Only some companies decided to reimburse the batteries to the telegraphers, who still had to buy the device at their own expense. Another limit was given by not finding a standard electromagnets coil impedance that was good for the electromagnets made by Martin and those of other emerging companies that produced similar devices. To overcome the limitations of cost, use of batteries and electrical compatibility, the manufacturers went towards a fully mechanical solution, a solution found by Martin in 1906 with the invention of the Vibroplex [2], which allowed the automatic generation of points with a fully mechanical system. Due to the success of the bug so far, there are various historical sources and operating manuals on it. In particular, the amateur radio operator Fabio Bonucci (IK0IXI) has realized probably the most complete collection about Vibroplex [3], putting together years of history and products made by Horace Greeley Martin and his historical Vibroplex Company Inc. The paper is divided into 5 sections. Section 1 presents and details the semi-automatic key. Section 2 presents the multi-body model of the bug and the main assumptions made. Section 3 is devoted to the validation of the model with experimental data taken on a real bug key. In Sect. 4 the multi-body model is used to test different setup configurations and the corresponding characteristics of the dits generated. Section 5 present the conclusions of this paper.

1.1 Semi-automatic Key

Figure 1 shows the drawing of the semi-automatic key as reported in the patent of Horace Martin [4] (commercially known as Vibroplex®Original Deluxe). With reference to the numbers in the figure, the key is divided in two parts: the main lever (4) rotating with respect of revolute joint (7). A second lever (16) carries one (33) or two (34) weights which can be shifted along the lever. The two levers are connected by a flexible spring (15). The movement of the key is driven by the operator by means of the lever of double knob (40), which can rotates with respect to hinge (7). Pushing the round knob (32), the lever rotates clockwise and the contact between (44) and (45) produces a line in the Morse code (*dash* contact). Pushing the elliptical knob (31), the lever rotates counterclockwise, moving also the main lever (4). The impact between the lever and the front stop screw (10) generates the oscillation of the second lever (16). A second contact spring (19) bends touching the contact screw (20). The elastic force generated by the contact spring moves back the lever keeping alive the oscillation an producing a series of *dit* in the short time of contact between (19) and (20). Springs (13) and (43) provide the contrast forces necessary to return the levers to their original position. Screw (9) acts as a stop of the main leverage in the clockwise direction, while elements (36) and (35) act as a damper to arrest the lever (16) once the operator releases the knob (also known as deadener).

The system takes the name of "mechanical semi-automatic key" since the dashes (lines) are manually generated by the operator, while the dits (points) are automatically generated by the dynamics of the leverage. The frequency of these points and the length of the signal can be modified by changing some parameters such as the distance of the mass. The main setup parameters responsible for the frequency of the dits are mainly three:

- Position of the front stop screw (10) (FSS): it defines the **impact distance** between the main lever (4) and FSS (10). This distance determines the initial conditions of the oscillating masses (id. all the parts supported by the flexible spring (15)).
- Position of the dit contact adjusting screw (20) (DCAS): it defines the **contact dits distance** between the spring (19) and DCAS (20). This distance determines the counteracting elastic force of the spring.
- Position of the weights (33–34) with respect to the FSS (20): it defines the **weight distance** (WD) responsible of the oscillating frequency and consequently of the speed of dits generation. In this paper only one weight will be considered.

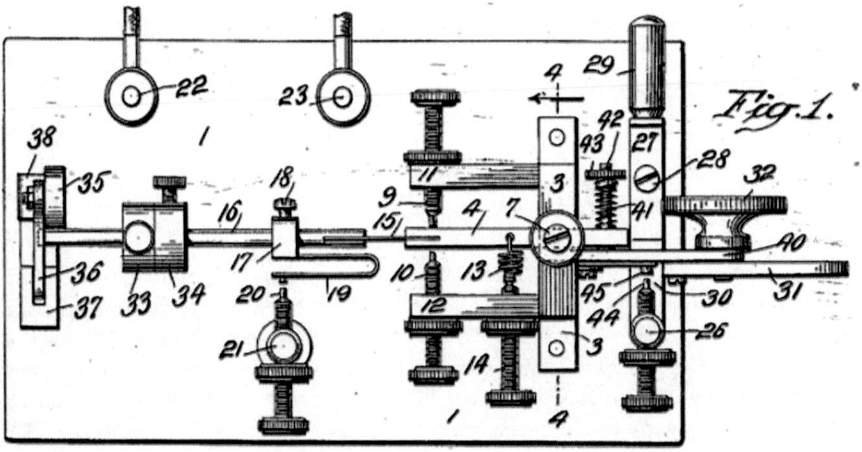

Fig. 1. Drawing from Horace G. Martin's February 13, 1923 patent [4]

1.2 Key Adjustment

Figure 2 shows a simplified drawing of the bug, and a list of the main dimensions taken from an Original Deluxe model owned by one of the authors. Only setup parameters relevant to the dits generation are considered as detailed in previous section: the contact dits distance, adjustable through the DCAS; impact distance, adjustable through FSS; weight distance, adjustable sliding the mass from the FSS. In Fig. 1, you can see other adjustment springs, for example the dits (13) and dash (41) tension springs that adjust the necessary forces that the operator must apply to generate dits and dash. The pre-load of these springs is a useful parameter to find the right working comfort for the operator but which does not affect the dynamics of the device and for this reason, they will not be considered as working parameters.

The reference values for these parameters has been defined following the guideline given by the U.S.A Department of the army and air force [5]:

- Impact Distance: adjust the FSS until to reach a recommended distance of 0.015 in./0.38 mm.
- Contact Distance: the best choice of this parameter is obtained operating the dit paddle counterclockwise and holding the lever in this position until the stop of the vibration of the reed. Adjust the DCAS until the dit contact just meet.
- Weight distance: if the dits are too fast (High oscillation frequency) move the weight in the direction of the deadener.

According to the American Departments of the Army and Air Force [5], the best operation of the semi-automatic key will be obtained when it is adjusted to send dits and spaces of equal length for the first 12–15 dits. Furthermore, the signal should generate at least 25 dits before stopping.

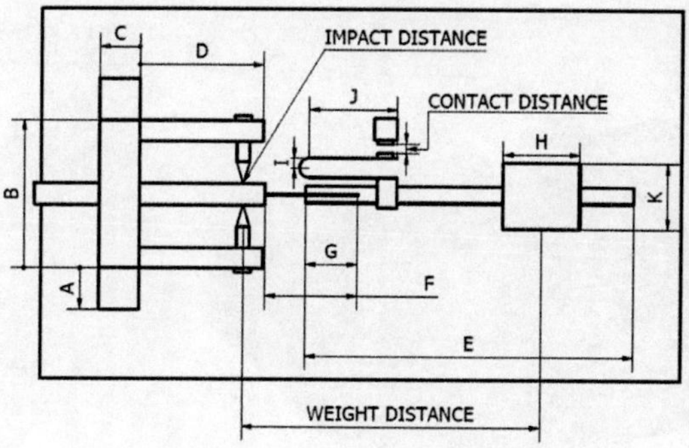

A	10 mm
B	36 mm
C	10 mm
D	31 mm
E	82 mm
F	23 mm
G	13 mm
H	19 mm
I	2.5 mm
J	22 mm
K	16 mm

Fig. 2. Technical drawing taken from a Vibroplex Original Deluxe key.

2 Multi-body Model

In this section, a multi-body model (Fig. 3a) of the bug is presented, reproducing the geometry and the material of a real Vibroplex®Original Deluxe and analyzing its dynamics by means of the Simcenter 3D motion software by Siemens.

Multi-body modelling refers to the dynamic analysis of a system modelled both by rigid and flexible bodies interconnected to each other. The approach to model a flexible body in Simcenter allows to combine the standard multi-body simulation algorithms with a representation of body flexibility using a set of deformation modes. This approach is the one most used by solvers and was first introduced by Craig and Bampton [6]. The method represents the deformation of a body in a dynamic regime starting from the superposition of its vibration modes. In particular, it considers eigenmodes of the system but also the constraint modes, which synthesize the constraint conditions of the body. The modal analysis of the flexible body was carried out using the Simcenter Nastran solver that require a finite element model of the component under study and its material properties.

The model can be simplified considering that the deformation of the system occurs only in two components - the central lamina (Fig. 3b) and the contact spring (Fig. 3c) - which are modelled as flexible bodies. Flexible bodies will be subject to deformation, this deformation is represented by the displacement of the flexible body mesh points. Flexible bodies are connected to rigid bodies that cannot deform, creating a constraint at the connection points for the deformation of the flexible body. To take into account this constraint on the flexible body Simcenter provides a function called *1-D connection* [7]. This function defines one-dimensional elements with stiffness and damping properties that connect each point of the flexible body mesh to a point on the rigid body simulating the constraint due to the connection.

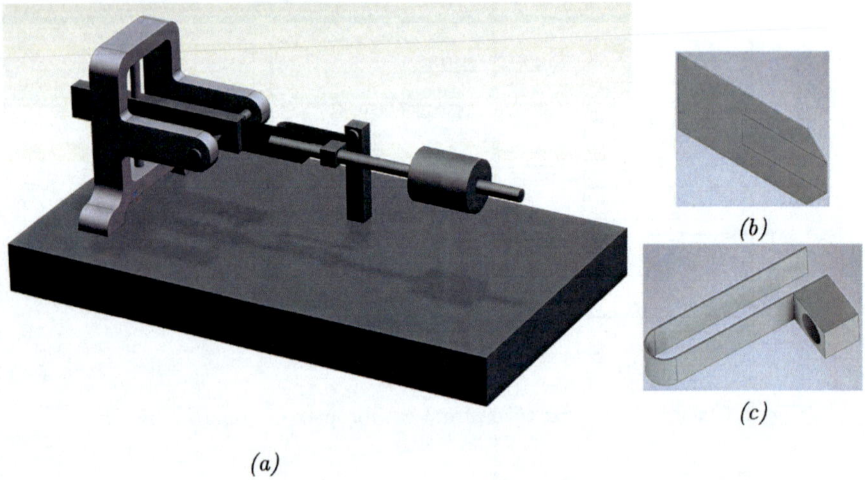

Fig. 3. (a) Multi-body model of the bug with details of: (b) central lamina, (c) contact spring

Neglecting the interaction between the oscillating lever and the deadener, the main contacts during the motion remain two: the one between the main lever and

FSS, the one between the dits spring and the DCAS. To model these contacts the *3D-contact formulation* was used, which is based on the cad geometry of the parts and generates a normal repulsive force between the bodies as a result of the interpenetration of the two geometries, according to the formula [8]:

$$f_n = max(f_{ns} + f_{nd}, 0) \tag{1}$$

where f_{ns} is the spring force, and f_{nd} is the damping force defined by Juhwan Choi [9]:

$$f_{ns} = K\delta^m \tag{2}$$

$$f_{nd} = C_{bpen}\dot{\delta} \tag{3}$$

where K is the stiffness of the material, δ is the penetration depth, calculated by the solver, m is the stiffness exponent or force exponent, a value greater than one indicates the contact force increases at a faster rate than a linear function of penetration. $\dot{\delta}$ is the penetration velocity (that is, the penetration differentiated by time) and C_{bpen} is the damping coefficient and it can be evaluated from the Eq. (4) by a function of a penetration depth [9]:

$$C_{bpen} = step(\delta, L_{init}, C_{init}, L_{bpen}, C) \tag{4}$$

Figure 4 shows the profile of C_{bpen} as a function of penetration depth δ, the step function profile starts from a value of C_{init} for a penetration equals to L_{init} and it stops at a value C for a boundary penetration depth equals to L_{bpen}. In this model, we assumed $L_{init} = 0$, $C_{init} = 0$.

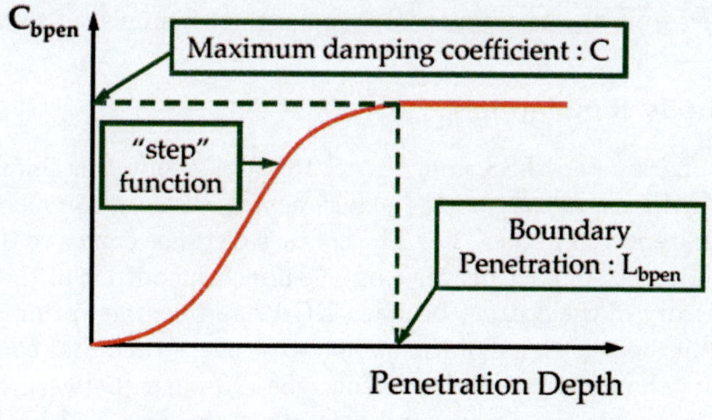

Fig. 4. Material damping coefficient [9]

For flexible bodies this formulation cannot be used, because the geometry of the body is modified during deformation, for this reason contact between DCAS and the tip of the flexible spring was modeled considering the so called *sphere-to-cad* modelling. Sphere-to-cad approximates a sphere in the contact area of the flexible body. The repulsive force generated is calculated in the same way.

The software Simcenter provides some default parameters from which to start to get the desired contact. Table 1 shows the parameters recommended by Siemens for steel and those chosen for each contact. The stiffness has been increased for the contact between the FSS and he main lever, since an extremely rigid impact is required to ensure that most of the energy reaches the moving mass. Instead, the contact between the flexible spring and DCAS requests an inelastic collision, consequently the stiffness was decreased. The other parameters were left unchanged except for the maximum penetration depth, which was decreased in both cases to avoid the onset of excessive and uncontrolled repulsive forces.

Table 1. Contact parameter

Parameter	Default	FSS contact	DCAS contact
K	100000 N/mm	10000000 N/mm	1000 N/mm
m	1.5	1.5	1.5
C	50 N * s/mm	50 N * s/mm	50 N * s/mm
L_{bpen}(Max)	0.1	0.02 mm	0.02 mm

3 Validation of the Model

The validation of the simulated model has been carried out comparing the real signal obtained from the experimental setup with the simulated signal.

3.1 Multibody Simulation

The distance parameters FSS and DCAS to carried out the simulation have been chosen in the same way as the experimental set-up recommended by the manual [5] and reported in Sect. 1.2. The contact distance equals to 0.84 mm was derived from a target impact distance of 0.38 mm. The output of the simulation is the time-history of the contact between DCAS and flexible spring. This signal reconstructs the electric signal of the bug because every time that contact occurs during the dits the circuit is closed. Hence the difference between contact and non-contact represents the closing and opening of the circuit returning a signal that oscillates from 0 to 3 V (Fig. 5) which is the supply voltage used for the experimental test (as will be detailed in the next section). The output signal has been obtained for a distance of the weight (WD) of 84 mm from the impact distance.

As we can see the signal reflects the directives given by [5], that the first 13 dits are defined and the signal stops abundantly after the first 25. With the same distance configuration, another two simulations have been made for two different mass distances one at the minimum point reachable by the weight and another in

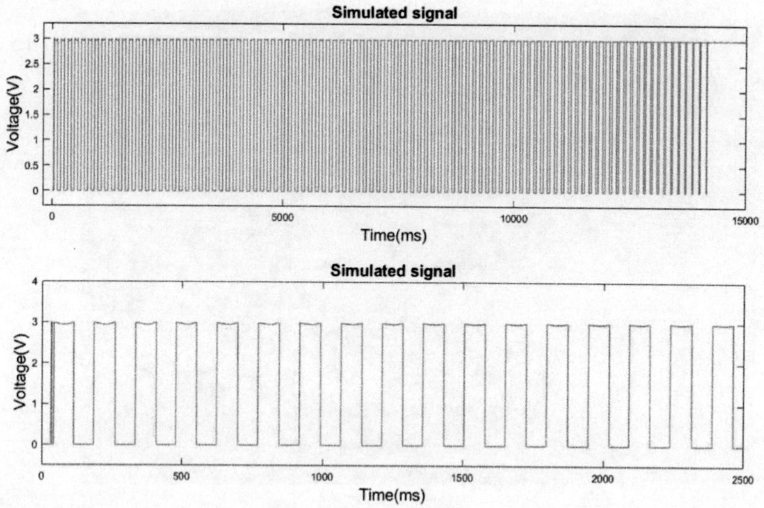

Fig. 5. Simulated signal of the key output: full signal (upper plot) and zoom of the first 25 s (bottom plot)

the intermediate point, the results are almost similar in terms of signal quality, there is just an obvious increase of the frequency when the distance of the weight from the impact point decreases, in fact the frequency is 6 Hz when the weight is at the maximum distance and reach the 10 Hz for the shortest distance. Further tests will be presented in Chap. 4.

3.2 Experimental Setup

The tests have been conducted on a copy of the Vibroplex Original Deluxe owned by one of the authors. A power supply providing 3 V constant voltage was connected to a National instrument NI-9162 DAQ board passing through the key. The signal was acquired with National Instrument's NI-9239 analogue input module (Fig. 6). A feeler gauge was used to set the FSS and DCAS distances in order to obtain an impact distance equal to 0.38 mm and the contact distance as suggested in the manual.

The first test was carried out with the mass at the maximum distance $84mm$, the knob is rotated counterclockwise to generate dits. The lever is kept in contact with the front FSS until the oscillation of the moving mass ends. The signal obtained is a square wave, with the high level represented by the contact (3 V) that progressively increases in duration until the oscillation stops and the contact remains constant (Fig. 7). The duration of a single test is approximately 15 s sampled 1200 Hz.

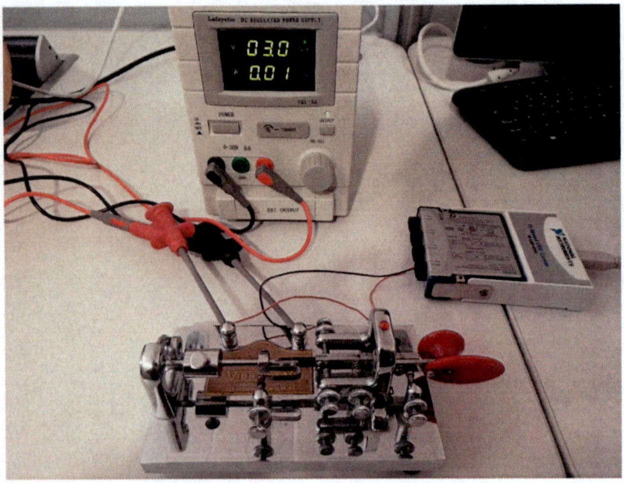

Fig. 6. Experimental acquisition made by a the Vibroplex key, a NI DAQ board and voltage supply.

Figure 8 shows the comparison between simulated and real signals. The signal obtained experimentally is very similar to the simulated one. This is confirmed by the spectra of the signals which show the same harmonic behaviour except for a few Hertz differences of the fundamental (6.94 Hz for real and 6.79 for simulated) (Fig. 8(a)). This difference has been results also for different distances from the mass. The similarity between the two signals can be better identified in the comparison of the first 16 dits (Fig. 8(b)).

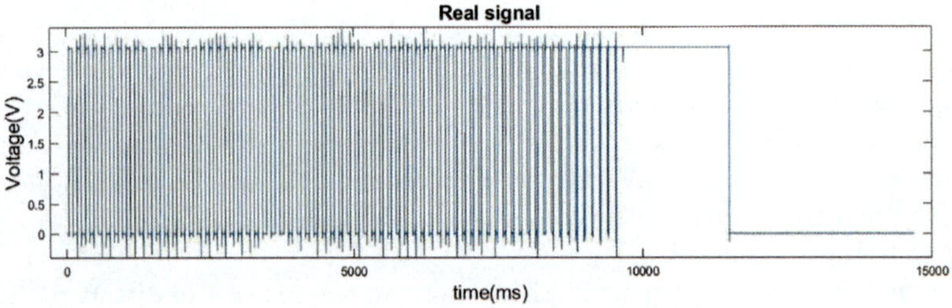

Fig. 7. Real signal from the experimental test

4 Setup Configurations

Once the model was validated, other tests were carried out for different setup configurations of the impact distance and the contact distance. For each of these

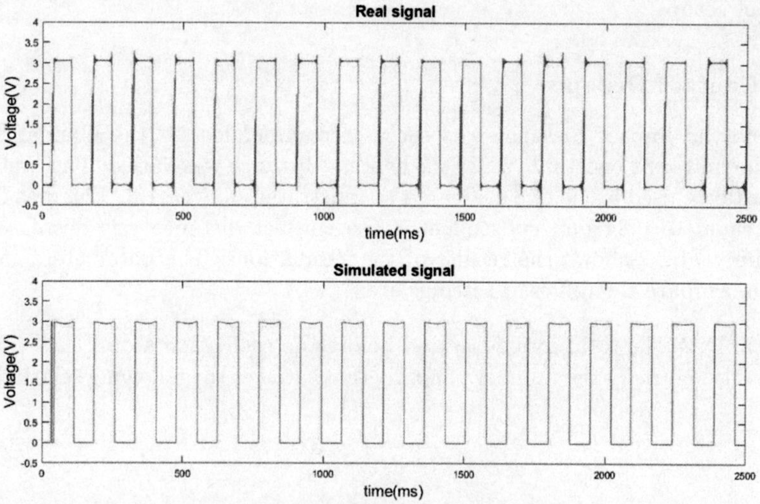

(a)First 16 dits comparison

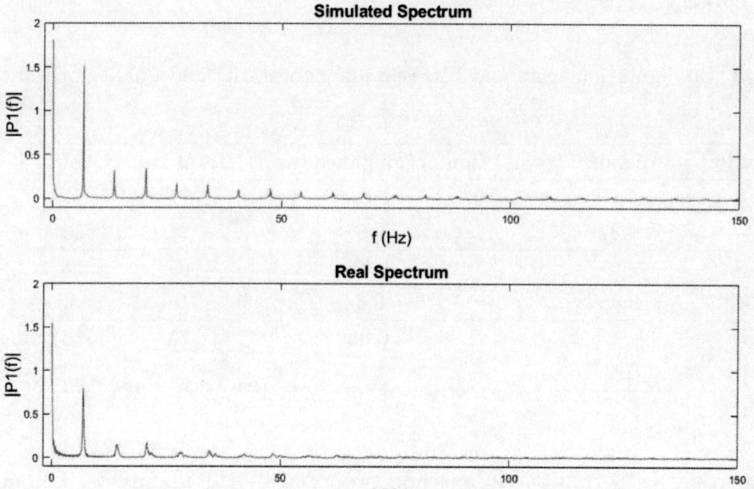

(b) Spectrum comparison

Fig. 8. Signal comparison

configurations 3 tests were carried out at different distances of the weight: 58, 71 and 84 mm. With reference to [5], the duration of the signal generated should last 25 dits at least before stopping. This simple parameter is used to compare different setups.

4.1 Contact Distance

Initially, the impact distance was set as recommended by the manual [5] to 0.38 mm and kept constant, while the contact distance was varied. The first contact distance used is the one evaluated as illustrated in Sect. 1.2. This was found to be equal to 0.84 mm, subsequently, the contact distance was decreased by 0.01 mm. Table 2 shows the results of the simulations. The parameters considered to compare the different settings are:

- Total Dits: the total dits generated before the oscillation stops.
- Quality: given by the number of points that satisfies the following relationship.

$$\frac{|t_c - t_b|}{t_c + t_b} < 5\% \tag{5}$$

where t_c is the contact duration, namly the time during (19) and (20) are in contact and t_b is the break duration, namly the time during (19) and (20) are not in contact.

Table 2. Configuration tests and corresponding total dits and quality of the output signal

Impact distance (mm)	Contact distance (mm)	Total Dits	Quality
0.37	0.81	71	21
	0.82	91	32
0.38	0.83	69	18
	0.84	97	30
0.39	0.85	73	15
	0.86	93	27

According with results, the method proposed by the manual is the one that give the best contact distance for parameters valuated.

4.2 Impact Distance

The other tests were carried out by modifying the impact distance ±0.01 mm the reference value of 0.38 mm. For each impact distance two values of contact distances are computed according to previous section. The ideal distances are highlighted in grey in the Table 2 for each impact distance test.

Table 3. Frequency of dits generation vs. mass distance

Mass distance (mm)	58	64.5	71	77.5	84
Frequency (Hz)	10.13	9.06	8.13	7.46	6.79

An increase in the impact distance recommended in the manual leads to a reduction of the total dits and quality, while decreasing the impact distance the total dits decrease but at the same time the qualities increase.

It is interesting to analyze the frequency trend in the various tests, in fact, the only parameter that influences the frequency is the distance of the weight as reported in Table 3. There are no variations of the frequency when the impact distance and contact distance are changed. Free weight tests were also performed, i.e. without contact between DCAS and flexible spring, and it was interesting to see that the frequency is not affected by the missing contact.

5 Conclusion

This article shows the dynamic analysis of a semi-automatic mechanical key for the manipulation of Morse signals. These devices have been used for several years to convey messages through intercontinental communications, and today are still used by thousand of ham-radio operators. The analysis was carried out on a specific model, the Vibroplex®Original Deluxe, usually referred to as bug. A multi-body model of the bug was created on Siemens Simcenter and subsequently validated by the experimental tests. The validation has taken into consideration the comparison between the contact signals obtained from the model's simulation with the experimental signal. Then the model was used to assess the sensitivity of the contact signal at the different setup configurations. The results show that the setting recommended by the manual [5] is the optimal configuration and that the signal frequency is independent on the impact and contact distance parameters but only depends on the position of the weight. The results obtained in the paper shown the efficiency of multi-body model in the reconstruction of the real model given to the users of this fascinating instrument a deeper theoretical knowledge of it. The virtualization of historical mechanisms is a well-known procedure (e.g. the 3D models of Leonardo Da Vinci's inventions), and it has the advantage to return a dynamic model that can be used to test structural modifications without building real prototypes, or to explain how it works for scientific dissemination purpose.

References

1. Martin, H.G.: Telegraphic transmitter (1902). US732648A
2. Martin, H.G.: Telegraph-transmitter (1906). US842154A
3. Bonucci, F.: Vibroplex: Storia, collezione e impiego dei famosi tasti telegrafici americani. Edizioni C&C (2012)

4. Martin, H.G.: Semiautomatic telegraphic sending machine (1923). US1445226A
5. Departments of the Army and Air Force. TM 11-459 International Morse Code (instructions) (1957)
6. Craig Jr., R.R., Bampton, M.C.C.: Coupling of substructures for dynamic analyses. AIAA J. **6**(7), 1313–1319 (1968)
7. Siemens. 1-D Connection. https://docs.plm.automation.siemens.com/tdoc/nx/1899/nx_help#uid:id628141
8. Siemens. 3-D Contact formulation. https://docs.plm.automation.siemens.com/tdoc/nx/1899/nx_help#uid:xid1128419:index_motion:xid429120:id563181:id638907
9. Choi, J.: A study on the analysis of rigid and flexible body dynamics with contact. Ph.D. thesis, School of Mechanical and Aerospace Engineering Seoul National University (2009)

Geometric Optimization of a Product Using the Design for Disassembly (DFD) Criteria

Pedro A. Ojeda-Escoto(✉)

Center for Design, Equipment Development and Industry 4.0 (CICMA), Technological University of Aguascalientes, Aguascalientes, Mexico
pedro.ojeda@utags.edu.mx

Abstract. Nowadays, product manufacturing has generated an evolution in manufacturing processes and in the transfer of information from engineering to mass production. It is also facing increasing competition and demand from end users with respect to the new product and, it is feasible to determine then, that the objective is not only to exceed the defined requirements, but it is possible to have feedback in terms of functionality from the end user for the formal consolidation of the product. Design for Disassembly (DFD) is a technique that assists in the design or redesign of products for reuse, recycling or ease of maintenance. Seeking consolidation and identifying opportunities for improvement, in this paper a redesign based on DFD criteria for production alignment was performed based on the new manufacturing process proposal outlined by geometry optimization. The theoretical framework was established based on DFD and Lean Manufacturing criteria, which allowed the successful redesign of an agave harvester (case study). Once the design is consolidated, a functional analysis is also shown between the use of DFD and a traditional methodology for the conformation of the final concept. Finally, the global procedure to determine the harvester redesign and the evaluation and analysis strategies to conform the final architecture under the geometry characterization and manufacturing cost reduction approaches are presented.

Keywords: Design for Disassembly · Redesign · Geometry optimization · Lean Manufacturing · Functional analysis

1 Introduction

The design and development of a new product consists of extracting the information from its specification by following a methodology, the architecture of the product in question and its procedural details in order to understand it. It is important to mention the methodology needed to be able to propose the solution to a need; that is, to formulate or propose a problem requires a previous analysis of the knowledge about its nature and, therefore, to have considered the important or relevant facts and concepts according to which it is possible to define in a synthetic way the problem in question. Most of the structured methodologies to generate redesign seek to renew products from a particular approach, to mention a few: assembly, manufacturing, production [1, 4,

M. Ceccarelli and R. López-García (Eds.): HMM 2022, HMMS 40, pp. 395–405, 2022.
https://doi.org/10.1007/978-3-030-98499-1_32

11]. Some design models are based on the implementation and evaluation stages or on the optimization of an initial alternative [12, 17, 18]. In this context, the initial solution option is evaluated and improved by taking into account different aspects such as: performance, cost, assembly, functionality, reliability and maintainability. Feedback is imposed, so that the initial concept can be modified; there is a method of redesign, known as the repeat and modify paradigm. It consists of repeating a previous design process and modifying actions whenever necessary and possible according to the original design intent for the new specifications [14, 16]. In the engineering context, disassembly can be defined as the use of assembly methods and configurations that enable the development of cost-effective products that facilitate the separation of components and materials from the product used to promote recovery and reuse. Disassembly is necessary for recycling, maintenance, remanufacturing, etc. The origins of DFD are based on factors such as: shifting trends towards sustainable development models; globalization of products; new product requirements (environmental requirements); high product innovation; product design through life cycle design [8, 10, 15, 20].

This paper reports the procedure and results of the redesign of an agave harvester (case study) structured with DFD criteria and using as approaches: geometry characterization and manufacturing cost reduction. Likewise, the results obtained from the finite element analysis (FEA) of the mentioned trailer are reported in order to optimize its final architecture. Finally, defined strategies to align manufacturing to production are mentioned.

2 Theoretical Framework

2.1 Product Life Cycle

The manufacturers have a high impact on environmental and they should play a significant role in fulfilling the requirements of consumers at the same time. It should be paying attention that sustainable development not only involves environmental protection but also significant economic issues. During the design and manufacturing process, the designer should apply the principle of DFD into the design. The introduction of disassembly systems aims "closing the loop" of a product at its life cycle. Modifications have to be made between the requirements and the actual possibilities considering the environmental requirements for production in the current technology and economy. There is a need for a comprehensive disassembly strategy concerning the recovery of industrial goods [7].

2.2 Design for Disassembly and Lean Manufacturing

Design for Disassembly is a well-known target design methodology which allows the easy separation of components in industrial products. It involves the selection and use of appropriate materials, the design of components and product architecture and the selection and use of joints, connectors and fasteners, which could be easily disassembled. In fact, it has been determined that the cost of dismantling is almost linear with the depth, while the revenue has a rapid initial increase and then settles down, without further growth as the depth increases. For this reason, it is recommended in the DFD to place the most important components in easily accessible areas where possible [2, 6]. There are three important factors which must be consider a designer that use this technique [5, 8]: 1) Material Selection; 2) Fasteners and Connections; and 3) Product Structure & Component design.

Lean Manufacturing (LM) involves several tools that help eliminate all operations that do not add value to the product/service and/or process, increasing the value of each activity performed and eliminating what is not required, reducing waste and improving operations. The LM system has been defined as a philosophy of manufacturing excellence, based on [3, 19]: a) The planned elimination of all types of waste; b) Continuous improvement; and c) The permanent improvement of productivity and quality. The implementation of LM is important in different areas, since different tools are used, so it benefits the company and its employees (see Fig. 1). Some of the benefits it generates are [9, 13]: a) Reduction of at least 45% in production costs; b) Reduction of inventories and lead time; c) Better quality and less labour force; d) Increased equipment efficiency and reduced waste; e) Overproduction and decrease in waiting time; and f) Transport, processes, inventories and movement of final product.

3 Conceptualization and Results Analysis

3.1 Case Study

To define a new harvester design, the following approaches were considered: geometry characterization and manufacturing cost reduction; based on the above approaches, several configurations were proposed and examined to provide a solution to the problem under study. Several configurations were conceptualized taking into account the defined approaches and finally the final architecture of the agave harvester was generated (Fig. 2).

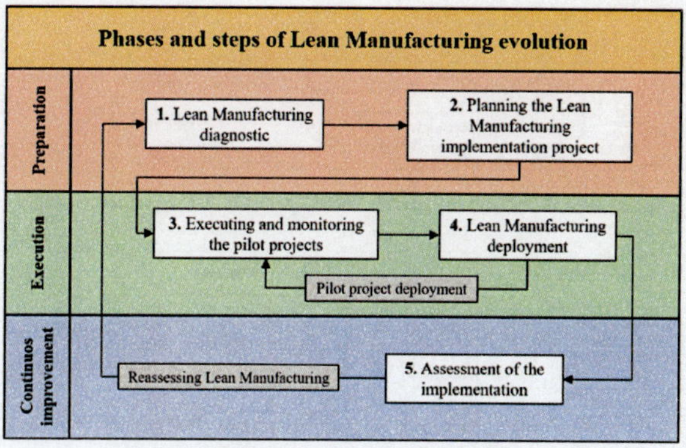

Fig. 1. Plan for lean implementation

Fig. 2. Final model based on DFD results

3.2 Methodology

The method used for the development of this research is shown in Fig. 3. The sequence of defined stages and the implementation of DFD for the redesign of the agave harvester can be observed. Is important to mention that the redesign also takes into account the manufacturing process for the optimization of costs and final time for the alignment of the harvester to mass production.

3.3 Solution Proposal

The agave harvester is a machine that is pulled laterally by a tractor and performs the work of harvesting and crushing the agave plant, the design contemplates the total or partial cutting of the plant at ground level. Once the final architectural geometry was selected, finite element analyses were performed to validate and optimize (for assembly purposes) that architecture. Table 1 shows a summary of the general conditions used for the management of FEA analyses.

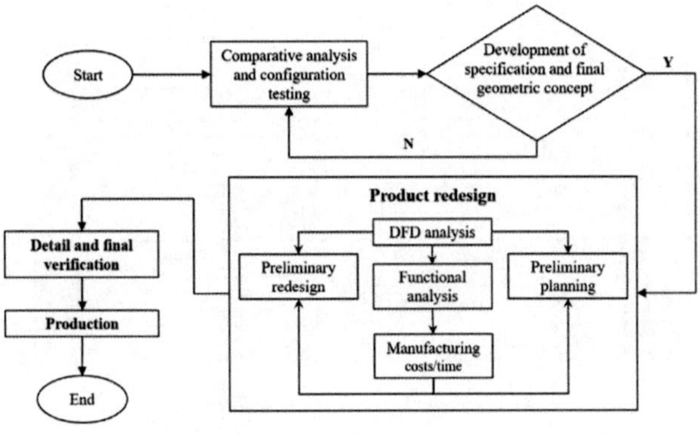

Fig. 3. Flowchart of the research process

Table 1. General conditions for FEA analysis

Model description	Type of analysis	Load group (ton force)	Type of mesh	Max/min. element size (inches)	Number of elements
Main chassis	Static	6.0	Solid	5.25/1.25	21354
Cutting head support	Static	3.5	Solid	1.75/0.875	26987
Material container platform	Static	4.5	Solid	6.5/1.375	15895
Feeder support	Static	3.5	Solid	1.75/0.875	28451
Suspension system	Static	4.5	Solid	5.25/1.25	34521
Harvester (iteration 01)	Static	7.5	Solid	8.5/1.75	69733
Harvester (iteration 02)	Static	7.5	Solid	8.5/1.75	69124
Harvester (iteration 03)	Static	7.5	Solid	8.5/1.75	68942
Harvester (iteration 04)	Static	7.5	Solid	8.5/1.75	69383

The mesh of the models was generated based on the existing dimensional relationships between the components of the assembly and the load conditions proposed for the analysis were determined based on the solicitations that are presented in the normal work of the harvester. Table 2 shows the general operating functions of the combine harvester, which is the subject of the present study.

Table 2. Harvester operating functions

Operating function	Harvester condition	Operating cycle
Drag	In motion	0
Cutting head positioning	Stationary	1
Material cutting and shredding	In motion	2
Material conveying	In motion	3
Material pick-up	In motion	4
Change of containers	Stationary	5
Cutting head lifting	Stationary	1
Drag	In motion	0

3.4 Geometry *O*ptimization

For the optimization of the geometry (part information to generate a sub-assembly) and architecture (general prototype assembly information) of the agave harvester, two paths were followed: on the one hand, the sub-assemblies of the machine were optimized using DFD criteria (Fig. 4); and on the other hand, finite element analysis was used to validate and optimize the geometry (for final assembly issues). The mesh of the models was generated based on the dimensional relationships between the components of the sub-assemblies and the applied loads were determined based on the stresses that occur in the natural work of the harvester (see Table 2).

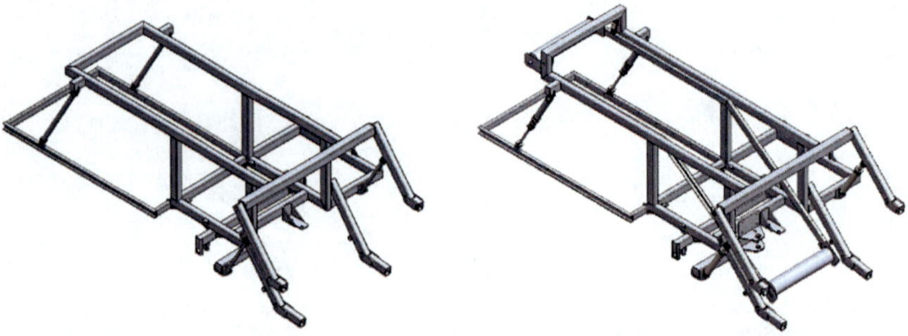

Fig. 4. Optimization of sub-assemblies under DFD criteria (reduction in the number of parts): *left)* second iteration; *right)* fourth iteration

The finite element analysis (FEA) was carried out in several stages, in each of which different criteria were used for the analysis and several stress application zones were defined according to the normal work performed by the harvester (Fig. 5). Table 3 shows the results obtained from the analyses performed.

Other analyses that were carried out to complement the study and application of DFD were first to analyze the drawbar and to define the general architecture of the agave harvester, in terms of the displacement it will have when it is towed and in the natural operation within the harvest field (see Fig. 6). On the other hand, an analysis of the different positions of the agave harvester was also carried out, as shown in Table 2. This

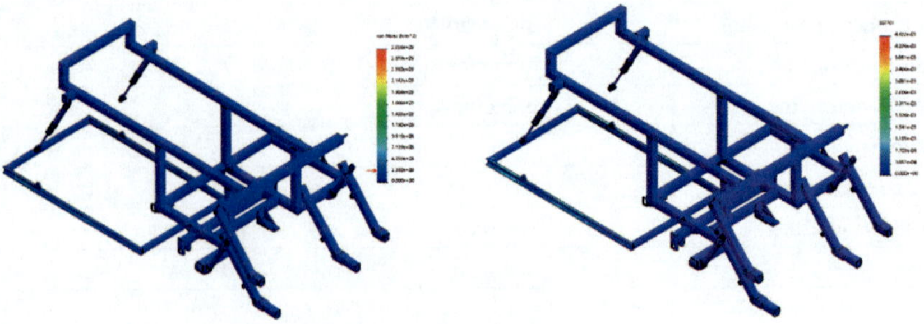

Fig. 5. FEA Results: final iteration

analysis defined the optimization of the weight obtained and supports the application of DFD (see Fig. 7).

Table 3. Result of static analysis

Iteration	Description	Type (unit)	Result
01	Stress	von Mises (MPa)	Minimum 0.000e+00 Maximum 3.661e+08
	Displacement	URES (mm)	Minimum 0.000e+00 Maximum 3.122e+08
	Strain	ESTRN	Minimum 0.000e+00 Maximum 1.452e−03
02	Stress	von Mises (MPa)	Minimum 0.000e+00 Maximum 2.943e+08
	Displacement	URES (mm)	Minimum 0.000e+00 Maximum 2.886e+08
	Strain	ESTRN	Minimum 0.000e+00 Maximum 1.289e−03
03	Stress	von Mises (MPa)	Minimum 0.000e+00 Maximum 3.182e+08
	Displacement	URES (mm)	Minimum 0.000e+00 Maximum 2.843e+08
	Strain	ESTRN	Minimum 0.000e+00 Maximum 1.334e−03
04	Stress	von Mises (MPa)	Minimum 0.000e+00 Maximum 2.380e+08
	Displacement	URES (mm)	Minimum 0.000e+00 Maximum 2.780e+08
	Strain	ESTRN	Minimum 0.000e+00 Maximum 1.155e−03

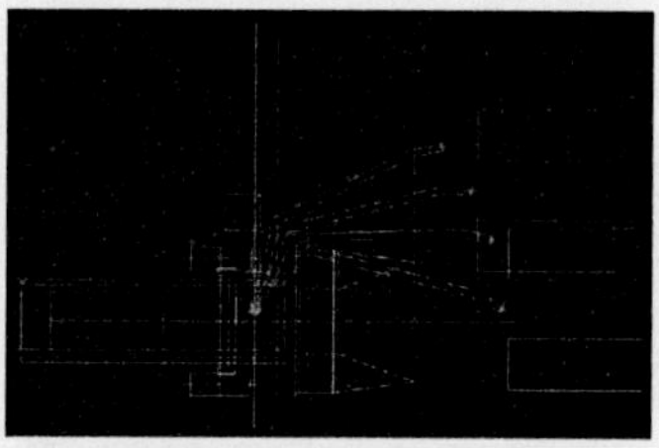

Fig. 6. Drawbar analysis

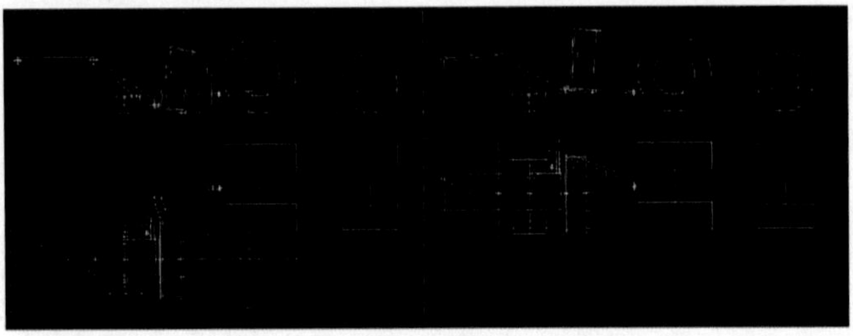

Fig. 7. General structure analysis

3.5 Analysis and Discussion

In the final process for the architecture definition, four iterations of the functional prototype were performed. Each of these iterations was performed based on the application of DFD for parts optimization and FEA for optimization of material usage and subsystem assembly.

The stress study carried out on the harvester model showed that each of its component elements had stress magnitudes lower than those allowed by the elastic limit of the material. It was even possible to optimize some elements and meet the design criteria that were taken into account to define the final architecture. The improvements obtained from the agave harvester redesign process yielded the following routes, which are directly linked to the DFD analysis: I) *Characterization of geometry to reduce the final weight of the product:* once the manufacture of the prototype was completed, an approximate reduction of 32% of the total weight of the agave harvester was achieved and II) *Reduction in design, engineering and manufacturing costs:* the new agave harvester design resulted in a 35% reduction in the total costs involved in the development of the new product.

Table 4. DFD design proposal: Main chassis

Description	Design					Ratio
	1st iteration	2nd iteration	3rd iteration	4th iteration	Redesign	
Main chassis sub-systems	8	7	7	6	6	25%
Components	22	19	19	17	17	22%
Disassembly time	124*	120*	120*	115*	107*	—
Manufacturing time	286	275	269	247	231	—
Total time	410	395	389	362	338	—

* Estimated times

Table 5. DFD design proposal: Cutting head support and material container platform

Description	Design					Ratio
	1st iteration	2nd iteration	3rd iteration	4th iteration	Redesign	
Cutting head and material container sub-systems	46	42	42	41	39	15%
Components	126	121	120	117	112	11%
Disassembly time	492*	484*	459*	442*	421*	—
Manufacturing time	624	631	602	593	581	—
Total time	1116	1115	1061	1035	1002	—

* Estimated times

Table 6. DFD design proposal: Feeder support and suspension system

Description	Design					Ratio
	1st iteration	2nd iteration	3rd iteration	4th iteration	Redesign	
Feeder support and suspension sub-systems	38	35	32	32	29	23%
Components	94	92	88	87	82	12%
Disassembly time	376*	370*	354*	339*	321*	—
Manufacturing time	521	511	502	483	476	—
Total time	897	881	856	822	797	—

* Estimated times

Tables 4, 5, 6 show the results obtained by applying DFD to the final configuration of the agave harvester, which served as the basis for optimizing the final assembly and production process.

4 Conclusions

Design is to establish and define pertinent solutions and structures for problems that have not been solved before or new solutions proposed in a different way for problems that have been previously solved. Advances in computational analysis and technology allow engineers and researchers to have effective diagnostic and simulation tools that facilitate, at any given time, the design, redesign or optimization of a mechanical system. In this work, the optimization of the geometry of an agave harvester and the results of the application of DFD criteria were presented as a case study. In order to define a new agave harvester design, several configurations were proposed and reviewed to solve the studied problem. Several configurations were also conceptualized taking into account the defined approaches and finally the product architecture was generated. Finally, the strategies defined for the adaptation of the new product developed for production are presented, taking into account sales projections, product image and marketing: i) Product development and prototype evaluation; ii) Marketing strategy; iii) Commercial viability strategy; iv) Market testing strategy; and v) Commercialization strategy.

Acknowledgments. This research work was carried out thanks to the technical, financial and logistical support of the Technological University of Aguascalientes.

References

1. Adam, A., Cosma, C., Cosma, A., Pop, A.: Redesigning a product using modern CAD–CAM software. Procedia Technol. **19**, 221–227 (2015)
2. Azab, A., Ziout, A., ElMaraghy, W.: Modeling and optimization for disassembly planning. J. Mech. Ind. Eng. **5**(1), 1–8 (2011)
3. Dombrowski, U., Mielke, T., Engel, C.: Knowledge management in lean production systems. Procedia CIRP **3**, 436–441 (2012)
4. Evbuomwan, N., Sivaloganathan, S., Jebb, A.: A survey of design philosophies, models, methods and systems. J. Eng. Manuf. **210**, 301–320 (1996)
5. Fang, H., Ong, S., Nee, A.: Product remanufacturability assessment and implementation based on design features. Procedia CIRP **26**, 571–576 (2015)
6. Frizziero, L., Liverani, A., Caligiana, G., Donnnici, G., Chinaglia, L.: Design for disassembly (DFD) and augmented reality (AR): case study applied to a gearbox. Machines **7**(29), 1–26 (2019)
7. Min, C., Mamat, T.: Design for disassembly in electric rice cooker manufacturing. Progr. Eng. Appl. Technol. **2**(1), 731–742 (2021)
8. Mule, J.: Design for disassembly approaches on product development. Int. J. Sci. Eng. Res. **3**(6), 1–5 (2012)
9. Mostafa, S., Dumrak, J., Soltan, H.: A framework for lean manufacturing implementation. Prod. Manuf. Res. **1**(1), 44–64 (2013)
10. Ojeda-Escoto, P.A.: Design for Disassembly (DFD): product redesign success case. In: Proceedings of the XXVI Annual International Congress of SOMIM, vol. 6 no. 1 (2020)
11. Ojeda-Escoto, P.A.: Structural evaluation and optimization as a strategy for redesigning a product. Spec. Issue Aristas Mag. Basic Appl. Res. **8**(15), 233–238 (2020)
12. Pahl, G., Beitz, W., Feldhusen, J., Grote, K.-H.: Engineering Design: A Systematic Approach. Springer, London (2007). https://doi.org/10.1007/978-1-84628-319-2
13. Powell, D., Alfnes, E., Strandhagen, J., Dreyer, H.: The concurrent application of lean production and ERP: towards an ERP-based lean implementation process. Comput. Ind. **64**, 324–335 (2013)
14. Smith, S., Smith, G., Shen, Y.: Redesign for product innovation. Des. Stud. **33**(2), 160–184 (2012)

15. Soh, S.L., Ong, S.K., Nee, A.Y.C.: Application of Design for Disassembly from Remanufacturing Perspective. Procedia CIRP **26**, 577–582 (2015). https://doi.org/10.1016/j.procir.2014.07.028
16. Ullah, I., Tang, D., Wang, Q., Yin, L., Hussai, I.: Managing engineering change requirements during the product development process. Concurr. Eng. Res. Appl. **26**, 1–16 (2017)
17. Ullman, D.: The Mechanical Design Process. Mc Graw-Hill, New York (2009)
18. Ulrich, K., Eppinger, S.: Product Design and Development. Mc Graw-Hill, New York (2004)
19. Upadhye, N., Deshmukh, S., Garg, S.: Lean manufacturing for sustainable development. Glob. Bus. Manag. Res. Int. J. **2**(1), 125–137 (2010)
20. Wang, P., Liu, Y., Ong, S., Nee, A.: Modular design of machine tools to facilitate design for disassembly and remanufacturing. Procedia CIRP **15**, 443–448 (2014)

Brief History of the Pulleys and Explanation of Chief Instructor Ishaq Efendi's Work on Pulleys and Pulley Systems

Irem Aslan Seyhan(✉)

University of Bartın, Bartın, Turkey
iaseyhan@bartin.edu.tr

Abstract. Pulleys and pulley systems have been used since the early times of history. The application of this simple machine has been available since the earliest civilisations. It was widely used in construction and transportation for lifting heavy loads on its own or as a part of another machine. These systems have a capacity of lifting heavy weights very easily, sometimes just by changing the direction of a tractive force in a fixed simple pulley and/or sometimes by reducing the tractive force considerably in the compound pulley system. The theory of this system grows much later than the application itself. The theory was considered first by Archimedes and many centuries later by Galileo. The application of this system was, of course, available in Ottoman land, but the subject's theory was attached to the official curriculum of the Military Engineering School in the 18-19th century. In the beginning, there were not any Turkish written sources. The first Turkish written treatise including this subject was Ishaq Efendi's (1748? -1835/36) *Mecmua-i Ulûm-i Riyâziyye*. It was a compendium written in order to introduce the Western-style mathematical sciences to the Ottoman students. The 3rd volume of this treatise was dedicated to physics and mechanics. There is a chapter about simple and compound machines in the mechanics part. After publishing this book, the Military Engineering School teachers started to use this treatise as a primary source of education on mechanics courses. This paper presents a brief history of the pulley, and the related chapters of Ishaq Efendi's *Mecmua-i Ulûm-i Riyâziyye* are reviewed. His treatise can be considered as fundamental significance in the education of the Military Engineering School of Ottomans.

Keywords: Ottoman military engineering schools · Chief instructor Ishaq Efendi · Compendium of mathematical sciences · History of pulley systems · Ottoman history of mechanics

1 Introduction

Ishaq (İshak) Efendi (1748?-1835/36) was one of the eminent chief instructors of Military Engineering School (Mühendishâne-i Berri Hümayun) in 19th century [2]. His four volumed *Mecmua-i Ulûm-i Riyâziyye (Compendium of Mathematical Sciences)* was a compendium written in order to introduce the Western-style mathematical sciences to the

M. Ceccarelli and R. López-García (Eds.): HMM 2022, HMMS 40, pp. 406–418, 2022.
https://doi.org/10.1007/978-3-030-98499-1_33

Ottoman students of the mentioned school. Military Engineering School's main aim was to educate students by means of modern military and engineering services. Therefore, the curriculum was arranged according to this particular need, and in that curriculum, there was a mechanics class [1]. The 3rd book of *Compendium of Mathematical Sciences* is devoted to physics and mechanics. In this book, between the pages of 180 and 266, there exists a chapter with the name of "dragging of inanimate bodies from the science of dragging of heavier weights". In this part, Ishaq Efendi mentions some simple machines. "As far as known, it is the first treatise mentioning the theory of simple machines in Ottomans in the Turkish language" [10]. The main aim of this chapter was to describe simple machines which make it possible to lift or drag heavier bodies by using less force. There were 15 sections in this chapter, and the 6th section was devoted to pulleys (between the pages 213–217) and the 10th section to the pulley systems (between the pages 231–234) [13]. Ishaq Efendi didn't discuss pulleys and pulley systems consecutively. But instead, he discussed inclined plain, clamp [9] and wedge in between the 6th and 10th chapters. The reason for this is Ishaq Efendi divided these machines into two groups as simple and compound machines according to a classical classification. He first described simple machines, and later, he described compound machines as a composition of the simple ones in accordance with the literature.

Before Ishaq Efendi's *Compendium*, pulley systems were taught in the classes of Military Engineering School by direct translations from the French books in the 18-19th century. The Ottomans were indeed using these simple machines in practice (in constructions, in naval related needs, supplying water from the wells, etc.) but, before the 18th century, it is unknown for us whether or not were there any theoretical essays on that subject thought to students in madrassas or not. As far as it is known, the first theoretical essays on mechanical science (cerr-i eskal or ilm-i mekanika) on pulleys were included in the curriculum from the late 18th and beginning of the 19th century. These essays -as we mentioned before- were generally translated -mainly- from French books. The eminence of Ishaq Efendi reveals itself at this point for Ottomans, and his *Compendium* was the first text written in Ottoman Turkish, including the theory of simple machines. After the publication of the 3rd book in 1832, this text was used as the main source in education in Military Engineering School along with the French books [1]. Although Ishaq Efendi's book did not bring any innovations on this subject's theoretical part, the fact that it was the first Turkish source of such kind makes it necessary to examine his work in terms of Turkish history of science and history of terminology.

2 Brief History of Pulleys and Pulley Systems

Pulleys are simple machines that provide force gain in moving objects and change the applied force's direction. They can be used singly or in combination due to the need for work. Pulleys consisted of a wheel, axle, and rope (chain, cable or belt). There is a groove on the wheel so that the rope can pass around it. Pulleys are considered as simple machines used for lifting heavy loads. There are two types of pulleys, fixed pulleys and moveable pulleys. Fixed pulleys are engaged to the fixed place (bar, beam, ceiling etc.). A fixed pulley is mainly used to change the direction of the force. It is used for extracting something from an unreachable depth, such as a well, or to carry something to an unattainable height, such as hanging flags. In such pulleys, there is no gain from the way: The further the rope is pulled, the higher the load will have lifted. The direction of application of the force does not matter. Fixed pulleys provide ease of doing work by changing the direction of the force. On the other hand, moveable pulleys are not engraved to any fixed place so they can actually move. One end of a rope is tied to a fixed position, while the other end of the rope is free for applying the force. In such pulleys, the weight is attached to the wheel, which locates between the rope's ends. When we apply force in such pulleys, both the weight and the wheel move in the same direction as the force.

Moreover, in such pulleys, the object's weight is shared between the fixed location and a person who applies the force. Therefore, as the number of pulleys increases, the amount of force required to do the work decreases. Accordingly, in order to raise the load to the height of h, the end of the rope -where the force applied- must be pulled 2h in a moveable pulley. In most cases, the weight of the pulleys and friction is ignored. But if the load P is balanced with the movable pulley of weight G, then the pulley's weight is added to the weight of the load (P + G), and calculations may pursue. There are also compound pulleys (block and tackle pulley) which are a combination of fixed and moveable pulleys. There is a considerable gain from the force in this system, which is balanced with the loss from the path.

Pulley was mainly used as a lifting device in history. According to Abbott Payson Usher, an expert on the history of technology, "The pulley in its simple forms had a most extended line of uses, from raising water from wells to its many applications on board ship" [20]. Pulley has been used on its own or as a part of another machine, such as a crane or inclined plane. It was used in constructing massive monuments, lifting monoliths as a part of capstans. The practical application was available even before the theory of this simple machine. The invention of the pulley can be dated back to Ancient Egypt [11]. It was first used for drawing water from a well. Also, it is thought pulley used in the construction of pyramids as well as transportation in Ancient Egypt. The oldest pulley belongs to Ancient Egypt, is dated the 12th dynasty period, was used to change the direction of the applied force (Fig. 1). However, archaeological findings demonstrate that the Ancient Egyptians have the information of pulley of changing the direction of pull and reducing the pulling force [8].

In the Hellen and Hellenistic periods, pulley systems were widely used to compensate for daily needs, transportation, and sailing. It is also known that pulley systems were used in the back stage of theatres for creating visual effects, dating from the 5th century BC. The expression "Deus ex machina" was first derived in Ancient Greek and Roman drama. When the story seems to be knotted at the end of the plot, the god appears and resolves the plot. This appearance effect was created by some cranes (including pulley) [14, 19].

The first investigation on theoretical and mathematical aspects of mechanics is attributed to the Archytas (c. 4th century BC), but no such treatise survived belonged to him. Therefore, this title is reserved for Archimedes (287–212 BC). He proved the property of levers mathematically, basing on certain postulates on the treatise *On Plane Equilibriums*. Sir Thomas Heath mentions Archimedes' related works, *On Balances* (or perhaps *On Levers*) and *On Centers of Gravity* are lost but, we are aware of his work

Fig. 1. The oldest pulley belongs to Ancient Egypt [8].

through Pappus of Alexandria (c.290 - c.350) and Simplicius (c.490 - c.560). It is uncertain whether these were the two separate treatises or parts of the same more extensive treatise on *Mechanics* [12]. It is also said that Archimedes used pulleys to haul a ship ashore.

In the Roman period, Vitruvius (1st cent. BC) mentioned pulleys in his *De Architechture*, the first book written on architecture. Pulleys are mentioned in chapter VIII of book IX- *Sundials and water clocks* and chapter XI of book X - *Balistae* [21]. Not far from that date, one of the most significant engineers, Heron of Alexandria (10 AD-75 AD), emerged in the scene of history. Heron mentioned pulleys in his several works such as *Pneumatica, Automata, Belopoeica* and *Mechanica. Mechanica* only survived in Arabic translation consisted of 3 books and has vast information on lifting heavy objects (the Baroulcus). This work was also cited by Pappus of Alexandria. *Mechanica* was based on the works of Archimedes. It includes a great deal of information related to pulleys, such as lifting and transporting heavy objects with mechanical devices [7]. As Schiefsky stated in the article on *Heron's Mechanics*, "This work provides an important precedent for the core of Heron's theoretical project: the attempt to 'reduce' the five simple machines or mechanical powers (the wheel and axle, lever, compound pulley, wedge, and screw) to the circle and, ultimately, the balance" [17].

In the Eastern world, pulleys were used in daily life, and some engineers used the knowledge of simple machines to make various devices. Al-Jazari (1136–1206) - the most famous engineer of the Eastern world- mentioned various mechanical systems, including pulleys, in his *Book of Knowledge of Ingenious Mechanical Devices*. Pulleys were mentioned several times in his treatise, such as water extraction devices, and they were used to balance weight and counterweight in weight clocks and candle clocks (Fig. 2) [16].

"Since early Renaissance, the subjects of Mechanical Engineering have attracted more and more interest both for practical applications and from a theoretical viewpoint in response to an increase of societal needs" [6]. In the age of humanism, Leonardo Da Vinci (1452–1519) studied the science of weights basic machines, including pulley and compound pulley [15]. In the late 16th century Simon Stevin (1548–1620) worked on

Fig. 2. Candle clock of Al-Jazari [18].

the mechanical advantages of simple machines. But, aside from all these works, Galilei's (1564–1642) treatise *Le Mecaniche* can be considered as a milestone. It was the first academic course on machine theory since it used modern concepts. In this treatise, Galilei inquires about the simple machines for lifting weights, namely lever, capstan, pulley, and screw. In this treatise, he considers these simple machines as an application of lever, which is regarded as a classical Archimedean approach to mechanical problems. Among the sources of Galilei there are Pappus of Alexandria (that he refers explicitly), Francesco Maurolico (1494–1575), Girolamo Cardano (1501–1576), Alessandro Piccolomini (1508–1578), Federico Commandino (d.1575), Giovan Battista Benedetti (1530–1590), Guidobaldo Del Monte (1545–1607) [6].

As we can see from this brief introductory historical presentation, pulleys' applications developed before the mechanical theories. Later, this subject was considered together with the force, friction, tension, and resistance subjects in physics.

3 Pulleys and Pulley Systems by Ishaq Efendi

Ishaq Efendi starts the 6th chapter by defining pulleys [13]. He defines the pulley that does not move as a "fixed pulley" and the pulley which can change its place and that moves with the weight as a "moveable pulley". After giving these definitions, Ishaq Efendi gives several theorems and results of these theorems. His first theorem is about the directions of force and weight on the fixed pulley. He explained that if an applied force and weight are equal on both sides of the rope along the pulley, there would be equilibrium. Then he explained this with an example and a figure related to that example (Fig. 3). Ishaq Efendi gives all the figures at the end of his book. The numeration of the pages, including the figures, was separately given.

As the first result of this theorem, he compares the type 1 class lever and the fixed pulley. He approached the fixed pulley as a kind of lever as an Archimedean approach

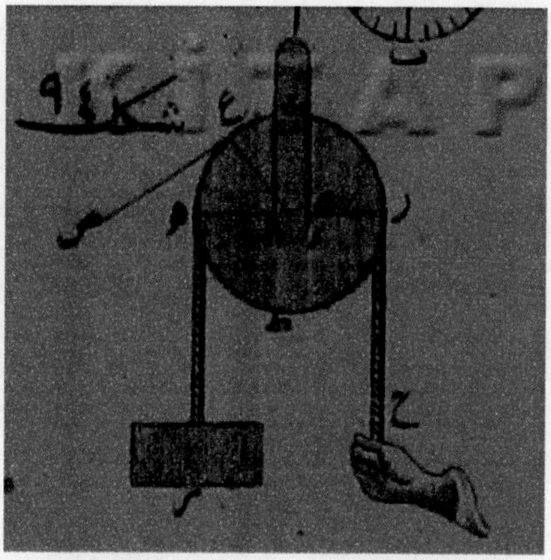

Fig. 3. Fixed pulley, figure 94 of Ishaq Efendi [13].

to mechanical problems. He considered the pulley's horizontal diameter as a scale of arm. Then he explained calculations varies according to the objects' arrangement, and according to the force and weight's quality. In the second result, he discussed the cases of friction and resistance. He first explained that the fixed pulley by itself does not increase or decrease the intensity of weight and/or force. Equality occurs with the equality of the ores (cevher). The impact intensity of the roller decreases due to the inertia of the moving wheel and the resistance of the contact. The inertia and resistance and the weight itself correspond to the force. However, the obstacle and delay are all resistance of the contact, and inertia corresponds to the wheel's movement before it starts moving. He concludes with the resistance of the contact being permanently ignored as in the scales, the rules described on the weight of the contact are followed. In his 3rd result, he explained that the direction of application of the force and the weight does not create any difference in the calculations in a fixed pulley.

In his second theorem, he explained moveable pulleys by explaining the figure 95 (Fig. 4). First of all, he explained the right pulley solely in this figure by stating that one end of the rope (the rope stands in the rightest) is tied to a fixed point, the other end of the rope is free for applying the force, and the weight is attached to the centre of the wheel. Then he explains if the proportion between the applied force and the weight is equal to the proportion between the radius to the diameter (i.e. 1/2), the system would be in equilibrium. In other words, that means in moveable pulleys, the object's weight is shared between the rope attached to the fixed location and the person who applies the force with the rate of 1/2.

In the first result of the second theorem, he explained that since the weight stands between the force and the fulcrum, such moveable pulleys may be considered the second type of lever. In the second result, he again stressed that if the proportion of the force applied to the weight, is equal to the proportion of radius to diameter, is equal to 1/2, and the system would be in equilibrium. Then he explained this fact with the help of the figure. In the 3rd result, he constructed the relation between the radius and force by explaining that it is regardless of the angle on the moving pulley, the applied force is proportional to the weight, even if the radius grows. In result 4, he descriptively explained that fixed pulleys provide ease of doing work by simply changing the direction of the force.

In the last result of this theorem, Ishaq Efendi mentioned the path-time-weight relationship that occurs in the system due to the conservation of energy. He has accepted that the distance taken is directly proportional to the time. He stated 1 step is taken in 1 min. In the continuation of this section, he stated that "since the force applied to lift the load will be halved, the distance must be doubled". In other words, for the force to go up 1 step, the load must go up 0.5 steps. He completed this part by reminding us that he will explain the (compound) pulley systems in the part of compound machines (alât-ı mürekkebe) in the following chapter of his book.

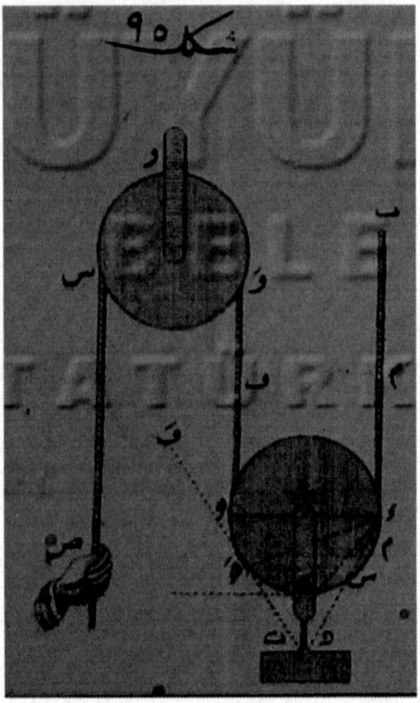

Fig. 4. Moveable pulley, figure 95 of Ishaq Efendi [13].

10th chapter dedicated to compound pulleys. The name of the chapter can be translated as "the compound pulleys from compound machines". Here, Ishaq Efendi defined the system as a combination of simple machines. According to his following statements, compound machines made of simple machines could take various forms depending on how they were used. Compound machines were formed by combining and arranging the most profitable multiple levers. The tools consisting of a lever or a clamp that continuously transfers the effect of the force to each other was such type of machine. He highlighted that these tools were the most profitable improvement in this subject. Furthermore, there were industrial tools, consisting of multiple pulleys, called compound pulleys. In such systems, each pulley moved and worked with its own rope around the wheel and one end of that rope was attached to the fixed to points, and on the other end, the force was applied. Or one rope might have passed around all the wheels, and this same rope might drive all the wheels. He described this system rhetorically with the help of his figure 103 (Fig. 5).

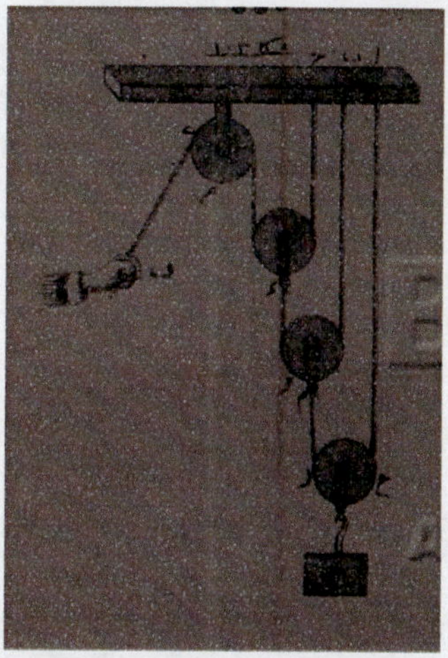

Fig. 5. Figure 103 of Ishaq Efendi [13].

After that point, he continued with explaining the *horizontal compound* (terkib-i ufkî) (Fig. 6) and *vertical compound* (terkib-i ‘amûdî) (Fig. 7) descriptively with the help of his figures 104, 105 and 106 in his book. Then he categorised the *vertical compound* as *terkib-i ‘amûdî* evel as in his figure 105 (Fig. 7 left), that was -as he described- if one end of the rope was at the applied force and the other end was attached to the inner wheel. And *terkib-i’ amûdî sânî* as in his figure 106 (Fig. 7 right) that one end of the rope was at applied force and other end was attached to a fixed point on the rod which was attached to the weight and the inner wheels.

As a result, in this subject, he concluded friction was negligible in the calculations, and the number of fixed pulleys didn’t change the proportion between the force and weight. Then he continued this section with two theorems. First of these theorems stated that the number of the parallel ropes in the vertical and horizontal compound was proportional with the ratio between the force and weight. Then he described that equilibrium with the help of his figure 103 (Fig. 5). He explained in detail how the load would divide accordingly to the pulleys on the figure, and he calculated the proportion between force and weight as 1/8.

Last but not least, in his last theorem on this subject, he discussed the relation between the force, weight, rope and, therefore, the wheel number. In the first result, he explained that in horizontal and vertical pulley systems (if the tractive force is on the fixed pulley) proportion between force and weight should be equal to the proportion of 1 to the number of pulleys. Then there would be equilibrium as in 104,105 (Fig. 6 and 7 left). In this theorem, he also explained the relation in detail that, while there was a gain in force, that would cause loss from the road. In order to move the load 1 unit on the movable

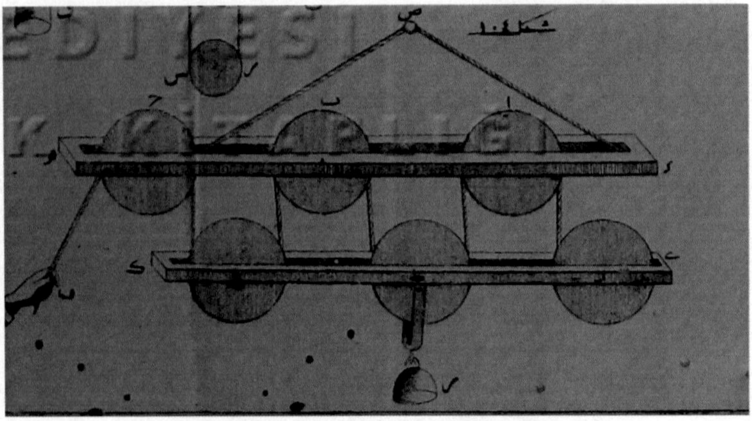

Fig. 6. *Horizontal compound* figure 104 of Ishaq Efendi [13].

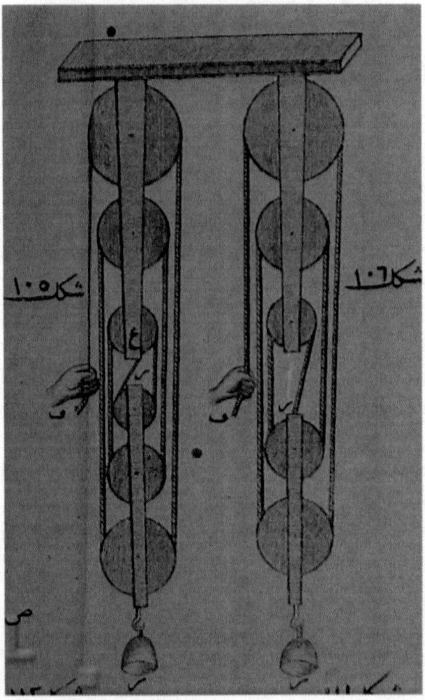

Fig. 7. *Vertical compound. Terkib-i ʻamûdî evel* figure 105 (left). *Terkib-i ʻamûdî sânî* figure 106 (right) of Ishaq Efendi [13].

pulley, each rope should move 1 unit, and it was necessary to pull the rope on tractive force 2 units (in total). So, if the weight goes up 1 unit in his example, the force should be applied 2 units in a downwards direction.

The second result of this theorem is about *terkib-i 'amûdî sânî* as he described above in his figure 106 (Fig. 7 right). He described that in that system, the proportion between force and weight should be equal to the proportion of 1 to the number of pulleys plus one (that is the case if the tractive force is attached to the moveable pulley). That can be expressed in modern terms (Eq. 1) when F is the applied force, P is the weight, and n is the number of the pulley then,

$$\frac{F}{P} = \frac{1}{n+1} \tag{1}$$

He also conducts this relation with the number of ropes by assuming that each wheel is attached to 1 rope from 2 sides. At that point, he warned the reader that the direction of the applied force in compound pulleys is irrelevant with the calculations.

In the third result, he investigated the path-force relationship on compound pulley systems in which the pulley's weight and friction are negligible. He explained that when there is more gain in force, there should be more loss in the path. In other words, he tried to explain that the distance travelled by the load can be calculated according to the work principle. As an example, for this result, he assumed a weight attached to a pulley which is attached to 4 ropes. He says if the weight wanted to lift 1 unit, then the ropes have to move 4 units (in total). We can describe his argument in modern terms as in Eq. (2), (3) & (4). Let the $4F = P$ in a system and force F moves h' and weight P moves h. According to the work principle

$$\mathrm{F.h'} = \mathrm{P.h} \tag{2}$$

$$\frac{P}{4}.\mathrm{h'} = \mathrm{Ph} \tag{3}$$

$$\mathrm{h'} = 4\mathrm{h} \tag{4}$$

In the last result, Ishaq Efendi reminded the relation of the number of the moveable pulleys, force and weight. According to his previous descriptions, when at least two of these unknowns are known, then the other would easily be calculated. He concluded that section with two examples for both types of pulleys [13].

4 Conclusion

Pulleys and pulley systems have been used since the early times of history. The application of this simple machine was known since Ancient Egypt and Mesopotamia. It is used widely in constructions, transportation for lifting heavy weights and water extractions from the deep wells. The theoretical aspect of pulleys was studied first by Archimedes. Then many centuries later, Galileo was the first scholar who handled this subject academically in modern terms, as is mentioned above. The knowledge of the pulley has been available in Anatolia since very early times of history. Ottomans were also using that simple machine for their daily needs. It is unknown for us whether or not there were any theoretical essays on that subject which was taught to students in madrassas, before 18th century [1]. As far as it is known, the first theoretical essays on mechanical science (cerr-i eskal or ilm-i mekanika) on pulleys were included in the curriculum from the beginning of the 19th century. These essays were generally translated from French books.

First teachers who included the simple machines in the curriculum was Yahya Naci Efendi (d. 1824) and Hüseyin Rıfkı Tamani (d.1817). In 1806, while Hüseyin Rıfkı Tamani was the chief instructor in the engineering school, it is well known that Sultan gifted many books to Military Engineering School. There were Abbé Charles Bossut's (1730–1814) books among these books. Around the 1810s, Yahya Naci Efendi taught physics from French books to the Military Engineering School. Around 1814–1815, it is known that he was teaching Bossut's book with the translated name *Cümlei ulûmi riyâziyye,* which includes mathematical sciences as well as mechanics. Teachers were taught these books either directly from the French originals or by simultaneously translating them. Later around 1834, Yahya Naci Efendi's son, Mehmed Nureddin, translated a mechanics book of Bossut's with the name *Kitabı Cerr-i eskal* [5]. Since these books were catalogued under the Ottoman name, further studies are needed to uncover which book of Bossut's were exactly used during these translations [1].

The first Turkish treatise was written about mechanics (therefore about simple machines and pulleys) was Ishaq Efendi's third volume of *Mecmua-i Ulûm-i Riyâziyye (Compendium of Mathematical Sciences),* which was written in 1832. Since it was the first of such kind of study, it was also important because it introduced the Ottoman terminology in simple machines and pulleys. Because of that reason, it was challenging to understand the Ottoman text during this study and physical interpretation is used while evaluating and explaining the text. It can be said that Ishaq Efendi explained the basic subjects of pulleys and pulley systems in great detail. He used quantitive examples, theorems, results and warnings on the subject. Friction and resistance cases were ignored as well as the mass of the pulleys, during the lecture on pulley systems. The main source of the 3rd volume was Étienne Bézout's (1730–1783) *Cours de mathématiques, à l'usage du corps royal de l'artillerie*'s 4th book [3, 9]. After the publication of Ishaq Efendi's compendium in 1832, this book started to be used as the main source in education in Military Engineering School since this subject was handled entirely in Turkish.

After this work, various books were written and translated on this subject (mechanics including the theory of simple machines and pulley), especially in the second half of the 19th century. One of the most important works written on this subject was by Ishaq Efendi's student Derviş Mehmed Emin Pasha (1817–1879). Derviş Pasha's work called *Usûl-ü Hikmet-i Tabiye (Introduction to Natural Philosophy),* considered as the first "physics" textbook taught in Ottoman Turkey by some historians. In the first article of that book, pulleys explained. This book was printed twice in 1865 and 1870 [22]. It can be assumed pulleys were explained to the students only theoretically and was not shown in practice in the first half of the 19th century, since Military Engineering School's list of tools did not include any pulleys in the years 1801, 1816, 1822 and 1836 [4].

References

1. Akbaş, M.: Osmanlı Türkiyesi'nde Modern Fizik (19. yüzyıl). Unpublished Ph.D. Dissertation, Istanbul University, Social Sciences Institute, Philosophy Department, Istanbul (2008)
2. Aslan Seyhan, I.: Mathematical instruments used in the military engineering academies of the ottomans. In: Zhang, B., Ceccarelli, M. (eds.) Explorations in the History and Heritage of Machines and Mechanisms. HMMS, vol. 37, pp. 243–258. Springer, Cham (2019). https://doi.org/10.1007/978-3-030-03538-9_20
3. Bézout, É.: Cours de mathématiques, à l'usage du corps royal de l'artillerie Tome quatrième, concernant l'application des principles généraux de la Mécanique, à différens cas de movement and d'équilibre, de l'imprimerie de Ph.D. Pierres, rue S. Jacques, Paris (1788)
4. Beydilli, K.: Türk Bilim ve Matbaacılık Tarihinde Mühendishâne ve Mühendishâne Matbaası ve Kütüphanesi (1776–1826), Eren Yayıncılık, Istanbul (1995)
5. Bossut, C.: Kitabı Cerr-i eskal (Tr: Ahmed Halil, Mehmed Nureddin b. Yahya Naci), Bulak Matbaası, Kahire (1834)
6. Ceccarelli, M.: Early TMM in Le mecaniche by Galileo Galilei in 1593. Mech. Mach. Theory **41**(12), 1401–1406 (2006)
7. Curley R. (ed.): Heron of Alexandria, The 100 Most Influential Inventors of All Time, pp. 25–26. Britannica Educational Publishing (2010)
8. Dieter, A.: Building in Egypt: Pharaonic Stone Masonry. Oxford University Press, Oxford (1991)
9. Dosay-Gökdoğan, M., Aslan-Seyhan, I.: İshak Hoca'nın Mecmua-i Ulûm-i Riyâziyye'sindeki Basit Makinelerden "Mengene" Bahsi. In: 19 Ulusal Makine Teorisi Sempozyumu Bildiri Kitabı, pp. 531–539 (2019)
10. Gökdoğan, M.D.: Mechanics (machines) in ishaq efendi, the chief instructor of military school in the ottomans. In: Zhang, B., Ceccarelli, M. (eds.) Explorations in the History and Heritage of Machines and Mechanisms. HMMS, vol. 37, pp. 3–16. Springer, Cham (2019). https://doi.org/10.1007/978-3-030-03538-9_1
11. Gorringe, H.H.: Egyptian Obelisks, pp.146–160. H.H. Gorringe (published by the author, 32 Waverly place), New York (1882)
12. Heath, T.L: Archimedes. Society for Promoting Christian Knowledge. Macmillan, London (1920)
13. İshak Efendi: Mecmûa-i Ulûm-i Riyâziye, c. 3, Bulak Matbaası, Kahire (1845)
14. Kadiroğlu, G.: Antik Dönemlerde Vinçler ve Kaldırma Araçlarının Kullanım Alanları, Unpublished Dissertation, Mehmet Akif Ersoy University, Archeology Department, Burdur (2014)
15. Pisano, R., Capecchi, D.: Tartaglia's science of weights and mechanics in the sixteenth century. History of Mechanism and Machine Science, vol. 28. Springer, Dordrecht (2016). https://doi.org/10.1007/978-94-017-9710-8
16. Romdhane, L., Zeghloul S.: Distinguished figures in mechanism and machine science. In: Ceccarelli, M. (ed.) History of Mechanism and Machine Science, vol. 7, pp. 1–22. Springer, Heidelberg (2010). https://doi.org/10.1007/978-1-4020-6366-4
17. Schiefsky, M.J.: Theory and practice in heron's mechanics. In: Laird, W.R., Roux, S. (eds.) Mechanics and Natural Philosophy Before the Scientific Revolution. Boston Studies in the Philosophy of Science, vol. 254, pp. 15–49. Springer, Dordrecht (2008). https://doi.org/10.1007/978-1-4020-5967-4_1
18. Sezgin F.: Technology and Science in Islam. Katipli Mum Saati (2017). http://www.scienceinislam.com/katipli-mum-saati/

19. The Editors of Encyclopaedia Britannica: "Deus Ex Machina". https://www.britannica.com/art/deus-ex-machina
20. Usher, A.P.: History of Mechanical Inventions. Harvard University Press, Harvard (1954)
21. Morgan, M.H.: Vitruvius: The Ten Books of Architecture. Harvard University Press, Harvard (1914)
22. Yinilmez-Akagündüz, S.: Osmanlı Devleti'nde Okutulan İlk Fizik Ders Kitabı: Usûl-ü Hikmet-i Tabiiye (Doğa Felsefesine Giriş). Turk. Hist. Educ. J. **2**(2), 58–77 (2013)

Past Iberoamerican Distinguished Figures, Experiences and Designs in MMS

Mechanical Engineering Devices in the America Colonization

Rubén Dorado-Vicente(✉), Rafael López-García, Juan Manuel Quero-Nieves, and Gustavo Medina-Sánchez

Department of Mechanical and Mining Engineering, University of Jaen, Jaen, Spain
rdorado@ujaen.es

Abstract. Due to the scientific gap between America and Europe at the XVI century, the first Europeans' settlements in the new world had to face different technological issues. Besides to expose the scientific and technological situation, this paper shows two singular mechanical wits representative of the technical advances introduced in America to overcome different problems. Firstly, a pump based on an alternative movement system through crankshafts and pistons used against the continuous flooding suffered in the Ciudad de México valley. Although this system did not solve the problem (it was solved in the XIX century), it was essential to the continuous depletion of water. Secondly, a port crane that due to its capabilities to handle cannons, military devices and constructive materials, was useful during Lima fortification in the XVII century.

Keywords: Port crane · Hydraulic pump · Reverse engineering · Industrial heritage

1 Introduction

The technological and scientific advances in Europe during the XV and XVI centuries were critical for the development of agriculture, metallurgy, warfare and naval transport [1]. Each improvement feedbacked the others and all together were boosted by the trade routes in the Mediterranean Sea as well as the opening of routes between East and West, and finally with the New World.

During the evolution of the aforementioned techniques, the printing press played an important role increasing the possibility of expanding the wealth of existing knowledge to extraordinary levels in a society that already had a large amount of graphic production.

Innovations in agricultural technology were unevenly lodged in Europe. These advances include the use of better water mills to drain the land, the use of drying techniques and obtaining a balance between crops and animals, with a method consisting of cultivating a specific part of the land to feed the animals, which in turn fertilized the land. The insertion in Europe of new crops from America and Asia, required new engineering knowledge, both to drain the land and to set up new farms.

From the middle of the XV century, Germany laid the foundations for the mining exploitation and subsequent metal processing, this by improving mining techniques,

M. Ceccarelli and R. López-García (Eds.): HMM 2022, HMMS 40, pp. 421–432, 2022.
https://doi.org/10.1007/978-3-030-98499-1_34

since at that time Germany had some of the richest mines. During the first half of the XVI century, a growing transfer of mining and metallurgical technology took place in two directions: in America, to exploit the silver mines of Potosí and Mexico, and in Central Europe [1].

Other metallurgical procedures were undergoing improvements, such as the melting of metals, contributing to later advances in Chemistry by providing experience on chemistry empirically, since to use this science was necessary to have an exhaustive and precise control of the metals and non-metals that intervene in it.

Following the above discussion, in the XV and XVI centuries, the Spanish settlement of colonies in America required an important scientific and technological effort to solve many technical issues [2]. Regarding the approaches implemented, it is noteworthy the lack of historical references that help to value and appreciate the engineering quality of these solutions.

Secrecy and vertical knowledge based on petitions from vassals to the king was the preferer way for scientific documentation in the Spanish empire [3]. This approach did not reduce scientific creation, but it has contributed to hide the large number of technical advances and scientific developments achieved by Spanish and Native Americans.

This work explains how a reverse engineering procedure [4, 5] can provide virtual prototypes of lost mechanical systems: a pump used to drain the Ciudad de México valley and a port crane used in the port of Lima. Despite of the difficulties raised by the secrecy pointed out above, the use of engineering methods supported by current CAD and numerical simulation tools can improve our knowledge about these ancient technical wits.

Regarding the manuscript structure, Sect. 2 describes the pump and the crane devices in the spotlight. Section 3 and 3.1 respectively explore the existing information about the pump and the crane, describe the recovering procedure applied, and show the resulting virtual mockups. Finally, the main conclusions are drawn in Sect. 4.

2 A Draining Wit

In 1521 the Spanish settled Mexico City in the 9 square kilometer area formerly known as Tenochtitlan. Unfortunately, the city is located in a valley that suffers continuous floods.

According to the writings [6], on October 14, 1555, the Cabildo (the responsible for the municipality's government) requested a solution to the water problem. Viceroy Velasco ordered rebuild the stones works destroyed in the past, but that solution did not solve the problem.

In the absence of short-term solutions, the viceroy proposed changing the location of the city, to which Madrid replied that it would cost more to build the capital from scratch than to solve the flooding problem. Therefore, it was needed to find a quick and effective way of drainage to continue facing the problem of the works. This task was designated to the German engineer Enrico Martínez, who had the ingenious idea of draining the valley by drilling it to leakage the formed lagoons' water.

After successive and unsuccessful attempts to continue with his initial idea, Enrico Martínez was accused of negligence and taken to prison. However, he was finally released to participate in the project again because of his experience. When Enrico Martínez died in 1630, Fray Andrés de San Miguel, a Spaniard living in Mexico, accepted the challenge. He had excellent knowledge of hydraulics and knew the previous solutions and works to be improved.

The viceroy, faced desperate by the flooded capital situation, ordered the Cabildo to promote a drainage system and boosted the project with 24 machines and the necessary men and animals for continuously evacuating water in strategic points of the city. The work promoted by the Viceroy took advantage of the open-pit excavations carried out by Enrico Martínez. However, the initial project, focusing on the surrounding lagoons, was never finished due to the continuous government changes and the non-continuity of the works.

The continuous extraction of water from the subsoil during the colonial period was critical compared to the failed project for controlling the lakes' water through channels. Thus it is interesting the recovery and technological analysis of a hydraulic pump used for draining water.

2.1 Description of the Pump

According to Fray Andrés de San Miguel, who was present at the drainage work in Mexico City during the viceroyalty, the hydraulic pump understudy, illustrated in Fig. 1, was used to drain water in Mexico City.

The hydraulic pump consists of two supporting pillars, two carriage-type wheels for manual movement, one wooden structure for the cylinder block, one lantern transmission mechanism, a very long crankshaft, eight connecting rods, eight cylinders contained in the same block, a system of valves operated by pressure difference, a system of conduits for evacuating water and a set of water outlet pipes.

The system, most likely based on the technical knowledge from Europe, works as follows (Fig. 1 and 2):

1. The movement of two donkeys runs a vertical wooden axis. The lantern mechanism transmits the axis circular movement to the crankshaft, which is mapped into an alternative movement.
2. The crankshaft moves the connecting rods, which generate a continuous reciprocating movement in the cylinders housed in the block, creating pressure differences in the cylinders, opening the suction valves, and closing the non-return valves, water entering the odd-numbered cylinders. At the same time, the pairs eject and vice versa in a continuous movement.
3. The water leaves the cylinder block through a system of tubes that leads it to a vertical tube which evacuates it to the desired drainage point.

2.2 Reverse Engineering: Pump Reconstruction

One of the main issues in obtaining a detailed design of an ancient device is dealing with and studying the current information and archeological remains. The starting point for the pump system under study was a drawing of Fray Andrés de San Miguel (Fig. 1). That illustration provided the information to identify the bill of materials and the main

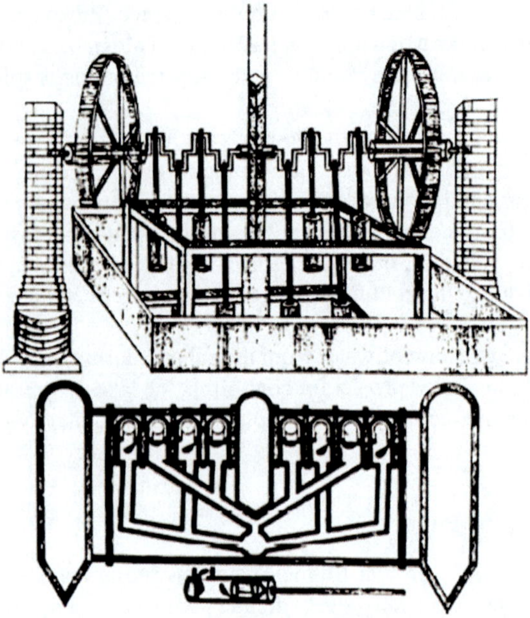

Fig. 1. Scheme of the hydraulic machine under study used for draining water [7].

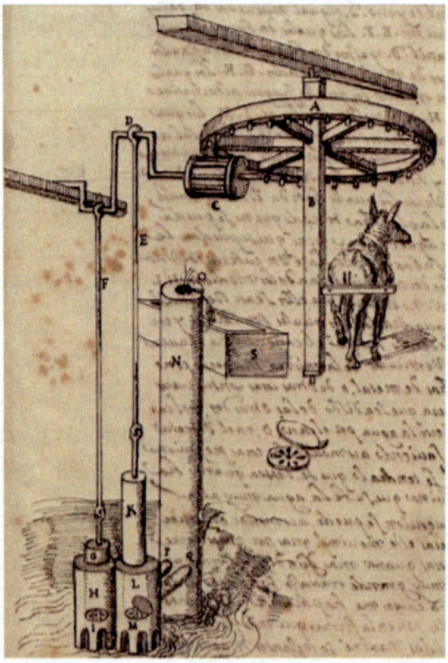

Fig. 2. Two-cylinder force pump from "*The twenty-one books*" [8].

dimensions of the system elements. It stands to reason considering that the parts were drawn to preserve the relation among component sizes. Besides, the bricks corresponding to the structure that supports the ends of the crankshaft are the reference used to scale parts dimensions. From a specific bibliography, the bricks used for purposes similar to that of the machine were about 45 mm thick.

Fray Andrés de San Miguel focused on the pump wit, but he did not care about how to power the system, and it is only known that it was blood-actuated. The approach considered in this work is based on the Ctsibio pump. It is operated in the same way as "*The Twenty-One Books*" suction machine (Fig. 2). Due to the enormous power requirement given its dimensions and the fact that it consists of 8 cylinders, most likely, the pump was animal-run by at least two donkeys which fit well with the symmetrical nature of the system.

To size the lantern gears that run the pump through the crankshaft, and to figure of flow rate, it is needed an estimation of the animal power $\dot{W}_{\text{Animal}}$, an estimation of the gear performance η_{Gear} as well as the input-output pressure difference ΔP, flow rate of water Q, and the performance of the pump system η_{Pump}.

The power provided by the two donkeys $\dot{W}_{\text{Animal}}$ depends on their average speed v when moving a specific load F; both data were obtained from specific bibliography (v ~ 1 km/h for $F = 1960$ N [9]). Taking into account the η_{Gear} (that for a lantern gears is usually considered equal to 0.5), the crankshaft received this power:

$$\dot{W}_{\text{Crankshaft}} = \dot{W}_{\text{Animal}} \cdot \eta_{\text{Gear}} = 2 \cdot v \cdot F \cdot \eta_{\text{Gear}}. \quad (1)$$

On the other hand, the power needed to suction water is:

$$\dot{W}_{\text{Pump}} = \Delta P \cdot Q \cdot \eta_{\text{Pump}}. \quad (2)$$

The relation of Eq. 1 and 2 and the hypothesis: $\eta_{\text{Pump}} = 0.6$, ΔP ~ 1 atm provide a way to estimate the flow rate Q. The flow obtained is 47.39 l/s. This value could not be obtained manually in any way.

Once known the pump volume, the flow rate, and the transmission rate (relation among crankshaft and animal rotational speeds r ~ 2.1), it is possible to size the lantern gears. Once the calculations have been made, it is possible to obtain a detailed 3D model of the system and simulate its movement (Fig. 3).

3 A Port Crane

Callao is and was in the past the largest port in Peru. Formerly, it was an indigenous fishing village called "*Piti Piti*," and its use as a pier and place of reception and departure of ships predated the formation of the city of Lima founded by Francisco Pizarro.

The first name it received was "*Puerto de la Mar*," and commerce was its primary use. Proof of this is the Cabildo license to construct a warehouse on March 6, 1537, to store the merchandise produced by the port's commercial activity [10]. Antonio de Herrera's work, "*Descripción de las Indias Orientales*", mentions the existence of a custom house and its large capacity and activity.

The above commercial placed the port in the spotlight of other nations, which attacked it several times during the 16th and 17th centuries. As a result, the port had to be fortified,

Fig. 3. Virtual mock-up of the pump understudy.

creating the first defensive structures under the mandate of Viceroy Francisco de Toledo in 1570.

Viceroy Juan de Mendoza y Luna built a platform in 1615 which had a defensive artillery system separated into three fortifications, whose shooters had a view over the entire bay. Given the danger of the continuous enemy expeditions in the South Sea, the port was walled in 1640 over the existing base. Callao became a fortified city port.

It is worth mentioning that the Callao pier was designed to avoid the cumbersome loading and unloading of the ships through unstable boats that usually overturned due to the load during the journey to the beach.

Port cranes were installed on the piers to load and unload military material, merchandise, and other heavy construction elements from boats. There is evidence of several testimonies of the first large crane installed: "Port cranes were placed that served to unload and load merchandise. They were rotating and moved by men who stepped on large wheels, and sometimes they also had multiplications based on pulleys and hoists… As a safety measure, they used ratchets [11].

Port crane design would be based on a technical manual such as "*The twenty-one books*" [8]. In this book, Juanelo Turriano described a crane to lift loads, which could turn on itself and had a treadwheel. It meets all the characteristics described in the testimonials and has a perfect geometry for port use, given its simplicity and the mechanical advantage it undoubtedly provides. Therefore, the builders of the system understudy could use the schemes portrayed in Juanelo Turriano to manufacture the crane. Years after its implementation, the crane was changed for improved solutions, which appeared in the "*Album de construcción naval del Marqués de la Victoria*" [12].

3.1 Port Crane Description

The port crane consists of treadwheel, treadwheel support structure, pillar with a coupled pulley, pillar structure for rotation, support the weight of the people who rotate it and finally stabilize the column to avoid buckling and unwanted horizontal displacements that would cause instability. Pillars are buried in a compacted land as the foundations of the houses of the time.

According to "*The Twenty-one Books*," which was the basis of the technique that probably travels to the new world, it is a construction crane, but it was used as a port crane by extending its size and reinforcing the structure for permanent use. It was installed around the year 1707, according to the testimonies.

The system works as follows:

1. Men walk inside the treadwheel made of wood and power its rotation producing a thrust torque.
2. The rotation of the wheel wraps the crane rope in the drum attached to its axis. This rotation tightens the rope raising the load connected to it.
3. The vertical movement provided by the treadwheel combined with the rotation of the leading crane pillar allows complex movements as required in port loading-unloading activities. Workers can push the timbers attached to the pillar or move the load by hand to achieve the desired lateral movement.

The pillar ends in a large metal bolt supported on its base, thanks to it is possible the rotation and at the same time prevents horizontal displacements. This mechanical

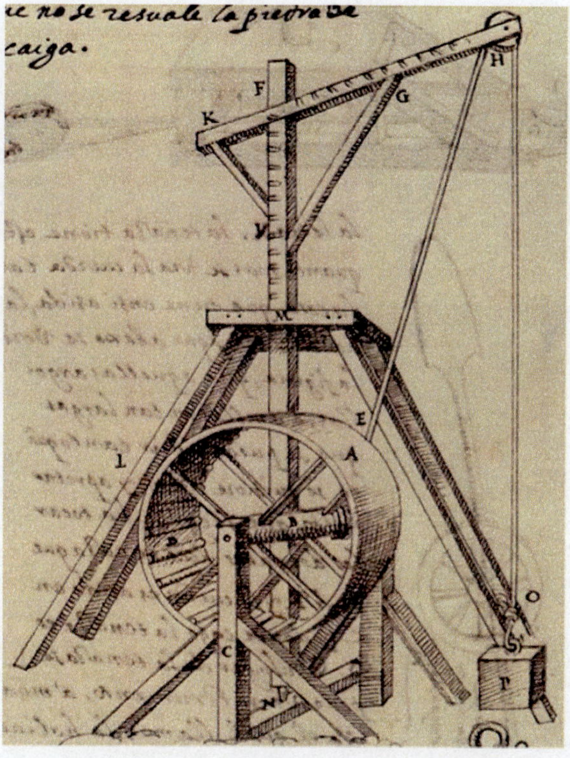

Fig. 4. Crane from "*The twenty-one books*" [8].

approach, lubricated with animal or vegetable grease, gives stability against buckling, limits lateral displacements, and reduces the bending of the pillar base.

4. Once the load is lifted and turned until it rests on the land side, the men are told to stop walking since the rope's tension is lost. Very thick logs were used as a ratchet if, for some reason, the wheel had to be stopped, supporting it between the internal logs of the wheel and the base log.

3.2 Reverse Engineering: Crane Reconstruction

The starting point to dimension the crane elements is the definition of the treadwheel size. It should be large enough that a man can walk comfortably inside it. Once this wheel is defined, the rest of the crane elements (Fig. 4) can be proportionally dimensioned and drawn using AutoCAD software. Considering an average man height of 1.70 m, it stands to reason taking a wheel radius of 3 m, which leaves more than 1 m between the wheel axis and the worker's heads.

It is necessary to establish the number of men needed to run the crane to analyze how it works. An energy balance provides the equations to estimate this number of workers. Taking into account the scheme portrayed in Fig. 5, the torque provided by N men that push the radius R wheel with a force E is equal to the torque transmitted to the radius r drum that winds the rope connected to the load F:

$$N \cdot R \cdot E = r \cdot F. \tag{3}$$

Regarding the maximum weight to lift for this crane, since it was installed a few years after 1700, the heaviest constructive or military element composed of a single piece would be a 24-pound cannon, used by the Spanish navy to defend fortifications from XVI to XIX century. Its weight was $F = 2500$ kg in total, of which almost 500 were part of the gun carriage.

It is interesting to calculate the mechanical advantage produced by the system or the factor that multiplies the force provided to overcome the resistance. Known r and R values:

$$F = R/r \cdot E \rightarrow F = 6.207 \cdot E. \tag{4}$$

Considering that an average man weighs $E \sim 70$ kg and Eq. (5), the number of workers needed to lift the maximum load is 6.

Dimensions and power input have been determined. The following step concerns analyzing the crane structure to verify that it can withstand the stresses to which it is subjected in operating conditions.

It is important to note that all the crane elements are mahogany wooden since it is a large tree to obtain the necessary pieces. Its wood is robust and is very abundant in the forests of Lima.

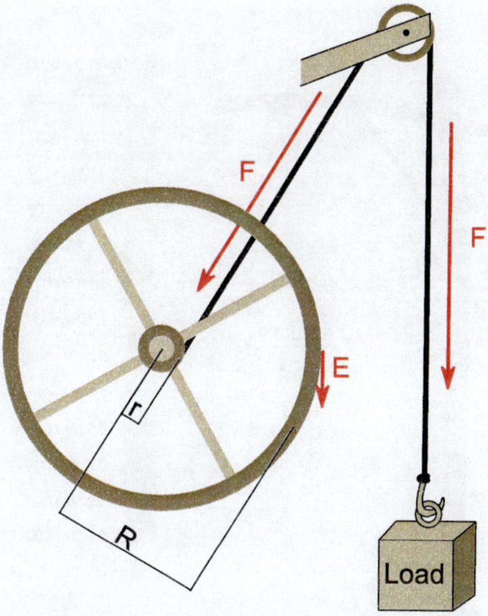

Fig. 5. Forces scheme for the crane system.

Finite element analysis of the pillar (Fig. 6).

The pillar rotates on itself, and it is connected to the load $F = 2500$ kg through a pulley. The joints are considered rigid since the woods are very well joined with each other. After conducting a static analysis by finite elements in the software Solidworks, the pillar that supports the load shows that it fully complies with the Ultimate Limit State for resistance. Being its maximum tension 22.13 MPa, 26.55% that of the material (83.35 MPa). So its service condition is safe and it possible to lift heavier loads with more than 6 workers. It is worth mentioning that the timber on the left has no structural function, so it is confirmed that it was used to rotate the crane.

Finite element analysis of the pillar base (Fig. 7).

The pillar supports are semi-buried and completely fixed to it. A force of 6867 N is applied on the upper face, the equivalent of having ten people weighing 70 kg simultaneously on the platform, an exaggerated figure but which serves to check the structure for resistance with more security.

As shown in Fig. 7, the base does not reach even 1 MPa, and the elastic limit is 83.35 MPa since its supports are very robust and there are four so that ten people for it does not suppose anything. This robustness is to stabilize the pillar that goes inside and allow the crane to rotate safely.

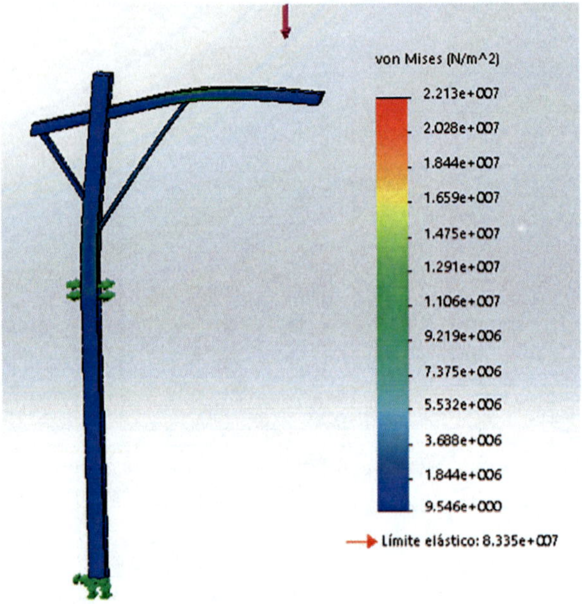

Fig. 6. Von Mises stress in the crane pillar.

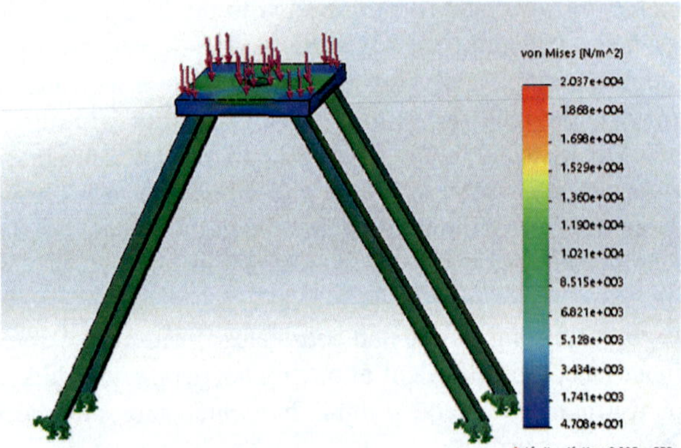

Fig. 7. Von Mises stress in the crane base structure.

Finite element analysis of the tread wheel support (Fig. 8).

From the volume and density of the treadwheel, its weight can be estimated at 12580.98 kg. A load of half the wheel's weight is applied to each hole that supports the wheel so that 6290 kg corresponds to each support, the equivalent of 61709 N per support. On the other hand, the supports are partially buried and wholly fixed to the ground as if they were cemented.

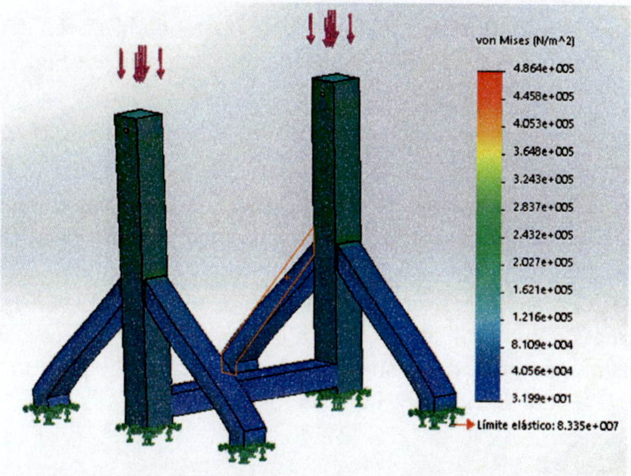

Fig. 8. Von Mises stress in the wheel support.

As a result of applying the conditions described above, it is observed how the maximum stress is reached in the joints with the diagonal supports, with a value of 0.4 MPa, far from the elastic limit, since this structure is oversized like the whole assembly in general with woods of great section and resistance.

It is observed that the crossbar inserted in the middle does not fulfill a structural function since the supports are embedded. Indeed, it was designed this way to compensate for possible movements of the supports due to the terrain, thus maintaining the structure's integrity with greater security against possible imposed displacements.

Fig. 9. Virtual mock-up of the port crane.

Once proven structurally viable and its whole operation has been determined, it is possible to make a complete 3D model and simulate its movement Fig. 9.

4 Conclusions

This work relives through virtual prototypes, built using reverse engineering, two mechanical wits used in the new world Spanish settlements from XVI to XVIII centuries. Specifically, this study focuses on a hydraulic pump used to drain water in the valley of Ciudad de México and a port crane installed in Callao's port (Peru). From limited historical information, the geometrical design, the engineering analysis, and numerical simulations fill the gaps found in the available information helping to accomplish a detailed CAD 3D model that can be simulated or printed.

References

1. Prieto Romero, J.J.: Ingeniería mecánica en el siglo XV y su contribución al descubrimiento de américa. (XV century mechanical engineering and its contribution to the Discovery of America). Tesis Doctoral, Universidad de Cádiz (2009)
2. Calvo, M.L., Enoch, J.M.: Introduction to the history of lenses and visual corrections: a reference to Spain and the Spanish colonies in the new world (XV–XVI C.). Revista Cubana De Fisica **22**(1), 3–12 (2005)
3. Canizares-Esguerra, J.: Bartolome Inga's mining technologies: Indians, science, cyphered secrecy, and modernity in the new world. Hist. Technol. **34**(1), 61–70 (2018)
4. Montes, J.C., López-García, R., Dorado-Vicente, R., Trujillo, F.J.: Mock-up of an eighteenth-century oil mill via rapid-prototyping. In: López-Cajún, C., Ceccarelli, M. (eds.) Explorations in the history of machines and mechanisms. HMMS, vol. 32, pp. 65–75. Springer, Cham (2016). https://doi.org/10.1007/978-3-319-31184-5_7
5. Torres-Garrido, S., Lopez-Garcia, R., Carou-Porto, D., Dorado-Vicente, R.: Historical and technological study of the 26 and 30 mechanisms of "The Book of Secrets" by Ibn Khalaf al-Muradi. In: Zhang, B., Ceccarelli, M. (eds.) Explorations in the History and Heritage of Machines and Mechanisms. HMMS, vol. 37, pp. 213–226. Springer, Cham (2019). https://doi.org/10.1007/978-3-030-03538-9_18
6. Catalá, J.: Ciencia y técnica en la metropolización de América. CSIC-CSIC Press (1994)
7. de San Miguel, A., Macías, E.B.: Obras de fray Andrés de San Miguel. Universidad Nacional Autónoma de México, Instituto de Investigaciones Estéticas (2007)
8. Turriano, J.: Los veintiún libros de los ingenios y máquinas de Juanelo Turriano: transcripción del manuscrito con prólogo de Pedro Laín Entralgo y reflexiones de José Antonio García-Diego. Fundación Juanelo Turriano (1996)
9. Chirgwin, J.C., Food and Agriculture Organization of the United Nations, de Roover, P., Dijkman, J.T.: El burro como animal de trabajo: manual de capacitación. FAO (2000)
10. Arrús, M.D.: El Callao en la época del coloniaje antes y después de la catástrofe de 1746: Posición real y efectiva de la ciudad, en relación con el antiguo presidio, Fundación de Bellavista. Terremotos y piratas, datos histórico-cronológicos de los virreyes y gobernadores del Perú y de los sucesos más notables desde 1530 á 1829. Imprenta de El Callao (1904)
11. López-Ayala, J.M.Z.: La Escuela de Fortificación Hispanoamericana. Consideración general. Las "academias" fundamentales. El gigantesco escenario de la fortificación hispanoamericana (1985)
12. Victoria, J.J.N., Naval, M.: Álbum del Marqués de la Victoria. Lunwerg (1995)

The Stamp Mill in Mexico's Antique Mines

Juan Carlos Jauregui-Correa(✉)

Universidad Autonoma de Queretaro, Querétaro, Mexico
jc.jauregui@uaq.mx

Abstract. This paper contributes to the understanding of the design and operation of antique mechanisms that worked in the silver mines before the industrial revolution. The work presented here describes the design principle and the kinematics and dynamics of a stamp mill. The stamp mill was used for breaking ore stones before extracting the silver with mercury (quicksilver). The kinematic and dynamic model were determined using the kinematic chain concept, the operation and dimensions were estimated from original drawings and technical data written in Europe and in Mexico. Nevertheless, it was not possible to find specific engineering data. The results from the analysis were verified estimating the production capacity reported in the XVI and XVIII centuries. Up to now, there are no archeological records of the dimensions and materials used in those days.

Keywords: Antique mechanisms · Reconstruction procedure · Kinematics · Identification techniques

1 Introduction

This paper is part of a series of articles devoted to describing the mechanism built in Mexico during the Colonial time [1, 2]. This study is a complementary answer to those questions regarding the construction and functioning of machines used in the silver mines (From centuries XVI to XVIII). It has been tough to find documents describing the technology of those days or the technical knowledge the people who operated and maintained the machinery had. It is known that there was a Mining Expert ("Perito Minero" in Spanish) and a Surveyor ("Agrimensor" in Spanish) who were appointed by a Mining Council [3] and were responsible for keeping and improving the equipment and the production processes. Since no records or documents explain those technologies, it is essential to reproduce their kinematics and dynamics with modern concepts.

The stamp mills were used in ancient Greece, the Eastern Mediterranean States, Rome, and China [4, 5]. Most of them were powered by water wheels, and a cam mechanism controlled the hammering effect. The stamp mill was used for breaking stones, whereas there were other mechanisms for grinding. In the Mexican silver mines, miners grinded ores with a particular machine known as "Tahona" [2]. Grain, fivers, and other materials were grinding with grinding wheels.

The stamp mills were widely used in Roman mining; they crushed ore from the mine veins for further process. Then, crushed material was grinded and smelted using charcoal bloomeries [6]. The Roman cam and stone anvil design evolved until medieval

M. Ceccarelli and R. López-García (Eds.): HMM 2022, HMMS 40, pp. 433–443, 2022.
https://doi.org/10.1007/978-3-030-98499-1_35

times. The Roman design was found in different gold and silver mines in Europe and the Iberian Peninsula. Most of them were powered by water wheels. Stamp mills were found in Samarkand, and they were widespread around the Islamic world, including Spain [7].

Archeologist have found mechanized hammers with the vertical pestle from different periods. They were applied for ore crushing, forging, stamping and other applications. They were represented by different artists in Europe and China. The stamp mills have different number of wooden hammers and the cams were out of face to have continuous impacts during the camshaft rotation [6].

For many years, understanding the functioning of ancient machinery and the reproduction of engineering data has been a research subject. Many publications have presented the Atikthera mechanism [8]. Yan and Lin [9] identified the joints and linkages and represented the mechanism as a kinematic chain. The "carroballista" is another mechanism that has been widely studied. The Romans developed this mechanism. Penta et al. [10] studied the "carroballista" with modern analysis tools, and they were able to confirm the historical evidence with engineering data. Azema et al. [11] presented a method for analyzing the welding process used by Romans to produce large bronze statues. They used ultrasonic phased arrays to study antique statues. Ceccarelli [12] described the most significant mechanisms that were developed during the Renaissance in Italy. Koetsier [13] described many mechanisms created from the ancient civilizations until the industrial revolution. Thomas [14] analyzed Leonardo Torres' endless mechanism. Hsiao [15] analyzed the synthesis of an antique crossbow, and classified three type of joints in ancient mechanism: similar joints as modern machinery, joints that are uncertain and joints that are rare in modern mechanisms. Chen et al. [16] analyzed ancient Chinese wooden horse carriage, and they applied Yan's approach for the synthesis of lost ancient Chinese mechanisms. Wauer et al. [17] analyzed the work done by Ferdinand Redtenbacher, who formally started the kinematic analysis of mechanisms.

The problem of analyzing antique mechanisms is the application of modern engineering tools to drawings without proper specifications or using archeological pieces with incrustations, corrosion, and post-corrosion. Veprek et al. [18] described a procedure for restoring archeological parts made of iron, silver, and bronze.

2 Silver Milling Process

During the Spanish domination in America (1492–1821) the economy depended on mining (silver and gold), agriculture, textile, and commerce [19]. Besides mining, most of the manufacturing processes were imported from Europe; thus, the machinery for producing gold and silver were developed locally. Unfortunately, few records support the engineering knowledge from that time. In 1557, Batolomé de Medina invented a new process for silver extraction. This process consisted on breaking, grinding and forming an amalgam of mercury and silver ore (known as backyard benefit or "beneficio de patio" in antique Spanish). After forming the amalgam, the silver was extracted by melting the mercury.

2.1 Description of the Silver Process

The ore was extracted from the mine veins; it was classified into three groups [20] almost pure ore, a mixture of ore and other materials without silver, and useless minerals. After classifying the ore, it has sent to the factory ("Hacienda de Beneficio"). Table 1 summarizes the process for producing silver from the ore.

Table 1. Silver extraction process [20]

Stage	Description
Grind	The grinding process had two stages: breaking the ore stones into small pieces with a hammer mill and grinding the ore with millstones and water. The output of this stage is a sludge formed with silver ore
Mixture	The sludge was deposited on a large backyard ("patio de beneficio") for producing an amalgam mixing the sludge with brine and mercury. The mixture was padded to ensure a good blend
Wash	Once the amalgam was formed, the mixture was washed with water. The washing process was done in mechanized washer tanks. The objective of this stage was to eliminate the salt and any remaining dust
Filter	The amalgam was filtered in leather bags that separate the water from the amalgam
Fusion	The mercury was eliminated by heating the amalgam. The mercury becomes liquid and flows at a specific temperature, leaving the silver
Foundry	Finally, the silver was melted in a ceramic foundry and cast as ingots

This paper presents the functioning of the stamp mill, which is described in the following section.

3 Description of the Stamp Mill

The stamp mill (Fig. 1) was adapted from similar designs produced in Europe. This type of machine was powered by mules, whereas in Europe, most were powered with water wheels. The main reason is that the silver mines in Mexico are located in desert areas.

Figure 2 shows an artistic representation of a milling yard. This area was part of the "Hacienda de Beneficio" (the name for the silver factories). The first stage for the silver production started at this yard. The process initiated breaking the ore stone into small pieces. The ore stones were broken with stamp mills. A complementary process was the selection of those stones that contained silver. Then, the selected material was weighted.

Figure 3 shows a sketch of the stamp mill widely used in the Spanish colonies. It was power by mules (blood engines) that pulled the bar (k) and rotated the vertical axel (i) and the gear (h). The gear (h) was a primitive version of a face gear. The rotating speed was incremented by the pinion (g) that transmitted the power to the camshaft (e). Several cams (f) were mounted out of phase (Fig. 1) to had a continuous hammering effect. The cams lifted the follower (b) and the shoe (c). The cam released the follower and, due to

Fig. 1. Schematic lithography of a stamp mill [21]

Fig. 2. Layout of the first stage of milling process

the gravity effect, the shoe hit the ore. The ore laid on the stone bed (a). Figure 4 shows a representation of the lifting mechanism (cam (f) and follower (b)).

Some improvements on the European concept were developed in Mexico. The miners suggested to use a copper plate for collecting the amalgam instead of having an

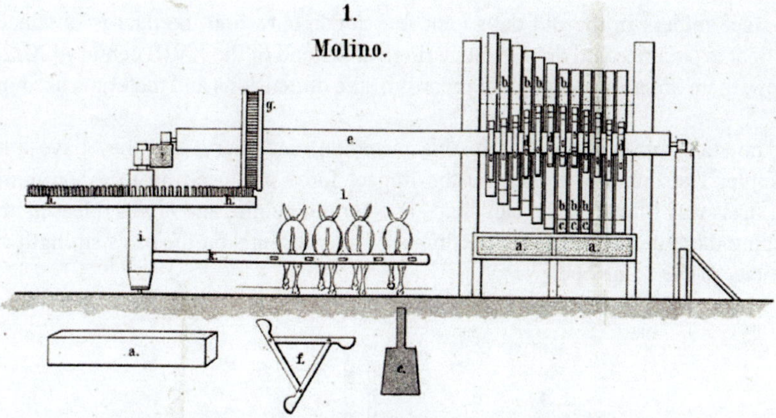

Fig. 3. Sketch of a hammer mill with technical data [22]

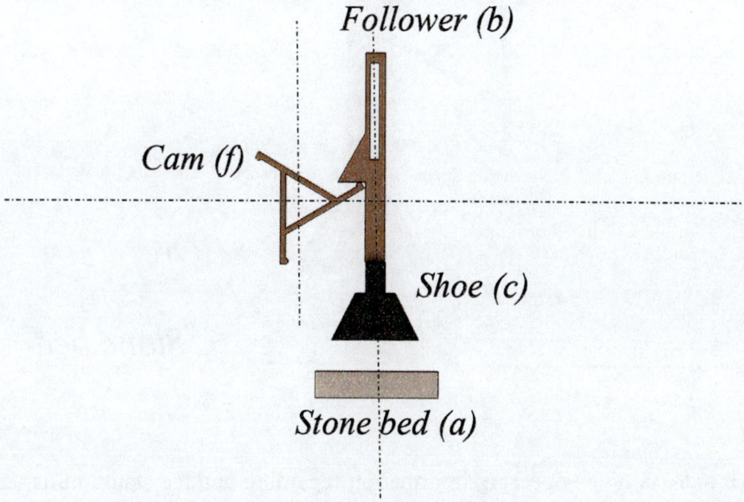

Fig. 4. Sketch of the stamp mill with technical data [22]

accumulation in the bottom of the vessels. This concept reduced the use of mercury, and it was named as the Cazo process. The Cazo process was adapted by many miners in Europe; nevertheless, nobody knows when it was invented.

The old technology is analyzed using the original descriptions, and the kinematics and dynamics behavior is reconstructed using modern concepts. The detailed analysis is based on current engineering principles, and the technical data is retrieved from the documents recovered in the XIX [19, 23]. The study is based on the original sketches and procedures that described silver production at that time; but, the reconstruction of the design concept, loading, and operation conditions is done using simulation models similarly to modern machinery. This approximation pretends to fulfill the information gap since miners in the old days kept few documents, and the data is obtained from practical experience and documents written at the end of the XVIII century [20, 24]. Up to now, there are no archeological records of the dimensions and materials used in those days.

The stamp mill evolved from this primitive machinery, and they have a similar principle. The ore was broken by the impact force produced by the shoe mass. The shoe mass was lifted (by the camshaft) at a certain height, and it was released and fell, crushing the minerals (Fig. 5). The impact force depends on the ore's strength and the elasticity of the stone bed.

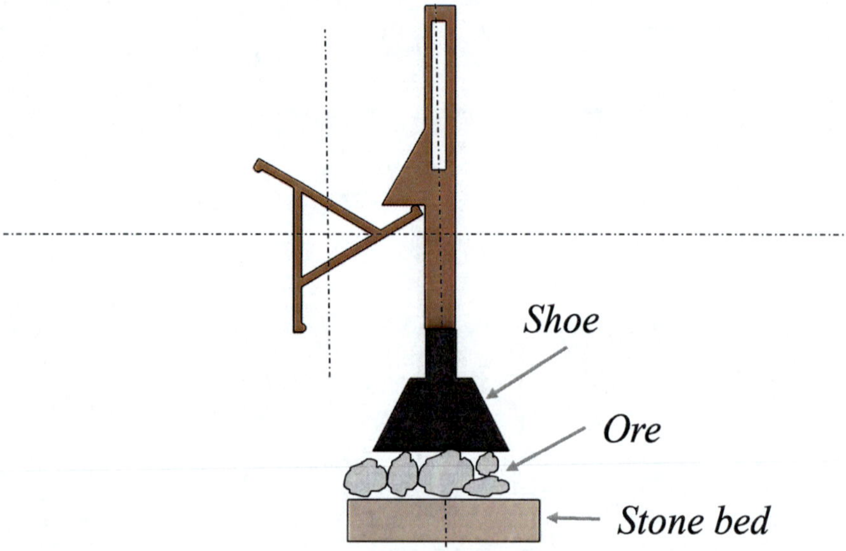

Fig. 5. Hammering effect

Since there is no engineering information regarding antique stamp mills, and the oldest publications that describe this equipment focus on overall parameters and production data, it is cumbersome to produce realistic models and results. An approximation of the kinematic and dynamic formulations of the antique stamp mill is described in the following section.

4 Kinematic and Dynamic Models

The loads on the hammers were determined from the description or the diagrams. From the original sketches, the model is determined based on Figs. 3. The kinematic model was selected approximating the old machine as a modern mechanism [15] (Fig. 8). The cam is represented as rotating and sliding joints. The other joints were easily represented as rotating pairs (Fig. 6). A kinematic model was derived using the geometric method, and the dynamic model was derived from the moments produced by the mules (blood engine), the pinion, and gear; the force was derived from Fig. 9. The external loading was estimated based on the data obtained from the operating manuals [24]; the power supply by mules is approximately 12 to 26 kW [25]; thus, the loading was calculated from the animal carrying capacity. The assumptions are: the number of hammers is 12, the impact load corresponds to the maximum stroke, and it depends on the rigidity of the stone bed; the production capability depends on the ore's strength. There are no specifications regarding the final dimensions of the broken minerals.

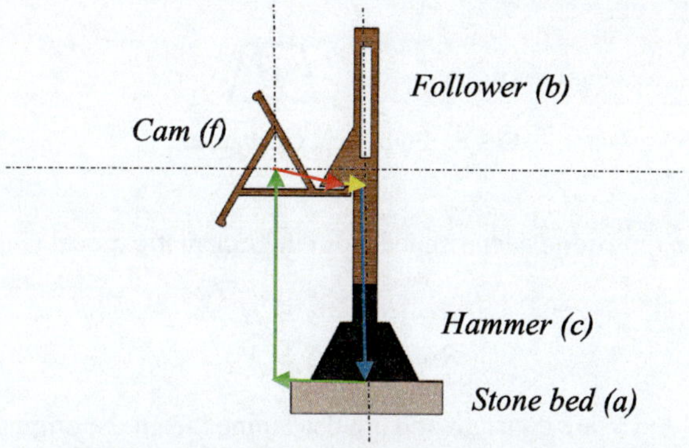

Fig. 6. Kinematic diagram

The kinematic chain is reconstructing using the vector representation of Figs. 6 and 7.

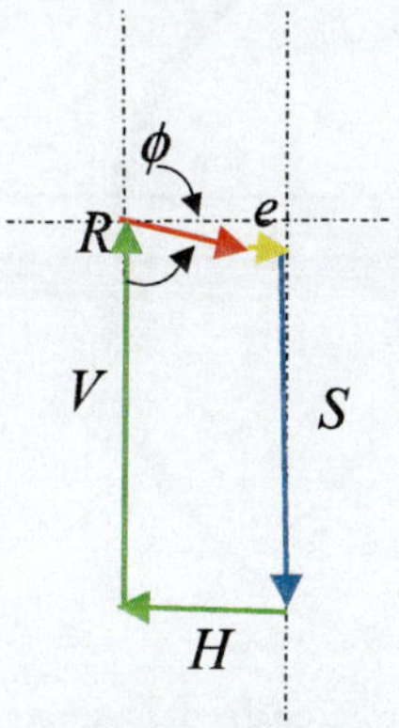

Fig. 7. Vector representation of the kinematic diagram

The equivalent mechanism is shown in Fig. 8. It is assumed that the kinematic motion is equivalent to a RRSS mechanism.

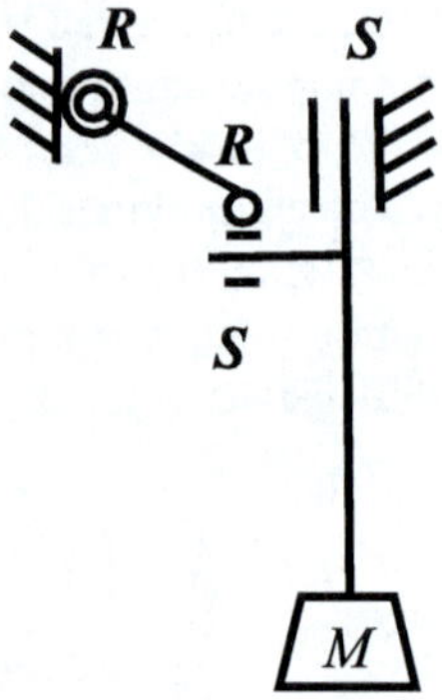

Fig. 8. Equivalent mechanism

The Mass (*M*) position is determined from the kinematic model (Fig. 7)

$$Rcos(\emptyset) + e = H$$
$$Rsen(\emptyset) + S = V$$

Where *R, H,* and *V* are constant and are determined from the original sketches. The solution of these equations is straight forward and it is out of the scope of this paper.

The dynamic model is determine using the diagrams shown in Figs. 9 and 10. The dimensions are estimated from the original sketch (Fig. 3).

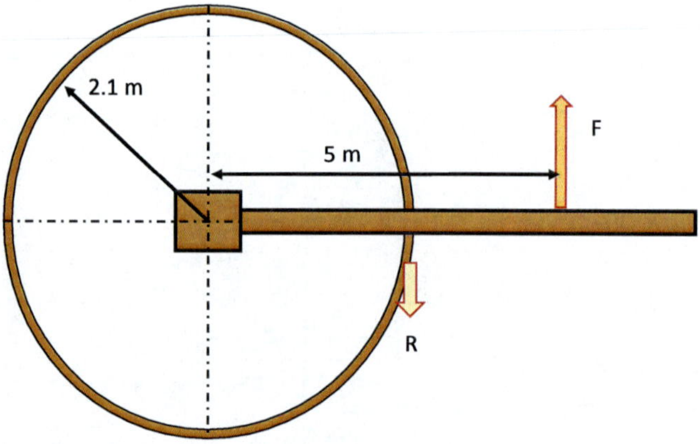

Fig. 9. Force diagram, upper view

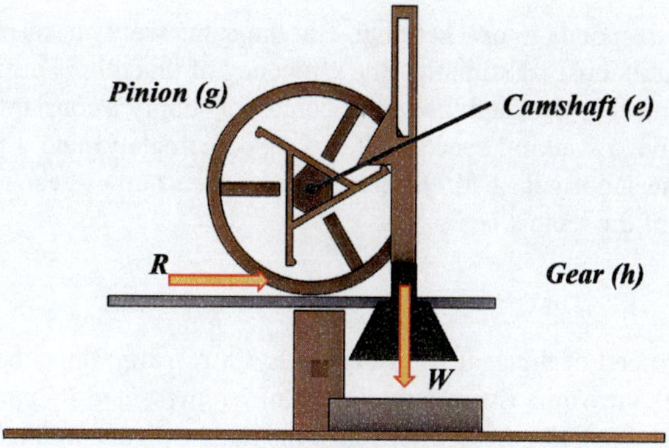

Fig. 10. Force diagram, side view

The maximum hammer displacement is also determined from the original sketches:

$$S_{max} = 2Rsin(60°) \tag{1}$$

This value is used for computing the maximum impact force that each hammer imposed on the ores. The impact force can be calculated with the method presented by [26]

$$F_{max} = 0.15M\left[\sqrt{\frac{g}{2S_{max}}}C + g\right] \tag{2}$$

Where g is the gravity constant and C is determined as:

$$C = \sqrt{\frac{(1-\nu)E}{(1+\upsilon)(1-2\upsilon)\rho}} \tag{3}$$

Where E is the Young's modulus for the stone bed (approximately 25 × 10^3 MPa), ν is the Poisson's ration (0.25) and ρ is the density (3 × 10^{-3} kg/mm^3).

5 Results

The recovered information can determine two parameters: the impact force and the rate of hits that the stamp mill could produce. The dimensions are estimated from the original sketching; thus, the equivalent impact force is calculated as:

$$S_{max} = 2sin(60°) = 1.75 \text{ m}$$

$$C = \sqrt{\frac{(1-0.25)25 \times 10^6}{(1+0.25)(1-0.5)3000}} = 100$$

$$F_{max} \approx 26M$$

This force corresponds to one hammer. But the cams were out of phase, so the ore was impacted continuously. Estimating the efficiency of this mill is difficult. The joints had many mechanical losses, and the mules could not supply a constant force.

The mules have a walking speed of 2 m/s (approximately) and a pulling force of 1600 N. [25]. The input gear had 40 teeth, whereas the output gear had 16; thus the speed increment of the teeth wheels was

$$u = 2.5$$

The rotating speed of the gear (h in Fig. 3) is 3.8 rev/min; thus, the camshaft (e in Fig. 3) rotated at 9.5 rev/min. Every turn of the mules represented 30 hammer hits. There are still many questions, but these values give the author a general idea of the capacity of the antique machinery.

The mass of each shoe is estimated from the kinetic energy supplied by the mules and the potential energy reached by the twelve masses. Using the maximum elevation, the estimated mass of a single shoe was 62 kg.

6 Conclusions

The engineering documentation has changed dramatically, and it is not easy to find technical information about the old machinery. Therefore, this paper presents the kinematic and dynamic analysis of the stamp mills used before the industrial revolution.

The design concept and the kinematic and dynamic analysis were developed, extracting the data from the original drawings and complementing the information with the technical books written in Europe and Mexico. The technical description of the mechanisms was obtained with the modern concepts of mechanisms. Although it is an approximation, the models can predict the production capabilities and the efficiency of those machines. Therefore, it is possible to represent the old mechanism with the kinematic chain theories; the significant limitation is the actual shape of the linkages, the unknown flexibility of the bars, the backlash, and the out-of-plane movements.

In the future, this work can be validated with a prototype since there is no evidence of an archeological sample.

References

1. Jauregui-Correa, J.C., Rodriguez-Zahar, G.: Dynamic reconstruction of a colonial mexican mechanism. In: López-Cajún, C., Ceccarelli, M. (eds.) Explorations in the History of Machines and Mechanisms. HMMS, vol. 32, pp. 87–100. Springer, Cham (2016). https://doi.org/10.1007/978-3-319-31184-5_9
2. Jauregui-Correa, J.C., Escamilla-Gonzalez, O.: Modeling an antique grinding mill of guanajuato silver mines. In: Zhang, B., Ceccarelli, M. (eds.) Explorations in the History and Heritage of Machines and Mechanisms. HMMS, vol. 37, pp. 101–112. Springer, Cham (2019). https://doi.org/10.1007/978-3-030-03538-9_9
3. Bustamante, M.V.: El trabajo en las minas de Guanajuato durante la segunda mitad del siglo XVIII. Estud. Hist. Novohisp., pp. 1–13 (2020)

4. Wilson, A.: Machine, power and the ancient economy. Soc. Promot. Rom. Stud. **92**, 1–32 (2002)
5. Liu, H.: The water mill and Northern Song imperial patronage of art, commerce, and science. Art Bull. **84**, 566–595 (2002). https://doi.org/10.2307/3177285
6. Gerogius, A., Hoover, H., Hoover, L.: De Re Metallica. Dover Publishers, New York (1950)
7. Luacas, A.: Industrial milling in the ancient and medieval worlds. Technol. Cult. **46**(1), 2005 (2005)
8. Moussas, X.: Antikythera Mechanism The Oldest computer and Mechabical Cosmos 2nd Century BC, 1st edn. University of Birmingham, Birmingham (2014)
9. Sen Yan, H., Lin, J.L.: Reconstruction synthesis of the lost interior mechanism for the solar anomaly motion of the Antikythera mechanism. Mech. Mach. Theory **70**, 354–371 (2013). https://doi.org/10.1016/j.mechmachtheory.2013.08.005
10. Penta, F., Rossi, C., Savino, S.: Mechanical behavior of the imperial carroballista. Mech. Mach. Theory **80**, 142–150 (2014). https://doi.org/10.1016/j.mechmachtheory.2014.05.006
11. Azéma, A., Angelini, F., Mille, B., Framezelle, G., Chauveau, D.: Ultrasonic phased array contribution to the knowledge of the flow fusion welding process used to make the Roman large bronze statues. Weld. World **57**(4), 477–486 (2013). https://doi.org/10.1007/s40194-013-0039-1
12. Ceccarelli, M.: Renaissance of machines in Italy: from Brunelleschi to Galilei through Francesco di Giorgio and Leonardo. Mech. Mach. Theory **43**(12), 1530–1542 (2008). https://doi.org/10.1016/j.mechmachtheory.2008.01.001
13. Koetsier, T.: On the prehistory of programmable machines: musical automata, looms, calculators. Mech. Mach. Theory **36**(5), 589–603 (2001). https://doi.org/10.1016/S0094-114X(01)00005-2
14. Thomas, F.: A short account on Leonardo Torres' endless spindle. Mech. Mach. Theory **43**(8), 1055–1063 (2008). https://doi.org/10.1016/j.mechmachtheory.2007.07.003
15. Hsiao, K.H.: Structural synthesis of ancient Chinese original crossbow. Trans. Can. Soc. Mech. Eng. **37**(2), 259–271 (2013)
16. Chen, F.C., Tzeng, Y.F., Chen, W.R., Yan, H.S.: On the motion of a reconstructed ancient Chinese wooden horse carriage. Mech. Mach. Theory **58**, 165–178 (2012). https://doi.org/10.1016/j.mechmachtheory.2012.08.007
17. Wauer, J., Moon, F.C., Mauersberger, K.: Ferdinand Redtenbacher (1809–1863): pioneer in scientific machine engineering. Mech. Mach. Theory **44**(9), 1607–1626 (2009). https://doi.org/10.1016/j.mechmachtheory.2009.05.007
18. Vepřek, S., Eckmann, C., Elmer, J.T.: Recent progress in the restoration of archeological metallic artifacts by means of low-pressure plasma treatment. Plasma Chem. Plasma Process. **8**(4), 445–466 (1988). https://doi.org/10.1007/BF01016059
19. Backwell, P.J.: Sivler Mining and Society in Colonial Mexico, Zacatecas 1546–1799. Cambridge University Press, Cambridge (1971)
20. Hermos, F.: Manual de Laboreo de Minas, Nueva Edic. Mexico: Librería de Ch, Bouret (1889)
21. Leon, D., Lara, A.: Guanajuato Herencia Minera. Universidad de Guanajuato (2018)
22. Tilmann, N.: Der Bergbau und das Amalgamations-Verfahren in dem Bergwerks. Distrikte von Guanajuato in Mexico. Münster (1866)
23. Sonnesghmid, F.: Tratado de la Amalgamacion de Nueva España. Londres: Martin Bossange (1925)
24. Sonnesghmid, F.: Tratado de la Amalgamacion de Nueva España. Martin Bossange, London (1925)
25. Vidal, M., Lopez, M., Herrera, O.: Diagnostico del Uso de Animales de Tiro en Una Zona Montañosa del Norte del Cauca. Acta Agron. **45**, 97–108 (1993)
26. Luo, F., Zhou, X., Wang, Y., Zhang, F.: Experimental study on rockfall impact force applied to frame shed tunnels. J. Highw. Transp. Res. Dev. **12**(2), 59–66 (2018)

Industrial Heritage of the City of Granada: The Water Well "Santa Clara" in "La Vega De Granada"

A. Dávila Ponce de León[4], M. J. Gámez[4], C. González-Sevilla[1], Rafael García de la Mata[2], and Lourdes Rubio[3](✉)

[1] Parque de Las Ciencias, Granada, Spain
[2] Restaura, Gestión Museística y Restauración Siglo XXI, Universidad de Granada, Granada, Spain
restauragmr@gmx.es
[3] Universidad Carlos III de Madrid, Madrid, Spain
lrubio@ing.uc3m.es
[4] Granada, Spain

Abstract. This paper presents the history and the recovery works of the water well "Pozo de Santa Clara". This construction dates back to 1923 and was intended to supply water to the agricultural area around the city of Granada, Spain, called "La Vega de Granada". The construction of the well is contemporary to the industrial development carried out in the city of Granada: the sugar industry based on the cultivation of sugar beet, the development of electric trams, the opening of the brewery "Cervezas Alhambra", the construction of the most important avenue of the city "La Gran Vía", etc. The water well had been abandoned since the 70's of the 20th century and the new owner, grandson of one of the first owners, started the restoration to protect the heritage and give it other uses than the one it was originally built for.

Keywords: Granada heritage · Water well · Vega de Granada · Old pump

1 Introduction

La Vega de Granada is an area located in the province of Granada in Andalucia, Spain (Fig. 1). The territory is surrounded by different towns and areas of the province: Los Montes to the north, the town of Guadix to the east, the Alpujarra Granadina to the southeast, the Lecrín Valley to the south, the town of Alhama to the southwest and the town of Loja to the west. It is made up of more than forty villages, generally small and with a population density higher than the average for the province. Of all of them, the most popular is the city of Granada, which is at the moment, and has been in the past, the historical capital.

Mainly La Vega is a flat area characterized by alluvial soils of excellent agricultural fertility thanks to the water of the Genil river and its tributaries. It presents architectural and technical vestiges of the industrialization started around the beetroot in the last

A. Dávila Ponce de León and M.J. Gámez—Private researcher.

M. Ceccarelli and R. López-García (Eds.): HMM 2022, HMMS 40, pp. 444–455, 2022.
https://doi.org/10.1007/978-3-030-98499-1_36

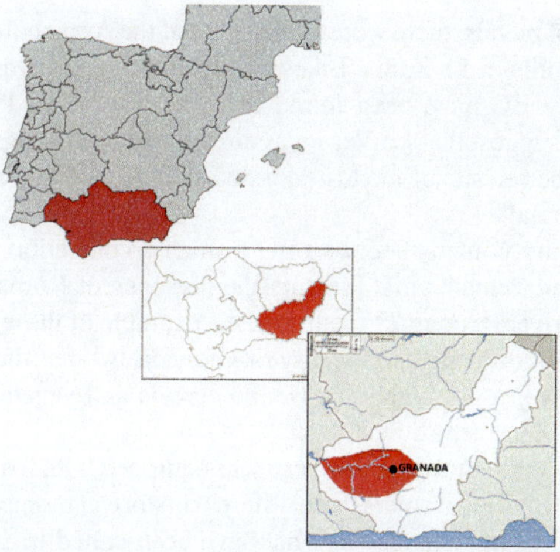

Fig. 1. Localization of La Vega de Granada in Andalucia, Spain

quarter of the 19th century and first third of the 20th century. It also presents archaeological remains and architectural heritage related to the different forms of agricultural exploitation throughout the centuries, from the Romans to the Arabs, from which it receives most of its legacy although there are studies of activities in the zone dated in the Paleolithic, Neolithic, and Age of Metals periods [1]. The fluvial genesis of La Vega composed of sediments giving permeability and drainage is the origin of good aquifers and natural reservoirs or warehouses used since ancient times to use the winter waters from Sierra Nevada during summer or dry periods by extracting them in springs, wells and waterwheels.

La Vega de Granada has been the economic engine that has led to human settlements, not only in the capital but also in the towns nearby. Due to the richness of the soil and the abundance of water different kinds of crops have been allowed. Among the crops that have contributed significantly to the economic development of the region are cereals and vegetables, very appreciated, up today, in the local gastronomy. Some other crops have been of great interest: hemp and flax, beets and tobacco. Special mention deserves the industrial plants used to supply other industries such as silk (18th–19th centuries) or the intensive cultivation of beets in the early and mid-twentieth century for the production of sugar. Unfortunately, these industrial crops declined after 1930.

2 Social-Historical Context

2.1 The Importance of La Vega for Granada. History of La Vega

"La Vega de Granada" and the territory that surrounds the Genil river, have been populated since Prehistory [2]. Thus, different vestiges can be found such as the site of "Las catorce fanegas" dating from the Neolithic. It is believed that these settlements were small huts that were inhabited only in summer and spring since it is assumed that the area was swampy and uninhabitable in autumn and winter.

Later, in the era of metals, there were settlements of the Argar culture in some areas of La Vega such as Armilla or La Zubia. Likewise, Phoenician remains, funerary remains, from the 8th century BC have been found like the mounds of Pinos Puente. These Phoenician remains suggest that La Vega was already generating agricultural resources. Proof of this is the Iberian site of the Mirador de Rolando (5th century BC), where farm devices have been found.

This agrarian activity intensified with the Roman colonization. There are archaeological and toponymic remains that indicate the presence of Roman villas in La Vega and along the Genil river (around 2nd century). An example of these remains is the town discovered in Los Mondragones, in the city of Granada. At that time, it is thought that olive trees were grown in the higher areas, and cereals and vegetables, fruit trees and wines were grown in "La Vega".

It was not until Islamic times that La Vega was equipped with irrigation systems. The irregularity of rainfall had led farmers from different historical moments to come together to collect water and organize its use in what have been called irrigation communities. The Muslims incorporated the knowledge and practices about irrigation into the local culture and established water infrastructures. In La Vega, one of these associations was created: the Community of Irrigators of the "Acequia Gorda del Río Genil" (Big Ditch of Genil River), which organized and managed the irrigation of a part of La Vega since the 11th century, although it is thought that the origin, and part of the infrastructures, could have been from Roman times. The Acequia Gorda continues operating [3] nowadays. Other ditches in La Vega were Arabuleila y Tarramonta, very much connected with the well of the work.

On the other hand, since the 13th century there is evidence of the existence of rules to regulate the distribution and use of the waters of the Genil River. These norms were translated and used later in the time of the Catholic Monarchs, after the Conquest of Granada in 1492. These norms, used by the Arabs, are still in force, with some minor modifications [3].

As a consequence of the need to supply more water to La Vega, it was necessary to carry out wells, some for the direct supply of crops, and others to incorporate water into the aforementioned "Acequia Gorda". As a proof of the interest in the existence of wells for supply is that today, the "Acequia Gorda" has the contribution of 8 wells with a depth of more than 150 m each [3].

2.2 Industrialization of Granada

The industrialization of the city of Granada is closely linked to LaVega del Río Genil. The industrial take-off of Granada took place in the last third of the 19th century and lasted until the first third of the 20th century and is mainly linked to the sugar industry [4].

It is in those years that electricity was introduced in the city (1892) and the sugar industries of La Vega were created and expanded, which in 1890 were seven of them [5]. At the same time, and as a support to the industries, the most important street, La Gran Vía, (High Street) known in its day, as Gran Vía del Azúcar (High Street of the Sugar) and later of Colón (1890–1925) was urbanized and put into service [4].

Likewise, as part of the modernization of the city, in 1904 the routes of the Electric Tramways of Granada [6] began. The trams were developed to connect La Vega and its towns, with the city, to allow the transfer of goods and people. Some other events took place around the industrialization and the modernization of the zone: in 1922 the Armilla aerodrome was created which, since 1925, has been called Dávila Aerodrome, in memory of Luis Dávila Ponce de León and which is still in service today as the Air Force Base [7]; in the same year, 1925, the Alhambra brewery (1925) began its operation among other industrial facilities in the city.

It was in this industrial environment that the construction and installation of the Santa Clara well began. Thus, the well was related to industry in two ways. Firstly, it was developed due to the need for water for the sugar industry, and later for the tobacco industry, and secondly, the industrial development of the region was used to incorporate an unconventional industrial equipment for the extraction of groundwater.

3 The Water Well of Santa Clara

The Santa Clara Well is an industrial agricultural construction dating from 1923. A stone tombstone with the inscription of the date of construction is at the main entrance of the building (Fig. 2).

It is a two-storey building that protects the Santa Clara Well and serves as a home and warehouse for the maintenance inside which, in the center of the room is the well. It is located in an area of La Vega de Granada close to Granada capital, about 4 km from the city. At the time of its construction, the well belonged to the "Cortijo del Pino" an agricultural estate belonging to both the town of Churriana de la Vega and the city of Granada as shown in the plans of Fig. 3. The "Cortijo del Pino" is a typical construction of La Vega and it is one of the oldest that has been preserved. The construction of the Cortijo dates from 1873 and has been owned by the same family since 1940, the year in which the current owner's grandmother bought it.

Fig. 2. The stone tombstone at the entrance of the house (owner's collection).

3.1 History of the Water Well

The "Pozo de Santa Clara" is a unique facility in La Vega de Granada, as traditionally irrigation was done through irrigation ditches and floods and not through well irrigation.

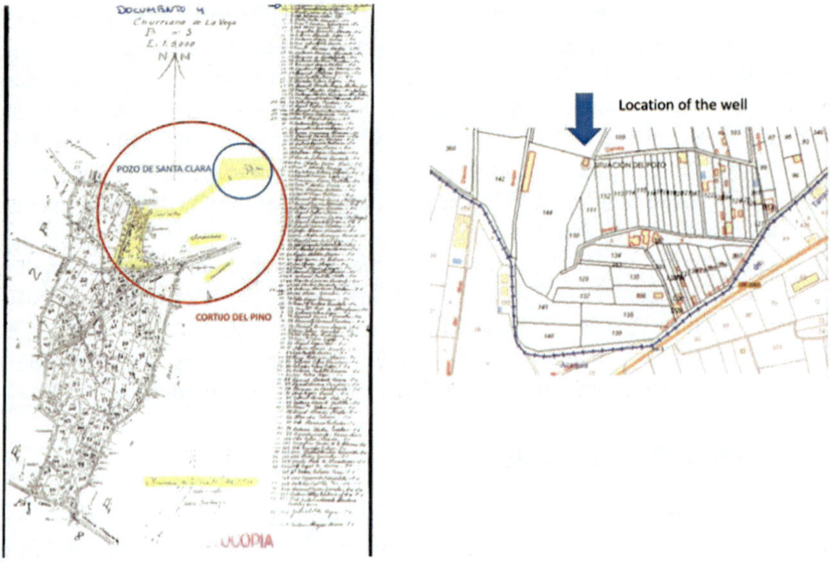

Fig. 3. Location of Cortijo del Pino and the Well Santa Clara in Churriana de la Vega and adastral representation and information of the "Pozo de Santa Clara" (owner's collection)

In fact, wells in the Vega became widespread with the introduction of diesel engines in the second half of the twentieth century, since before that there was hardly any quantitatively significant extraction of groundwater. Therefore, the Santa Clara well, dated 1923, is one of the oldest documented wells in the Vega of Granada and its existence must be related to the sugar industry. As mentioned above, the well is part of the "Cortijo del Pino" which was purchased in 1940 by the grandmother of the current owner of the well. The well was in operation until 1970's of the past century when it finally dried up [8].

It is an emblematic place that has been the subject of paintings such as the one shown in the image in Fig. 4 by the painter Francisco Carreño. It has not only been an emblematic place but also an enigmatic place since there is even a legend about the well and the man in the well that the tradition of the town of Churriana remembers and that

Fig. 4. "Pozo de Santa Clara" by Francisco Carreño (owner's private collection).

has given rise to a story-legend (Fig. 5) by the writer César Elvira [9] and to an escape room game (Fig. 5).

Fig. 5. Cover of the book "La leyenda del pocero de Santa Clara" and poster announcing the escape room "La leyenda del pocero".

3.2 Building and Civil Engineering

The building consists of a rectangular open plan of one room with a sloping gabled roof with ceramic tiles. In one side, a turret stands for housing the original electrical transformer (today located in the outer turret). Figure 6a shows the current state. The area occupied by the building is 92.40 m^2. The house is built with 50 cm wide enclosures made up of mixed stonework masonry and solid brick walls framing gaps and giving strength to the walls. The sloping roof is made up of poplar logs beams, planking, insulation and ceramic tile (Fig. 7). The building follows the vernacular architecture of agricultural spaces, such as farmhouses.

Fig. 6. The House of the Well. a) View from the rear part of the building. b). Inside the building, in the center, the well (owner's collection).

Inside, close to the center of the room, appears the well. It is made of masonry and is protected by a locksmith railing. As it can be seen in Fig. 6b it is a huge well with 3 m of diameter and up to 45 m deep. In order to do works of maintenance, an access to the bottom is provided through a helical staircase (Fig. 8) of 50 cm wide. The staircase descends 30 m to the platform where the electric motor of the water extraction system is located. The building containing the well was thought for maintenance and to be the living place of the well man. Just above the well a skylight allows light to enter the building.

The initial depth of the well was 35 m (about a 9-storey building) but in the 50s of the 20th century, due to the decrease in the water table of "La Vega", the former owner of the well and father of the current owner, began to deepen it to the current deep of 45 m (12-storey building). The work was carried out manually. The hollows of the scaffolding used for its excavation and interior reinforcement can be still observed.

Fig. 7. Detail of the wooden roof and the skylight. (owner's collection).

Fig. 8. The access to the well and the helical staircase. From the inside of the well (owner's collection).

An iron cast beam has been located bellow the wooden roof and over the well clamped in two opposite walls of the building in order to support a small bridge crane (Fig. 9).

Fig. 9. Clamped beam to support the bridge crane. (particular collection).

Regarding the inputs, something that is really important for an industrial set up, the distribution of electrical energy both for the house and for the engine of the water extraction system has been installed from the moment in which the "Casa del Pozo" was built. Likewise, the building has a drinking water network for human consumption.

3.3 Mechanical-Fluid Engineering

The raison d'être of this building was to supply irrigation water to the lands of the "Cortijo del Pino" and other lands adjacent to it. As mentioned above, "La Vega de Granada" bases its wealth on traditional crops, mainly vegetables and also crops for supplying industries, all of them with a great need for water. Taking advantage of the impulse of the industrialization of the region, in the first third of the last century, the owners of the Cortijo, proceeded to the installation of an innovative system in the agricultural sector of the time and in the region, introducing a pump extraction system equipped with an electrical engine. The idea behind this installation was to reach a great depth to achieve a large water flow from the aquifer, impossible with the procedures and technology used so far. In this sense, the use of industrial equipment such as the engine, generally reserved for other activities such as transport, in an agricultural installation, was a milestone in "La Vega".

The industrial equipment consists on a pump and the correspondent engine and the pipe. To install these elements, an iron cast bridge crane supported by the clamped-clamped beam was used. All together are the mechanical-hidraulic part of the water well.

The bridge crane, Fig. 10, is mechanical a device, quite simple in concept, but was a very important part of the well by the time of the construction as it was a unconventional element in the agricultural environment of La Vega that help to the installation of the rest of the elements. We can say that it was a landmark in the recent history of La Vega.

Fig. 10. Details of bridge crane. (owner's collection).

The old engine of the well can be seen in Fig. 11a. This part is located at about 30 m of depth from the ground floor of the building and is accessible by the mentioned staircase. Figure 11b shows the metal plate with the main properties of the engine. Although we do not know exactly when the pump was installed, the plate gives us some information. The manufacturer of the motor was the company CENEMESA, created in 1930 in Córdoba (Spain). Therefore, the engine has to be dated after 1930, but no later than that to the owner's knowledge. As can be seen, although the plate is quite deteriorated, it is an asynchronous motor with a power of 100 HP, usually used in other types of industrial assemblies.

Fig. 11. a) Details of the engine and the extraction pipe and b) the dataplate of the engine. (owner's collection).

In Fig. 12a, the long pipe for the extraction of the water can be observed going up in the center of the well. In the past, when the well was in operation, once the pipe reached the ground floor of the building it got out of it following the direction drawn, in Fig. 13a, in red. Outside, the pipe continues to the final part of the installation as shown in Fig. 13b. Also a detail of the pipe outside the building can be seen in Fig. 12b.

3.4 The Restoration

The restoration and rehabilitation of both the well and the house began in 2010 at the initiative of the current owner.

The restoration lasted 2 years and consisted on the recovery of the house and its reconditioning as a museum site and location, on the one hand, and the recovery of the civil part of the well, on the other. In Fig. 14 the architectural drawings of the building are shown. Figure 15 shows the comparative dimensions of the well and the house.

Regarding the recovery of the well, vertical masonry and locksmith work was carried out. As for the mechanical-hydraulic part, the motor was cleaned up and refurbished, although it has not been put into operation due to the low groundwater level and the fact that the well is currently dry.

Fig. 12. The extraction pipe of the well a) seen from the bottom b) outside de house. (particular collection).

At present “La casa del Pozo” is in the process of being declared “Bien de Interés Cultural (BIC)” (Asset of Cultural Interest) by the Junta de Andalucía, as it is an agro-industrial installation almost one century old. According to the “Ley de Patrimonio Histórico de Andalucía” [10] (Andalusian Historical Heritage Law), both the well house and the well fall within the cases of the classification of article 25 relating to the cultural interest of the Patrimonio Inmobiliario (Immovable Heritage): Sites of Ethnological Interest and Sites of Industrial Interest. In order to achieve the status of BIC, a report on

Fig. 13. Direction of the pipe a) inside the building and b) outside the building

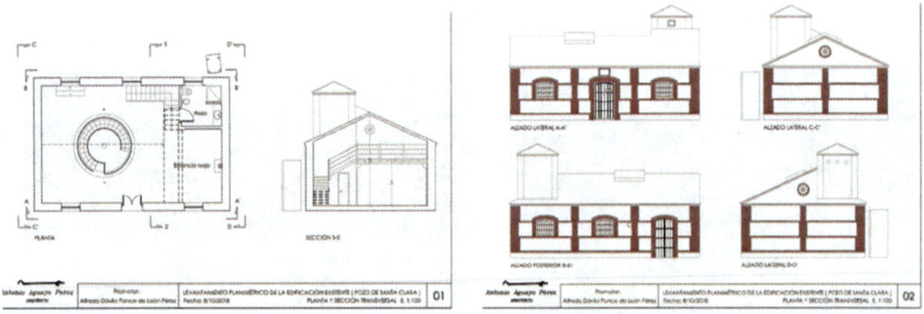

Fig. 14. Architectural drawings of the building for the restauration

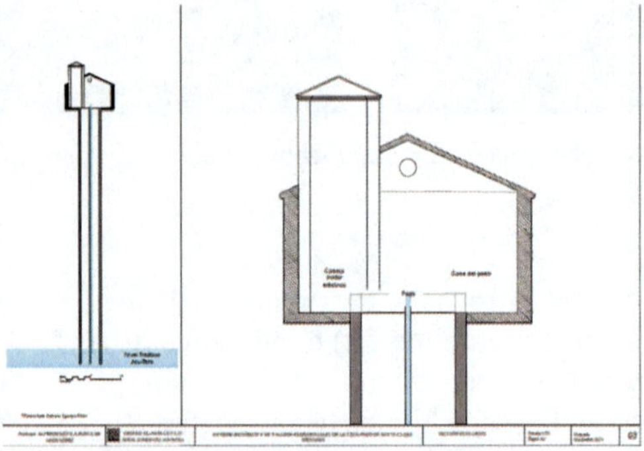

Fig. 15. Comparative dimensions of the well and the house

the historical and heritage values of the well has been commissioned by the owner, part of whose conclusions are reflected in this work [11]. The "Santa Clara well" will join, in this way, other ethnographic and industrial "Bienes de Interés Cultural" of Granada, such as the "Casa Molino del Marqués de Rivas" and "La Azucarera de San Isidro y el Ingenio de San Juan", a flour mill and a sugar factory, respectively [12].

4 Conclusions

This paper presents the history and the recovery works of the water well "Pozo de Santa Clara". In the first part, an introduction to the importance of "La Vega" for the city of Granada thought the history. The well, dates back to 1923 and was intended to supply water to the lands of the "Cortijo del Pino" and the surrounding area. It has shown to be contemporary to the industrial development carried out in the city of Granada: the sugar industry. The new owner, grandson of one of the first owners, started the restoration of the well, abandoned in the the 70's of the 20th century, to protect the heritage and give it other uses than the one it was originally built for. In the paper we also describe the building and mechanical-hidraulic equipment.

References

1. Historia de la Vega de Granada. http://www.culturandalucia.com/GRANADA/VEGA_DE_GRANADA/Historia_de_la_Vega_de_Granada.htm. Accessed December 2021
2. Notas sobre la Vega de Granada antes de los árabes. https://www.elindependientedegranada.es/cultura/notas-sobre-vega-granada-antes-arabes. Accessed December 2021
3. http://www.acequiagorda.es/. Accessed December 2021
4. Martín Rodríguez, M.: Cambio económico y reforma interior urbana: La Gran Vía de Granada 1890–1925. Ciudad y territorio (1986)
5. Salmerón, P., Seño, F., Pérez de la Torre, R., Garzón, D.: Estudio del paisaje de la industria azucarera en la Vega de Granada estrategias para su preservación y uso sostenible. In: II Congreso Internacional de Patrimonio Industrial y de la Obra Pública (2018)
6. Péris Torner, J.: Los tranvías eléctricos de Granada. http://euroferroviarios.net/inadex.php?name=Reviews&req=showcontent&id=4747. Accessed December 2021
7. Biografía de Luis Dávila Ponce de León. Real Academía de Historia. https://dbe.rah.es/biografias/45486/luis-davila-ponce-de-leon-y-wilhemi. Accessed December 2021
8. Dávila Ponce de León, A.: Technical documents. Private documents of the owners (2010)
9. Elvira, C.: La leyenda del pocero de Santa Clara. Ed. Ciempiés (2021)
10. Ley 14/2007, de 26 de noviembre, del Patrimonio Histórico de Andalucía. BOE
11. Rodríguez, A., Dávila Ponce de León, A.: Informe histórico y de los valores patrimoniales "Casa del Pozo Santa Clara". Private document of the owners (2021)
12. Catálogo de Bienes de Interés Cultural de Andalucía. https://www.juntadeandalucia.es/organismos/culturaypatrimoniohistorico/areas/bienes-culturales/catalogo. Accessed December 2021

Check for updates

Bartolomé Sureda: Mechanic, Inventor and Artist. A Character of Enlightened Knowledge

Patricia Zulueta Pérez(✉) and A. Sánchez Lite

University of Valladolid, Valladolid, Spain
{pzulueta,asanchez}@eii.uva.es

Abstract. The main objective of this work is to present the figure of Bartolomé Sureda y Miserol (1769–1851), a mechanic, inventor and illustrated artist, in relation to his contribution to scientific progress. The next goal is to demonstrate the extent to which graphic representation techniques were decisive during the aforementioned process, techniques that the enlightened scientists of Spain, the so-called "experimental natural philosophers" among whom was Sureda, used masterfully to develop their plans and drawings of machines. The scientists of Las Luces constituted a circle of technology professionals whose training came from various fields: artistic, analytical and technical. In the absence of specific educational centres, academies were created during the 18th century that were conceived to promote knowledge and were born from private gatherings. Bartolomé Sureda was trained in these centres and took an interest in both technique and art, and although he applied great geometric rigour in the representation of machines, his artistic will was never absent.

Keywords: Bartolomé Sureda · Engineering-Illustration in Spain · Graphic representation · Aquatint · Lithography

1 Introduction

At the beginning of the eighteenth century, scientific activity had very little social influence and practically no institutional support, and thus, research activity was almost nonexistent.

During the first half of the century, a period prior to the Enlightenment, Spain was immersed in changing historical situations that represented clear disagreements with the Old Regime. With the establishment of the Bourbon dynasty in Spain, voices began to be heard that, in addition to criticism, demanded the search for solutions to catch up with the intellectual activities that were being conducted in Europe.

Decades later, towards the middle of the century, within the intentions of progress advocated by enlightened reformists, scientific advancement and teaching of the sciences were predominant, and both were considered essential elements for facilitating the development of industry and commerce. Within those intentions of prosperity was

M. Ceccarelli and R. López-García (Eds.): HMM 2022, HMMS 40, pp. 456–469, 2022.
https://doi.org/10.1007/978-3-030-98499-1_37

also the manual or servile arts, i.e., arts or mechanical trades excluded from the set of the seven noble or liberal arts that made up the Trivium –grammar, rhetoric and dialectic – and Quadrivium – arithmetic, geometry, astronomy and music. The social attitude began to change as a result of the publication of the French Encyclopaedia, which included equality among the sciences, arts and crafts.

Spanish Enlightenment constituted a unique stage that generated fundamental events in Spanish history in its attempt to transform the world. In addition, established a line of thought that introduced the foundations of modernity and that led to a structure of relationships between scientific knowledge and technique. In many parts of Europe, scientific and technical progress was conceived as something intimately linked to political and social progress, and faith in science and technology was a basic element in the fight against superstition and irrationality.

In eighteenth-century Spain, one of the most characteristic events that occurred was the attempt, carried out by a small section of the population - the enlightened minority - to place Spain at the same level as the rest of Europe. With this objective, science and technology were promoted, and a series of urgent measures were adopted aimed at training new scientists. Among these measures was the hiring of foreign experts and the granting of scholarships to study outside of Spain, resulting in great achievements for our cultural landscape. Trips to Europe and expeditions to America were also promoted, and fundamentally, specialized institutions that promoted scientific research were founded. However, of all the measures described, the main mechanism of creating innovation in Spain during the Enlightenment was scientific missions. Thus, instead of prohibiting studies abroad, as had been done since the Counter-Reformation, scholarships were granted to expand training in other countries. This measure, which began at the beginning of the century, became common during the second half of the century [16]. The measure was strengthened by the monarch Carlos III (1716–1788), with regard to the training of experts abroad, turned out to be of more importance than the hiring of foreign specialists. First-hand learning was much more effective than imported knowledge [25].

Regarding the field of engineering, one of the primary objectives of the Spanish Enlightenment was the implementation of teaching and research in the new disciplines of hydraulics and mechanics to facilitate the development of future public works. To achieve these goals, individuals such as José Moñino (1728–1808), Count of Floridablanca, provided the necessary impetus so that in the spring of 1784, the Canarian engineer Agustín de Betancourt (1758–1824), a paradigm of Spanish enlightened scientists, was graced by the Secretariat of Indias with a pension to settle in Paris, an offer that was endorsed by the Royal Academy of Fine Arts of San Fernando, of which he was an honorary member. At first, his learning was directed to the study of underground architecture, but due to vocational inclination, he subsequently went to *École des Ponts et Chausseés*, a school of civil engineering in Paris and at that time the most prestigious in the world. With Betancourt at the helm, playing a managerial role, the "Hydraulic Team" was established in Paris in 1785, a compact group of Spanish pensioners from different branches of science among whom Bartolomé Sureda would be part of a few years later. The studies in the aforementioned school and the relationship with scholars, including Gaspard Monge (1746–1818), gave rise to a series of fundamental events associated with the subsequent development of Spanish engineering. In this context, various devices and machines were conceived, the steam engine was reinvented, and the optical telegraph was perfected.

During this stay in Paris, there was an event of great relevance, i.e., the continuation -by Betancourt and the distinguished scientist José María de Lanz (1764–1839)- of the unfinished work of Gaspard Monge, creator of descriptive geometry, materialized in a curricular treatise entitled *Essai sur le composition des machines*. The work, from 1808, was pioneering in industrial kinematics, and in its pages, the authors assured, based on certain scientific principles, that "form" and "function" are two directly related concepts [13].

Regarding the founding of scientific institutions, one decisively influenced the development of engineering in our country because of its origins and the group of people and series of events that made its founding possible. The Royal Cabinet of Machines opened its doors in Madrid in 1791 under the direction of Betancourt and the subsequent incorporation, in 1793, of Juan López de Peñalver (1764–1834). During their stay in the French capital, the members of the Hydraulic Team developed and compiled plans, models, and scientific reports representative of the avant-garde engineering in Europe [2]. The complete collection, i.e., 359 plans, 270 models and 99 memoirs, arrived from Paris to Madrid in the summer of 1791 and was installed in the rooms of Buen Retiro Palace. The Royal Cabinet opened its doors on April 1, 1792; however, it took until 1802, the year in which a new educational centre, i.e., the General Inspection of Roads and Canals, was created, for Betancourt to transform the Royal Cabinet into the School of Hydraulic Studies, acquiring notable importance as an educational centre considered among the best European scientific centres.

Following the French occupation in 1808, the pilgrimage of the machine collection of the Royal Cabinet to various points of Madrid began, from the drawing rooms of the Royal Academy of Fine Arts of San Fernando to Buen Retiro Palace. After the assault on this palace by Napoleonic troops, what remained of the collection was sent to Buenavista Palace, where it remained in obscurity until 1815. In that same year, the Royal Matritense Economic Society of Friends of the Country carried out an inventory, the results of which revealed the irreparable losses suffered. The collection remained there until 1824, when the Royal Conservatory of Arts was founded; it was then, after the last transfer to this institution, when the remaining machine collection of the Royal Cabinet almost completely disappeared. These documents undoubtedly constitute one of the most important and best testimonies of Spanish engineering during the Enlightenment [21].

The Royal Conservatory of Arts became the Royal Industrial Institute in 1850, a centre in which a new type of civil engineer began to germinate, i.e., industrial engineers [20].

2 Biographical Notes: Training, Professional Activity, and Inventions

Bartolomé Sureda was born in Palma de Mallorca in 1769 into a family of carpenters. When he was a child, he demonstrated an inclination for the arts and skill in drawing, entering the School of Noble Arts of the Royal Mallorcan Economic Society of Friends of

Fig. 1. Francisco de Goya. Portrait of Bartolomé Sureda (fragment). 1803–1804. National gallery, Washington

the Country at the age of 16, where he earned awards on two occasions [5]. The classes taught by Juan Muntaner (1744–1802), engraver, teacher, and director of the school allowed the young Sureda to recognize the importance of instruction in the technique of engraving, which was one of his contributions to scientific dissemination during the Enlightenment [24]. His education was nourished by writing books and treatises in addition to other texts that contributed to his technical training; however, his training also focused on the application of new techniques, including the then emerging intaglio engraving and lithography.

In 1791, with the help of the Mallorcan Tomás de Verí (1763–1838), a pensioner in Paris and a member of the Hydraulic Team, he moved to Madrid and entered the Royal Academy of Fine Arts of San Fernando, which then had a pedagogical facet, the following year. Through Verí, he came into contact with Betancourt, who, in view of his artistic gifts, incorporated him into his work team and invited him to make a to England on a scientific mission associated with industrial espionage. The main objective of the trip was to learn about the "hydraulic press," a new invention by the English engineer Joseph Bramah (1748–1814). Sureda remained in England between 1793 and 1796, and as Betancourt's right hand, he took trips and conducted industrial and engineering studies on the design and construction of machines, perfecting his knowledge of drawing and engraving and undergoing training in the manufacture of English pottery. Among the chalcographic engraving techniques he learned were the "aquatint" and "aguada" varieties of etching; thee techniques consisted of generating various shades on a copper plate, giving it a certain pictorial character.

In March 1797, the year they returned to Spain, at the Royal Academy of Fine Arts of San Fernando, both scientists presented the new hollow engraving technique learned by Sureda in England, a technique that he later showed to Francisco de Goya (1746–1828), who used the technique in his *Caprichos* collection made between 1798 and 1799 [6, 10]. Pedro García-Ormaechea write about the presentation of these techniques before the Board of Academics, among whom was Goya:

"… a new method of painting on glass, executed in England, a picture engraved by Don Bartolomé Sureda in a way that imitates the washing of Chinese ink, another engraved by a method that imitates a pencil and is executed on the sheet as soon as it is drawn on the paper, and another engraved on a board; all of these were presented by the Honorary Academician Agustín de Betancourt as evidence of what is being advanced in the arts in that part of Europe" [9].

Apparently, in gratitude for the teaching received, between 1803 and 1804, Goya painted portraits of Bartolomé Sureda (Fig. 1) and his wife Thérèse Louise Chapronde Saint Armand; these paintings are displayed in the National Gallery of Washington.

On the other hand, the Valencian artist José Antonio Jimeno y Carrera (1757–1818) was one of the first cultivators of aquatint engraving in Spain. Thus, the possibility has raised that Jimeno initially showed Goya the new technique [6, 12].

In August 1796, a scientific expedition left for Cuba with the objective of building a canal from the Güines Mountains to Guantánamo Bay [11]. The Cuban authorities requested the presence of Betancourt, Lanz and Sureda to execute this project and direct the construction of machines for sugar mills, which had operated until then by oxen and slaves. Sureda would have been responsible for creating plans and levelling. However, the aforementioned mission was never carried out because the ship in which they were travelling was seized, on June 9, 1797, by the British frigate Boston, and the surveying instruments were seized. The construction of the canal was replaced with the construction of a railroad that linked Havana with Güines, the first railroad to be built in the Spanish-American world. Subsequently, Sureda joined the Royal Cabinet and dedicated himself to engraving representative plates of machines for informative purposes under the direction of the engineer López de Peñalver (Fig. 2).

Fig. 2. B. Sureda. (a) Nail making machine. (b) Crane for the port of Mahón. Descripción de las máquinas… 1798 [15]

In 1800 and with the constant support of Betancourt, Sureda travelled again to France to undergo training in the construction and handling of textile machines under the guidance of the machinists Charles Albert and James Collier, who spun cotton, wool, linen and hemp in Coye-la-Fôret (Oise). The purpose of this apprenticeship was to take charge of the Royal Cotton Factory of Ávila, which had been managed and operated by Betancourt since 1799. From France, the Mallorcan sent detailed plans of the structure and operation of the new machinery; Betancourt, based on the plans, was having the machines built with support from the Madrid Cabinet.

In 1802, he moved to Paris to learn porcelain manufacturing techniques at Manufactory Nacional de Sèvres, whose industrial secret had not been revealed. Upon the death of Felipe Gricci, director of the Royal Porcelain Factory of Buen Retiro, and at the express request of the Spanish crown, Sureda returned to Madrid in 1803 to take over the company.

In the history of Spanish porcelain, Bartolomé Sureda will always remain linked to Buen Retiro Manufacturing, the company founded by Carlos III in 1760 and in operation for almost half a century. The discovery of the first Spanish hard porcelain during Sureda's mandate marked the beginning of a process of technology transfer in Spain in the early nineteenth century [22]. Chemical and mineralogical analyses of a selection of fragments of crockery, tiles and sculptures that were found during the archaeological excavation of the factory in 1996, confirmed the production of a new type of hard porcelain in which a magnesium clay, sepiolite, was used instead of kaolin. The steps that he had to take before being able to manufacture hard-paste porcelain are recounted in a report that he wrote in November 1808, when he was about to leave the factory [18]. During the five years of his tenure, he developed a porcelain different from other European productions, i.e., "Porcelanas de Madrid," even producing pieces of a higher quality than that manufactured in Sèvres. The Royal Factory, known as the China Factory, was occupied as a barrack by the French in 1808 and burned by the English in 1812, apparently with more commercial than political interests.

In the middle of 1809, Sureda travelled again to the French capital, now with the objective of following Betancourt, who had been in Russia. However, Napoleon (1769–1821) did not allow him to continue his journey and ordered him to remain in France, where he dedicated his time to improving his knowledge in the field of porcelain and the textile industry. During this stay, he began to develop his skills as an inventor, patenting, together with the mechanic Etienne Calla (1760–1835), a machine for spinning all types of filamentous materials (Fig. 3).

After five years, he returned to Palma de Mallorca, where he collaborated in several projects supported by the Mallorcan Society of Friends of the Country, for example, the mechanization of wool weaving in workshops associated with the hospital. However, his stay there was brief; a few years after his arrival, he was assigned as a technical director to the Royal Cloth Factory of Guadalajara, again under the command of López de Peñalver. He was entrusted with the establishment of workshops to spin wool using machines; he was able to accomplish this because of the knowledge acquired in his previous stays in France.

In *Essai sur le composition des machines* by Lanz and Betancourt [13], Sureda is again described as the inventor of some mechanisms: an improved textile machine (Fig. 4), a machine for polishing windows (Fig. 5), a machine that applies continuous circular motion at a uniform velocity that can be changed to an alternative rectilinear motion (Fig. 6) and a machine for mincing leather to make cards (based on the motion of the previous invention) (Fig. 7), among others.

In 1820, he was assigned to the Royal Porcelain Factory of La Moncloa, which was created to replace the former Buen Retiro factory.

A few years later, in 1824, he was appointed director of the Royal Glass Factories of San Ildefonso, undertaking its renovation with the construction of new mills and the

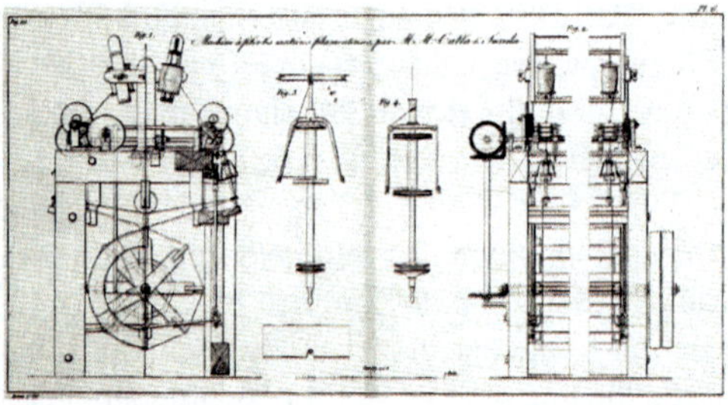

Fig. 3. B. Sureda and E. Calla. Patent: machine à filer les filamenteuses. 1812 [5]

restoration of the manufacturing and polishing of large mirrors using the aforementioned machines he invented (Fig. 5). In that same year, he was appointed director of the construction and repair of machines of the Royal Conservatory of Arts. Subsequently, Sureda was part of the jury for the first public exhibition of Spanish Industry products held in Madrid in 1827.

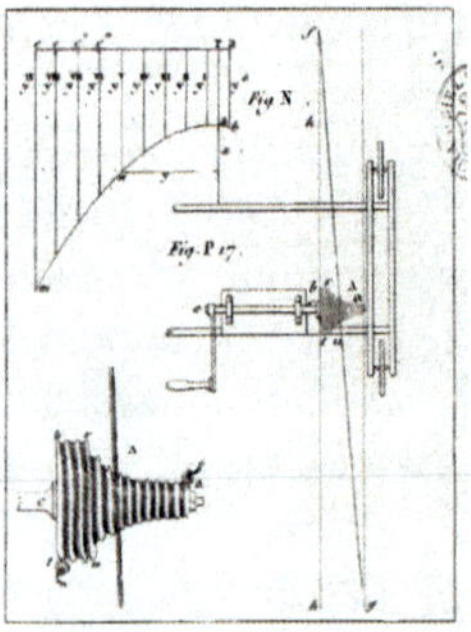

Fig. 4. B. Sureda. Invention: Improved mule-jenny spinning machine. Essay sur … (Mechanical description and application of the invention in facsimile of the first English edition of 1820, P17, p.149) [13]

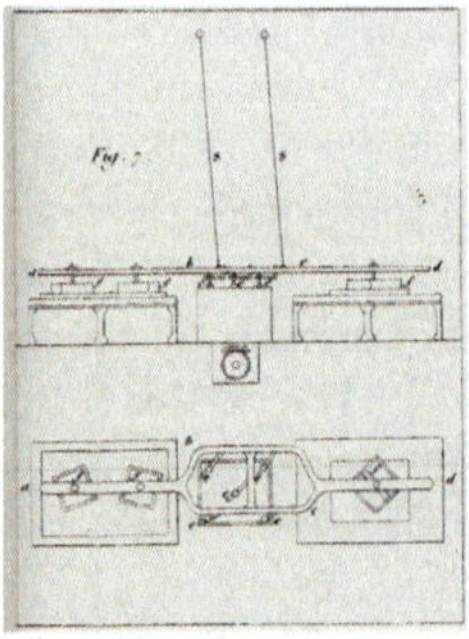

Fig. 5. B. Sureda. Invention: glass polishing machine. Essay sur… (mechanical description and application of the invention in facsimile of the first French edition of 1808, F.7., p.30) [13]

Around 1829, he returned to Mallorca. In recognition of an artistic career that he had never abandoned, he was appointed director of the Academy of Noble Arts by the Mallorcan Society of Friends of the Country, a position he held from 1837 to 1849, taking charge of the remodelling of the drawing and sculpture studios. In those years, he travelled with his followers to Barcelona to participate in the preparation of the plates for the *Historical-artistic Panorama of the Balearic Islands,* the work of Antonio Furió (1798–1853), a lithographic instalment by the painter and engraver Francisco Muntaner (1743–1805). The presence of Sureda in the Academy was an impetus for the application of new artistic techniques, i.e., daguerreotype and lithography.

Bartolomé Sureda died in Palma de Mallorca in 1851.

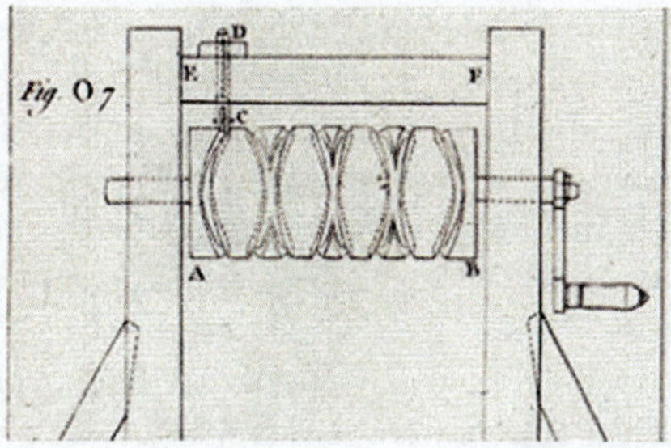

Fig. 6. B. Sureda. Invention: transformation of circular motion into rectilinear motion. Essay sur… (mechanical description and application of the invention in facsimile of the first French edition of 1808, O.7., p.36) [13]

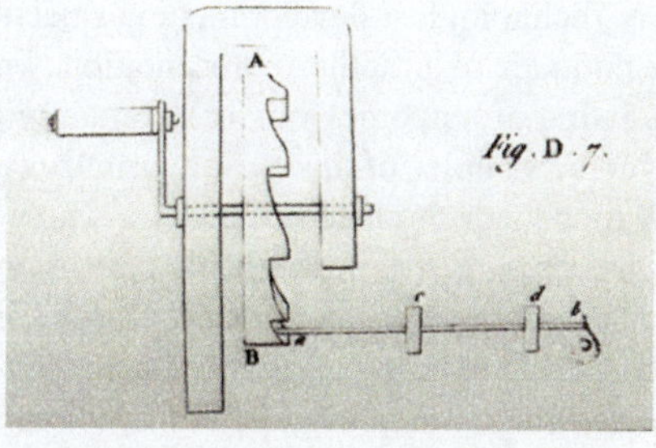

Fig. 7. **B**. Sureda. Invention: application of motion transformation to a leather chopping machine to make cards. essay sur… (Mechanical description and application of the invention in facsimile of the first French edition of 1808, D.7, p.29) [13]

3 Bartolomé Sureda, an Encounter Among Science, Technique and Art

In the extensive entry dedicated to Sureda in the Historical Dictionary of Illustrious Professors of Fine Arts in Mallorca, the first reference makes mention of his training in drawing, which together with mathematics constituted the pillars of the thought of Las Luces [7].

Bartolomé Sureda, as an enlightened person, was immersed in the great changes that occurred during the Enlightenment in conjunction with various technological and sociocultural phenomena that changed the character of life in the last decades of the eighteenth century. However, in his role as an architect of the graphic representation of machines, he actively participated in a fundamental moment, a before and after, marked by the historical significance of the Age of Reason and the incipient postulates of descriptive geometry. In this way, this period became an essential stage between the real and figurative staging that had begun in the medieval era and the normalized schematization of the contemporary age.

The drawing of machines in the 18th century contributed, to a great extent, to the characteristics of drawing in general. Enlightened scientists used a certain coding in their graphs that was accompanied by figuration and artistic rendering of the objects represented. It was then, in the late eighteenth century and as an innovative resource, when technical records were becoming popularized. The staged reading typical of previous moments was maintained, but gradually, a more technical definition was incorporated into the available graphic space, in such a way that the drawing of machines was progressively losing its plastic quality for the sake of industrial production in the eighteen hundreds. The new geometric postulates, together with serial manufacturing and the birth of the principle of interchangeability—direct antecedents of standardization in industry—contributed decisively to the creation of a graphic language of engineering that, as such, was acquiring a set of characters and symbols—a syntax—incomprehensible to unqualified readers [26].

Technological development has been greatly influenced at all times by the different techniques of graphic representation. From the Middle Ages to the Renaissance, the drawing of machines was based mainly on the staged representation of the object; the graphic samples of this period, usually embedded in technical texts, generally consisted of models of the mechanism as a whole with a merely illustrative function and sometimes presented some hints of conventionalism. However, in the Renaissance, technical representations began to exhibit certain recurring graphic aspects, and together with the theatrical rendering of machines and their environments, typical of the artists, other types of drawings were developed in which mechanical objects were shown with a more precise description, through some schematization and attenuation of the surrounding space. Later, in the eighteenth century, the transformation of social relations was decisive and of great depth, making it necessary to meet the demands of a new social system by resorting to inventing unprecedented technical procedures. A more intense relationship among science, technology and industry was established, leading to the birth of a true machine cult that led to a transfer of the craft to the industrial plane. In the 1700s, the production of images related to machines acquired a particular purpose that responded to the demands emanating from modifications in the social structure. During that century, technical records became popularized in the graphic space, which until then had only been artisanal; factories and workshops were shown as theatres in which activities were conducted. In this first phase of machinery, mechanical designs materialized in plans and drawings, in which washing or intaglio engraving was masterfully expressed in the communication of a certain type of information destined for technological development and formed the basis of an ideology of technicality. In short, beginning in the seventeenth century, and more formally in the eighteenth century, the same demands of the social system that had led to the division of labour were seen in the drawings of machines that sought decomposition in phases of the artisanal manufacturing processes [27].

In the first attempts to rationalize information and as an innovative fact of great importance – although with roots in some Renaissance machine theatres – the compilation of analytical and descriptive methods was linked, in the late eighteenth century, with a bipartition in the images of machines superimposed in the same work with comparative figuration of the whole and specifications of the compositional parts. The staged representation of previous moments was maintained, but gradually, technical records were becoming popular in the figurative representation of objects, and a more technical definition was incorporated into the available graphic space, with which the compositional parts of contraptions began to be exhaustively described. Finally, with the progressive arrival of industrialization at the beginning of the nineteenth century, the samples described above did not coexist in the same graphic space and continued, from that moment, as well differentiated currents. Bartolomé Sureda, along with other illustrious figures, were the last Spanish scientists interested in both technique and art. This affirmation is evident in the plates, drawn and engraved on paper by himself with the new chalcographic techniques learned in England, for the already named work directed by José López de Peñalver in 1798, *Description of the machines of more general utility that are in the Royal Cabinet, established in Buen Retiro* [15] (Fig. 8).

The century par excellence in terms of intaglio printing was the Enlightenment. The reforms undertaken during this period resulted in engraving being one of the most effective ways of disseminating ideas; engraving was conceived, mainly, as a means to create illustrations in scientific books. In Spain, all the technicians of the time were trained in hollow engraving in the recently founded Royal Academy of Fine Arts of San Fernando. Through this academy, technical advances and machines facilitated the progression of art, linking scientific studies to artistic development. In this way, the Academy played a fundamental role in scientific development by promoting decisive techniques for the dissemination of science: drawing and engraving [1].

During the seventeenth century, copper plates with engravings displaced wood blocks engraved with fibre, thus transitioning from relief techniques to hollow engraving to represent machines. One of these techniques, "aquatint," is a chalcographic method derived from etching in which resin is used in addition to varnish and acid.

Another new technique that Sureda learned in England was "gouache", "lavis" or "brush bite," which imitated the pictorial effect of an ink wash due to the direct application of water with a brush.

The aforementioned work by López de Peñalver, *Description of the most useful machines in the Royal Cabinet,* was developed with the active participation of Sureda and can be considered the first edition of a Spanish technical collection in which the modernizing of industry is illustrated. The prints, engraved with etching, gouache and burin, were made by Sureda and by the engraver Vicente Mariani (1765–1819) In the prints prior to the 19th century, the engraver included a series of Latin words that helped the reader understand the mechanism [19]. In the illustrations elaborated in gouache and illuminated later, as explained in the book itself, the following phrase is shown at the bottom of the sheet "Sureda drew it and engraved it" or *"Sureda lo dibuxó y gravó,"* which indicates that he was the only artist who worked on the plate. However, if we analyse the prints made by Mariani, created with burin, the caption that appears is *"Mariani sculp"* or *"Mariani sc,"* which means that he only engraved the plate and was not the creator of the original drawing (see the signature in Fig. 8 b).

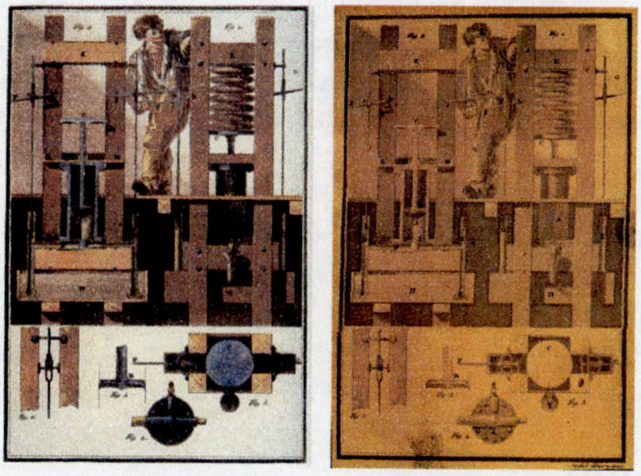

Fig. 8. B. Sureda. Hydraulic press of Mr. Bramah. (a) Stamp. (b) Copper plate. Descripción de las máquinas… [15]

During his stay in England, Sureda also learned about and trained in a new method of wood engraving that had been developed by Thomas Bewick (1753–1828) in 1775. Given the limitations of wood fibre carving, this new method involved engraving in relief. The process consisted of cutting the wood perpendicular to the axis of the trunk, countergrain, counterthread or testa, as it was called in Spanish, or gravure sur bois debout or wood engraving in French or English, respectively. This technique allowed working in a similar way to that of burin on metal.

Some years later, in the first decades of the eighteen hundreds, there was discussion in Spain about a recently invented image reproduction technique. It was an additive technique for planographic printing in which the matrix was obtained by adding material and not subtracting it, as in previous engraving processes. Lithography, given its name because it involves stones, gave rise to a matrix without incisions; therefore, it should not be included as an engraving technique, as it is sometimes designated [3]. The inventor of the aforementioned technique was the German Alöis Senefelder (1771–1834), author of plays, who, in 1796, in an attempt to print his own works and not having sufficient economic means to use copper, resorted to the use of stones from the quarries of Solnhofen (Bavaria), which led to the discovery of lithography. It was a flat printing technique that allowed stamping of an image drawn on a stone with ink or a grease pencil. Senefelder patented his invention in London in 1800.

The first report of lithography in Spain was the translation of the writings of Marcel de Serres (1783–1862) "News on lithography or the art of printing with stone moulds" included in the *Memory of Agriculture and Arts* published by order of the Royal National Board of Government of Commerce of Catalonia, Barcelona, by Antonio Brusi (1775–1821), in 1815–1816. Regarding its impact in Spain, lithography revolutionized graphic techniques used until then, primarily because of its ability to reproduce originals [23]. After a period of training in Munich, Bartolomé Sureda, Carlos Gimbernat (1768–1834) and José María Cardano (1782–1843) were the pioneers of lithography in Spain, with the intention of using the new technique as a means of scientific reproduction. In 1818, under the direction of Cardano, the first lithographic establishment was inaugurated in Spain and created for the management and processing of maps, charts and plans. Cardano himself showed Francisco de Goya the new technique [4]. Some years later, in 1825, Sureda assumed the task of advising José Madrazo (1781–1859), in charge of the Royal Lithographic Establishment, on the use of the lithographic presses in the Hydrographic Depository [17].

To conclude, with the newly implemented graphic techniques, Bartolomé Sureda was one of the first to learn and use daguerreotype. In the aforementioned work *Optical-historical-artistic panorama of the Balearic Islands* of 1840, Francisco Muntaner makes careful use of the lithographic press and the machine invented by Daguerre that ensured the reality of the representations of subjects [8] (Fig. 9).

4 Final Considerations

The extensive specialized training and the extensive and fruitful professional life of Bartolomé Sureda, a figure of knowledge and representative of the ideals of the Enlightenment, has barely been detailed in this work. Among the numerous facets of his activity,

Fig. 9. B. Sureda. Optical-historical-artistic panorama of the Balearic Islands. Molino de la Beata. Lithography, 1840 [8]

we wanted to note, specifically, his formidable contribution to the graphic representation of machines in their interaction among science, technique and art.

It is necessary to remember that, during the Enlightenment, due to new scientific postulates, fundamental changes occurred in the industry and originated the progress of modern engineering. In this sense, the similarities with the current moment in which a new industrial revolution is taking place are evident, now due to profound and emerging technological and methodological changes.

Engraving and lithography techniques played a fundamental role in the diffusion of technology; industrial advances have always been linked, in each stage or historical period, to graphic representations through these artistic techniques. War machines, hydraulics and a multitude of devices and machines, culminating with the steam engine in the nineteenth century, have been carefully drawn and reproduced, contributing to the dissemination of knowledge. Engraving, although initially considered a minor art, experienced a notable boom in Spain during the seventeenth century. As art at the service of the Enlightenment, engraving contributed to demonstrating the achievements made in areas as diverse as art, literature and science [14]. The most widespread reproduction engravings drew the eyes of Europe towards its own cultural assets. This technique was widely used in scientific dissemination, and it is an indisputable fact that one of its most notable creators was our character.

Through the graphic representation of machines across the centuries as an active subject of technicality, it seems evident that images themselves have held a preponderant position within the complex framework of machinery. It would not be possible to think of the industrial advances of the nineteenth century without the application of descriptive geometry to the science of machines, unifying the criteria in regard to capturing the achievements of the century.

Therefore, understanding and identifying the procedures by which these images were obtained is relevant because when thinking about their iconographic analysis, prior knowledge of the sources and familiarity with specific topics and concepts is required. To achieve this understanding, investigate the cultural tradition and scientific knowledge of the drawing of machines and, finally, pause and examine the essential work of Bartolomé Sureda y Miserol.

References

1. Bédat, C.L.: Real Academia de Bellas Artes de San Fernando (1744–1808). Fundación Universitaria Española. Real Academia de Bellas Artes de San Fernando, Madrid (1989)
2. Betancourt, A.: Catálogo de la colección de Modelos, Planos y Manuscritos que, de orden del Primer Secretario de Estado, ha recogido en Francia Dn Agustín de Betancourt y Molina. Madrid. (Biblioteca Palacio Real sig. DIG/II/823_A) (1792)
3. Blas, J., Ciruelos, A., Barrena, C.: Diccionario del dibujo y la estampa: vocabulario y tesauro sobre las artes del dibujo, grabado, litografía y serigrafía. Madrid. RABASF. Calcografía Nacional (1996)
4. Bonet, A.: Origen de la litografía en España. Boletín de la Real Academia de Bellas Artes Betancourt, A. de, de San Fernando, nº 73, 73–77 (1991)
5. Cantarellas, C., Sierra, J.: Cronología. In: Bartolomé Sureda (1769–1851): arte e industria en la Ilustración tardía. Madrid, Museo Municipal (2000)
6. Carrete, J., Checa, F., Bozal, V.: Summa Artis Historia General del Arte XXXI, El Grabado en España (Siglos XV–XVIII). "El Aguatinta". Espasa Calpe. Madrid (1987)
7. Furió, A.: Diccionario histórico de los Ilustres Profesores de las Bellas Artes en Mallorca. Por D. Antonio Furió, individuo de la Real Academia de Buenas Letras de Barcelona y de otros cuerpos literarios y científicos. Por Gelabert y Villalonga sócios. Palma (1839)
8. Furió, A.: Panorama óptico-histórico-artístico de las Islas Baleares. Edición facsímil de la de 1840. Prólogo de L. Ripoll. Ed. Luis Ripoll. Palma de mallorca (1988)
9. García-Ormaechea, P.: Betancourt y la Academia de Bellas Artes III: La actividad académica de Betancourt hasta 1800. Revista de Obras Públicas, CXII, nº 2992 agosto, pp. 937–945 (1964)
10. González, I.: Los Técnicos y el Arte, Betancourt, Sureda, Goya. In: Betancourt: los inicios de la ingeniería moderna en Europa. Ministerio de Obras Públicas, Transportes y Medio Ambiente; Centro de Publicaciones, D.L. Madrid (1996)
11. González, I.: Betancourt y los sacarócratas. El viaje frustrado a Cuba. In: Betancourt: los inicios de la ingeniería moderna en Europa. Ministerio de Obras Públicas, Transportes y Medio Ambiente; Centro de Publicaciones, D.L. Madrid (1996)
12. Lafuente, E., de Goya, L.C.: Edición facsímil. Ed: Gustavo Gili. Barcelona (1977)
13. Lanz, J.M., de Betancourt, A.: Ensayo sobre la composición de las máquinas. Facsímile de primera edición francesa de 1808. Facsímile de primera edición inglesa de 1820. Traducción al español de la edición francesa de 1808. Prólogo de J.A. García-Diego. Comentario al texto. J.M. Bogoliúbov. Colegio de Ingenieros de Caminos, Canales y Puertos de Madrid. Madrid (1996)
14. Larriba, E.: L'art au service de la divulgation scientifique: le rôle des gravures dans le Semanario de Agricultura y Artes dirigido a los Párrocos (1797–1808). El Argonauta Español (2005).https://doi.org/10.4000/argonauta.1185
15. López-Peñalver, J.L.: Descripción de las máquinas de más general utilidad que hay en el Real Gabinete de ellas, establecido en el Buen Retiro. Madrid en la Imprenta Real, por D. Pedro Julián Pereyra. (B.N. Sala de Investigadores Cervantes R/39554) (1798)
16. López, J.M.: Impulso y desarrollo de la actividad científica. In: Carlos III y la Ilustración 1788–1988, 2 v. Ministerio de Cultura, Barcelona (1988)
17. Origen de la litografía en España: el Real Establecimiento Litográfico. Fábrica Nacional de Moneda y Timbre, Madrid (1990)
18. Pascual, C., et al.: La porcelana de sepiolita de Bartolomé Sureda (1802–1808). Investigación arqueométrica sobre la Real Fábrica de Buen Retiro. Boletín de la Sociedad Española de Cerámica y Vidrio **50**(6), 311–328 (2011). https://doi.org/10.3989/cyv.402011. noviembre-diciembre

19. Rubio, M.: Ayer y hoy del grabado y sistemas de estampación: conceptos fundamentales, historia y técnicas. Ediciones Tarraco, Tarragona (1979)
20. Rumeu, A.: Ciencia y tecnología en la España Ilustrada. Colegio de Ingenieros de Caminos, Canales y Puertos de Madrid. Ed. Turner, Madrid (1980)
21. Rumeu, A.: El Real Gabinete de Máquinas del Buen Retiro: origen, fundación y vicisitudes: una empresa técnica de Agustín de Betancourt. Ed. Castalia, Madrid (1990)
22. Sureda, B.: Cuaderno de notas sobre cerámica (1802–1806). Edición al cuidado de José Sierra Álvarez e Isabel Tuda Rodríguez. Museo Municipal de Madrid (2000)
23. Vega, J.: Bartolomé Sureda y las técnicas gráficas. In: Bartolomé Sureda (1769–1851): arte e industria en la Ilustración tardía. Madrid, Museo Municipal (2000)
24. Vega, J.: Ciencia, Arte e Ilusión en la España Ilustrada. Gobierno de España, Ministerio de Ciencia e Innovación. CSIC. Madrid (2010)
25. Zulueta, P.: La Mirada a Europa de los científicos españoles de la ilustración. 18th INGEGRAF International Conference, Barcelona (2006). http://ingegraf.es/?page_id=11
26. Zulueta, P.: El dibujo de máquinas: sistematización de un lenguaje gráfico. In: Técnica e Ingeniería en España VI. El Ochocientos. De los lenguajes al patrimonio. Real Academia de Ingeniería, Institución "Fernando el Católico", PUZ, Ed.: Manuel Silva Suárez. Zaragoza (2011)
27. Zulueta, P.: Los ingenios y las máquinas. Representación gráfica en el período ilustrado en España. Colegio oficial de Arquitectos de Castilla y León Este, Secretariado de Publicaciones e Intercambio Editorial, Universidad de Valladolid (2007)

Inheritance Value of the Machinery of the Factory "Azucarera del Pilar" in Motril and the Sugar Industry in Eastern Andalusia

José Antonio Saez-Calvo[1,2](✉)

[1] Polytechnic University of Valencia, Valencia, Spain
jose.saez@intec.edu.do
[2] INTEC- Technical Institute of Santo Domingo, Santo Domingo, Dominican Republic

Abstract. The Pilar factory is an example of the survival of steam engines in the 20th Century. It is currently the most extensive mill train preserved, not only in Spain but also in Europe. In terms of dimensions, it is only comparable to a steam steel rolling mill. Up to 11 steam engines are preserved for different uses, from the movement for grinding to the elevators and cane drivers. The mill train is the heart and the essential part of the symbolic sugar factory in Motril, which should be transformed into an industrial sugar museum. The machinery has the declaration of Asset of Cultural Interest. At the same time, the chimney and the rest of the enclosure that is conserved are protected because they are part of its context. Steam engines, manufactured in foundry and representing the last period of the sugar industry, are the only reference of this type in Europe as a set of parts representing an assembly. In this paper, there will be a technical and historical review, with figures, of the five mill trains that remain in Spain, curiously in the provinces of Granada and Malaga. Emblematic places where a robust sugar industry developed at the end of the 19th Century, based on steam engines, presented different configurations, and grinding sugar cane and beets.

Keywords: Mill train · Sugar cane industry · Industrial heritage · Steam machines · Granada development

1 Introduction

Historical aspects of the development of the province of Granada and some of the regions of Eastern Andalusia cannot be understood if it is not mentioning its almost thousand-year-old sugar tradition. Although in the "Calendar of Cordoba" in 961, the existence of sugar cane is cited on the coast of Salobrena and Almunecar, the botanical studies where the cultivation in the area stated dates from the end of the 11th Century with Ibn Wafid or the first decade from 12th Century with the Granada botanist Al Tignari in his work "Treaty of Agriculture."

Sugar in Granada was a lever in developing the urban nucleus and of infrastructures, especially from the end of the 19th Century when the overseas provinces of Cuba, Puerto Rico, and the Philippines were lost from the Spanish Crown.

M. Ceccarelli and R. López-García (Eds.): HMM 2022, HMMS 40, pp. 470–486, 2022.
https://doi.org/10.1007/978-3-030-98499-1_38

Fig. 1. Aerial overview. Azucarera del Pilar [5]

It was significant potential in the tram or railway network in the characterization of the landscapes of the buildings—currently, one of the top-ranked in the cataloging of cultural assets [12].

These transformations necessarily introduce social and ecological conditioning since the hydraulic and landscape organization of the environment is affected by the socio-economic and environmental structure. Agricultural technical issues reflect that the crop requires abundant irrigation in summer and unique climate and location conditions.

The basis of the current paper focuses on "Nuestra Senora del Pilar", Fig. 1, in Motril. It shows the sugar production and storage system from an industrial point of view, and the result of the technological development that sugar is industrialization entailed focused on steam engines.

During the 19th Century, the sugar industry became the recipient of technological transformations such as the application of steam engines in different applications, such as moving to mills or internal transportation for the raw materials and final products Fig. 2.

Starting in 1882, an essential beet-sugar industry was created in Spain. Therefore, it must bear in mind the phenomenon of the development of technology applied to beet, becoming an apparent competition to the sugar cane.

2 The Sugar Cane Industry in the South

2.1 The Crop as an Industrial Origin and Development

Sugarcane is a grass of the genus "Saccharum" that grows optimally in tropical and subtropical territories, ranging in height between two and five meters. The stem, three to five centimeters in diameter, is made up of cylinder joints separated by knots, in variable

Fig. 2. Cane square and overhead cable laying [7]

numbers depending on the height of the stalk (from 40 to 60), covered with long leaves with serrated edges alternately arranged. The stem is covered with hard tissue. Its interior is composed of fibrous material or spongy pith, in whose interstices microscopic cells containing sucrose are housed; This, separated from other impurities, constitutes sugar.

These characteristics are reflected in the layout design of the industrial plant. A transport line is necessary where the trunk of the cane enters a sequence of mills, where the grinding takes place and the start of obtaining the so-called juice.

Botanists have described more than 20 species of the genus Saccharum, among which the Creole cane (called in Andalusia "Cana de la Tierra," Doradilla, or Algarrobena) and the yellow cane of Tahiti (called American cane in Andalusia) stand out. Its genetic instability makes it susceptible to degeneration. Still, at the same time, it allows producing crosses, new varieties destined to increase its saccharin richness or improve its resistance to more extreme climatic situations [16].

The optimal climatic and water conditions for its development range between 20 and 25 degrees Celsius in average temperature and 1,200–2,000 mm. of precipitation. The coastal plains of the Andalusian Mediterranean had a moderate temperature, and a thermal amplitude at the limit of these requirements, and the insufficient rainfall could be compensated by artificial irrigation.

2.2 Location of the Industry, Near the Crop Development

Due to the microclimate configured on the coast from Adra to Velez Malaga, known in its longest distance as Costa Tropical, it meets the climatic, orographic, and latitude conditions for the sugar crops to acclimatize and be productive.

In addition, isolated to the north by the mountain ranges and bathed in the south by the Mediterranean Sea, a microclimate is created that protects from the rigors of the continental climate.

Sugarcane was the alternative that was followed in the 19th Century. There is now a commitment to other tropical crops, such as kiwi, mango, custard apple, kakis, avocados, papayas, guavas.

It was established to consider being an exotic crop in Europe and a specific luxury item, which led to industrial plants associated with farmland, creating activity, developing technology, and productive techniques. Irrigation works business initiatives had to be faced, together with the workforce's learning, soil conditioning, and the worst part, the destruction of mountains, to obtain wood for the energy supply of the boilers.

Specifically, the "Vega del Rio Guadalfeo" is a large coastal plain that is climate protected and meets the requirements for developing sugar cultivation. It results from the silting process of an old bay and with the sedimentary contributions of the river and other channels such as the Ramblas de Molvizar, Puntalon, Alamos, and Las Brujas.

And it is here where the first sugar factory with steam engines was in Almunecar and the Pilar factory.

3 The Industrial Process

Since each building is handled individually, it may be designed to conform with any given set of requirements, made necessary by conditions existing at the factory site.

The width of the building is determined by the space occupied by the tandem, gearing, power machinery, hydraulic accumulator, and the conveyor for transportation.

To obtain the sugarcane juice and industrially design the extraction, within a continuous process to increase the performance faster, the industry substituted both the animal and water traction for the force generated in a steam engine associated with the milling machine Fig. 3.

A series of operations were performed, including selecting and preparing the sugar cane, cutting, milling to obtain the bagasse, cooking, and crystallization. In particular, the technology applied to Pilar's factory focuses on milling the sugar cane using a gear train driven by a steam engine, whose movement is communicated to cylinders that produce the milling (pressing) of the sugar cane.

3.1 Industrial Design is Based on the Type of Crop and Technology Available

When the cane is cut (mature) to be driven to the mills, these must be of such capacity and in such good condition that they allow the total tonnage to be continuously ground in the shortest period possible.

By grouping several mills into a train with their units, a profitable extraction of the cane juice is ensured in one plant. As a result of this high extraction, a minimal percentage of sucrose remains in the bagasse after the last mill. And bagasse contains so little moisture that it is ideally suited as fuel.

The possibility to join the new technologies coming from the Industrial Revolution, together with the option of wood to feed the boilers, made the incorporation of steam

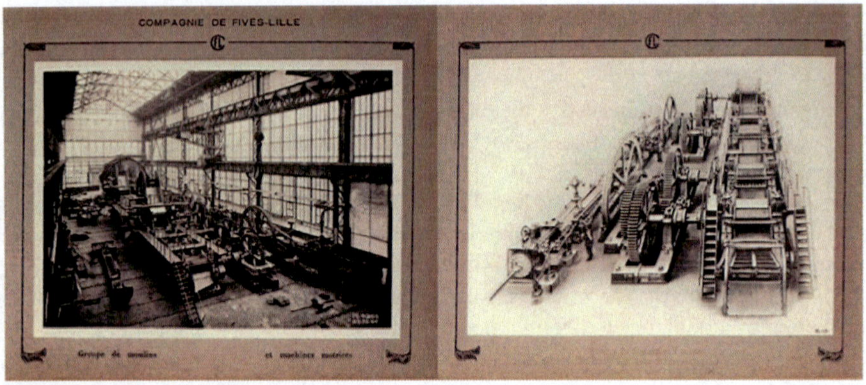

Fig. 3. Left: Test bench at the Fives-Lille factory for a mill train with shredder and drive machines. 1924. Right: Steam-powered cane mill train, built by the Fives-Lille company. 1921. Miguel Gimenez Yanguas Collection [4]

engines, associated with the mill, that can propel the mill, along with hydraulic pumps, powered by steam boilers through a connection of pipes.

3.2 Machinery Description

All the mechanical components are fixed in bases made of cast iron due to the stress to which the machinery is subjected.

The side covers are made of semi-steel and are made so that they hug the side heads and are also provided with a spring for adjusting screws that allow an easy and fast adjustment of the cane and bagasse hubs while they are grinding.

The top caps are cast steel and are provided with hydraulic pistons to regulate the pressure in the hubs. All thrust bearings are phosphor bronze. The crowns are tough to cast steel, turned cast, and well-tempered. The axes of the hubs are provided with double crowns, adjusting perfectly.

Iron plate guards cover the crowns for protection, preventing bagasse from clogging them and collecting excess oil that escapes from the bearings.

A steam engine moved the conveyor belt, and from the cane square, where there is a well from which the conveyor belt starts, it fed the cane mill train from the top.

First, it entered the defibrillator to make fiber from the cane, and then it entered the three mills, whereby mechanical pressure, the cane was squeezed, and the juice obtained Fig. 4.

There is a transfer of the reeds, where steam engines operate the shafts that the mill moves the mass, and the stakes have trunnions, which can weigh 1000 kilos.

The mills have hubs or fluted cylinders, which go in an equilateral triangle, and with gears that join them to rotate in unison.

The tops of the mills have hydraulic hubs, which, if any stone or unwanted element penetrates, hydraulic pressure accumulators such as oil presses, and a machine or steam cylinder for a pressure meter. The closures must be caught next to the gears, and it carries large bronze windings that correspond to the defibrillator Fig. 5.

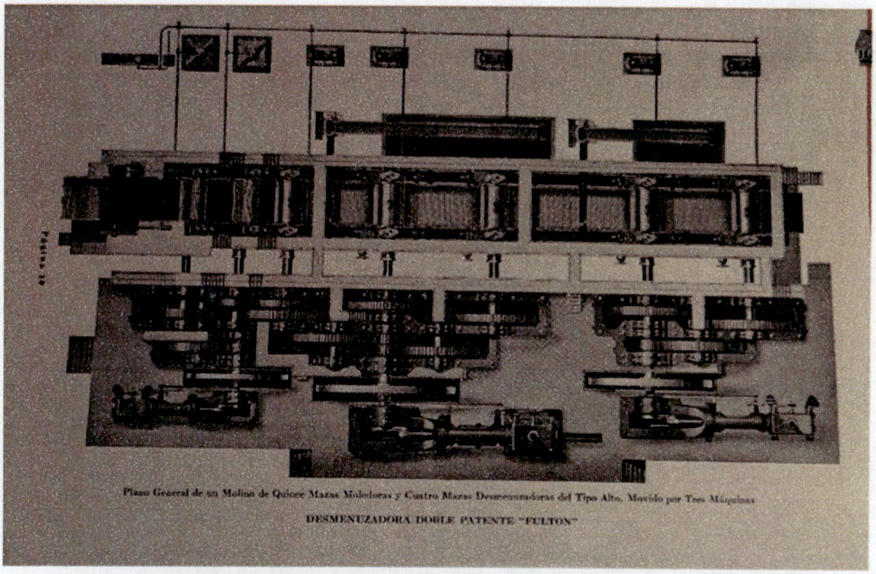

Fig. 4. Drawing. Layout triple mill train. Fives- Lille [4]

The oil enters through the upper part and is injected by the pump. There are conveyor belts, which go up to the upper floor from the bottom of a mill.

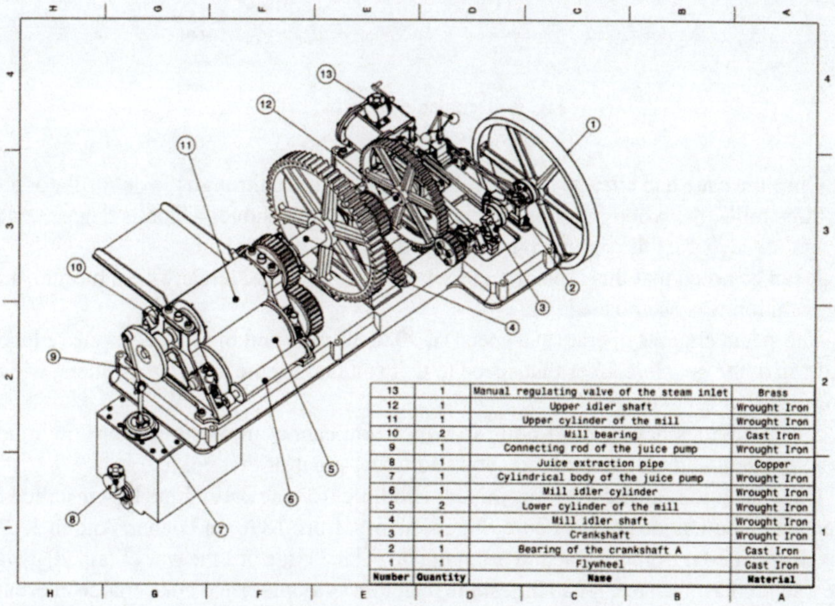

Number	Quantity	Name	Material
13	1	Manual regulating valve of the steam inlet	Brass
12	1	Upper idler shaft	Wrought Iron
11	1	Upper cylinder of the mill	Wrought Iron
10	2	Mill bearing	Cast Iron
9	1	Connecting rod of the juice pump	Wrought Iron
8	1	Juice extraction pipe	Copper
7	1	Cylindrical body of the juice pump	Wrought Iron
6	1	Mill idler cylinder	Wrought Iron
5	2	Lower cylinder of the mill	Wrought Iron
4	1	Mill idler shaft	Wrought Iron
3	1	Crankshaft	Wrought Iron
2	1	Bearing of the crankshaft A	Cast Iron
1	1	Flywheel	Cast Iron

Fig. 5. Mill and steam machine assembly plan [14]

The milling or pressing of the sugar cane requires considerable compression force. However, the steam engine generated very high speeds that were unsuitable for milling, and it was, therefore, necessary to transform speed into power employing a compound gear train Fig. 6.

Fig. 6. Detailed gear train [14]

Once the cane had already been shredded, it had passed through the defibrillator and the three mills, the woody part of the cane, the resulting residue, which is bagasse, was going to feed as fuel the special burners that the steam boilers had.

It can be noted that this type of factory was ecological and already used biofuel, and cogeneration was established [6].

The steam engines operate at a speed of 70 to 80 rpm, and the speed-torque reducer, made up of the gears, reduces that speed to that suitable for the hubs or cylinders, which is of the order of 8–10 rpm Fig. 7.

Steam engines have a dead point in which you cannot move no matter how much steam entered, and with the ratchet, you had to get it out of that point.

Once the juice came out, it was mixed with lime from a nearby lime kiln installed in Granada in the Duque de San Pedro Sugar Factory, from 1890, in Lachar. And in 1929, they dismantled it from Lachar and brought it to Pilar. Oxide of lime was obtained, it was like a small blast furnace, it had limestone (calcium carbonate) and coke coal in alternate layers, it was made to burn, that is, it burned, and the heat made the calcium carbonate decompose into oxide of lime and carbon dioxide. The lime oxide came out from the

Fig. 7. Steam engine without steam-chamber side cover [14]

lower part and went to a trommel that quenched the lime oxide with water, and calcium hydroxide was obtained. The calcium hydroxide was mixed with the juice from the mill train. Then it made a series of chemical products precipitate, mainly monocalcium phosphate, which is soluble, and tricalcium phosphate was formed. As lime was added in excess to carbonate the juice, in the manufacturing hall, making the carbon dioxide that was extracted from the upper part of the lime bubble, it was sucked up with pumps, moved by steam engines, bubbled through the juice with each paid and the excess passed to calcium carbonate.

In other words, we have tricalcium phosphate, which precipitates, calcium carbonate that we have precipitated in excess, with carbonic anhydride. In the press filters, through some cloth, the juice was passed with the pressure in horsepower because of the pumps, and the solid phase remained in the fabrics, which, when the filter press was opened, fell into the collection hopper.

As soon as the juice was purified, it had to be concentrated under a vacuum, and for this, we have an evaporation set. The fluid enters the first boiler with indirect steam from the steam engines. Evaporation of the juice takes place, these vapors are conducted to a second boiler to feed the exchanger of the double boiler, the fluid continued to evaporate, and thus it was passed to the third, and from the third to the fourth. From the fourth, the steam went to the vacuum column, 10 m high, were at the top was a water shower. Then the moisture condenses and produces a vacuum.

Of course, heating the juice has reactions, not only physical, if not chemical, of decomposition of the liquid, where methane, carbon dioxide, and carbon monoxide were obtained. If it is not removed, the vacuum that is created is ruined. There were two auxiliary steam engines, one from 1900 and the other from 1884, which moved vacuum pumps that removed the non-condensing gases from evaporation. All this in a quadruple effect. The gases were released into the atmosphere, with a very typical smell Fig. 8.

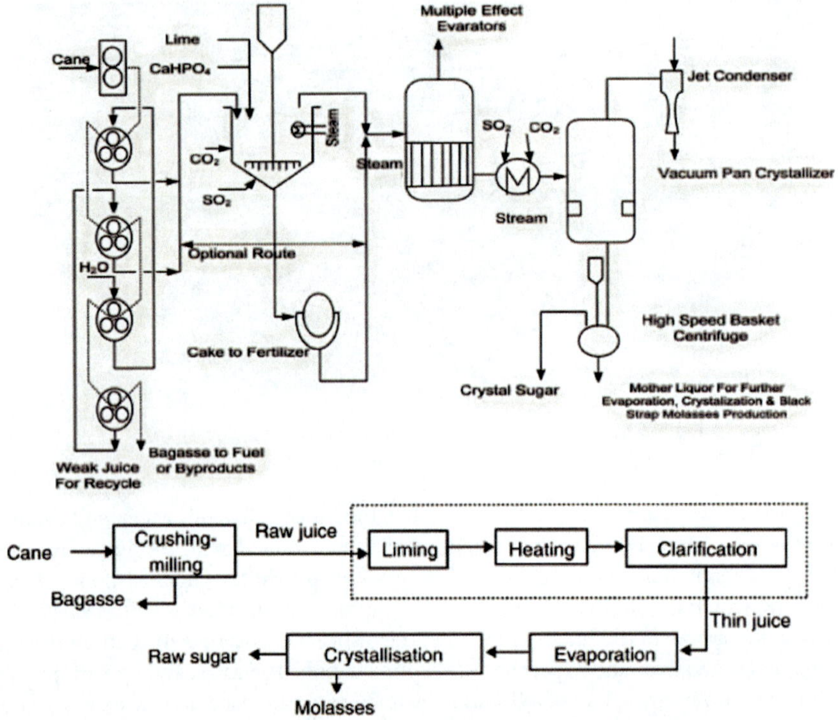

Fig. 8. Simplified flow diagram. Sugar physical-chemical process. Author

3.3 Why Do Steam Engines Survive in the Sugar Industry Compared to Electric Motors?

The system works in cogeneration. The industry needs steam for its operation and with a heat exchanger. Not over direct heat as at the beginning but steamed.

Producing steam is expensive, but in the sugar industry, it is necessary to assure of the processes [13].

The circuit was fed at a pressure of 10 kg./cm^2 of motive force for the cane milling. The water of about ten ° C must be raised to 100 degrees; the specific heat of the water is high. Once 100 ºC is reached, we have the latent heat of vaporization, but when we have steam and reheat it, the specific heat is small as the pressure increases. We put it in the steam engine, and instead of letting it escape, we send it at 2 kg. of pressure to the rest of the manufacturing process Fig. 9.

The production of the factories in Motril was concentrated in El Pilar. In 1929 it was reformed to move about 1,000 tons of cane per day. In the reform, they removed the centrifuges, where the sugar is separated from the honey. They made cane syrup concentrate here and sent it to the Atarfe distillery, directed by the engineer Zubiela. It was then sent to a small distillery in Salobrena at the San Francisco factory.

When Sociedad General Azucarera dedicated itself to making alcoholic beverages and created the Bermudez Rum Winery thus, in the last year of operation of El Pilar, they devoted themselves to manufacturing sugarcane syrup concentrate to feed the alcohol factories [7].

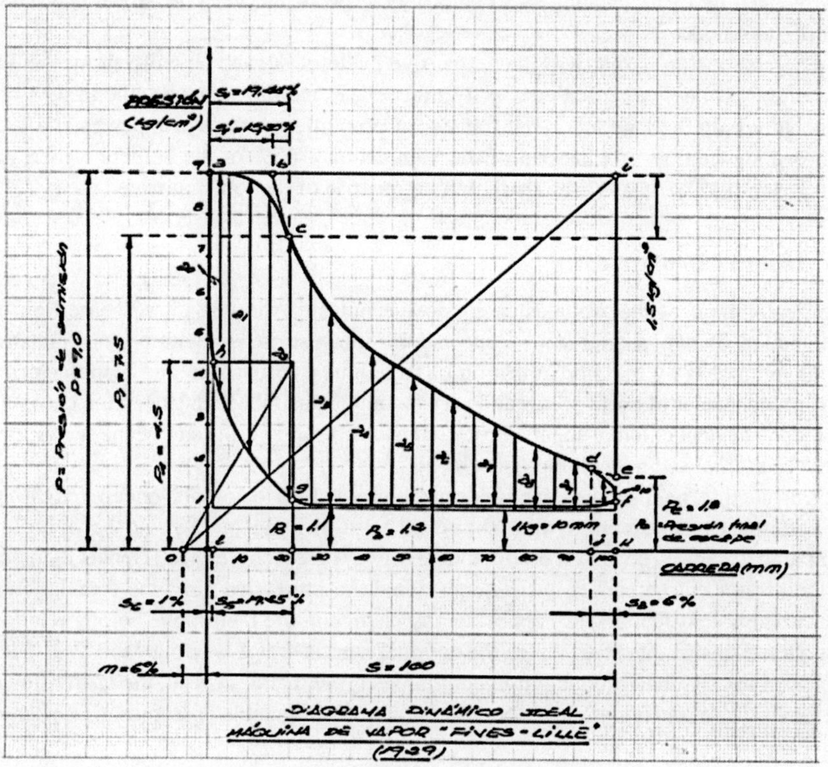

Fig. 9. Dynamic diagram Fives- Lille [4]

4 Local Deployment and Development

4.1 Crop Expansion

During the six months that a mill worked per year, it needed to have the ovens constantly lit, consuming large amounts of firewood annually. With the Royal Provision of Carlos II, it has come to prohibit the entry of cattle in the sugar lands because firewood was necessary for the milling of the cane.

According to "Simon de Rojas," the cultivated area of the "Vega de Motril" could reach 800 Ha in the middle of the 16th century; according to other sources, in 1561, it could be around 1,200 Ha. In the middle of the s. XVIII is when there are more precise files, with the Ensenada cadaster, which included between registered and no man's land around 970 Ha in 1752. Later, in 1785, 1,600 Ha of cultivation area were recorded;

and they are no more because the flooding of rivers and floodplains are not controlled. Already in 1830, 40% of the surface of the Vega continued being unproductive [15].

It is with the impulse to cultivate sugar cane when in 1860 the unproductive land was reduced to 19%, reaching 1930 with only 4.2% of the Vega de Motril and 6% of that of Salobrena.

The conclusion is that with the expansive cycle of sugarcane cultivation, the surface of the productive agricultural soil improved. Regarding property, the impact of a monoculture on the land led to the transfer of properties to sugar companies and local land-owning families. At the end of the 19th century, 65% of the holders owned less than 1 Ha. In 1930, it was 74%, participating in 16% of the total cultivated area.

4.2 Industry Deployment

The deployment of the Andalusian sugarcane industry from 1845 was based on the conjunction of two phenomena: a spectacular increase in agricultural and industrial yields, promoted by the modernization of techniques, cultivation, and manufacturing, and circumstances of fiscal policy and very favorable tariffs, which gave it the necessary protection against colonial sugars. Protected by such support, the sugarcane sector grew to the limit imposed by geography itself [2].

Despite this, peninsular sugar yields never came close to colonial ones, even though the differences in outcomes were shortened until arriving in 1881, to be only 20% compared to Cuba. The lowest productivity is always the Achilles heel. Hence, industrial and cultivator entrepreneurs always considered the protective tariff the foundation of their existence, systematically opposing any change in the metropolis-colony relations that altered this fragile status since the national market was a way out each time. More irrelevant for a Cuban production in continuous growth and, instead, the tariff allowed the survival and consolidation of a peninsular industry incapable of representing a danger [3].

This scenario became complicated around 1880; The tense relations between the colonies' interests and the Hispanic metropolis were joined by the birth and rapid development of the Spanish beet industry from 1882. From the Vega de Granada, the cultivation of beet would spread very quickly throughout the lands of the Ebro Valley and Castile, achieving at the end of the century to produce almost all the sugar consumed by Spain; Without geographic limitation and with the support of industrial productivity based on its greater saccharin wealth, the production of beet sugar tripled that of cane in a short time [9].

The national market was organized as a protected market. An oversized sector would have to supply and be composed of diverse production capacity, technological modernization, and production costs (whether sugarcane or beet); This made it necessary to carry out a drastic reconversion of the sector, resulting in the constitution of the "Sociedad General Azucarera of Spain" and the closure of the least profitable factories. In these circumstances, the worst part was taken by the sugarcane facilities. The attempt at a sugar business monopoly, in any case, did not manage to stably order the sector, having to resort to the prohibition of setting up new factories [11].

Thus, after reaching a historical record in 1900 - with the milling of 357,000 tons of cane and obtaining almost 34,000 tons of sugar - the sugarcane sector faced a drastic decline throughout the first quarter of a century, minimizing its presence in the total production of national sugar - from 35.7% in 1900 to 4.2% in 1925- and radically limiting its entire show, which in 1925 did not exceed 25% of that obtained in 1900 [10].

The fall in agricultural and sugar production constitutes the most obvious example of the industrial dismantling carried out in the first years of the century and the progressive loss of profitability in the farms, accelerated by the fall in prices and sugar cane yields. In 1900, 29 sugarcane factories were in operation. Ten were in the province of Granada, 17 in Malaga, and 2 in Adra; During that campaign, nine factories were grinding in the Vega del Guadalfeo, of which we mention Nuestra Senora del Pilar [1].

5 "Nuestra Senora del Pilar" Factory

5.1 Description

Designed by the Granada-born architect Francisco Jimenez Arevalo, the construction of the factory and warehouses began in 1882. The building complex-built brings us to the factory and colony characteristics of the sugar world. The typology of its industrial buildings results from extensions, modifications, and destruction carried out over a century.

The current factory is located to the east of the city of Motril, in the foothills of a hill included in Rambla de las Brujas; Originally, the factory was located on the outskirts of the town, although today it has been practically integrated into the urban complex.

At a lower level, the old canal is located, which serves as a dividing boundary between the old plain and the upper lands not traditionally subject to irrigation; This site has an approximate area of 7 hectares, forming a kind of theater open to the Southeast and at the base of which the facilities themselves are located.

In 1984, after 102 years of almost uninterrupted operation, the Nuestra Senora del Pilar factory permanently shut down its boilers and was waiting to be dismantled.

The factory plant consists of:

- Four industrial buildings with a three-story elevation.
- The distillery is separated from the body of the factory. It consists of a height of three floors.
- Warehouse for tools and appliances, formerly it was the chapel and stables.
- It has a central plan and an octagonal shape to which another building with a rectangular plan is attached.
- Sugar warehouse, rectangular. Inside there is a second floor.
- All these units have gabled roofs.
- The industrial area of the triple mill train consists of a plant of significant proportions to implement the vast layout Fig. 10.
- Manor and Management House has a rectangular plan with a three-story elevation attached to the four original buildings.

Fig. 10. Mill train- industrial building N.S. El Pilar. Current situation. Photo: Author

5.2 Machinery

Regarding the most valuable set of machinery that is preserved in Spain, from "Ntra. Sra. Del Pilar", it has the following eleven steam engines, Fig. 11, which are worth mentioning [8]:

- 1 Aktien- Machienenfabrik machine, which acts as the driving force of the electric generator, with a flywheel and simple slide valves.
- 1 Fives-Lille engine from 1889 gave the sugarcane cutting machinery motion power, with two twin flywheels and single slide valves.
- 2 Fives-Lille machines from 1929 provide the driving force for the mill train. The machine's shaft is coupled to a speed reduction gear train, which, in turn, connects with the mill cylinders. It has a centrifugal speed regulation system with hydraulic damping. It represents the technological cutting edge of the time, with a power of 200 CV.
- 1 B.M.A. machine. (Braunschweigische Maschinenbrau Anstalt) From 1893, which provides a driving force to a carbon dioxide compressor that, coming from the continuous lime kiln, is installed to feed the factory carbonation. It has inlet and expansion valves through a simple slide system and a Porter-type centrifugal speed regulator.

- 1 Etabs machine. Carson- Delmotte from 1898 provides the driving force to a carbon dioxide compressor coupled to its shaft, with intake and expansion valves using a simple slide system and a Watt-type centrifugal speed regulator, with four spheres.
- One machine from "Talleres Castanos" of Granada in 1900 supports an air compressor for the sulfidation process, vertical counter pressure, and admission and expansion valves by simple slide, lacking a speed regulator.
- 1 Fives-Lille machine from 1889 gives motive power to the cloth washing machine, with two twin flywheels, speed regulation by Watt system, and single slide valves.
- 1 Fives-Lille device from 1889 gives motive power to a vacuum evaporation pump through a simple slide system, Watt type centrifugal speed regulator Fig. 12.
- 1 Fives-Lille machine from 1900 provides the driving force for an evaporation vacuum pump using a simple slide system, a Watt-type centrifugal speed regulator, and an inertia flywheel.
- 1 S.A. machine from "Constructions Electriques" de France, from around 1930, gives motive power to turbines and malaxers. It has an inertia flywheel, hydraulic-type speed regulator, and Sulzer-type intake and expansion valves.

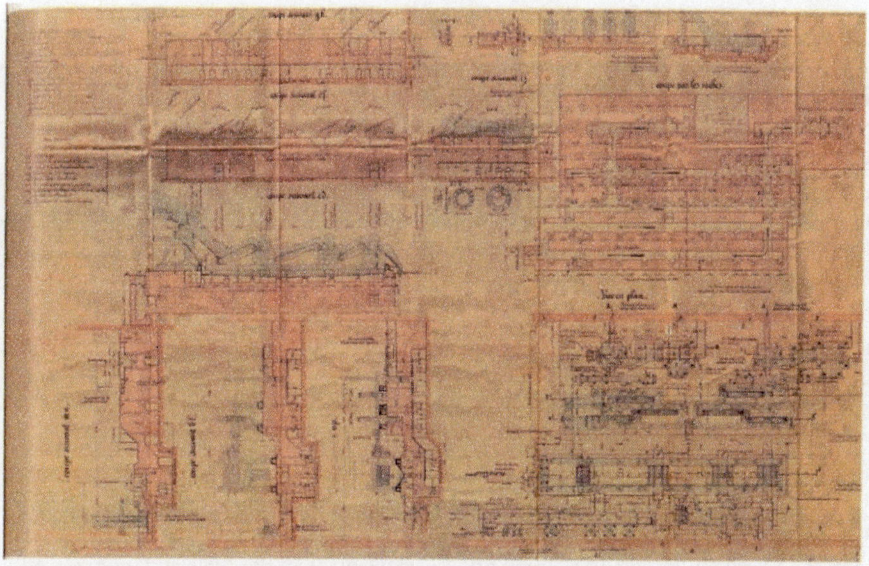

Fig. 11. Installation drawing of the mill train of the «Nª Sª del Pilar» sugar factory. 1929 ca. Miguel Gimenez Yanguas Collection [deposit]

Fig. 12. Steam machine. N. S. El Pilar. Fives- Lille catalogue [4]

6 Conclusions

An industrial remnant must be valued for its technological exceptionality, not only for its outdated and antiquity, nor vintage but also for its survival character coming from a disappeared industrial activity. And bearing in mind, adding value has impacted society and development.

The Pilar sugar factory is the most complete and has the oldest triple mill train in Spain. Its construction dates from 1882. In 1984 it carried out its last milling works, being definitively closed out. It is the most valuable vestige of the industrial relics that are preserved linked to sugar production. The architectural integrity of the complex is combined with the inside machinery built between 1880 and 1929.

Therefore, the main protagonist of this article is mentioned as the sugar factory of "Nuestra Senora del Pilar" and the remains of the other four mills that remain in the provinces of Malaga and Granada, which refuse for disappearing. It is a believe this architectural and industrial complex matches with the same considerations that UNESCO used to declare the "Voelklingen" steel plant in the Saarland, considered a unique testimony of the steel industry during s. XIX and principles of the s. XX. Such mention that "Nuestra Senora del Pilar" would also deserve, in addition to the museum project.

Unlike architectural structures, which are much more resistant to the passage of time, the historical machinery linked to sugar production has undergone a considerable decline in recent decades. A certain number of pieces have been rescued and restored, but the destruction of many others is now irreversible. Despite this, there is still a rich movable heritage with a very uncertain future unless steps are taken to acquire and preserve it.

The value of an operation of this type is none other than trying to put together pieces that, regardless of their intrinsic value, can jointly give a detailed overview of how sugar was produced at the end of the last century and how the sector evolved technologically. Therefore, it is not about the conservation and exhibition of isolated machinery but constructing complete industrial sets Fig. 13.

The comprehensive rehabilitation of this factory is the best service that can be done to safeguard the historical and technical heritage that we have inherited from the sugar industry that allowed the social and urban development of the province of Granada.

Fig. 13. Steam engine machine [4]

Acknowledgments. I want to acknowledge the support received from Industrial Engineer and UGR professor Miguel Jimenez Yanguas, for his large machine and documentation collection, as a mentor, advisor, restorer, and entirely devoted to the conservation of the industrial heritage, especially the steam machines and the Granada's sugar industry. Also, to Professor Miguel Angel Sebastian from the UNED Industrial Engineering School.

References

1. Rozua, J.M.B.: Reforma urbana y destruccion del patrimonio historico en Granada. (Urban reform and destruction of the historical heritage of Granada) University of Granada (1998)
2. Perez, J.C.: Factores de la estructura socioeconomica de Andalucia Oriental. (Factors of the socioeconomic structure of Eastern Andalusia) Archivum. Universidad de Granada (1993)

3. Evangelista, L.: Tratado de fabricacion del azucar de cana y remolacha. (Treatise on the manufacture of cane sugar and beets) Madrid (1895)
4. Fives-Lille Catalogue for Sugar Mills (1925)
5. Yanguas, M.G., Santos, S.P.: Motril y el azucar. Paisaje, historia, patrimonio. (Motril and sugar. landscape, history, heritage) Brochure Expo Millennium Granada' Kingdom. pp. 75–91 (2013)
6. Yanguas, M.G.: Importancia Tecnologica de la Industria del azucar de cana. Motril y el azúcar. Tradicion y modernidad. (Technological importance of the cane sugar industry. Motril and sugar. Tradition and modernity) Granada (1989)
7. Yanguas, M.G.: El patrimonio industrial azucarero en la Costa Granadina: origen, situación y perspectivas. (The sugar industrial heritage on the Granada Coast: origin, situation and perspectives) Jornadas Ibericas de Patrimonio Industrial y la Obra Publica. Sevilla (1994)
8. Samos, J.P.: Yanguas, M.G.: La fabrica azucarera Nuestra Senora del Pilar. (The sugar factory Nuestra Senora del Pilar) Motril (Granada). Valoracion y propuestas de actuación. VIII Congreso Internacional para la conservación del patrimonio industrial. Actas. Madrid (1995)
9. Rodríguez, M.M.: Azucar y descolonizacion. Origen y desenlace de una crisis agraria en la Vega de Granada. (Sugar and decolonization. Origin and outcome of an agrarian crisis in the Vega of Granada) University of Granada (1982)
10. Martinez, A.P., Rosado, A.: El fracaso del cartel la Sociedad General Azucarera de Espana: Un análisis desde la legislacion sobre este sector industrial (1899–1931). (The failure of the cartel the Sociedad General Azucarera of Spain: An analysis from the legislation on this industrial sector) XII International AEHE Congress. Salamanca. (2017)
11. Pascual, A., Rosado, A.: Creación, quiebra y remodelacion del trust azucarero español: la Sociedad General Azucarera Espanola (1903–1916) (Creation, bankruptcy and remodeling of the Spanish sugar trust: the "Sociedad General Azucarera Española"). Investigaciones de Historia Económica. AEHE (2019)
12. Samos, J.P., Yanguas, M.G.: Motril y el azucar. Del paisaje industrial al patrimonio. (Motril and sugar. From industrial landscape to heritage) Asukaria Mediterranea (1996)
13. Sola, J.I.R., Marin, J.R.U.: The steam engines in sugarcane production in Spain. DYNA Review #171 (2012)
14. Sola, J.I.R., et al.: An approach to the study of a 19th-century sugarcane mill and steam engine through CAD techniques and mechanical engineering. Machines throughout History for Technological Development: Analysis from Engineering Graphics. Jaen (2021)
15. Gandia, M.A.R., Yanguas, M.G., Mesa, J.M.R.: Patrimonio Industrial en Granada. (Industrial Heritage in Granada) Obra Social Caja Granada, pp. 171–187 (2003)
16. Sagra, R.: de la. Informe sobre el cultivo y la fabricacion de azucar en las costas de Andalucia (Report on the cultivation and manufacture of sugar on the coast of Andalusia) Azucarera Peninsular Company. Madrid (1845)

Industrial Heritage in Granada and Almería - The Recovery of the Zurgena Railway Station as an Example of the Reinterpretation of the Lorca-Baza Railway Line

Rafael García de la Mata[1(✉)], C. González-Sevilla[2], A. Dávila Ponce de León[3], M. J. Gámez[3], and Lourdes Rubio[4]

[1] Restaura, Gestión Museística Y Restauración Siglo XXI, Universidad de Granada, Granada, Spain
restauragmr@gmx.es
[2] Parque de Las Ciencias, Granada, Spain
[3] Sevilla, Spain
[4] Universidad Carlos III de Madrid, Madrid, Spain
lrubio@ing.uc3m.es

Abstract. In this paper we present the history of an old railway line in Andalucía: the Lorca - Baza railway line. In the paper we recall the beginning and development, the rise and fall of the British company that kept the line alive from the late nineteenth century until the Spanish Civil War, and the exploitation by RENFE until 1984. The railway line was of enormous importance in the transport of iron ore and goods from Andalusia to the port of Águilas (Murcia), and in the supply of all types of goods throughout the provinces of Almeria and Granada. The most important station, that of Zurgena, was restored at the beginning of the 21st century by the Town Council of Zurgena. Currently, a project of rehabilitation and reconstruction of all the railway structures, workshops, turntable and a set of tracks is being developed to reinstall a historic train using the original rolling stock, as a way to recover old techniques and knowledge and offer, in a new way, a real vision of this incredible train line.

Keywords: Granada heritage · Train · Railway line · Zurgena station

1 Introduction

The Great Southern of Spain Railway Company Limited (GSSR), a British company, was incorporated on 15th December 1885 to build a railway from Águilas and Lorca to Granada via the Almanzora Valley, passing through Baza, thus connecting the provinces of Murcia, Almería and Granada (Fig. 1). The history of this railway line, is a fascinating story of trials and tribulations, riches to rags, the triumph of hope over experience and the final painful death by civil war.

A. Dávila Ponce de León and M.J. Gámez---Private researcher

M. Ceccarelli and R. López-García (Eds.): HMM 2022, HMMS 40, pp. 487–498, 2022.
https://doi.org/10.1007/978-3-030-98499-1_39

Building the line was a daunting prospect due to the mountainous nature of the terrain, thus requiring many bridges, tunnels, cuttings, embankments, culverts and walls to be built. This employed some two thousand four hundred work-men in the late 1880s. As in every story, there has to be a hero and that was Gustave Gillman who was a relatively lowly member of The GSSR staff but who saved the company on many occasions mainly by selling The GSSR's ser-vices to the mines in the area, and by that token was almost single handedly responsible for the explosion of iron, lead, talcum etc. mining in the area. The railway functioned up until 31st December 1984 under the state railway operator Renfe. In the 1980s due to the parlous state of the railway system and the lack of money, many lines were closed including The GSSR (however, with support from Águilas council the section Águilas to Lorca and Murcia was kept open up until recently).

This is a terrible shame, as the line would have been a perfect tourist route for steam enthusiasts and would have brought much needed revenue to the Almanzora Valley. This would have been possible before the track was removed, as a steam locomotive still exists.

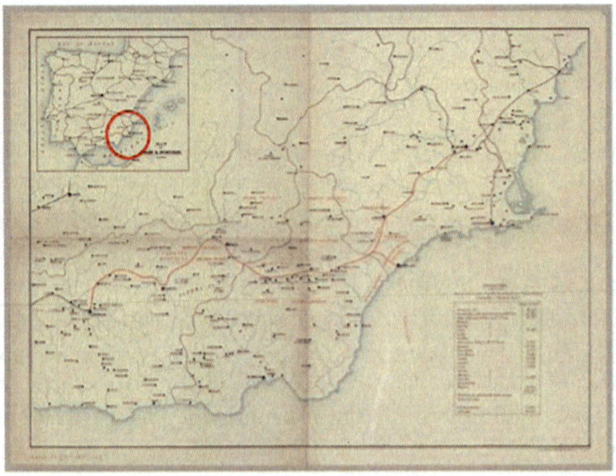

Fig. 1. Localization of Lorca-Baza line, Spain [2]

2 A Short History of the GSSR

2.1 The GSSR Until Spanish Civil War

In 1879 the government auctioned a concession to build a line between Granada and Murcia. A British retired ship-ping merchant called Edmund Sykes Hett was the only bidder and so, on 15th December 1885, the GSSR was incorporated and construction started.

In October of 1885, the sum of £1,250,000 was raised (approximately £200 million today - not a lot compared to Hereford- Sheffield line for example considering that the distances are almost the same –299 km), later to be increased to £1,684,357 for the construction of the GSSR. In addition, the Spanish government subsidy was only £532,000 (£70 million at today's rates, making the grant £150,000 per mile). However, The GSSR wasn't the construction company, which was 'Hett, Maylor & Co Ltd' (Edmund Sykes Hett & John Maylor) to be paid a fixed sum of £2,089,184 (£270 million today) on completion of the project, they do not seem to have done much work, as the main construction was subcontracted to the Marquis de Loring (Jorge Enrique Loring y Oyarzabal), who had been successful in the past as a railroad builder, and to August Lecoq, an expert in metal construction.

The original Board of Directors comprised:

- The Right Hon. Sir James Fergusson, Bart., M.P., G.C.H.L., K.C.M.G., 24 Grosvenor Gardens, S.W. (Chairman).
- Sir George Russell, Bart., M.P., Swallowfield Park, Reading. Director of the South-Eastern Railway company.
- David Davies, Esq., M.P., Llandinam, Montgomeryshire. Deputy Chairman of the Barry dock and Railway Company.
- Edward K Hett Esq., Eltham, Kent. Director of the Great Western of Brazil Railway Company (Limited).
- W. Goodwin Barnes, Esq., Beaufort Lodge, Cambridge Park, Twickenham. Shipowner.
- Perry F. Nursey, Esq., 161, Fleet Street, London, E.C. President of the Society of Engineers.
- Alfred Fernandez Yarrow, Esq., Isle of Dogs, Poplar, London, E. Engineer.

Each of the above bought 100 £10 shares in the company, as did:

- Howard Webb, Cardiff. Colliery prospector.
- Robert Mercer, Bromley, Kent. Civil engineer.

Nearly everything needed for the construction of the line was brought from Great Britain, France or Antwerp by ship.

Considering the difficult topography, construction went on at an impressive pace, the section between Almendricos and Águilas, having been started in October 1887, opened in April 1890. As an example of the topography from the point of view of a railway engineer, 133 km (41.8%) of the track between Granada and Lorca has inclines of 1 in 66 or steeper. The gauge was 1.668 m (5 ft 5¾ in.) as opposed to the British 1.435 m (4 ft 8½ in.).

The amount of work involved was enormous, there being more than three kilometres of construction by way of bridges, culverts, tunnels etc.

On March 24, 1890 the inauguration was held in which locomotive No. 2 "LORCA" took all the dignitaries to each of the stations on the line (Pilar de Jaravía, Pulpí and Almendricos during which Gustave Gillman took photographs. Figure 2 is thar corresponding to Almendricos station.

As mentioned, the line included the difficult section Pulpí to Pilar de Jaravía part of which was the Puerto de los Peines, that is still in service (Fig. 3). However, a few

Fig. 2. Inauguration of the line in Almendricos [2].

months later, by the end of 1890, Hett, Maylor & Company were declared bankrupt. This was mainly due to the gross underestimation of the difficulties of constructing a railway in a countryside such as that in Andalusia: the elevated costs of buying land, the lack of a road network making it difficult to transport heavy construction equipments and thus the need of access roads before stations could be built, track laid or bridges constructed. Also, Hett, Maylor & Co were being ripped off by sub-contractors. On top of all of this was the endemic bureaucracy in Spain at the time. The Hett, Maylor & Co bank-ruptcy had dire consequences for The GSSR.

Fig. 3. Puerto de los Peines (section between Pulpí and Pilar de Jaravía). (author's collection)

Notwithstanding, The GSSR pressed on with the construction of the rest of the railway (Fig. 4).

The bankruptcy of Hett, Maylor & Co left the GSSR in very bad situation as creditors, especially George Loring, successfully filed claims in court to seize all GSSR-owned material and freeze its assets. This despite the fact that the debt was owed to HM and not The GSSR. He also organized the placement of blockades on the Huércal-Overa to Zurgena section that was ready to open, as mentioned in El Boletin (the newspaper of The GSSR) [1].

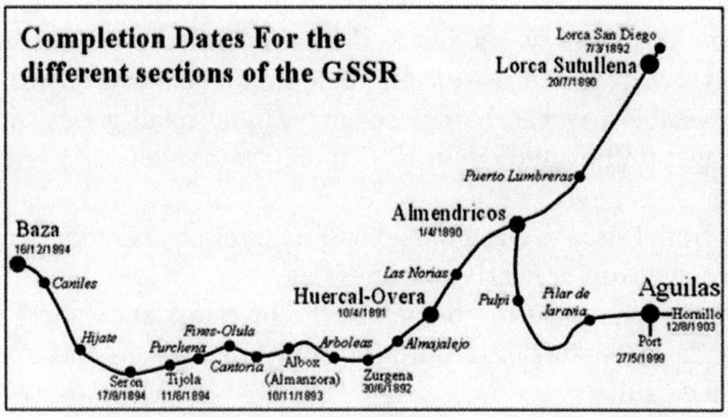

Fig. 4. Completition dates for the different sections of the GSSR [2].

2.2 Restructuring of the GSSR

The GSSR was in serious trouble, and wasn't even close to finishing the Baza to Granada section and they didn't have enough money to do this. They had, in fact, already built Deifontes and Piñar Stations and possibly much more. The answer was to restructure the company by first going into voluntary liquidation. Their main creditor (created in 1888 to build railways in Spain) was 'The Spanish Railways Trust and Investment Company' which re-arranged the concessions so that The GSSR only had the concessions for Lorca - Baza and Almendricos - Águilas/Hornillo (Table 1).

Table 1. Companies operating the different sections of the line

Section	Company
Granada-Guadix	South of Spain Railway
Guadix-Baza	The Granada Railway
Baza-Lorca	The GSSR
Lorca-Alcantarilla	Alcantarilla-Lorca Railway
Alcantarilla-Murcia	Madrid-Zaragoza-Alicante Railway

Other section, that corresponding to Baza and Guadix was taken on by 'The Granada Railway Company Limited' (TGR) which paid to The GSSR for services rendered. TGR also had almost insurmountable problems with construction which is why that section didn't open until 1907. It seems that The GSSR had brought in quite a lot of hardware for the Baza - Granada section which they proceeded to sell to The Granada Railway, and this probably helped the accounts somewhat.

The GSSR had to go back to their investors for more money. As the track was already laid between Lorca - Baza and Almendricos - El Hornillo, they were only asking for a small amount to keep the company solvent until the fares rolled in! (they didn't).This was mainly because the section between Baza and Granada hadn't been finished, so there was little incentive for passengers to use The GSSR.

However, now that there was a link to the north of Spain via Lorca/Murcia and to the port of Águilas there was a reasonable trade in merchandise such as esparto grass, marble, fruit, vegetables, wheat, barley, sugar beet and retail goods. In addition there was some passenger traffic although in 1897 it only warranted one passenger train each way per day.

The journey from Lorca to Baza took 6 h (1 h 30 min by car today) and a third class ticket cost an average worker nearly two days pay.

Anyway, the company had to go back to the investors and cancel all investments issuing new bonds, debentures etc. referring only to Lorca - Baza/Águilas and not Murcia – Granada/Águilas.

In fact, there was another restructuring of The GSSR in 1904 due to the need to purchase 150 40 tonne mineral wagons as a result of Gustave Gillman's amazing efforts to encourage mining along the route of The GSSR as well as more rolling stock so The GSSR restructured again. The fact that they were able to do this shows that the investors knew that without the mining transport their shares, debentures etc. would be totally worthless. So, in 1904 the company managed to get its rolling stock and so a new era began for the GSSR.

2.3 Operation in the Early 20th Century

The income from passengers at the turn of the 20th Century was very poor as the section between Baza and Granada hadn't been completed, so there was little interest for passengers. Thus, for example, in 1902 there was only one passenger train in each direction per day. This was a mail train with first, second and third class passenger cars. It left Lorca San Diego at 13.24 h and stopped at all stations. It passed Almendricos at 15.03 h, arriving at Baza at 19.58 h, averaging 20 km/h (12 mph). In 1908 the first train from Alicante to Granada passed through The GSSR and ran until it was suspended by RENFE in 1970 whereby they instigated an express between Barcelona, Granada and Almería.

The transport of goods was somewhat more promising although business started slowly. Most freight was marble (Fines-Olula), esparto grass, animals, minerals from mines near to Almendricos and Jaravía etc. nearly all of which went to Águilas Port for loading onto ships. It has to be said that The GSSR was certainly saved by the opening of the mines in Las Menas, Bacares and El Tesorero and the pier El Hornillo in Águilas as well as government grants. However, matters improved in the early years of the twentieth century with a total of 124,396 tonnes of ore from the Sierra de En-medio (Almendricos Station being in the foothills) as well as 403,272 tonnes from the Sierra de Los Filabres representing 1,257 trains of 12 wagons that year. The transport costs to the mining companies were significant, being approxi-mately half of the costs of extraction and transport to the loaders. Due to the topography between Serón and Águilas there were stretches where double and triple headers were needed (two locos at the front, one rear).

It is interesting to note that in 1915 (when there was a reduced market in iron ore due to the First World War) the chairman of the board, Mr. Thomas Harrison said, when talking about sales in 1914: "Minerals account for nearly £30,000, goods for £4,000, passengers £1,000".

There was a large variety of goods carried, including coffins, livestock, citrus, tomatoes, olives (and their oil), almonds, potatoes etc., wood, pine bark (for tanning), fertilizer, copper, calamine, sulphur, talcum, barium, gypsum, ochres, porphyry, lead, tin, silver and gold.

Agricultural products such as melons and tomatoes continued to be transported from Alicante to Águilas right up until the 1960s.

2.4 The Line Under RENFE Rules up to the End in 1984

The death of The GSSR came at the beginning of the Spanish Civil War when, on 17th July 1936 the Railway Worker's Council took control of the running of The GSSR, all its track and stock. This was part of the Republican side's activities and the Republicans ran the railway from this point. Eventually, the Nationalists under General Francisco Franco took control and set up a national railway network called RENFE which still functions today as Renfe Operadora.

The GSSR board of directors said this as part of their 1936 report to shareholders:

> "The directors regret to report that on 5th September 1936, control of the railway was assumed by The Local Railway Worker's Council with the indulgence of the state representative. Subsequently, a decree of 26th December 1936 enacted that the state provisionally should take charge of the working of the remainder of the railways not comprehended in an earlier decree of 3rd August 1936, under the terms of which the workings of the large systems had already been taken over by The State and confided to the Railway's Exploitation Committee."

In fact, RENFE was served a poisoned chalice as much of the rolling stock and track had been destroyed during the war, although The GSSR stock survived better than most of the Spanish railways as Almería and environs remained in republican hands until the bitter end, thus preventing much destruction to rolling stock and track. The other problem for RENFE was lack of investment over many years such that the situation in 1963 was as follows (Table 2):

Table 2. Length of the line

Period	Km/miles track
Over 45 years old	2891/1796
35–45 years old	4430/2752
25–35 years old	2335/1450
10–25 years old	1705/1059
Less than 5 years old	2063/1282
Total	**13424/8341**

Just for comparison, the total length of track in Great Britian is 15,811 km (9,824 miles). Additionally, during The Spanish Civil War 1,309 (42%) locomotives, 32,000 (40%) goods wagons and 3,700 (70%) passenger wagons were destroyed.

The GSSR gamely carried on as a company based in London until on 4th February 1955 the company was dissolved. In the restructuring of 1904, many of the investors nearly everything. As Bladó said in [1]: "Those that had invested in the company came out very badly. The money lenders received debentures for their money and paper of dubious value for the interest. The debenture holders lost 50% of their capital with the remainder in shares that today (1918) are quoted on the stock exchange at 10% of their face value. The shareholders lost 90% and 95% respectively of their money. So, the first shareholders (if they had not already burnt their certificates in desperation) now owned 5 pesetas for every 1,000 pesetas invested. Probably they had sold them as scrap paper".

Although one might think that the GSSR management were swindlers, the reality was different. Touring the rail line allows one to see that the work that was done, without the aid of machinery, only with pick and shovel, was incredibly arduous and required thousands of workers, to the tune of 2,400. As far as is known, the company was overwhelmed by the problems created by the topography of the area and was forced to overcome the problems of gorges, hills, rivers, etc. Only the management of Gustave Gillman, allowed GSSR to continue. He convinced the investors to invest in the mines of Los Filabres, in the construction of a loader in Serón, of a pier nin El Hornillo for the transport of millions of tons of iron ore.

The end came finally on 31st December 1984 when Renfe closed the line between Guadix and Lorca, ripping up track in the late 1990s. Thanks to the efforts of Águilas Town Hall, the line between Águilas and Lorca (is still open with a limited service of two trains each way per day.

Some town halls have tried to maintain the station buildings: Zurgena, Serón, Cantoria, Almanzora and Tíjola being notable examples. Other buildings are in danger of falling down, especially Gor, Gorafe, Fines-Olula and Arboleas. Almanzora is currently (2019) being renovated again.

What is left of the track is mainly ballast although certain areas have been converted to green routes. Either way, they present excellent opportunities for walkers and mountain bikers to see the sublime Andalusian countryside first hand, and to appreciate the enormous amount of work that went in to constructing The GSSR. And, more recently, with the recovery project of the railway facilities of the Zurgena train station, the laying of a section of railway on the original route and the creation of a railway industrial heritage restoration school with the support of the Zurgena City Council and the Almanzora Association of Municipalities we are facing a reinterpretation of this railway line of capital importance in the communication of Andalusia and the Spanish Levante that will mark the line to be followed by similar projects in the coming years.

3 Works of Recovery

3.1 The Zurgena Railway Station, a First Class Train Station in Progress to Return to the Service for Historic Trains

This station was very important from the point of view of the running of the railway. Apart from the station, there were engine sheds, workshops, supplies departments and offices. It was a storage point for goods trains to stay overnight and to continue their journeys to Águilas/El Hornillo or Guadix in the early hours. In Fig. 5 an old drawing of the station can be seen.

Although it had no track machinery assigned to the station like Águilas and Lorca-San Diego did, it was the site for the track machinery to be kept when not in use. However, it did have permanent engineering staff. There was, also, a turntable (see Fig. 6) and the engine sheds had inspection pits for the locos. In addition, there were staff to manage sleep-overs for the train crews and their charges.

The main reason for all this activity was that this station is where the triple header iron-ore trains would leave for Huércal-Overa, labouring up the 40 m in 11 km incline and then two of the three returning.

Zurgena was also a commercially important station, both in passenger and freight traffic. This was not due to the population of Zurgena, which was quite small, but because it also served other towns in the region such as Lubrín, Cuevas de Almanzora, Vera, Garrucha, Los Gallardos, Mojácar, etc. The installation was completed with two additional private lines for two factories: Carlos Tortosa (marble) and Campsa (fuel).

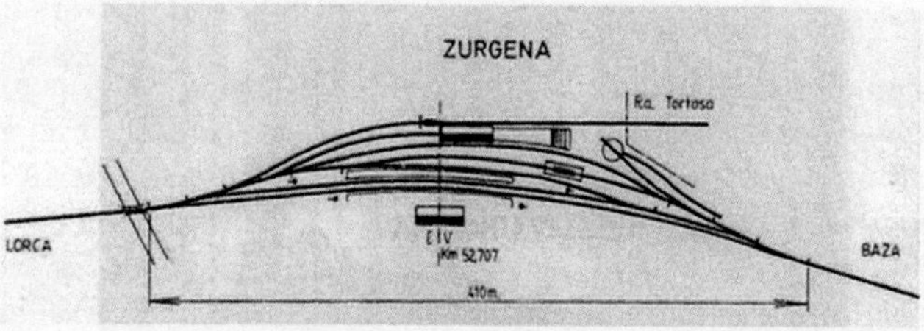

Fig. 5. Drawings of Zurgena railway station by Pedro Pintado Quintana.

The actual shape of the station is, in general, good but the tracks and sleepers, buried under a 40 cm layer of ground last 20 years, are in regular or poor condition. An "escuela-taller" promoted by Zurgena's Town Hall at the end of last century, garantized the integrity of the main buildings (see Fig. 7 and Fig. 8) much better than other stations of the line, in very poor condition or, directly, destroyed.

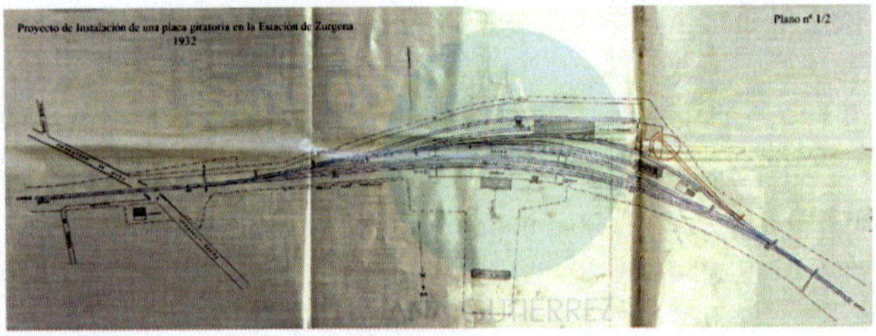

Fig. 6. Drawings of the works to install the turntable (in red colour) in 1932 (Railway Historic Archive of Spain)

Fig. 7. Works of rehabilitation of the interior of Zurgena station (author's collection)

This fact, moreover, is the real will of the City Council of Zurgena to improve its industrial heritage. Proof of this is the subscription in early 2021 of an agreement to collaborate in this project with the technology company of the University of Granada "Restaura, Gestión Museística y Restauración Siglo XXI", experts in the conservation of industrial heritage. Thus, at this moment there is an advance of the recovery works of the entire railway infrastructure of the station, the turntable, the conservation workshops of railway material and the laying of a railway line between the towns of Zurgena and Arboleas, in a first phase, within what was the line Lorca - Baza. The objective is the exhibition of historic railway material and the creation of a training school in classic railway material, in order to train technical personnel for the maintenance of this historic line and its facilities and to collaborate with other historic lines in Spain and abroad. This type of school is almost non-existent at present, and there is a great demand for qualified technical personnel in this field.

Fig. 8. Passengers terminal of Zurgena Railway station (author's collection)

4 The Rolling Stock for the New Railway Line. Units to Restore and Put in Service. New Adaptions of the Original Engines and Technologies to the New Enviromental Requirements

The rolling stock that will equip the historic train line that will circulate in the Zurgena train station facilities and the section of track that is currently being installed is intended to be a representative sample of the units that circulated on this railway line during its entire operating life, that is, it will have steam locomotives and diesel traction locomotives, with the adequate adaptation of their fuels to meet the criteria of low or zero polluting emissions and that they do not pose fire risks in the summer season, so common in steam locomotives that use coal as fuel.

Already in the 19th century there were shunting locomotives in the sugar industries of the Vega de Granada that moved exclusively with fuels of vegetable origin (cotton seed, marc, almond shells, wood, etc.). Both Granada, Almería and Murcia have been and are traditional producers of rainfed almonds and the almond shell is a fuel with a very high energy value and zero kilometer, affordable and with close and immediate availability, being the main fuel that will be used in the steam locomotives that will be added to the historic train, specifically the steam locomotive 020 WT, brand Orenstein and Koppel from 1912 (Fig. 9), with manufacturing number 5,894 and the steam locomotive

Fig. 9. Locomotive 020 WT Orenstein and Koppel in 1923 [2].

020 T from 1902 (Fig. 10), brand Ateliers de Tubize, former Andalusians 05 after their restoration period in the workshops of the Zurgena station.

Fig. 10. Locomotive De Tubize – Andaluces 05 [2].

5 Conclusions

In this work we intend to recall part of the industrial heritage of southeastern Spain showing the historical evolution, from construction to dismantling, with all the drawbacks, technical and economic, found in the Lorca-Baza railway line. In the first part of the work we offer a historical summary of the railway line being of capital importance in the communication and for the transport of raw materials. In the second part, we briefly present the recovery works of the main station of the line, the Zurgena station, as well as the recently initiated restoration works of some rolling stock, wagons, locomotives and other old railway equipment. The intention of this project is a new interpretation that restores part of the splendor that the line had for almost 100 years with a reopening project that represents a before and after in the new uses of industrial heritage in Spain.

Acknowledgments. The authors would like to thank Maxwell Adran Kite and Donald Gaunt for their contribution to the GSSR history, to Zurgena Town Hall for their contribution to the project and for the documental information and, finally, to the association of railway friends of Zurgena.

References

1. Baldó, M.L.: El Boletín. Periódico de la Great Southern of Spain Railway Company Limited (1917)
2. Gaunt, D.: Almeria and the Great Southern of Spain Railway (the GSSR). Ed. Gaunt (2006)

Correction to: Evolution History of Non-circular Gears

Olga V. Egorova[1](✉), Nikolay N. Barbashov[2], and Roman M. Kiselev[2]

[1] V.I. Vernadsky Crimean Federal University, Simferopol, Russia
tmm-olgaegorova@yandex.ru
[2] Bauman Moscow State Technical University, Moscow, Russia
kiselevrmbmstu@gmail.com

Correction to:
Chapter 30 in: M. Ceccarelli and R. López-García (Eds.): HMM 2022, HMMS 40, pp. 372–381, 2022.
https://doi.org/10.1007/978-3-030-98499-1_30

In the original version of the book, the following belated corrections have been incorporated: References [23] and [24] have been updated in the reference list in Chapter 30 and Figure 1 caption has been revised. The book and the chapter have been updated with the changes.

The updated version of this chapter can be found at
https://doi.org/10.1007/978-3-030-98499-1_30

M. Ceccarelli and R. López-García (Eds.): HMM 2022, HMMS 40, p. C1, 2022.
https://doi.org/10.1007/978-3-030-98499-1_40

Author Index

A
Ambrosetti, Nadia, 345
Angeloni, Renato, 134
Aslan Seyhan, Irem, 406
Aterini, Barbara, 330

B
Barbashov, Nikolay N., 71, 372
Barmina, Natalya A., 24
Bustos, Alejandro, 120

C
Calderón, David, 197
Callegari, Massimo, 134
Caparrós-Mancera, Julio Jose, 41, 184
Carbonari, Luca, 134
Carrasco, A., 197
Carretero Díaz, Antonio, 266
Castejon, Cristina, 120
Castilla-Gutiérrez, Javier, 184
Castillo-Martínez, Agustín, 173
Ceccarelli, Marco, 3, 10, 24, 360
Chen, Yu-Hsun, 97
Chiavoni, Emanuela, 360
Chiu, Yu-Chieh, 97
Cigola, Michela, 295, 360
Clini, Paolo, 134
Cocconcelli, Marco, 314, 382

D
D'Alessio, Mirco, 134
Dávila Ponce de León, A., 444, 487
Dávila-Martín, Jose M., 184
Diacodimitri, Alekos, 360
Dorado-Vicente, Rubén, 84, 106, 421
Đorđević, Miloš, 210

E
Egorova, Olga V., 71, 372

F
Fernández-Liétor, Israel, 106
Fonte, Cosimo, 382
Fortes-Garrido, Juan Carlos, 184
Fuentes, Demetrio, 197

G
Gallozzi, Arturo, 305
Gámez, M. J., 444, 487
Gašić, Vlada, 210
Gasparetto, Alessandro, 10
García de la Mata, Rafael, 444, 487
Gokdogan, Melek Dosay, 49
González-Sevilla, C., 444, 487

H
Hanafi, Issam, 197
Hernández-Torres, José Antonio, 41, 184

J
Jauregui-Correa, Juan Carlos, 433

K
Kiselev, Roman M., 372
Koetsier, Teun, 279

L
López-García, Rafael, 84, 106, 421

M. Ceccarelli and R. López-García (Eds.): HMM 2022, HMMS 40, pp. 499–500, 2022.
https://doi.org/10.1007/978-3-030-98499-1

M
Mansilla, Luis, 197
Marco-Lozano, Sixto Manuel, 149, 160
Martínez Muneta, Luisa, 266
Mata, Francisco, 197
Medina-Sánchez, Gustavo, 84, 106, 421
Meneses, Jesús, 120
Monjas Hebrero, Oscar, 266
Moon, Francis C., 57
Moreno-Buesa, Jorge, 84
Mottola, Giovanni, 314

N
Navarro-Arcas, Abel, 149, 160

O
Ojeda-Escoto, Pedro A., 395
Oliveira, Agamenon R. E., 241

P
Padole, Pramod M., 255
Pisano, Raffaele, 227, 241

Q
Quero-Nieves, Juan Manuel, 421

R
Robarts, Julie, 227
Rodríguez-Pérez, Angel Mariano, 41, 184
Rubio, Higinio, 120
Rubio, Lourdes, 444, 487

S
Saez-Calvo, José Antonio, 470
Sánchez Lite, A., 456
Sanz Bobi, Juan de Dios, 266
Scoccia, Cecilia, 134
Shiwalkar, Prashant B., 255
Strollo, Rodolfo Maria, 305

T
Tikhomirov, Grigoriy, 71
Trubachev, Evgenii S., 24

V
Velasco-Sánchez, Emilio, 149, 160

Y
Yan, Hong-Sen, 97

Z
Zrnić, Nenad, 210
Zulueta Pérez, Patricia, 456